# *Encyclopaedia*
# *of*
# BIOLOGICAL SCIENCE, TECHNOLOGY AND ENGINEERING

# *Encyclopaedia of* Biological Science, Technology and Engineering

Volume 1

Dr. Albert Shawn

ANMOL PUBLICATIONS PVT. LTD.
NEW DELHI-110 002 (INDIA)

**ANMOL PUBLICATIONS PVT. LTD.**
**Regd. Office:** 4360/4, Ansari Road, Daryaganj,
New Delhi-110 002 (India)
Ph.: 23278000, 23261597
**Branch Office:** No. 1015, Ist Main Road, BSK IIIrd Stage
IIIrd Phase, IIIrd Block,
Bangalore-560 085 (India)
Tel.: 080-41723429
*Visit us at*: www.anmolpublications.com

*Encyclopaedia of Biological Science, Technology and Engineering*



First Edition, 2009

ISBN 978-81-261-4121-0 (Set)

PRINTED IN INDIA

---

Printed at Paras Offset Pvt. Ltd., New Delhi

# Contents

## Vol. 3

# Preface

Biology is the study of life and a branch of the Natural Sciences which studies living organisms and how they interact with each other and their Environment. It examines the *Structure, Function, Growth, Origin, Evolution, and Distribution* of living things. Also, it classifies and describes Organisms, their Functions, and how species come into existence. Four unifying principles form the foundation of *Modern Biology*: *Cell Theory, Evolution, Genetics and Homeostasis.*

Biology as a separate science was developed in the Nineteenth Century as scientists discovered that organisms shared fundamental characteristics. Biology is now a standard subject of instruction at schools and universities around the world, and over a million papers are published annually in a wide array of biology and medicine journals.

Most Biological Sciences are specialized disciplines. Traditionally, they are grouped by the type of Organism being studied: *Botany, Study of Plants; Zoology, Study of Animals; and Microbiology, Study of Microorganisms*. The fields within biology are further divided based on the scale at which organisms are studied and the methods used to study them: *Biochemistry* examines the fundamental chemistry of life; *Molecular Biology* studies the complex interactions of systems of *Biological Molecules; Cellular Biology* examines the basic building block of all life, *Cell*; Physiology examines the *Physical and Chemical Functions of the Tissues and Organ Systems of an Organism*; and *Ecology* examines how various organisms and their environment interrelate.

Author

# Chapter 1

# Matter and Energy in Life

Primitive man dissociates his idea of motive agent or impelling instrument from that of the body set in motion. Direct observation in fact teaches us that for a body to start moving some force must be applied to it. This idea was first scientifically stated by Aristotle and upon it was founded the rational mechanics of Galileo and Newton. Man does not intuitively perceive that movement may be but one of many properties of matter, inherent in the manner in which such matter exists.

Animate beings are indeed distinguished through their very capacity for motion. An animal moves and gives off heat or even light in some form; every living body is capable of growth.

Not so very long ago, in 1907, Strecker in his Der Kausalitätsprinzip in der Biologie (Cause and Effect in Biology) was still able to say: "Every organic body is a form of activity, whilst a purely chemical or physical object can do nothing more than be set in motion." Since it is impossible for a thing to become animated by itself alone, some motive principle special to living matter must be in existence.

Throughout the ages Biology has been vitalistic. Looking back through the history of physiological knowledge we see the continuous passing-on of the idea of a breath, spirit, ether or animating principle acting upon pure matter, that is, upon a passive immovable substance possessed of weight but incapable of action by itself alone. In the course of time these ideas have taken many different forms, though all are essentially similar.

Whence springs the activity present in live organisms? One of the most important steps in the progress of Biology was signalized by proof that the physical laws governing living matter are the same as those applying to the inorganic world.

## COMBUSTION

It is essential to take an objective view of matter and of the Universe in general. Certain bodies burn up. The Greek philosophers, Aristotle for one, considered fire to be one element out of the four — air, earth, fire and water — which, among them, covered the whole field of existence as then known. Combustion was held to be the liberation of the element fire from ignited bodies. According to the alchemists this element contained sulphur, or as they called it, brimstone. Becher in 1699 and Stahl in 1702 suggested a similar explanation for combustion, namely, that a hypothetical substance, which they called phlogiston, escaped from burning bodies and diffused into the air wherein it disappeared completely.

Subsequent to Scheele, Priestley recognized the fact that air did contain oxygen. Priestley showed that oxygen aids combustion and interpreted the phenomenon on the assumption that such oxygen was dephlogisticated air which favoured the absorption of phlogiston.

Had this theory been correct, and then the products of combustion should have weighed less than at the start of the process, since phlogiston is supposed to be liberated. Actually the opposite takes place, as in all cases of oxidation. Lomonossov in 1756 put a quantity of tin in a hermetically sealed flask and heated it up until the tin turned to a white powder. On opening the flask, air rushed in, thus proving that a part of what had originally been in the flask had disappeared. The weight of the white powder was greater than that of the original tin, so something must have combined with the metal — something which had previously been present in the air. Lavoisier in 1775 confirmed these facts and gave a definite explanation thereof. He heated quicksilver for four days inside a closed retort, at a temperature slightly below boiling point, and a red crust then formed on the surface of the metal. This product on being heated still more intensely separated into quicksilver, which could be recovered, and a gas.

The volume of such gas equalled that lost by the air in the first part of the process and its properties were identical with those of the gas Priestley obtained when using mercuric oxide as his starting point. The weight of the product obtained on heating mercury was greater than that of the original quantity — exactly as Lomonossov had observed — and the increase tallied with the weight of gas disappearing from the air enclosed in the retort and which was given off when the oxide had disintegrated.

The oxidation process must consist therefore in fixing one of the components of air on to a definite quantity of a substance, and Lavoisier decided to call such component oxygen, that is, acid producer. Lavoisier founded modern chemistry by the use of the balance, following up the conversion of substances by studying their weights. He established the Law of the Conservation of Matter, which states that the properties of substances vary as they react upon one another but that the total quantity of matter remains unaltered. Nothing is lost and nothing gained in a reaction or in a series of reactions. Combustion is therefore, in general, a form of oxidation.

The burning of coal in a grate is the same as oxidizing mercury, for instance. The laws governing each process are the same. Carbon dioxide, the product of the combustion of coal, weighs more than the coal itself, and the increase in weight is exactly equal to that of the oxygen consumed. Ideas on combustion at this stage have stridden a long way ahead of the old phlogistic theory.

## RESPIRATION

Let us look back once more. In 1667, Robert Hooke demonstrated to the Royal Society the need for " fresh air " — these are his actual words — "for maintaining life." Mayow in 1674 proved that the component of air necessary for breathing purposes was the nitro-aerial spirit, oxygen, which had already been found to be the substance essential for promoting combustion. He assumed, furthermore, that such nitro-aerial spirit must penetrate into the blood where it makes itself indispensable to the vital activities, by combining with the sulphurous particles in the blood in the same way as would happen in burning the substance.

In 1774, Priestley noticed that if mice were left for some time in a closed chamber, continuance of life was impossible because the air became impure. Living plants introduced into the chamber made the air breathable once more. Soon afterwards, Jan Ingenhousz in 1780 discovered that this occurred only if the plants were exposed to light:

"The sun's light induces in the leaves that activity which generates dephlogisticated air."Black in 1755 had proved that animals produced, when breathing, a gas different from air, which he called fixed air.

Lavoisier soon showed that this gas is the same as that obtained when coal is burnt, namely carbon dioxide. Jean Senebier in 1780 gave the explanation of the occurrences noted by Priestley and Ingenhousz, thus:

"The leaves of plants convert fixed air into dephlogisticated air," that is, they turn carbonic gas into oxygen. A long time was still to elapse before the isolation and discovery of the properties of chlorophyll by Pelletier and Caventon in 1818! Lavoisier arrived at systematized and precise concepts regarding respiration.

Beings living in air consume oxygen and exhale carbon dioxide in practically equal amounts. Combustion takes place within the organism and, as in the physical world, heat is generated. Lavoisier resorted to Laplace to get some idea regarding the means necessary to enable such produced warmth to be measured. He made the first calorimeter for biological use and completed the first calorimetric determinations ever carried out on living matter.

## CONVERSION OF ENERGY

Several years of political convulsion were to pass in Europe, the Napoleonic Wars with all their various consequences, and thus, nothing of note came to fight in our field until 1824, when Sadi Carnot published his celebrated work Réflexions sur la Puissance Motrice du Feu (Reflections on the Motive Power of Fire).

In due course there had arisen the peculiar paradox that though the steam engine was becoming more important because of its enormous value in industry, its invention had come about from purely empirical observation. It was of supreme interest to know the principles in virtue whereof such extraordinarily potent results could be obtained. Carnot studied the physical conditions obtaining in steam engines at work and discovered the so-called Second Law of Thermodynamics, namely that a difference of temperature occurs in various parts of an engine and that heat must pass from a warmer to a cooler portion, like water falling to a lower level and turning a millwheel. Heat in this case thus behaves like a material element.

In spite of the importance of Carnot's law, light did not yet break openly on the subject. Some twenty years later, an obscure German doctor, Julius Robert Mayer, wrote Die Mechanik der Wärme (Mechanics of Heat) in 1841, but his thesis did not achieve publication as it conflicted with the ideas then reigning in the world of physics. Among his points was the following statement:

"We must bring ourselves to think of two such dissimilar items as mechanical work and heat as two forms of the same thing." Heat can be converted into work and vice versa, such conversions being just as quantitative as transformations of matter itself. In changes of energy, in physics, as in chemical reactions, there is neither gain nor loss of energy throughout the whole series of possible conversions. There is a law of the conservation of energy just as valid as that of the conservation of matter.

Very soon, Mayer's ideas were confirmed by the work of James Prescott Joule, an amateur — like so many English scientists at the start — who in 1843 published his first thesis on heat and work. A great scientist, the physiologist Helmholtz, in 1847 proved conclusively and with a sure hand the arguments above mentioned. He was at that time but twenty-six years old and for the moment did not possess sufficient reputation as a scientist. Like Mayer, he met with difficulty in getting his work published and his ideas accepted.

## METABOLISM AND ENERGY

Mayer had already asserted that the Law of the Conservation and Conversion of Energy is applicable to the activity of live bodies. In 1845 he published his treatise Ueber die organische Bewegung in ihrem Zusammenhang mit dem Stoffwechsel, in which he showed that from the physical point of view, vital phenomena are indistinguishable from what occurs in the inorganic world. Helmholtz definitely set forth his views in his much-discussed work Ueber die Erhaltung der Kraft, published in 1847.

A group of young physicists, mathematicians and physiologists had founded The Berlin Physical Faculty, and whilst Helmholtz's reception on the part of the older diehards of the Academy of Sciences had been extremely cold, he aroused in the younger group such an enthusiasm as has not been quenched even to this day.

Complementary to these discoveries was the codification of chemical and thermochemical energy. Marcellin Berthelot in 1875 wrote his celebrated treatise demonstrating that mechanical energy is reproduced in chemical processes. In any chemical reaction conversions of matter occur simultaneously with displacements of energy. In this way exothermic or endothermic reactions are developed.

The calorimetric measurements taken in live bodies by Berthelot himself, Pettenkoffer, Atwater, Benedict, Du Bois, and many other physiologists, make it quite plain that the laws of thermochemistry govern nutritional processes in animals and plants just as they do in inorganic bodies, and that assimilation is a combination of synthetic reactions whose energy balance is endothermic, as for instance in the great assimilative mechanism represented by the action of chlorophyll, where energy supplied by the light of the sun is effectively utilized.

On the other hand, disassimilation in its more distinct forms is an exothermic breaking-down or release of heat and this energy appears in life under many different guises, such as motion, warmth, electric power, light, syntheses coupled with disassimilative reactions, and the rest. In the living world the laws of the conservation of matter and of energy recur. Energy in living beings is a special manifestation of the universal energy of the Cosmos.

Let us return to Lavoisier who was the first to express the whole thing with clarity. When coal burns in a stove, fixing oxygen and liberating carbon dioxide, such chemical change gives out heat, namely the heat of the fire. Animals and plants breathe, consuming oxygen and exhaling carbon dioxide gas. Combustion of carbon takes place within their organism, as well as many other biochemical processes and exothermic disassimilative reactions.

Here also the combustion of carbon, just the same as in the stove, sets a certain amount of energy at liberty. The same thing occurs in the other reactions taking place at the same time in the living creature, and such energy appears in the form of heat and in other shapes such as activity in the functions, or other living processes. Looked at from the energy side of things, every living organism is a thermochemical machine.

## ANTOINE-LAURENT LAVOISIER

The substance which combines with metals subjected to heat and increases the weight thereof is none other than the purest portion of the air surrounding us and breathed in by us, and which by such process passes from a state of flexibility to that of fixation.

When charcoal is burnt, it disappears in gaseous form, but the resultant gas is greater in weight than the original charcoal consumed, because this latter has combined with the pure portion of the air.

This is identical with what happens when quicksilver is burnt. Conversely, a strong heating-up of the burnt product allows quicksilver to precipitate and air is then given forth which can be breathed freely.

When charcoal is used for reducing metals such as quicksilver and the rest, it completely disappears whilst the metal returns to its original state, fixed air (carbon dioxide) being generated at the same time. From this we may deduce that the substance which has been called fixed air is a combination of the pure substance, breathable oxygen from the air, with charcoal. I propose to analyse this conclusion in detail in the following series of notes.

**Experiments on the Breathing of Animals**

Among all the phenomena of animal economy none is more striking nor more worthy of attention on the part of philosophers and physiologists, than those occurring in the act of breathing. We know very little in regard to this function; merely that it is necessary for the preservation of life and cannot therefore be suspended for any length of time without exposing living creatures to the risk of instant death.

The experiments carried out by certain philosophers, especially Hales and Cigna, have thrown a certain light in this connection. Doctor Priestley in an important work has recently widened the range of our knowledge. He has tried to show by means of certain extremely ingenious, novel, and delicate experiments, that animal breathing has the property of phlogisticating the air in a way similar to that seen in the combustion of certain substances and in other analogous chemical processes. The air then becomes unfit for breathing purposes and.this occurs when it is overloaded and eventually saturated with phlogiston.

However plausible the theory of this celebrated philosopher may appear to be at first sight, and however well conducted may have been the large number of experiments upon which their author relies, I must confess that I find some of the phenomena described to be so mutually contradictory as to give rise to certain doubts on my part. Consequently, I have myself experimented by different methods and the results of my researches have led me to conclusions distinctly at variance with his. Air which has been used for the combustion of metals contains nothing more than the noxious residue from atmospheric air. The part thereof formerly fit for breathing becomes combined with quicksilver during combustion. The air breathed in, once separated from the fixed air, is exactly the same as that entering into combustion. In fact, by combining with the breathable residue, somewhere about a quarter of its volume of dephlogisticated air (oxygen) extracted from the mercury consumed, I was able to restore the air to its original state perfectly suitable for respiration, combustion, etc.

The same occurs if air vitiated by combustion of mercury be used. The addition of a proper amount of the pure principle obtained by long and intense heating of the said calcinated mercury, restores to the air its qualities of breathability and its capacity for combustion.

From these experiments it is evident that in order to restore air which has been rendered noxious by breathing, we must —Firstly, get rid of the fixed air contained therein, using quicklime or caustic alkali for such purpose, and Secondly, add an amount of pure breathable or dephlogisticated air equal to what has been lost.

Respiration gives the converse effect.In fact, in my paper read at the Public Session last Easter, I pointed out that dephlogisticated air can be converted into fixed air by burning powdered charcoal in it, and that there are various other means of obtaining a similar result. It is possible that respiration possesses the same property, namely that dephlogisticated air taken in by the lungs may be returned in the form of fixed air. Pure air may combine with the blood and that might be the reason for the red colour distinguishing the latter.We may take it for granted:

- Breathing changes only the pure or dephlogisticated air contained in the atmosphere. The residue or noxious portion enters the lungs and is expelled in the same state without change or modification.
- The burning of metals in a given quantity of atmospheric air takes place proportionately to the dephlogisticated air contained therein, and combining with the metal.

- Similarly, an animal confined in a given quantity of air which is not renewed, will die after having consumed the breathable portion of such air, and converting it thus into fixed air which is useless for the purpose of breathing.
- The noxious air, not fit for breathing, remaining from the calcination of metals is identical to what remains from respiration, once the fixed air has been separated.

The noxious air in one case or the other can be restored to the atmospheric state if a quantity of dephlogisticated air equal to what disappeared in the burning or breathing process be incorporated therewith.

## MARCELLIN PIERR-EUGÈNE BERTHELOT

Initial attempts to analyse organic substances led to results so strange that the problem itself, approached with certain tenacity from the seventeenth century to the beginning of the eighteenth, seemed to many at that time to be impossible of solution.

Later researches showed, however, that the component parts of animals and plants, like the products deriving therefrom, are the only substances capable of giving oils and products of combustion on distillation.

This feature distinguishes them from mineral compounds which never give rise to such products of combustion. Thus, once for all, the distinction between mineral chemistry and organic chemistry was established.

Furthermore the way was thereby paved for the discovery of the constituent elements of animal and plant matter. It had been observed that all plants are formed from the same elements, since they all give similar products on decomposition, namely water, oils, carbonaceous earths and the rest.

The same applies with even more force to animals, as animal elements are up to a point identical with those found in plants. Indeed, the decomposition of animal matter gives substances identical with those obtained from the decomposition or distillation of plants, with the addition of considerable proportions of volatile alkalis, all of which seems to indicate some factor characteristic to animal chemistry. The presence of such volatile alkali is precisely the thing which from the chemist's viewpoint distinguishes animal matter.

When later the nature of water, of oils produced by heat, and of the volatile alkali became known and it was demonstrated that all these bodies were combinations of single elements, it was possible to deduce that carbon, hydrogen, oxygen and nitrogen are the four basic elements of living bodies.

Here we have an all-important result which suffices to show that these analytical studies, for all their inevitable imperfections, were of real value. Thanks to work on the chemistry of the air, interpretation of these facts suddenly acquired greater clarity. The discovery of the gases, the theory of combustion, the proof of the constancy in weight of bodies in chemical reactions, and of the stability of single bodies or elements throughout the whole range of conversion — in short, the sum total of knowledge definitely obtained by Lavoisier and his contemporaries — placed the science of chemistry on the solid foundation on which it still stands today. They gave a positive analytical basis to organic chemistry, as had previously been the case with mineral chemistry.

It was now possible to show that plants are formed mainly from carbon, hydrogen and oxygen and that animals contain these same elements with the additional feature of considerable amounts of nitrogen. Such simplicity in the chemical composition of five beings is so much the more interesting since the same elements are also met with in inorganic mineral compounds. Towards the middle of the eighteenth century, ideas regarding the composition of organic substances became somewhat more precise.

Taking into account the differences observed between the properties of products from analysis and those of the animal or plant substances analysed, chemists began to concern themselves with the question of isolating in their natural state not only the elements but also the "immediate principles" which made up the substance of living bodies.

A fair number of these immediate principles were isolated, recognized and identified in the same century. Certain acids, such as tartaric, oxalic, malic, citric, lactic, uric, formic, gallic, benzoic and the rest, and other substances like urea, sugar of milk, cane sugar, cholesterine or wax from the biliary calculi, not possessing the property of acidity, were labelled.

The foundation of organic chemistry, as M. Chevreul says, is the exact definition of the species of immediate principles in animals and plants. Every research into living phenomena requires an exact knowledge of the immediate principles which make up the tissues and liquids within the organism, where the phenomena in question occur.

## HERMANN LUDWIG FERDINAND VON HELMHOLTZ

Since the present discussion had from its nature to be addressed essentially to physicists, I have chosen to present its principles purely in the form of a physical hypothesis independently of any philosophical argument, developing their consequences and comparing these with the empirical laws governing natural phenomena in the various branches of physics.

The derivation of the principles established can be set about from two points of view, namely, the postulate that it is impossible to acquire unlimited mechanical power through the action of any one combination of natural bodies on another, or the assumption that all activities in nature may be attributed to forces of attraction and repulsion, whose intensity depends solely on the distance between their points of application. That both these theories are identical is demonstrated right at the very start of this discussion. However, they still have an essential significance for the final and real aim of the physical sciences in general, which I shall endeavour to set forth in this separate introduction.

The problem of science is to discover once for all the laws whereby individual events in nature may be traced back to general laws and be in turn determined from these latter. Such laws, for instance, as those of the refraction or reflection of light, Boyle's and Gay-Lussac's laws for the volume of gases, are obviously nothing but general notions, by means of which all phenomena appertaining thereto are gathered together. The search for these is the business of the practical side of our sciences.

Theoretical science, on the other hand, seeks to discover the unknown origins of the processes by considering their visible effects, seeking to comprehend them by application of the principle of cause and effect. We are led to and entitled to such practice through the axiom that every change in nature must have its own adequate cause.

The proximate causes, which we attribute to natural phenomena, can well be invariable or variable; in the latter case, the same principle compels us to seek in their turn other antecedent causes for such modification, and so on, until we eventually arrive at ultimate causes operating in accordance with an immutable law which consequently at all times produces the same effect under the same external conditions.

The ultimate aim of pure science is therefore to seek out the final unalterable causes of the processes of nature. If indeed all processes are to be traced back to such causes, if therefore nature is bound to be entirely comprehensible, or if there be modifications in it which elude the law of compulsory causality, and belong therefore to the sphere of free chance, this is not the place to discuss such possibility; in any event, it is clear that science, whose mission it is to understand nature, must proceed on the assumption of its being intelligible and according to such intelligibility make inquiry and inference until becoming compelled, perhaps by irrefutable facts, to acknowledge its own limitations.

Science considers the affairs of the universe according to two abstract principles, in the one case from their mere existence apart from their effects on each other or on our organs of perception, designating them in such case as matter. The existence of matter in the abstract is to us therefore something passive and inactive; we discern in matter its distribution in space and its mass, which are laid down as being constant in perpetuity. We should not ascribe qualitative differences to matter as such, for if we speak of heterogeneous substances, we then fix their dissimilarities only all the more in the dissimilarity of their effects, that is in the forces they exert. Matter in itself cannot, for that reason, admit the slightest change other than a spatial one which is movement. Objects in nature are not, however, inactive, for we generally arrive at our knowledge of them only through the activities impinging from them on to our perceptive organs, whilst from such effects we infer an effective cause.

If we want, therefore, to apply the notion of matter to reality, we may do so only by adding thereto still a second abstract idea, which in the first instance we were seeking to exclude: to wit, the ability to exert effects, which is the same thing as ascribing forces to such effects. It is clear that the two ideas of matter and force applied in nature may never be separated.

Pure matter would be of little moment for the rest of nature, since it could never establish a modification therein or in our organs of perception; a pure force would be something which ought to exist and yet on the other hand have no actual existence, for "matter" is the name we give to what is existent. It is equally erroneous to seek to define matter as something concrete, and force as a mere idea not corresponding to anything actual; both are, on the contrary, abstractions, each formed in exactly the same way from reality; we can in fact only perceive matter from the very action of its forces, and never in the abstract.

We have seen above that natural phenomena should be taken back to immutable ultimate causes; this requirement thus turns out in such form that in the temporal sense invariant forces have to be found as ultimate causes. Substances with invariant forces (indestructible qualities) have been defined by science as (chemical) elements.

If however we consider the Universe to be composed of elements possessing invariant qualities, then the only changes still possible in such a system must be in spatial position, which is to say movement, and the external conditions by which the effect of forces becomes modified can themselves only be spatial in their turn, so that forces are only forces of motion or impetus, dependent for their action only upon spatial relationships.

Or, to put it more clearly, natural manifestations have to be reduced to movements of matter whose fixed powers of motion depend solely on spatial circumstances.

Motion is a modification of relative positions in space. Spatial relationships are possible only in comparison with bounded quantities of space and not as against undifferentiable empty nothingness. Motion can therefore occur only in practice as a change in relative position of at least two material bodies in relation to each other; kinetic energy as its cause will therefore also only be inferable from a comparison of at least two bodies as against each other and is thus to be defined as the effort of two masses to change their position in relation to each other.

The force, however, which two integral masses exert on each other must be resolved into the individual forces of each of their material particles, i.e. of their molecules. Such points of space filled with matter have, however, no spatial connection with each other apart from the distance between them, since the direction of their line of union can be defined only from a comparison with at least two other points.

Any impulse, which they exert on one another, can therefore give rise only to a change in the distance between them and is therefore a force of attraction or of repulsion. This also validly follows directly from theory. The forces which two masses exert on each other must necessarily be determined by their size and direction immediately the position of the masses is completely defined. However, only one single direction is completely determinable through two points, to wit, that of the line joining their centres of gravity; consequently the forces they exert on each other must lie along this line and their intensity can depend only upon the distance between them.

Thus, the ultimate task of the physical sciences is to reduce natural phenomena to invariant forces of attraction and repulsion whose intensity is dependent on their distances apart. The resolution of this problem is at the same time the necessary condition for a complete understanding of Nature. Pure mechanics has not hitherto admitted this limitation for the concept of kinetic energy, firstly because it was not itself clear as to the source of its principles, and secondly because it is important to be able to calculate the results of complex forces in those cases where their resolution into simple ones has not yet been attained.

Moreover, a considerable portion of the general principles of motion of compound systems is demonstrable only in the case when these act on each other through fixed forces of attraction or repulsion; that is, the law of virtual velocities, the conservation of the plane of the principal axis of rotation and of the moment of rotation of unconstrained systems, the principle of the conservation of energy.

Only the first and the last of these principles are in the main applicable under terrestrial conditions, since the others relate only to completely free systems; the first again, as we shall show, is a special case of the latter, which therefore appears to be the most general and most important consequence of the deductions drawn.

Pure physics, therefore, if it is not to stick half way along the road to intelligibility must arrange its views, on the nature of the simple forces and their consequences, in harmony with the challenge laid down. Its task will only be accomplished when the reduction of phenomena into simple forces has finally been consummated and when it can at the same time be shown that this is the only possible deduction permissible under the circumstances. This then might prove to be the essential concept of the constitution of nature, to which the quality of objective truth would also thus be ascribable.

## WILHELM OSTWALD

All living organisms build up their energetic system in the first place upon the basis of chemical energy. The reason for this is that chemical energy is the most concentrated and most conservable of all the forms of energy available. We have already observed that the stock of free energy, to which all living creatures owe their existence, is being constantly renewed and delivered to us from the sun. The sun, however, only shines half the day on the average and stays beneath the horizon for the other half.

For any organism, therefore, to enjoy a permanent supply of such energy, there must be some kind of arrangement whereby it may not only absorb a proper amount for its immediate needs but garner in also a sufficient stock to take care of its existence for the other half of the day. For, if it so happened that the constant flow of energy through the unchanging organism, which is the very essence of its life, were interrupted for any appreciable period, then the machine would as a rule be absolutely unable to set itself going again, even though free energy were once more to become available. As a matter of fact it becomes increasingly evident in the preponderating majority of organic beings that even a very transitory stoppage may be sufficient to bring things to a permanent standstill and thus cause death.

Every provision must therefore be made to convert the daily supply of energy into some suitable permanent form for preserving the flow at least during the night, and even throughout the winter also, without the creature being reduced to extremities. Of all the forms of energy known to us, none fulfils this requirement better than chemical energy. We see this from the fact that chemical energy serves us as the stock for all other purposes of existence and industry.

Our foodstuffs consist of chemical energy, and the great source of energy for all industry up to the present has been almost exclusively bituminous coal reconverted into chemical energy. It is only quite recently that we have begun to use to any great extent the mechanical energy in the form of elevated masses of water put at our disposal by the sun.

Owing to the general state of affairs on this earth, the constitution of our organic life depends even to its smallest details on the necessity of founding the energetic economy of organic beings on the chemical form of energy. Our muscles work on chemical energy, and similarly the working of our nerves, still something of a mystery to us, is closely bound up with this form of energy. In particular, however, a special property of all organic phenomena depends also in all probability on chemical relationships, and that is the phenomenon of consciousness or intelligence in the most general sense, as was first discovered by E. Hering.

For instance, organic beings frequently demonstrate the property of being able to repeat with much greater ease such processes as have already happened to them one or more times than to carry out something which is entirely new to them. This is by no means a general. thing in nature, and is in fact so rare that in the inorganic world the contrary is more often the rule.

An electric cable does not carry the current any the more efficiently just because it has already served for that purpose, any more than a kettle comes more rapidly to the boil because it may already have been heated up some thousands of times. In other words, inorganic structures do not retain the least vestige of previous experience or history within themselves; a repeated occurrence is for them ever just as new an experience as it was the first time.

Only in exceptional and fairly complicated cases does something of a similar nature become evident, as for instance, when a machine grinds down its moving parts through wear and tear and runs thereafter more smoothly than at the beginning. But even in a case like this, we could reduce such a running-in period to zero, if we polished the working parts beforehand to maximum smoothness. It is a question, therefore, of chance rather than intrinsic similarity.

On the other hand, such behaviour is common in the organic world. Reference has previously been made to the remarkable fact that children become extremely like their parents, even when removed from direct influence of the latter for the greater part of their adolescent years. Among the possibilities as to how parental characteristics are transmitted, the most likely is the chemical theory, that is the hypothesis that such peculiarities are caused by corresponding peculiarities in the germ substance. For we call to mind that, in addition to spatial energy (under which classification I include the energy of motion, gravity and conversion in volume and area), chemical energy alone has proved to be a permanent and inseparable component of ponderable matter. The boundless multiplicity of animate matter can, however, be ascribed only with difficulty to the relatively invariant energy of space; therefore, of the types of energy known, only chemical energy remains as a carrier for these diverse qualities. Of course, we do not exclude the possibility that still other forms of energy, hitherto unknown, may come into the question, though up to the present no signs whatever of these have shown themselves.

Now, chemical arrangements, possessing the same property of recollection, can be imagined and invented, to carry out a repeated process easier than at first shot. I am quite willing to admit that the resemblances are mainly of a purely superficial nature; yet on the other hand it must be remembered that it is but little more than a decade since the field of chemical phenomena now under consideration in this connection, namely the catalytic processes, has been subjected to systematic treatment, and consequently only small portions of its ramifications have meantime come to light.

It is perhaps as well at this point to make some general remarks on questions of a cognate nature. The problems of applied science resulting when any imminent intricate natural phenomenon has to be interpreted on the basis of our knowledge of the laws of nature, are of a particularly treacherous character. Nature obviously has at her disposal the whole stock of possibilities and expedients which are in any way applicable, whereas our attempts at interpretation have to be made with the aid of such little science as we may possess for the moment and of which nothing is more certain than that it is absolutely incomplete.

Let us take but one example from recent practice: up to the present not even a relatively satisfactory idea could be found to explain the generation of heat from the sun, since the known sources of energy all flow too meagrely to supply the enormous needs of the sun's radiation. By the discovery of radium, which represents a source of energy some million times greater in concentration than those heretofore known, the problem has now been decidedly transferred to the realm of the explicable.

It has not indeed been proved that the sun's radiation actually originates in a process of the nature of spontaneous disintegration of radium, but we no longer have any need to say that no known source exists able to meet that requirement, since we do have such sources at our disposal and may even say with some degree of probability, on the basis of this experience, that further sources may exist different in quality but similar in productivity.

Living organisms by their very activity and existence thus set us thousands of unsolved problems, amongst which many are doubtless beyond solution with the means at present at the disposal of science.

The resources of science, however, expand from day to day, and what today not only seems but actually is still unintelligible may on the morrow be a very easily resolved question.

Thus, the information that this or that vital phenomenon is not susceptible to interpretation not only should never be regarded as an indication that it will always remain so, but should indeed be taken as a sign of the very reverse, to the effect that — whoever seeks to interpret all these kinds of phenomena solely on the basis of today's scientific knowledge has not yet properly appreciated the basic principles of the matter. The same holds, too, for those demanding something of a like nature from the champions of scientific biology. This last is said for the ears of those who allege that the phenomena of life are basically unintelligible.

# Chapter 2

# Cell Theory

The microscope was invented in 1590 by a Dutch spectacle-maker, one Zacharias Jansen, a resident of Middelburg. It was some threequarters of a century later, before Robert Hooke was to publish in 1665 his Micrographia embodying the results of his researches in this field. Hooke himself was originally an architect and professor of mathematics addicted to the study of optics and interested also in what was at that time the virtually insuperable problem of building a flying machine. He produced a compound microscope which, in comparison with Jansen's and the model which Leeuwenhoek was using at about the same period, was strikingly perfect.

Hooke gives a description of the structure of cork in the eighteenth section of Micrographia. Cork contains, says he, numerous cavities -which he denominates "pores" or "cells" — reminiscent of the structure of a honeycomb, sponge, or pumice stone.

Almost contemporaneously, Marcello Malpighi pointed out in 1675 that cell structure is observable in many plants and in certain animal organs. For all that, neither Hooke nor Malpighi arrived at the conclusion that the basic factor was actually what was contained in the cell itself and not the material by' which it was surrounded.

Around that same time, in 1673, Leeuwenhoek discovered certain free or independent cells, to which however he did not attribute any specific biological significance; these were the red blood corpuscles, spermatozoa, infusoria, yeast cells and others. Leeuwenhoek, though, was but an amateur not ranking particularly high in the opinion of his contemporaries; nevertheless his discoveries were of very real importance for the advancement of science. These various observations remained unmentioned for more than a century because of the predilection for organistic standards in vogue at that time. The organ, both in plants and in animals, was held to be the source of all functions. At the beginning of the nineteenth century, however, Xavier Bichat in 1802 saw, in a flash of inspiration, that the organs are made up of tissues and that it is these latter which determine the physiological properties of a body and which are subject to disease.

Morphology, that is the study of structure and form, had meantime borne fruit with fresh discoveries — none, however, of startling importance. Corti asserted in 1772 that the viscous liquid which Malpighi had stated as contained in cells, was in fact an ever-present constituent thereof. Somewhat later, in 1835, Dujardin confirmed the existence of a gelatinous substance inside the cells and proposed to give it the name of sarcode. Johannes Purkinje in 1840 and Mohl, the botanist, in 1846 brought the name protoplasm into favour.

In 1781 Fontana was able to give a description of a nucleus and nucleolus in cells and, in 1833, Robert Brown confirmed the existence of a small differentiated "nucleus" in plant cells and suggested that its function was of considerable importance.

## THE CELL BECOMES A BIOLOGICAL UNIT

In 1826 Turpin wrote as follows: "The cell is the fundamental element in the structure of living bodies, forming both animals and plants through juxtaposition." However, the cellular theory did not take proper shape until the famous meeting between Matthias Jacob Schleiden and Theodor Schwann. The former was a lawyer who gave up his profession in order to devote himself to the study of plant structure and physiology.

The latter was an anatomist specializing in zoological studies. In October, 1838, there took place a historic meeting in the course of which the two sages exchanged ideas with one another. Afterwards, Schleiden visited Schwann's laboratory and was shown certain slides of animal tissues which enabled him to verify the similarity between animal structure and plant structure; he opined that the cells are the structural units of the plant, and in the following year, 1839, Schwann asserted that the same holds also for animals. "The elemental parts of the tissues are the cells, similar in general but diverse in form and function. It can be taken for granted that the cell is the universal mainspring of development and is present in every type of organism. The essential thing in life is the formation of cells."

Later on, in 1861, Max Schultze showed that plant cells are in fact identical with those forming the lower animals. He enunciated what is now called the "Protoplasmic Theory," namely that the basic unit of organization in every living being consists of a mass of protoplasm surrounding a nucleus, and that this system of organization is common to all classes of living bodies.

## CELLS IN PHYSIOLOGY AND PATHOLOGY

At the same time, the physiologists and pathologists were increasing their field of morphological knowledge. Of prime importance is Rudolph Virchow's great work Cellular Pathology published in 1858. "Cellular theory applied to all living bodies leads us to a cellular physiology and cellular pathology each based upon histology, that is, upon the anatomical knowledge of the elements.

The seat of illness must be investigated in the cell and symptoms of disease are nothing more in the final analysis than evidence of certain reactions of the component cells of the organism in the presence of the causes of the illness." And forty years later, in 1895, he repeats: "'The essence of disease is found, in my view, in some affected part of the organism, in some cell or conglomeration of cells, some tissue or some organ."

Physiologists like Johann Müller and Claude Bernard, botanists like Strasburger, recognized the truth of this theory which became universally accepted. All biologists since that time speak in terms of the cell and its components. From other quarters we see the emergence of the immortal discoveries of Louis Pasteur. Beginning with 1857 we proceed first from the discovery that the agents of fermentation are live bodies, then this idea is extended to embrace the causes of infectious diseases, which are independently living cells such as Koch was able to cultivate in 1867 outside the body itself, whilst Brücke was able definitely to formulate the general law that "every live body is a cell or a combination of cells."

Cells reproduce to form new individuals or new cells in the same individual. As Virchow wrote: "Every cell must come from a cell" (Omnis cellula e cellula,) combating those many who since the time of Schwann held that the cell coagulated in some manner from plasmatic liquids. This is indeed how the somatic cells which form the individual exist singly or in bunches and how reproduction becomes possible by the division of specialized cells, such as the germ cells, whether sexed or not. One lives and dies by means of the cell.

Since that time cellular physiological and morphological knowledge has increased considerably. Schleicher in 1878 and Flemming in the same year were able to give a description of karyokinesis, the behaviour of nuclear elements in the mitotic division of the cell. Strasburger in 1876 notes, moreover, the centrosome as the main factor in karyokinesis.

Previously, Hertwig in 1875 had studied the fertilization process using ovule and sperm from the sea-urchin. Van Beneden in 1883 and Boveri in 1886 obtained important data whereby it is proven that fertilization is merely the combination of two cells, the male gamete with the female, which promotes the segmentation of the egg already previously observed by Rusconi in 1826 and Newport in 1853.

From all points of the compass we get fresh confirmation that the cell is the real unit of life, that living bodies may consist of one single cell or of a generally most involved and abundant complex of cells. We learn that functioning cells make up the tissues and that through combination of the tissues we get the organs whose sum total amounts to the individual being. Also by multiplication of the cells which are a part of the individual, growth and the eventual reconstitution or repair of injuries is effected, and the reproduction of the individual is fostered so that the continuance of the species is in short nothing but a cell function. It is the cell, moreover, which becomes ill and dies causing the death of the whole body of the individual.

## SUBCELLULAR AND SUPERCELLULAR ORGANIZATIONS

The validity of the cell theory has been under fire of late. It is indeed possible that certain manifestations of life do exist which we may call infra or subcellular. Possibly molecular aggregates of sufficient size may form biological units. It is the so-called filtrable ultra-viruses and similar chemical and physico-chemical combinations which are the cause of certain extraordinary phenomena and lead us to think along these lines.

The living molecules or combinations thereof making up the micelle have their particular structure, chemical composition, and features which remain constant throughout all the processes of nutrition. The micelle or the molecule, each consists of living matter with vital properties distinguishing it as such.

It is further probable that micelles can be divided in the same way as the entity we call the cell, and that in this way they enjoy an independent private life, whence springs their ability to engraft in certain cells which thus become a living culture medium for them. This certainly may be the case with some of the viruses and perhaps also with the genes. The micelle, however, always comes from an organization of molecules and the cell from an organization of micelles.

Let us not forget on the other hand that the organisms come before our eyes as individuals, in the form of a biological supracellular whole. Polyplastid cells do not lead independent lives but must be subjected to the demands of rigid cohabitation whence they derive certain very important and inevitable vital properties. None of this, however, can be taken to mean that the functional and morphological element in life is anything other than the cell. The discovery of this truth, that the cell is the essential element, was one of the most outstanding and important steps forward in the progress of biological science and has enabled human knowledge to conceive ideas of incalculable value for the future.

## ROBERT HOOKE

I took a good clear piece of Cork, and with a Pen-knife sharpened as keen as a Razor, I cut a piece of it off, and thereby left the surface of it exceeding smooth, then examining it very diligently with a Microscope, me thought I could perceive it to appear a little porous; but I could not so plainly distinguish them, as to be sure that they were pores, much less what Figure they were of.

But judging from the lightness and yielding quality of the Cork, that certainly the texture could not be so curious, but that possibly, if I could use some further diligence, I might find it to be discernible with a Microscope, I with the same sharp Pen-knife, cut off from the former smooth surface an exceeding thin piece of it, and placing it on a black object Plate, because it was itself a white body, and casting the light on it with a deep plano-convex Glass, I could exceeding plainly perceive it to be all perforated and porous, much like a Honey-comb, but that the pores of it were not regular; yet it was not unlike a Honey-comb in these particulars.

Next, in that these pores, or cells, were not very deep, but consisted of a great many little Boxes, separated out of one continued long pore, by certain Diaphragms, which represents a sight of those pores split the long-ways. I no sooner discern'd these (which were indeed the first microscopical pores I ever saw, and perhaps, that were ever seen, for I had not met with any Writer or Person, that had made any mention of them before this) but me thought I had with the discovery of them, presently hinted to me the true and intelligible reason of all the Phaenomena of Cork; As, First, if I enquir'd why it was so exceeding light a body? my Microscope could presently inform me that here was the same reason evident that there is found for the lightness of froth, an empty Honeycomb, Wool, a Spunge, a Pumice-stone, or the like; namely, a very small quantity of a solid body, extended into exceeding large dimensions.

## THEODOR SCHWANN

Though plants outwardly exhibit an immense variety of forms, the simplicity of their internal structure is no less remarkable. Their extraordinary variations in form arise solely from the various types of combination of the simple elementary structures, namely the cells which, though showing modifications to a greater or less degree, are nonetheless inherently similar.

The cellular plants consist entirely of cells, some being composed exclusively of single cells, some of groups of cells homogeneously bound together. Similarly the vascular cryptogams in their primitive state consist only of simple cells, whilst the pollen-grain which, according to Schleiden's discoveries, is the basis of the new plant, is in its essence nothing but a cell. In perfectly developed vasculars, the structure becomes somewhat more complex, and up to a comparatively short time ago, their elementary tissues were differentiated into cellular or fibrous, coils or vessels.

Investigations into the structure, and particularly into the growth of such tissues, showed that such fibres and coils were, however, nothing but elongated cells, the coiled fibres being merely deposits in spiral form on the internal surface of the cells. Thus the vascular plants consist also of cells, some of which have reached only one step higher in development. The milk-tubes are the only structure not yet broken down into cells, and further research is necessary into the question of their growth. According to Unger, in his Aphorismen zur Anatomie und Physiologie der Pflanzen (Principles of Plant Physiology and Anatomy), Vienna, 1838, these also are composed of cells whose dividing walls dissolve into nothing.

Animals present a much greater variation in external form than is to be found in the vegetable kingdom and, particularly in the higher more perfectly developed classes, exhibit also a much more complex structure in their individual tissues. How far exactly does the distinction go between muscle and nerve, between this latter and cellular tissue which by the way is like that of plants only in name, or between flexible and horny tissue, and so on?

When we look into the question of development of these tissues, however, it becomes evident that all their manifold forms likewise originate solely from cells, from cells in fact which are absolutely similar to plant-cells and in remarkable accord with them in some of the vital phenomena they produce. The object of the present work is to prove this thesis by a series of observed facts.

First, however, we must give some account of the vital phenomena occurring with plant-cells. Within certain limits, every cell is an individual, independent unit. The vital manifestations of one are repeated wholly or partly in all the others. These individuals are not, however, drawn up side by side to form a simple mixture, but work together in some way unknown to us so as to produce a harmonious whole. The processes which run their course in the plant cells may be reduced to the following main heads:

- The production of fresh cells.
- The expansion of the existing cells.
- Conversion of the contents of the cells and the thickening of cell walls.
- The secretion and absorption effected by the cells.

Firstly, regarding the production of fresh cells. According to Schleiden, in the phanerogams, except in so far as cambium cells are concerned, this always takes place inside the already ripened cells, and in a really remarkable way from the cell nucleus. In view of the importance of this latter in connection with animal organization, I am including at this point a shortened version of Schleiden's relevant description.

The section of an onion, for instance, contains R. Brown's areola or cell nucleus, or what Schleiden calls the cytoblast, varying from elliptical to circular in outline according to the passage from lenticular to perfectly globular shape of the solid it forms. It is mainly yellowish in colour, though sometimes it turns to an almost silvery white, which because of its transparency is often scarcely to be distinguished.

If stained with iodine the shade varies from a pale yellow to very deep brown, whilst its size also varies greatly from 0-0001 in. to 0-0023 in. according to age, to the particular plant and to the particular part of the plant in which it lies. Its internal structure is granular, though the granules forming it are not clearly separated one from the other. Its texture varies from being soft enough to dissolve easily in water to a stiffness capable of undergoing considerable pressure without distortion of shape.

Further to these peculiarities of the cytoblast, already published by Brown and Meyen, Schleiden has brought to light a small corpuscle in its interior which in the fully developed cytoblast looks like a thick gasket or thick-walled hollow globule. The appearance seems, however, to differ in different cytoblasts. At times it is possible only to distinguish the outside sharply defined periphery of the ring, with a dark spot at the centre, or more frequently only a sharply defined spot. In some cases his spot is infinitesimal and so minute as not to be in the least bit distinguishable.

As it will be necessary often to mention this body in the following discussion, I will call it the nucleolus for short. According to Schleiden, sometimes two or even maybe three, or as he has told me personally as many as four such nucleoli may occur in the cytoblast. Their size varies considerably, ranging from the half-diameter of the cytoblast down to practically nothing.

Having done with these preliminary observations, we now pass on to the animal world. The similarity between certain individual animal tissues with those of plants has already been mentioned a number of times. Nevertheless, as is logically correct, no inference has been made from such individual similarities.

Every cell is not necessarily analogous in structure to some plant cell, and as to its having a polyhedral shape this necessarily arises from the cells being closely compacted against each other and is obviously no more a token of similarity than the mere chance caused by dense crowding together. We can only draw an analogy between the cells of animal tissues with plant structures of similar elementary identity on the following grounds:

- By demonstrating that a large part of the animal tissues originates from or consists of cells each having its own particular wall, when it may be possible that such cells do correspond to the cellular elementary structure present in all plants; or,
- By proving in connection with any particular animal tissue made up of cells, that in addition to its cellular structure, forces similar to those operating in plant cells are at work in its component cells; or, since this cannot be proved directly, by showing that the action of such forces in the shape of nutrition or growth proceeds in them in a way similar or identical to that which occurs in the plant cells.

I was considering the matter from this point of view last summer, when in the course of my experiments on nerve-endings in the tails of frog larvae, I not only saw for myself the beautiful cellular formation of the dorsal cord in these larvae, but also discovered the nuclei in the cells. J. Müller has already proved that the dorsal cord in fish is composed of separate cells provided with definite walls and packed closely together like the pigment in the choroid.

The nuclei, so similar in form to the usual flat nuclei of plant cells, might well be mistaken for these, and thus supplied a further point of similarity.

Müller had proved, in connection with the cartilage corpuscles discovered by Purkinje and Deutsch, that because of their gradual change into larger cells they must be hollow and therefore cells in a wider meaning of the term; Miescher also points out a special class of spongy cartilages cellular in structure.

Similarly, nuclei were observable in the cartilage corpuscles. I next succeeded in actually seeing the true wall of these corpuscles, first in the branchial cartilages of frog larva and later also in fish, and the concordance between all of them thus proved that all cartilages possessed a cellular structure in the restricted sense of the word. During the growth of some of the cartilage cells, a thickening of their walls was also perceptible, thus supporting still further the idea of similarity in the process of vegetation in animal and plant cells. Dr. Schleiden at this juncture most opportunely advised me of his admirable investigations on the origin of fresh cells from nuclei inside the parent cell in plants, thus clearing up for me the previously baffling contents of the cells in the branchial cartilages of frog larva, for I now recognized these for infant cells provided with a nucleus. Meckauer and Arnold had already discovered fat vesicles in the cartilage corpuscles.

As, soon after this, I succeeded in ascertaining the origin of young cells from nuclei inside the parental cells in the branchial cartilage, the matter was settled. Cells in the animal body showed themselves with a nucleus whose position with regard to the rest of the cell, shape and variants were similar to those of plant cytoblasts, while thickening of the cell wall took place and infant cells were formed within the parent cell from a similar cytoblast, thus proving the development thereof without the aid of vascular connection.

This concordance was still further proved by many other details and in this way, so far as concerns such individual tissues, the necessary evidence was obtained to show that these cells did indeed correspond to the elementary cells of plants. I soon surmised that the cellular formation might be a widely embracing, perhaps universal, principle governing the formation of organic substances.

Many cells, some with nuclei, were already within our knowledge; for instance, in the ovum, epithelium, blood corpuscles, pigment, and so on. It was an easy step in logic to gather all these known cells into one head and compare the blood corpuscles for instance with the epithelial cells, and consider these — and likewise the cells of cartilages and plants — as tallying with one another and each as facets of the main common principle.

This was the more likely, as many points of similarity in development of such cells were already known. C. H. Schultz had already demonstrated the pre-existence of nuclei in the blood corpuscles, the formation of a vesicle around them, and the gradual distension of this vesicle, Henle had observed the gradual expansion of epidermal cells from the lower layers of the epidermis towards the top.

The growth of the germinal vesicle observed by Purkinje also served in the beginning as an example of the growth of one cell within another, though afterwards it became more probable that it was not a cell but rather a cell nucleus in its nature thus supplying a demonstration of the fact that not everything of cellular form need necessarily correspond with plant cells. An exact term for those cells corresponding to plant cells ought to be adopted, say either elementary cells or vegetative cells.

On still further investigation, I continually found this principle of cellular formation coming into its own. The germinal membrane was soon found to be composed entirely of cells, and, not long after that, cell nuclei and later on cells themselves were discovered to be at the origin of all tissues in the animal body, whence all tissues consist of cells or are formed from cells by various means, thus supplying an alternative proof of the analogy between animal and vegetable cells.

## RUDOLF ALBERT VON KOELLIKER

Our knowledge of the elementary structure of plants and animals is the fruit of work carried out during the last two centuries. This study began with Marcello Malpighi and Anton van Leeuwenhoek, in the period when the microscope, then very simple in design, was first placed in the hands of observers. However brilliant the advances made by science, thanks to the labours of men like Ruysch, Swammerdam and others, such progress had no positive basis on which to establish itself.

On one hand, investigators were not yet sufficiently practised in the technique of microscopic observation, and on the other, the cultivation of the less delicate branches, particularly physiology, embryogeny, and comparative anatomy, absorbed too great a proportion of their attention. Thus in spite of certain isolated observations, valuable from many points of view, made by Fontana, Muys, Lieberkühn, Hewson and Prochaska, histology reached no decisive heights during the eighteenth century, and remained at that time nothing more than a hotchpotch of uncoordinated facts scattered here and there.

Not until 1801 was the genius of one man able to give proper shape to General Anatomy alongside the other branches of anatomical study. This man though certainly he did not enrich histology with any great discoveries did in actual fact bring to fruition something which no one had attempted before him.

He classified the material accumulated by his predecessors and considered the physiological and medical application of what was known up to that time and this with so much success that the science of histology can be definitely dated from his work. The tissues are not examined solely from the morphological point of view, but the relationship between them and their functions both in the physiological and in the morbid sense is also discussed.

Considerable impetus has been given to the immense progress made hi the field of histology in the present century by continuous improvement of methods of research and of the microscope, and, what is more, the ever-increasing zeal of students. It is not to be wondered at, therefore, that during the past fifty years histology has gone far ahead of what was accomplished in the second half of last century, which may be considered as the initial date of its existence.

Over the last thirty years in particular, discoveries have come thick and fast and at the same time it has been possible to link them together in line with their natural logical relationship in such a way that there is now no longer any danger of microscopic anatomy becoming lost in a welter of tiny details, as was formerly the case.

I must particularly point out the year 1838, when Dr. Theodor Schwann proved the unity of constitution of the animal organism, showing that it was originally and uniformly composed of cells and that all the tissues, even the highest morphological structures, were derived from such elements.

This is a fruitful theory combining all observations made up to that time, and acquires greater importance as a starting point for further researches.

If Bichat, by the creation of his system, founded the theoretical basis of histology, Schwann with his practical investigations placed it on a foundation of positive facts. Advances in histology, commencing with Schwann, have indeed been of the greatest importance for physiology, medicine and pure science. Certain points glanced at or only briefly mentioned by Schwann afterwards became subjects for more intensive study, for instance, the origin of the cells, the significance of the nucleus, the development of the higher tissues, the chemical relationships between cells, and many others.

There still remains much to be done, even after all that has already been achieved. Without pretending to prophesy, I may say that the present morphological state of histology will continue until it is no longer possible to penetrate more deeply into knowledge of organic structures, when it is no longer possible to discover fresh facts to show that what we today think of as a simple element is in fact complex. .

If it should ever be possible to discover the molecules which make up the cells, the fibrils of the muscles, the axial fibres of the nerve, etc.; if one could actually fathom the complete laws of the juxtaposition of parts, the laws of their development, of growth, and of the functions of these supposedly constituent parts of the cells; a new era in functional histology would be on the way.

In such an event, the discoverer of the "law of cell-origin," or of a "molecular theory," for instance, would be as celebrated, and rightly so, as those who founded the theory of the cellular composition of all animal tissues, by definitely proving that "all tissues, both plant and animal, are formed and proceed from the cell."

## RUDOLF VIRCHOW

At the beginning of the present century, Bichat established the principles of General Anatomy, and new horizons were opened to medical science. But the progress which histology owes to Schwann has not been greatly developed nor sufficiently applied to pathology. It is a matter for surprise to see how little has been achieved in the period since that time.

Each day brings forth fresh discoveries but it also opens up fresh matters for uncertainty. Is anything positive in histology, we have to ask ourselves. What are the parts of the body whence commence the vital actions? Which are the active elements, and which the passive? These are queries which have given rise to great difficulties, dominating the field of physiology and pathology and which I have solved by showing that "the cell constitutes the true organic unit," that it is the ultimate irreducible form of every living element, and that from it emanate all the activities of life both in health and in sickness.

This manner of holding life to be a special process may be thrown in my face, and many people may incline to accuse me of some type of biological mysticism, impelling me to set life apart from the great agglomeration of natural occurrences and discount the sovereign laws of chemistry and physics. The object of this course is to show that it is not possible to possess more mechanical ideas than those I myself profess, provided that one seeks to explain what happens in the elementary forms of organism.

There is no doubt that the molecular changes occurring inside the cells are referred to some part or other composing them; the final result is however due to the cell from which the vital action started and the living element is not active except in so far as it presents us with a complete whole enjoying its own separate existence.

It has many times been objected that there is no general agreement as to what must be understood by the term "cell." These difficulties date from the work of Schwann himself, who, founding his system on that of Schleiden, interpreted his observations as a botanist in such a way that theories of plant physiology were applied to animal physiology without taking into account that even within the general structure there were plant cells presenting peculiar characters not safely to be identified with the characters exhibited by animal cells. Robert Brown's discoveries contributed greatly to the unification of concepts: I refer to his detection of the nucleus in animal and plant cells. An unimportant role was, however, attributed to the nucleus in the preservation of the cell whilst in contrast its effect in the development of cellular elements was greatly exaggerated.

On the other hand the nucleus, in elements of development, frequently includes an important formation: the nucleolus. According to Schleiden, whose opinion was later adopted by Sahwann, the development of the cell takes place in the following manner.

The nucleolus becomes the first vestige of tissue, coming to life in the innermost recesses of the formative liquid, blastemes or cytoblasteme, and rapidly acquiring definite volume; small granules separate out from the blasteme and settle around it whilst a membrane is formed to cover these elements. Once the nucleus is completed, a new mass starts to surround it. Some time after, a second membrane starts to form at a definite point on the surface of the nucleus, the well-known "watchglass" form. From this is developed the protoplasm and then the cell membrane.

This theory, admitting the development of the cell at the expense of the free blasteme and requiring the formation of the cell to precede that of the nucleus or cytoblast, is generally known under the name of the "Cell Theory." It would be more precise to call it the theory of the free formation of cells.

Nowadays it has been almost entirely abandoned. In spite of most careful investigations, starting with study of tissues under the microscope, there has never been observed any part which, being ready to grow, or multiply physiologically or pathologically, has not contained nucleate elements as the starting-point of the growth process.

In all cases, the first important modifications have their root in the nucleus and this can be confirmed by examining the appearance of the nucleus when an element is about to become transformed. Nothing fresh is created, neither in the shape of complete organisms nor individual elements.

Similarly, the digestive juices do not form the tapeworm, nor do infusoria, algae or cryptogams arise from organic plant or animal waste; in the same way in histology, we deny the possibility of forming a cell at the expense of a non-cellular substance.

The existence of a cell presupposes the prior existence of some other cell — omnis cellula e cellula (every cell must come from some other cell) — just the same as a plant cannot occur except it be derived from some other plant, or an animal from some other animal.

Throughout the range of living beings, plants, animals, or the constituent parts of one or the other, there rules the eternal law of continuous development. Development cannot cease to be continuous, because no particular generation can start a fresh series of developments.

We must reduce all tissues to a single simple element, the cell. The whole of an individual made up of distinct tissues is the result of cellular proliferation. There is only one way to form cells, that is by fissiparity; one element is divided after the other. Each new generation proceeds from some preceding generation.

It is not possible to look on an elementary granule, globule, or fibre as the starting point of histological development; indeed there is no reason to suppose that living elements arise from unorganized parts, any more than it can be assumed that certain substances, liquids, plastic elements, matter or blastema give rise to the formation of cells. The formative materials are to be found within the cell, they are neither extracellular nor precellular.

Such are the observations which seem to me to constitute the startingpoint for any theory of biology. A single elementary form makes up the living body and remains ever constant. This brings us to a consideration of the higher formations, the plant and the animal, as the sum of the greater or less number of like or dissimilar cells each possessing all the characteristics necessary to life.

It is not, however, in a point of superior organization, in man's brain for instance, that the character of vital unity is to be found, but rather in the regular and constant co-ordination of isolated elements. Thus, the individual is the result of a type of social organization, an agglomeration of individual existences dependent one on the other.

But this dependence is of such nature that each element has its own special type of activity and even when other parts produce some motivation on it, some impulse or excitation whatever they may happen to be, the function does not emanate the less from the element itself, because of that.

# Chapter 3

# Significance of Stimulus and Excitation

## IRRITABILITY

"Irritability is the property in muscular fibres of perceiving an irritation and reacting thereto. Thus irritation causes contraction and when such irritation ceases the muscle fibres relax. This is identical with natural perception through irritation of sensitive fibres. The power of movement is a general property of the organs and inherent to life itself . . . Thus irritability is the prime cause not only of muscular movement but also of life in general.

The same thing becomes evident similarly in a state of sickness. Every part of the body suffering discomfort seeks to get rid of it. This effort may be called irritation, whilst the parts able to perceive the damage and react to it should be labelled as capable of being irritated."

Almost a century later, Albert Haller made a distinction between muscular and nervous excitability. "Incitability," he wrote, "is a property of the muscular fibres whilst sensitivity belongs exclusively to the nervous or muscular system containing nerve fibres.

The muscles have the property of contracting quite independently from the nerves, but sensitivity depends definitely on the nervous system." Haller proved by his various experiments that muscle contraction can occur either by exciting it direct or by exciting its motor nerve. The nervous impulse provoked by excitation of the nerve descends through the latter and on reaching the muscle promotes in its turn excitation of the muscle itself and consequent contraction thereof.

"All living matter can be excited," was the axiom laid down by John Brown in 1788. "Life is excitability and we say living matter is excitable because it can be affected by outside forces." According to whether excitability increases or diminishes, the resultant infirmities are called aesthenic or asthenic.

## ELECTRICAL STIMULUS

Luigi Galvani's discovery in 1791 was an important scientific achievement; he found that when two dampened metals, in contact, touch a muscle or motor nerve they cause the muscle to contract. This was a chance discovery, made during other experiments; as Galvani himself explains: "I had dissected a frog and having duly prepared it for quite a different purpose placed it on the top of a table on which stood an electrical machine.

The frog was not in contact with the conductor of the machine in question and indeed was a fair distance away from it. One of my assistants chanced to touch the leg nerves of the animal with the point of a scalpel and immediately the lower members contracted as if they had been attacked by violent tetanic spasms.

This happened only when the conductor was made to spark." Subsequent to this observation, Galvani found it was sufficient to make a circuit with two different metals closing contact with the animal in order to cause a contraction of the muscles. A few years after this important discovery, Galvani was deprived of his professorship as a reward for his political integrity and died practically in penury.

Baron Humboldt gave the name of "galvanic" to these results, thus honouring their discoverer, and observed that through excitation direct or on a nerve, not only were muscular contractions produced but sensations also were aroused which fomented the working of the glands. This he assumed to be due to what he called "galvanic flux." Alexander Volta, professor at Pavia, discussed the nature of these observations with Galvani to decide whether they were due to a cause originating in the animal's body or whether this arose from the contact of the two metals reproducing the effect of the electrical apparatus. This latter idea which finally won the day led Volta to the invention of the electrochemical cell which now bears his name.

The electric current is produced by the interaction of two different metals or of one metal and carbon — "metals or carbonaceous bodies" a and this is the current which stimulates the organ. This is an important fact, but years later it was found that the converse also is true, that living tissues in action give rise to variations in electric potential.

In 1802, Volta read before the French National Institute in Paris a paper proving the identity of the galvanic fluid with electric flux and also mentioned the physiological properties thereof.

The Great Proconsul awarded him a special gold medal bearing the terse inscription "To Volta." Thus was discovered the agent which on application to the organs causes excitation in them. Nevertheless half a century went by before general use was to be made of electricity as an exciter so easy to handle, so accurate and of such great importance in physiological research.

## IRRITABILITY AND ILLNESS

Whilst this was going on, the doctors still indulged in their theoretical speculations on excitation or irritation and its normal manifestations particularly in the pathological field. Rasori in 1807 distinguishes between aesthenic and asthenic illnesses, namely those which he holds to be due to constriction and those which may be due to relaxation.

Aesthenia arises from an excess of stimulus and is combated with counter-irritants, such as sedatives, antimony, opium or bleeding. Asthenia is itself a counter-irritant and its remedy is therefore some stimulant such as gamboge, belladonna, strychnine, ipecacuanha or nux vomica. These ideas were inspired by Francis Joseph Victor Broussais, Bretonborn like Laennec, though of vastly inferior calibre to the latter. Broussais was an army doctor in Napoleon's forces, assistant professor at the Military Hospital at Val-de-Grâce, and in 1830 professor at the Faculty of Medicine in Paris. He was a pupil of Bichat and thought himself sufficiently well up to run the Physiological Medicine courses. He lacked, however, the genius of his young teacher.

Broussais was an inveterate theorist who, writing certainly with great eloquence, acquired enormous prestige whilst his works became unaccountably acclaimed, thus causing havoc throughout the world particularly in the nations of Southern Europe and America.

His influence extended almost to the end of the nineteenth century. "Life," he says, "depends on the stimuli setting sensitivity and contractility in motion. Increase in these factors gives rise to irritation which can extend its influence to other parts of the body by means of the sympathy exercised through the nervous system.

As irritation increases so does the body heat, stimulating the chemical processes of life. Increase in irritation is inflammation, and subsequent increase in heat is fever. The part of the body most sensitive to irritation and most liable to receive and transmit the sympathetic action is the stomach and intestines. Fevers and pyrexia are manifestations of gastrointestinal inflammation and this being so the basic disorder of the human body is gastro-enteritis.

Thus any sickness must be treated with diet and bleeding." Broussais's work was responsible for the fact that all patients for quite a hundred years were atrociously subjected to diet and that the use or rather abuse of bleeding was practised with zest by nearly all doctors, quacks or otherwise.

In those cases where bleeding was not thought to be absolutely necessary, the least that was done was to inflict on the sufferer some dozens of leeches. As Castiglioni reminds us, the collection and breeding of leeches was one of the most prosperous industries in Europe thanks to the influence of Broussais.

## THE PHYSIOLOGY OF EXCITATION

These outrageous ideas were opposed by the scientific judgment of Claude Bernard, who, by means of his physiological researches, sought repeatedly to bring some order and arrangement into the ideas then current.

He published his celebrated book L' Irritabilité (Irritability), and in 1878 brought out his Leçons sur les phénomènes de la vie, communs aux animaux et aux végétaux (Lectures on reactions common to animals and plants). Irritability is a synonym for excitability.

"Every anatomical element, that is to say the protoplasm intrinsic in its make-up, possesses the property of being set in operation and of reacting in a particular way under the influence of outside stimuli." The concept of life is inseparable from that of excitation.

"Every living thing is irritable because anything which lives reacts to excitations." There is no vital occurrence which is not the consequence of some type of excitation. "Irritability is an intrinsic factor in perception and motion." Stimuli may be of various types, mechanical, physical or chemical. Claude Bernard was constantly making use of electrical stimulus in his researches, though on occasion he did avail himself of mechanical means when, for instance, he discovered the effects of the glycosuric puncture, whilst in other cases he used stimuli of a chemical nature.

Excitation usually produces some increase in functions. In 1846, however, the brothers Weber had already noticed that electrical stimulus of the vagus or tenth cranial nerve in the neck could cause the heart to fail.

They found other instances at later dates. Excitations can therefore be of a stimulating nature or on the other hand they may cause inhibition.

## EXCITATION, FUNCTION AND NUTRITION

It was a great step forward in the theory of excitation to connect it with the condition of metabolism, and it was in Claude Bernard's works that the idea first appeared that excitation is disassimilative and a destroyer of living matter which it thus converts into excremental products. This idea is established pithily in the phrase: "Life is death."

However, up to the time of Hering in 1878, excitation was not identified with disassimilation nor inhibition with assimilation. This concept was crystallized by him in his theory of colour vision. Years later in a famous lecture in 1889, published by the Prague medical periodical, Lotos, Hering confirmed and extended his idea of the nature of excitation and inhibition to cover widely different functions.

In 1886 Gaskell gave a detailed account of these ideas and suggested the terms anabolism and catabolism to distinguish the two phases, the assimilative and the disassimilative, of metabolism.

Verworn also in his Allgemeine Physiologie (General Physiology), which had such a formative influence on physiologists at the beginning of the present century, finally codified the theory with some well-chosen examples.

Every effect of excitation is accompanied and conditioned by acts of disassimilation, of great intensity in the muscle but on the other hand very light in the nerve. These actions are assuredly the two most distinct and characteristic. The effect in the muscle is well known through the evidence of motional and thermal discharges of energy and by the metabolic changes which are inherent to excitation.

The nerve effect is a matter of some considerable discussion. Thunberg showed in 1904 that the nerve breathes just like other organs, and Tashiro in 1913 confirmed that gaseous interchange in such nerve is intensified on excitation. An increase in heat production consequently takes place within the nerve, but so negligible that at first physiologists in general thought that no such thing did in fact occur.

Later research by A.V. Hill, Gerard, and Zottermann has enabled such functional heat to be measured, on the scale of one millionth of a Centigrade degree for each nervous impulse. One ten-millionth thereof corresponds to the initial heat and the remainder to the heat of recuperation. In 1938, Bronk gave a clear sketch of the development of these important phenomena.

## PHYSICAL CHEMISTRY OF EXCITATION

Certain modifications in the physico-chemical structure of an organ occur under excitation alongside the chemical effects mentioned. Du Bois Reymond in 1848 discovered the negative variation, that is that the field surrounding a part under excitation becomes electrically negative in relation to the remainder of the organ.

Hermann in 1867 describes the "demarcation" or critical current, at which any injured portion of the surface or inside an organ is also negative in relation to the uninjured surface. It is easy to infer from these results, that wherever molecular simplification, disorganization, decomposition or catabolism occur, there will be a zone of negative potential as against the unaffected portions.

The source of currents due to action, the negative variation which is observed in bipolar conduction, is referable to that of the demarcation currents. Which again confirms that every excitation is accompanied by catabolic discharges. This entails alterations of the ionic and colloidal balances in those physico-chemical systems represented by the living tissues. Hermann in 1879 assumes that electric charges of one sign or the other, or as we now call them ions, are distributed on both sides of a membrane, impermeable or practically so, whence electrical polarization arises. Stimulus increases the permeability of the membrane in such a manner that the ions hitherto held back may pass through it. This is what we know as excitation, as already previously described by Pfeffer in 1873. Nernst in 1908 propounded a similar excitation theory.

The process of excitation would presuppose the concentration of certain ions on one side or the other of the membrane. As Helmholtz stated in 1853: excitation demands a minimum concentration of such ions, otherwise it will not take place. Both theories agree in considering that the basic condition in the process of excitation is a modification of the respective concentration of ions on either side of the membrane.

These theories were modernized in mathematical interpretation by A.V. Hill in 1910 and Lorente de Nó in 1946.

In the excitable elements, particularly the nervous elements, there is a state of electric polarization of the cytoplasm in the surface of the cell, whether nerve-cell or some other type, where manifestations develop in the same way as if there were a membrane.

Such polarization is tied up with the life of the cell, and when the latter dies, polarization vanishes. There is therefore in every living element a difference between the electric charges residing in it and those of the surrounding medium, which difference is maintained by the living matter forming the surface or membrane of the cell.

Excitation is accompanied by instantaneous depolarization whether it be a case of natural excitation, say of a muscle by nervous impulse, or of an artificial excitation, say by electrical stimulus. Such depolarization is arrested in the very act through a reciprocal repolarization process which restores the equilibrium of the electrical charges and brings them back to a state of rest. Obviously such changes must be reversible.

Variations in electric charges have to be pretty rapid to promote excitation before the cell becomes repolarized, and to determine an impluse or activation which is transferred along the cell from the nerve. Yet, on the other hand, they cannot be absolutely instantaneous, since in that case there would be no efficient employment of exceedingly slow changes in electrical potential, that is the electric stimulus, nor of changes too great in speed of the high-frequency currents.

The gradient of rise and fall of electric potential and the duration of the change in potential or stimulus constitute, as we have already seen, very important factors in the study of excitation. The physical and physico-chemical changes caused by stimulus on an excitable cell are immediately translated into metabolic effects.

## MATERIAL FACTORS IN EXCITATION

From the foregoing we arrive at two important facets of the question. On one hand, there is the possible participation of material factors in excitation, a problem which is still occupying the minds of contemporary physiologists and which at the present time has reached a high stage of development. Langley in 1906 played with the idea of a "receptor substance" intervening when an impulse is transmitted from nerve to muscle; then there was Elliot's "myoneural connection," "intermediate material," and so on.

The latest evidence of the effects of adrenaline in the case of certain organs, of acetylcholine in others, and of the potassium ion and other electrolytes, certain hormones, parahormones and drugs of particular classes is proof that material factors do play a part in the excitation processes.

## EXCITATION AND TIME

A second aspect comes from consideration of the time factor conditioning such occurrences. In point of fact, the main object of Nernst's theory was to explain why high-frequency alternating currents did not act as stimuli. J. Müller in 1846 held the view that "the time needed for the transmission of a sensation from the periphery to the brain is infinitesimal and therefore incommensurable."

Helmholtz in 1850 showed that this was not necessarily so, that nerve impulses are not transmitted instantaneously, but quite the contrary, they travel at the relatively low speed of slightly less than 100 ft/sec in cold-blood creatures such as the frog, and around three times as fast on the average in warm-bloods such as man, the rapidity of conduction varying with the nature and diameter of the nerve fibres, physiological condition, etc.

This relative slowness in action requires that stimulus, if it is to be effective, shall be applied to the excitable organs with sufficient suddenness. This also, however, has a limit. The writer named did not in his researches take into consideration the minimum time the stimulus would have to last, i.e. "the characteristic time" for each tissue as Waller called it.

Hoorweg and Weiss in 1907 showed that if the period of actuation were too short the stimulus had no effect. The work of the Lapicques culminated in the notion of "chronaxy" embodying two parameters, the intensity of stimulus and its period of application in the process of excitation. It was known from previous work that if the stimulus fell below a certain intensity, excitation was not produced, such minimum excitation threshold intensity being denominated by Lapicque as the "rheobase."

When therefore it is a question of determining the influence of time in excitation, we have to use a stimulus which is quantitatively sufficient, for instance an electric current of twice the threshold intensity. It is then necessary to measure the minimum effective time, by shortening the period of stimulus gradually, until we reach the minimum or "chronaxy." Obviously a body will be the more excitable the lower its chronactic number.

Modern researches have shown how excitation develops in relation to time, such periods being reckoned in thousandths of a second. In the first place, excitation does not commence simultaneously with the application of the stimulus, but only after a period of latency which usually is fairly short. Einthoven thought that there was no such latent period, and that the delay observable in the response of a muscle to direct excitation or to a nerve impulse depended upon purely mechanical conditions such as the inertia of the system or the elasticity of the organ or some such thing.

Exact measurements by Cooper and Eccles in 1930 showed that in a muscle the actual latency period is onethousandth of a second; similar values were observed for nerves. In muscles, impulses travel slightly more slowly than in the nerves. The excitation process continues splitting into various phases according to the metabolic changes taking place. Some considerable time after Marey had studied the phase of inexcitability in the heart, Gotch and Burch in 1899 found that there are also moments of refractory condition in the nerve; first an absolute period of total inexcitability, then a relative period of progressive recuperation from the excitability.

Adrian and Keith Lucas discovered that after the refractory condition there developed a period of supernormal excitability followed finally by normal repose. It is not difficult to approximate these facts to respective metabolic conditions.

Inexcitability in the course of full excitation must be attributed to the disassimilative flow produced in those particular instants. Supernormal excitability is connected with the assimilative reaction which balances the preceding functional catabolic phase. Researches by the aforesaid Lucases in 1912, by Lillie in 1922 and of other still more recent investigators confirm this.

## SIGNIFICANCE OF EXCITATION IN LIFE

We are gradually getting to know more exactly what excitation may really be, but the process still possesses many unknown features and presents itself in many guises affecting all forms of animate matter. In life there appears a certain functional tonality, a primary biological feature or property allied with a particular metabolic tone. Tonality can be modified in either direction: increase signifies excitation, and decrease, inhibition, producing corresponding modifications of the metabolic tone, namely dominance of catabolism or of anabolism.

Such variations occur in a wavy ever-moving line resulting from the influence of stimuli which previously had been thought to be always outside the animate being, but which as we now know may also be internal to it. This is why the life "quantum" is not uniform but varies constantly in accordance with external and internal conditions.

This possibility of suitably exaggerating or diminishing the various functions, a possibility which may be applicable to all animate manifestations, allows the adaptation of living bodies — animal principally — to changing conditions to which they are ever subject, and is a prime, positive and essential condition for the maintenance of existence.

## GIOVANNI ALFONSO BORELLI

The nerve fibres are not solid, hard-packed and impermeable bodies, neither are they hollow empty tubes like a reed, but they are conductors made up of a spongy matter similar in structure to elder pith. Thus the marrow of fibre may be easily moistened by volatile juice from the brain with which such fibres are in communication and can moreover be saturated until they swell up like a sponge soaked in water.

All writers now admit that the nervous juice is not an accumulation of wind or of inhaled air but a liquid of the consistency of spirits of wine. This nervous juice even though spirituous and active is always of a corporeal nature and is unable without physical contact to intensify or lower the animal spirits. It is by means of its corporeal presence that on mixing with the animal spirits which also are of a corporeal nature it heightens the action of such spirits or drives them out or changes them in some way.

It is impossible to imagine that nervous action may be developed without some local movement of the nervous juice, passing along the nerve in one direction or the other, either from the surface to the brain generating sensations, or from the brain to the muscles thus causing movement. In this latter direction, the nervous humour as conceived by Willis fulfils a double function. It not only gives rise to visible movements through use of the muscles but also exerts a variable power of nutrition whereby the foodstuffs supplied through the blood are converted into living flesh.

If one of the ends of a nerve is compressed, nipped, struck or punctured, the disturbance in the form of an impulse or wave is communicated to the opposite end, since in proportion to the nearness of the parts arranged in ordered series the impulse keeps flowing successively throughout the whole nerve. From this it follows that the fibres or spongy ducts of certain nerves distending with spirituous liquor can be vibrated through the gentle motion of the spirits by means of which the brain communicates to the outer reaches the actions impelled by the mind.

The impulse will descend through the whole length of the nerve and then by irritation of the end there will be driven out or discharged from that extremity a few droplets of spirit into the corresponding muscle causing a type of explosion or ebullition which makes the muscle contract and become tense.

Again, if one extreme of a sensory nerve terminating in the skin, nose, ears or eyes is compressed, pinched or tickled the spirituous juice of such nerve transmits the wave or impulse to the spot in the brain where the nerve ends. And here, the power of the sensitive pith in relation to the region of the brain affected together with the force of the disturbance and the manner or quality of the nervous movement, will produce an assessment of the object causing such movement.

## FRANCIS GLISSON

Now I come to the second reason which shows that animal spirits are not the immediate cause of the feelings and movement. It is the exception in nature for the same conduit to carry matter in both directions, flow out and flow in. In breathing, the air, inhaled as well as exhaled, certainly travels through the same organs, but the cartilaginous pipes forming the breathing passages are made for that special purpose and the thorax expands and contracts alternately. Also in the viscera peristaltic and antiperistaltic motions take place but these are necessary for the preparation and conveyance of the chyle. In the case of the nerves, it would be difficult to conjecture why the spirits flow sometimes forward and sometimes backward.

The sensation produced by a hard object lasts for many hours in certain cases. How can we imagine that all this time the animal spirits might be going to and fro from the sense-organ to the nerve-centres and that the sensation should remain continuously throughout such time? This problem is beyond me.

It is said, moreover, that these spirits cause the fibres to swell in motion. Certainly the muscles contract in length but they also thicken at the same time. The experiment I shall describe below contradicts the idea that muscular contraction is something of the nature of an explosion or outburst, that is to say a distension or swelling of the muscle.

Take an oblong glass tube of suitable size and shape, and to its mouth affix another upright tube ending in a funnel. Get a strong muscular man to thrust his bare arm into the large tube and stop the mouth of the tube immediately with clay around the start of the arm near to the shoulder, in such a way that there will be no escape of the water with which the space between the walls of the tube and the arm is to be completely filled by means of the funnel at the top end of the vertical tube.

The water must reach a certain level in the vertical tube. Instruct the patient first to contract his arm muscles strongly by clenching his fist and then to relax them right away. It will be seen that during contraction the level of the water falls and that it rises again on relaxation. Clearly, therefore, during contraction the muscles do not distend but quite the contrary they decrease in volume.

This shows that the fibres contract by reason of their intrinsic vital activity and not because they stretch and consequently become shorter from a flow of vital or animal spirits bearing orders given out by the brain and causing such voluntary movements.

This intrinsic property of the muscle is similar to that of the gall bladder which though constantly receiving bile only empties itself at particular instants. This is the property of being irritable and such property of certain organs may be called "irritability."

## WILLIAM CULLEN

While the doctrines of Stahl were prevailing in the University of Halle, Dr. Hoffmann, a professor in the same university, proposed a system that was very different. He received into his system a great deal of the mechanical, Cartesian, and chemical doctrines of the systems which had appeared before.

But, with respect to these, it is of no consequence to observe in what manner he modified the doctrines of his predecessors, as his improvements in these respects were no ways considerable, and no part of them now remain; and the real value of his works, beyond what I am just now going to mention, rests entirely on the many facts they contain. The merit of Dr. Hoffmann and of his works is that he made, or rather suggested, an addition to the system, which highly deserves our attention. Of this I cannot give a clearer account than by giving it in the author's own words.

"Defects in the microcosmic movements of solids may explain, through nervous affections, a certain type of internal malady. Alterations which run their course through the nerves or nerve membranes will give rise to anomalies outside the usual course of nature, sometimes serious, sometimes of little consequence. Careful observations confirm the reality of unhealthy motions which have their origin in the alteration of influences emanating from the nervous organs, which by way of the nerves, extend their action to all parts of the body.

It is these conductions of the subtle juices which, alternating in condition from systaltic to diastaltic, arrive in the gut, belly, gullet and anus, in the arteries and bile, salivary and urinary ducts, muscles and subcutaneous regions and the very membranes of the brain and medulla, in the sensory organs and ligaments which enfold the bones and form the joints. There is no pain, inflammation, spasm, loss of power, fever, alteration of the humours or of the secretions in which such influences do not play some part."

But, in these writers, it was either not extensively applied to diseases, or was still so involved in many physiological errors, that they had attracted little attention; and Dr. Hoffmann was the first who gave any tolerably simple and clear system on the subject, or pointed out any extensive application of it to the explanation of diseases. There can be no sort of doubt that the phenomena of the animal economy in health and in sickness, can only be explained by considering the state and affections of the primary moving powers in it.

It is to me surprising that physicians were so long of perceiving this, and I think we are therefore particularly indebted to Dr. Hoffmann for putting us into the proper train of investigation; and it every day appears that Physicians perceive the necessity of entering more and more into this inquiry. It was this, I think, which engaged Dr. Kaaw Boerhaave to publish his work entitled Impetum faciens; as well as Dr. Gaubius to give the Pathology of the Solidum vivum.Even the Baron van Swieten has upon the same view thought it necessary, in at least one particular, to make a very considerable change in the doctrine of his master, as he has done in his Commentary upon the 755th Aphorism. Dr. Haller has advanced this part of science very much by his experiments on irritability and sensibility. In these and in many other instances, particularly in the writings of Mr. Barthez of Montpellier, of some progress in the study of the affections of the Nervous System, we must perceive how much we are indebted to Dr. Hoffmann for his so properly beginning it.

## LBRECHT VON HALLER

Not only in the animal kingdom but also among the plants there is to be found a widely spread contractile force which causes the fibres to approach one another. This appears to us to be the cause of cohesion in general, which is evidenced when a fibre stretches or tightens and then relaxes returning to its original length, the cause being an elastic force of both a physical and a cohesive nature.

On the other hand a live or dead fibre is also given to shrinking when heat is applied or if it be submitted to some other treatment. Such behaviour is to be seen in all living bodies which are not of too soft a nature, such as the cerebral pulp, or too hard, like the bones or teeth.

Alongside this, however, we must consider a special type of contractile force inherent to the muscles. In a living animal or in one which has just died there frequently occurs in the muscular tissue a rapid and striking contraction, during which the end of the muscle is brought nearer to its middle and after relaxation pushed away therefrom.

This contractive motion cannot normally be produced spontaneously except when brought about by specially applied irritations in the live animal. The force causing contraction in the live state cannot help but be something different from that provoking retraction in the dead body, since each of these processes differs in the laws which govern it, both as to duration and as to the point at which it develops.

The force at play in the live being may be called "inherent force" (vis insita) and the tissues possessing it may be called, as Glisson suggests, irritable. This inherent force, quite different from elasticity and from that causing retraction in dead tissues, both of which latter are common to all the fibres, is characteristic of the muscular fibre and constitutes a particular feature thereof distinguishing the physiological character of such fibre, which is rendered distinct from the very fact of its being irritable. We are dealing therefore with a special force different from any other, which may be classified among the causes of motion, the ultimate source whereof is unknown. It is inherent in the fibre itself and does not reach it from any other part remote therefrom.

Therefore the heart and the intestine containing the muscular fibre are by their very nature irritable; they are able to function by themselves alone without any influence reaching them from outside.

Some writers are inclined to call this the "vital force," but such term does not seem justified in my view, since it is possible to demonstrate effects caused by it even after life has departed. I prefer to call this the "inherent or intrinsic force" of the muscle. We have seen that it comes into action through the application of various stimuli, but the very interesting fact also comes out that a very small stimulus may provoke a movement of great intensity.

Alongside this force there is another which may be exerted on the muscle. It is carried from the brain through the nerves and sets the muscles in motion. This we call the "nervous force" (vis nervosa). Like the inherent force it survives the death of the body, as may be demonstrated in cold-blooded animals recently dead, wherein without either sensations or voluntary movements, contractions are produced to irritate the nerves ending in the muscles, provided such muscles remain damp and intact. The same may also occur in warm-blood creatures.

The ancients attributed to the nerves a subtle and refined "humour" of great fluidity. The word "humour" however suggests the idea of something indolent, lazy, sluggish, slow, dawdling or heavy, for which reason we prefer to call the active agent "spirit," that is, something movable, invisible and powerful like the air. These theories persisted for many centuries. At last, like so many others which were good enough for antiquity, they threatened their own destruction. Thus it is that an important school like Stahl's came to think that the soul might act directly on the whole body without needing any intermediate instrument to transmit orders from a distance. In point of fact, the existence of certain spirits, which would be nothing more or less than material principles, is somewhat doubtful.

It has been put forward, as proof of the non-existence of such a class of spirits in the nerve, that the latter does not stretch or swell above a constriction which would not allow such spirits to pass on to their fate.

This objection is invalid, since the fluid might be so tenuous that the constriction would be no obstacle to its transmission. There are no proofs, on the other hand, that these hypothetical vital spirits are of an alcoholic, acid, sulphurous, albuminoid or combustible nature.

Other writers have assumed that they were aerial in nature and during the present century the name "ether" has become fashionable to such an extent that it is customary to attribute to the ether everything whose causes not being susceptible of practical demonstration remain unknown, such as light, the force of gravity, and magnetism among others.

In line with this tendency certain people assume that the spirits, the "nervous fluid," are of an ethereal nature and compounded of ether. This idea does not seem justified to me, nor that such agents should be credited with electrical properties.

Of what nature, then, can these spirits be? A spirit must be an element with its own essential properties distinct from everything else, of so tenuous a nature that it cannot be perceived by any of our senses, but coarser than fire, ether, electric or magnetic flux, since it can be contained in ducts, restricted by binding and produced and maintained by food.

All of which allows us to assume, since light is different from fire and the material of a magnet differs from one to another, whilst air and ether are two entirely different things, that what we call spirits, the inherent force as well as the nervous force, may be some special element peculiar to living organs and that we may know them only by their effects! A property of living matter.

## CLAUDE BERNARD

Protoplasm, the agent of the phenomena of organic creation, not only possesses the power of chemical synthesis which we have previously examined, but, furthermore, exhibits for the purpose of setting such power in play, the faculty of being "irritable," and consequently through its irritability the power of motion. It can, in fact, react by contracting under the action of external excitants. Happenings in life are not spontaneous manifestations of any internal vital principle.

They are on the contrary the result of conflict between living matter and external conditions. Life is manifested through the reciprocal connection between these two factors in the phenomena of sensitivity and movement which are considered in general to be of a high order, as well as by those due to physico-chemical action.

This continuous relation between organic substance and its surrounding medium is a general feature of organic life, just as it is of the socalled animal life. Nutrition, like sensitivity or movement, shows in more or less complex form this property of response to excitation. This faculty, which is an essential condition of all life phenomena in plants and animals, exists in its simplest form in the protoplasm. It is "irritability."

In generic terms we may define irritability as "that property possessed by every anatomical element, that is to say protoplasm, of being set in motion, and of reacting in a particular way under the influence of excitants." Every vital manifestation demanding the combination of set conditions or external excitants is a manifestation of "irritability." Sensitivity or perception which is in the highest degree a complex phenomenon cannot be other than a particular case of irritability, which is the only fundamental biological property common to both living kingdoms.

We must consider irritability a simple form of perception and, reciprocally, in perception we find a very high form of irritability, that is of the property common to all live beings of reacting according to their particular nature under alien stimuli. Linnaeus made perception the distinguishing feature of the animal kingdom, saying: "Plants exist, animals feel".

If, however, we examine this sensibility, seeking to acknowledge it as a higher attribute of animal life, we soon become convinced that it is not a simple "property" but a complex vital manifestation corresponding to a "function." Moreover it is necessary to distinguish between the functions of a live body and the properties of the organic matter composing it. Perception may be a complex phenomenon inherent in certain organisms, but it arises from a simpler general source.

Broussais did not accept any essential property of organic substance other than irritability, which implied consequently perception, contractility and other secondary actions. Virchow likewise held that vital phenomena possess the essential feature of irritability, a generic term comprising all other vital properties.

This theory is to be found in its incipient state even in Bichat, who used the word "sensibility" or "perception" under all circumstances, thus causing considerable confusion; nevertheless what he understood by sensibility was what we now call irritability. In his time no distinction had yet been made between the two concepts. Bichat recognizes in animals an "animal sensibility" and on the other hand a "vegetative" or "unconscious sensibility" residing in the vegetative organs and translatable into visible actions on the part of the organs when they are aroused by an internal stimulus.

It may, however, happen that such reaction to physiological or artificial excitants is not translated into any visible sign or movement, and yet nevertheless the reaction does develop, being lost intrinsically in the nutritional motions which do not come into evidence except by their subsequent effects. This is what happens particularly in the case of plants, wherefore Bichat attributed to plants and certain parts of animals an "imperceptible sensibility," that is, something not evidenced by any visible sign whatsoever.

Whatever opinion one may hold regarding the choice of these terms—conscious perception, unconscious, unknowing, imperceptive or senseless sensitivity— it must be recognized that they stand for actual facts and tally with a proper feeling of reality. All actions of the organism are acts provoked by external or internal, physiological, normal, abnormal or artificial stimuli.

For their fulfilment they require perception; if this word is not to cover anything other than the power of reacting to stimulus, let us call it "irritability." It is of course correct that in such reaction there may be various stages, from "purely nutritional or trophic reaction" – imperceptible – up to "motor reaction" which is clearly appreciable, and finally the highest stage of "conscious reaction."

## Chapter 4

# Biocatalysts

### CATALYSIS

In 1812 Kirchhof discovered that if starch were heated with sulphuric or other strong acid solution it was converted into sugar without any change or loss in the acid itself. Thénard noticed that hydrogen peroxide solution decomposed, giving off oxygen in the presence of undissolved bodies, such as finely divided platinum, silver, manganese dioxide or blood fibrin, not entering into the reaction, whose mere presence was sufficient to produce this effect. Sir Humphry Davy showed that platinum in the powdered or spongy state, and therefore of greatly increased surface area, can oxidize alcohol vapour exposed to the air without the application of heat.

Berzelius in 1837 called general attention to the meaning of these facts: "In reactions of this nature there is present a chemical force quite distinct from any hitherto known," and he proposed to call such operations catalytic in view of their being absolutely different from analytic actions resulting from the use of those reagents normally used in chemical research. In ordinary analytical processes the reagents come in according to the laws of affinity of normal chemical relationships, whereby they are transformed whenever they are present. On the other hand, in catalysis, the factors giving rise to the action have their effect merely by being present and in themselves suffer no intrinsic change whatever.

Prior to this basic statement by Berzelius, which is one of the cornerstones in the history of chemical science, Dubrunfaut in 1830 had already made one important discovery, namely that fermented barley extract or malt converts starch into glucose even in the cold state, similarly to the way an acid acts in the presence of heat as above mentioned. Shortly thereafter, Payen and Persoz in 1832 succeeded in separating the amylolytic principle from malt by extraction in water and precipitation with alcohol, giving it the name diastase.

They considered this agent identical with the natural ferments, which cause souring in wine or milk. Robiquet consequently showed that bitter almonds contain a protein substance which breaks amygdalin up into prussic acid and glucose; this active protein was christened emulsin by Liebig and Whöler.

The fermentative properties of the gastric and pancreatic juices came under consideration practically at the same time. Pepsin was discovered by Eberlé and Schwann, and trypsin by Corvisart. These newfound substances were to be of the greatest value in bringing the study of digestive chemistry nearer to that of the catalysts. Thus it is that the digestive diastases at that time came to be the most typical examples of physiological catalysis, and that is how diastasis and catalysis acquired their greater renown as factors in the vital processes. The idea got abroad that certain animal organisms contained or produced diastases in just the same manner as did certain vegetable tisses.

## FERMENTATIONS

Pasteur's researches on the processes of fermentation extended from 1857 until 1870. Some score years previously, in 1838, Cagniard de La Tour asserted that fermentation was due to microscopic beings. Berzelius opposed this idea that fermentation should be one of the vital processes. "The case of fermentation," he wrote, "is similar to that of the decomposition of hydrogen peroxide in the presence of platinum, silver or fibrin. A catalytic action like any other."

Pasteur's great work definitely proved the effect of various yeasts and bacteria in different fermentation processes and determined the conditions in which such processes could develop. "Fermentation is the work of living organisms," he wrote, and very soon proved the fact that not only are single-celled micro-organisms possible factors of fermentative change, but also that cells from metazoic or metaphytic tissues existing communally in polyplastid organisms might also play their part in the process. Fermentation might therefore be shown to be a vital property inseparable from the structure and functions of the cell.

Lebig, from 1839 onwards, maintained that ferment is a soluble matter of protein nature working through catalytic action, and that structural or functional life was not essential to demonstrate its effects and properties. Ferment, he said, was neither an organized nor an organic product with all the complexities which organized life would entail. Liebig's "soluble ferment" was the same as Payen and Persoz's "diastase" — later on Duclaux was to propose for general acceptance the generic use of the word diastase to distinguish soluble ferments — it is the same thing as Kühne's "enzyme" or "zymase," etc.

## DIASTASES AND PHYSIOLOGY

The labours of Traube (1838), Berthelot (1860), and Hoppe-Seyler (1876) finally facilitated the correlation of these various theories, whether life or mere catalysts, were responsible for fermentation. Numerous diastases were discovered formed from the most diverse types of cell. That is how cells make diastases and how the fermentative effects of such cells become evident through the diastases. Life is, therefore, catalytic through the action of its diastases originating from vital actions and comes about at the same time through catalysis; which suggestion does not, however, prevent the parallel existence of catalysts not endowed with life and even of an inorganic nature.

These concepts found definite confirmation as a result of the researches made by Buchner and published in 1897; yeast extracts totally free from cellular matter even in fragmentary form, containing therefore only dissolved matter formed from living cells and separated entirely from cell plasm, were capable of developing alcoholic fermentation if they came into contact with glucose. The diastases of alcoholic fermentation are manufactured by yeast cells, that is living organisms, in the same way as pepsin, trypsinogen, amylase or lipase are manufactured by the gland cells of the stomach or pancreas. Diastases are therefore merely products of metabolic action or cell functioning.

These ideas became still more positive when the intervention of the diastases was speedily discovered to be a factor in all nutritional processes and an influence in the development of all functions. Diastases became accordingly something more than mere digestive factors. Wittich in 1873 proved the diastatic nature of the evolution of glycogen in glucose in the liver.

He isolated the diastase by solution in water and glycerine and precipitation with alcohol. Claude Bernard in 1877 after having many doubts on the matter, confirmed Chauveau's assumption of 1856 that glucose was consumed by the tissues. Similar consumption occurs in the blood, which when protected from putrefactio n at room temperature in the laboratory rapidly loses its sugar content. Lepine in 1890 and Arthus five years later attributed the glycolytic diastase or diastases present in the blood to be of leucocytic origin, but immediately pointed out also that the tissues in general and the muscle in particular produced similar diastases in greater quantity.

Starting from this point, discoveries of numerous and distinct diastases from all the tissues came thick and fast. Each act of metabolism arises from the action of the corresponding diastase or diastatic group, and as time goes on we are daily increasing our knowledge of fresh details in regard to the intimate relationship between nutritional reconversion and the functioning of the organs.

The study of diastases thus covers a very wide field, particularly during recent times. As Fearon said in 1940: "Since 1900, research on the diastases has reached such a stage of development that each single diastase or a single feature of a particular class of diastase is sufficient to occupy fully an entire group of research workers."

## THE NATURE OF CATALYSIS

Ostwald, in 1896, defined catalysis as an acceleration of the corresponding chemical reaction without modification of the state of balance of such reaction and without affecting therefore the amount of energy liberated or absorbed thereby. From that time on, researches have been repeatedly started for the purpose of investigating the nature and course of diastatic actions.

Among the many scientists concerned, we may mention Jacobus Henricus van 't Hoff, Arrhenius, Armstrong, Duclaux, Sabatier, Födor and others who busied themselves with the question of diastatic catalysis and conditions leading thereto. They were able to assert once more that diastasis is one particular case of catalysis.

According to Hughes a diastase is any substance which by its mere presence allows certain molecules to undergo chemical change through the effect of a critical quantity of energy less than what would be necessary were the catalyst not present.

There are many types of catalysts at work in living matter; sometimes they are the hydrogen, hydroxyl or iodine ions and their like, at others the catalyser may be a dissolved organic particle, metabolite, hormone or vitamin, an inorganic colloid like the metallic suspensoids or lyophobic rolloids, distinct organic particles, or diastases of some special type; and perhaps, finally, combinations energetic in action rather than of a specific material effect, for instance, porous walls (Becquerel, Giraud), radiations (Victor Henri, Achalme), or membranes or other cellular structures.

## THE BIOLOGICAL SIGNIFICANCE OF CATALYSIS

It is through catalysis that vital phenomena are developed in living organisms, since only by this means can one explain how otherwise impossible chemical reactions take place. There is no single physiological process which is not under the effect of many vital catalysts.

Diastases, hormones, parahormones, harmosones, vitamins and certain ions govern the activity of the organism as a whole as well as each of its parts, and control the appearance and development of the most diverse phenomena of life.

The vital catalysts are countless in number and act in a systematic manner co-ordinated rigorously within the framework of the being, their intrinsic potentiality being most intense. Haldane for instance has estimated that a molecule of catalase can decompose somewhere about 200,000 hydrogen peroxide molecules per second. Conversely, each molecule of the substrate will come into contact with a large number of diastase molecules and thus be converted. Haldane's calculations show that the average atom undergoing oxidation in the course of its metabolic life through the cell encounters no less than 100 catalysing molecules one after the other.

These facts show that the immense metabolic activity present in the living state can be explained by the presence and working of suitable catalysts. It is not possible to guarantee that these figures completely represent the true state of affairs, but, even though they be mere approximations, the extent and intensity of the biocatalytic function can thus be inferred.

Such catalysts must unfailingly be present in all living bodies, whether plant or animal, and it is because of them that life continues to exist. Without catalysts there would be no metabolism, and without metabolism there would be neither function nor morphogenesis. It is impossible to conceive of any vital manifestation of any type without the assistance of the relative catalysts.

## CATALYSTS AND LIVING CONDITIONS

Catalysts enter into the appearance and development of the vital manifestations. Conditions most favourable to them occur naturally in circumstances which are likewise best for the maintenance of life: moisture, for instance, since biological catalysts must be dissolved in water — the substance cannot act unless dissolved; warmth, since they act best and with greater intensity at the body temperature of warmblooded creatures; particular osmotic pressure and a predetermined acid-base balance individual to each one of the various groups of diastases; ionic surroundings and the presence of certain substances also help.

In this way complex diastases may be observed, and the functions of co-diastases, activating or inhibitive agents, and antidiastases must also be borne in mind. Diastases may have their source in the inactive form of prediastases and then be set in motion by various means.

Other vital catalysts, like the hormones or vitamins, demand the help of particular circumstances, or contrariwise find their physiological effects opposed by the presence of particular types of disturbing factors. Various metabolites, of which we could mention many, such as glutathione, diphenyloxylic acids, adenylic acid, flavin, etc., exert powers of fermentation analogous to those of the diastases, probably without any diastase whatever having to intervene. As may be seen, the catalysing agents in organic life are of a varied character, the wealth of their activity belongs to life itself and the need thereof is equally intrinsic to requirements for the continuance of life.

In little more than a single century immense progress has been made in theory, particularly as far as concerns biological catalysis and the real importance of those phenomena, originally thought to be nothing more than individual simple applications of organic chemistry of no greater interest than, say, their value in industrial application; the fermentation processes, for instance, have been effectively developed. It is obvious that a lot more still remains to be done; catalyst mechanisms have been shown to be indivisible and of the same essence as all the processes of life. "Life is ferment," as Paracelsus wrote in the sixteenth century.

## A FORCE DIFFERENT FROM ANY HITHERTO KNOWN

It is indeed surprising that the blood-stream supplying all organs alike should, without the help of any other liquid whatsoever, produce saliva, milk, urine, and so forth. In 1812 Kirchhof made a discovery which gave us a first glimpse at the life processes though remaining still far from a full explanation thereof. Kirchhof observed that starch could be converted into glucose by the action of dilute sulphuric acid, the latter remaining unchanged and unconsumed in the process. Thénard showed subsequently that hydrogen peroxide can be decomposed by finely divided platinum and other substances in suspension. Platinum itself in the spongy form is capable of oxidizing alcohol at room temperature.

It is common knowledge that the conversion, for instance, of sugar into carbon dioxide and alcohol takes place during fermentation under the action of an insoluble body which we call ferment, and that animal fibrin, coagulated albumin obtained from various plants, cheese and similar substances behave like such ferment though with less intensive effect. Such reactions cannot be explained by the laws of chemistry, as a double decomposition taking place between the ferment and the sugar or any similar reversible reaction.

It may perhaps be better to compare such conversions with inorganic processes such as the wellknown decomposition of hydrogen peroxide in the presence of platinum, silver, fibrin, and so on, from which we must assume that the action of the ferment may be of similar character. I do not by any means imagine that we are dealing with chemical forces different from those exhibiting themselves as properties of matter. In some cases it is quite possible to explain reactions of the type under consideration by availing ourselves of the well-known laws of chemistry and physics.

Therefore I propose calling the power of substances causing the conversions above mentioned under the conditions indicated catalytic, contrasting the word catalysis with the term analysis, so much in vogue for signifying the separation of the components of a body by means of chemical affinity in the normal sense.

Catalytic power seems to consist in the faculty certain substances exhibit for setting in action potential affinities and chemical activities which may be lying dormant at a particular temperature and are presumably something quite different from the real affinity between the bodies, since their action occurs through their mere presence.

There are well-justified reasons to suppose that in living animals and plants, thousands of catalytic processes take place between the organic fluids and tissues, whence we get the formation of countless heterogeneous chemical compounds, starting with plant saps or animal blood whose composition does not change through the fact that so many and various products are manufactured.

It is likely that some time in the future we shall find out the cause of it all is the catalytic power of the tissues forming the organs of the living body.

Every solid living tissue placed in contact with a liquid seems capable of acting chemically upon the substance in solution, whilst at the same time the dissolved matter can react without any interchange of elements Catalysis is both chemical and physical in its nature.

Attention has recently been called to the fact that a body may exert influence by mere contact upon some other body which is thereby decomposed or combined. Platinum, for instance, does not decompose nitric acid neither at the boiling point of the acid nor in the finely divided non-reflecting state (spongy platinum), but an alloy of silver and platinum is easily dissolved in nitric acid.

The oxidation suffered by silver gives rise to a similar change in platinum. In other words, platinum in the presence of silver undergoing oxidation, acquires the property of decomposing nitric acid. Copper does not decompose water even when it is treated with boiling dilute sulphuric acid, but an alloy of copper, zinc and nickel dissolves under such conditions, hydrogen gas being liberated.

These and similar reactions cannot be attributed to "special forces" differing from those governing chemical affinities, forces which would be set in motion by a touch as light as a feather. The unstable substances which are so easily decomposed give the greatest number of instances of this type of phenomenon. The components of these bodies are found in such a state of tension that the slightest modification of their atomic balance overcomes the effects of chemical affinity. They exist only by the force of inertia and the slightest shock or disturbance is sufficient to destroy the mutual attraction of their components.

Hydrogen peroxide belongs to this class of substance; it is decomposed by any product capable of attracting the oxygen which goes into its composition and even by the mere presence of certain substances such as finely divided platinum or silver, which themselves do not enter into combination with any of its constituents.

The cause of the sudden separation of the elements in hydrogen peroxide has been held to be quite distinct from that giving rise to a common chemical decomposition, and has been called the "catalytic force." It should be remembered, however, that the presence of platinum or silver has no other effect than to accelerate the decomposition of the hydrogen peroxide, which would take place even without the intervention of those metals but at a very much slower rate.

Certain organic bodies seem to enter spontaneously into a state of fermentation or putrefaction, especially those containing nitrogen. It is worthy of note that very small quantities of such substances in that state possess the power of bringing about similar transformations in unlimited quantities of substrate. Thus a small sample of grape-must in fermentation added to a large amount of the same liquid in the still state causes the whole lot to ferment. Similarly the smallest amount of putrefying milk, meat extract or paste added to fresh milk or meat will soon cause the latter to putrefy in the mass.

All these changes differ apparently from those ordinary decompositions which are developed through the effect of chemical affinity. These are chemical actions, conversions or decompositions which become stimulated by substances in the same condition. It is demonstrable that inorganic compounds made up of the simplest molecules are the more stable and resist decomposition, whilst others more complex in structure are more liable to decompose.

The cause seems clear: the greater the number of atoms forming a molecule, the greater the number of ways in which the said atoms are mutually drawn together. Such complex balances are practically unstable and persist only through inertia. In this way, complicated chemical combinations, like the sugars, for instance, become the object of such decompositions or fermentation.

We have seen that some metals acquire the power, not inherent in them, of decomposing certain bodies, such as water or nitric acid among others, through mere contact with other metals in the act of chemical reaction. We have also seen that hydrogen peroxide and other peroxides which decompose easily can transmit this quality to other substances present. This is the same thing that happens, as we may also recall, in the case of fermentation or putrefaction brought about by organic matter which itself is fermenting or putrefying.

Different bodies, in the act of combining or decomposing are found to be in a position to induce such processes, that is to cause changes in balance between the component atoms of the molecules, particularly of organic molecules which are highly complex, and causing these atoms to become arranged in other states of balance of greater stability, i.e. in the form of smaller moleculed bodies, in accordance with their natural affinities.

Among the substances especially active in this respect must be reckoned in the first place those organic substances containing nitrogen, particularly the albumins.

## LOUIS PASTEUR

In the first of his disclaimers, Liebig formally states I cannot have produced ferment from beer nor furthered alcoholic fermentation in a sweetened mineral medium wherein an extremely small amount of yeast may have been sown.

Here is the touchstone of truth or error. Liebig, in fact, holds fermentation, like putrefaction, to be a phenomenon similar in nature to death, if one may say so. Certain substances, and particularly those which he calls albuminoids, such as albumin itself, fibrin, casein and others or the organic liquids containing them, namely blood, milk, urine, etc., possess the property of imparting to the molecules of fermentescible matter the motion which exposure to the air determines in them.

This fermentescible matter is divided into fresh products without taking anything away from the inductive substances and conversely without giving them anything either from its own make-up. I maintain on the other hand, that fermentations are correlative with life, and think I have proved beyond cavil that a substance capable of producing fermentation never enters into fermentation unless there is some continuous chemical change going on between its living cells which increase by assimilation of a part of such fermentescible matter.

Liebig's theory was in full swing when I was able to show that in any fermentation special organisms are to be found and that on considering that the phenomena produced were due to the presence of dead albuminoid substances, no account is taken of the fact that life develops alongside and parallel with the fermentation process.

I was able to prove, moreover, that all these fermentations become impossible even in contact with the air, if it so happens that the air does not bring right up to the substances in question the organized germs which it ceaselessly carries from one part to another in the strata nearest the earth.

Nevertheless, and this is another of my results, fermentescible mixtures which do not ferment, because of the absence of the germs responsible, may be oxidized by mere contact with pure air, suffering appreciable chemical changes which are however quite distinct from those which develop during fermentation.

I made up fermentescible media with the substance to be fermented, suitably selected mineral salts and the figurate germs. Fermentation took place immediately in each case. And, in accordance with these facts, I assert, for instance, that every time wine becomes turned into vinegar it is through the action of a film of vinegar plant, Mycoderma aceti, which develops on the surface.

Nowhere on earth does there exist a single drop of vinegar whereby wine may have soured on contact with the air, without the yeast spore mycoderma having been present. This microscopic plant has the property of condensing oxygen from the air, in the same way that platinum black or red blood-corpuscles do, fixing such oxygen in the subjacent materials available.

Liebig denies the correctness of these assertions: "With dilute alcohol used for the speedy manufacture of vinegar, elements nourishing the mycoderma are excluded, and vinegar is produced without its help." Liebig says, moreover, that he discussed this with the manager. of one of the most important acetic acid factories in Germany, Riemerschmied, who affirmed that in his works, alcohol in the course of its conversion received not the slightest external addition and that apart from the air, wooden surfaces of the vats, and coal, nothing is able to act on the alcohol.

He adds even that Riemerschmied places no credence in the presence of mycoderma. Very well then, I would suggest to Mr. Liebig that he gets from the works a few wood shavings, drying them quickly and sending them to Paris for examination by an impartial committee.

Mycoderma would certainly be found on the surface of such shavings. I would also suggest a further experiment to him, namely, that he gets Mr. Riemerschmied to lend him one of his vats, which has been in use for some time with such great success, that he puts into it daily an amount of vinegar equivalent to a bit more than half a gallon of absolute alcohol, filling it up with boiling water which should not be left in more than half an hour, after which he should start the vat running again. According to his ideas, it should work just as well as before the scalding, whilst I, on the other hand, maintain that the vat so treated will give no vinegar, not, at any rate, for a fairly long time, that is to say until there will have developed on the inside surface of its staves fresh colonies of fungus, i.e. mycoderma, taking the place of those killed by the boiling water.

## SIR FREDERICK GOWLAND HOPKINS

Some years ago I was endeavouring to discover if vitamins were to be found among sulphur-containing compounds, and was led part of the way towards the separation of the substance now described. A little later, acting on the suggestion that acid formation in muscle is a necessary factor in contraction I wished to discover if by chance in the absence of carbohydrate, acetoacetic acid from fat might function instead of lactic acid. This led me to apply the nitroprusside test to tissues. At this time Arnold's papers had not yet appeared and I was ignorant of Heffter's publication.

There can scarcely be a doubt that the substance to be described is the "Philothion" of de Rey Pailhade. In 1888 this author showed that yeast cells and aqueous extracts of yeast have the property of reducing sulphur to hydrogen sulphide. Later he showed that many animal tissues possess the same property.

Throughout a long series of communications upon the subject he has courageously maintained the view that the labile hydrogen thus shown to exist in living cells has important respiratory functions. His views as to the probable' nature of the hypothetical substance (named as above) which he supposed to carry this labile hydrogen, have been modified from time to time. In his latest writings he speaks of it as the hydride of a protein.

## Chapter 5

# Metabolism

Growth in the animal and vegetable kingdoms is a process from which no individual is exempt. Growth, that is increase in size and weight of the living matter forming the individual, results from assimilation.

By assimilation, matter similar to that of the organism itself is built up from the foods consumed, which latter are moulded and converted to form the specific substance of which the whole individual is composed. Despite their general similarity of composition, both the living substance and the greater portion of the organic material in various organisms are different and distinctive, certainly for each particular species and even for each individual within any particular species.

The basic property of five matter is its power of building up substances similar to itself. Consequently, synthesis of living substances always presupposes the influence of some other five matter, and synthesis can only take place when such matter is present.

Synthesis comes about in such a manner that fresh living matter makes its appearance complete with chemical, histological and anatomical formations identical to those of the organism entering directly and immediately into the assimilative process.

In this identity between the products of assimilatory synthesis and the organized substances which are the instruments of the synthetic process, consideration must be paid to the specific assimilation applying to each organism, to each cell thereof, and even to each differentiable portion of such cells. Thus, the growth of an individual with all its chemical and morphological characteristics comes about through progressive assimilation and the number of individuals forming a species is likewise increased.

Assimilation is the process of manufacturing living matter proper, that is to say, the essential cell plasma, residing and living within the cell, whilst at the same time it produces inclusions, paraplastic materials, nutritional reserves or functional product$ playing a special part in metabolism and in the body functions.

In short, through assimilation, nature is able simultaneously to create both the machine and its fuel. These assimilative and anabolic operations are of course endothernmic, whence it happens that assimilation must integrate not only matter by itself but energy as well. Typical of assimilative changes are those brought about by the action of chlorophyll.

It is known that this substance causes water to combine with carbon to form one of those self-same paraplastic substances just referred to, in this case starch, which forms a food reserve capable of subsequent use for countering the many necessities of the individual. To the substance, in this instance inorganic, namely water and carbon, energy in the form of sunlight has to be added so that organic matter may be synthetized. Organic molecules, being much bigger and more complex than inorganic particles, contain more heat than the latter.

The greater the complexity of the number of atoms forming them, the more the energy within the molecules. Assitmilation, the combined whole of the synthetic process, demands fixation of the energy supplied to the material elements making up the large particles which, in the first instance, are organic, and then become alive.

It must therefore be borne well in mind that assimilation does not consist merely of the manufacture of complex material edifices or particles of vast structural complexity, but that it must also supply such systems with the amount of energy necessary for their synthesis.

## ASSIMILATIVE DISASSIMILATION

The source of this energy is sunlight. Once the glycides are produced, through union of bodies low in energy such as water and carbon, such glycide reserves will be able to supply the energy necessary for further assimilative changes because of their disassimilative decomposition.

Energy will pass from a system in disassimilation to another in synthesis, as shown by Pasteur in 1875, when he wrote: "The energy supplied by oxidations typical of disassimilative operations can be used for completing synthesis within the organism." Therefore assimilation and disassimilation develop in indissoluble conjunction one with the other.

Coupled catabolic and anabolic reactions frequently occur as instanced by Meyerhof in an interesting. example, namely, the resynthesis of glycogen from lactic acid in the muscle, by the use of the energy liberated from a portion of the lactic acid. Such energy of chemical origin being applied to the unconsumed portion of the lactic acid raises it by assimilative operation to a glycide. Such processes are now known as "the Pasteur-Meyerhof reactions," and occur repeatedly in animal and plant life.

Plants, besides being able to convert carbohydrates into fats, also manufacture proteins by incorporation of nitrogen with ternary molecular structures. This stands for synthesis and by the same token a need for energy, which is introduced by decomposition, most certainly oxidation, of a fraction of the glycides resulting from the chlorophyll action. In animals too, the immediate organic principles making up the foodstuffs have to rise to the level of more complex living matter with a greater heat content than the feeding matter itself.

This is effected by consuming a part of the food components in disassimilative operations in such a way that they supply the energy necessary for such assimilation. In fact, the three main objects of dynamogenic feeding within the limits of animal metabolism are: maintenance of kinetic energy, generation of heat, and output of energy necessary for assimilatory synthesis.

## FUNCTIONAL DISASSIMILATION

Once living matter takes shape, once the peak which is the giant living molecule has been reached, that disassimilative exothermic disintegration commences whereby the vital processes are sustained. The functions get their energetic nutriment from these disintegratory processes which always finish up with a credit balance.

The consumption of dynamogenic matter does not take place solely at the expense of living matter properly so-called, as on the whole it uses up reserves stored within the cells, forming the fuel referred to above.

Living organisms, from this point of view, may be compared to a heat engine burning a special type of fuel but also partly consuming certain portions of the engine itself, which portions are immediately replaced. The flow of energy in the case of live organisms cannot be compared to the current of a river as seen from the banks unless we take it that a portion of the banks themselves are carried away by and in the stream itself at the same time.

## LIVING MOLECULES IN THE STATIONARY STATE

Metabolism, an essential feature of life, is a continual self-destruction and self-recomposition of living matter. Cuvier called it a "living whirlpool." Goethe before him had written "life is a flame," whilst, according to Buffon, it was "a minotaur devouring its own substance." It is in fact a constant flow of matter and energy; living systems are complexes in a stationary condition (Ostwald), seat of continuous changes of matter and energy under the cloak of stability.

Pflcüger in 1898 portrayed the living molecule as a figure in threedimensional space made up of rays darting upwards, downwards, rightwise and leftwise, forward and backwards from a common centre reaching towards the surface of an ideal sphere, such rays being composed of groups of radical atoms corresponding to the various immediate principles, proteins, glycides and lipids forming an extremely complicated particle which has a very high molecular weight despite the low atomic weights of its separate elements.

To this concept of the living molecule Verworn gave the name biogen. From the biogen linear atomic groups of greater or less complexity, proteins, carbohydrates or fats, or portions of such radicals, emerge into the open, these being the "lateral chains" which Ehrlich in 1904 sought to use as an explanation of the phenomena of immunity.

These lines of atoms ceaselessly falling from the biogen into the disassimilative processes would leave atomic remnants in the remainder continuum of the biogen, upon which by assimilation further chains would build up similar to those already broken away. We must imagine the biogen, therefore, as the seat of constant renewal of its constituents, some disappearing whilst others make their appearance, and that, despite all this, the chemical identity of the whole is preserved. There takes place in it — and this is the essence of life — an uninterrupted flow of matter and of energy; assimilation makes up for losses caused by disassimilation. Disassimilation liberates energy whilst assimilation on the contrary demands it. Thus it is that the stream of matter in metabolism entails a corresponding flow of energy.

Bertalanffy names the aforesaid uninterruptedly continuous physico-chemical vital process an "open system." The organism is a system or combination of unified systems in an open steady state of rhythmic constancy, to which the living being tends ever inexorably to return whenever disturbances become apparent. This rhythmic recurrent balance is a basic characteristic of existence, and when it is interrupted, life ceases.

Biogens of various types go into the formation of cell plasma, cells, tissues and organs. Biogens are arranged according to their physicochemical structure, and consequently result in cytological and histological structures which make up the complete anatomy. Nutrition is a property of biogens and therefore of the cells. Living matter attains its animate condition through its nutritive capacity; metabolism must be therefore one of the principal factors in life.

Gautier wrote in 1892.: "In the structure and organization of chemical molecules forming the protoplasm and in their manner of association one with the other lies the fountain-head of the whole range of elementary phenomena of life. Variation of the integrating molecule causes changes also in the style of reaction throughout the whole organism; vital processes are the direct consequence of the chemical functions of the constituent particles."

## METABOLIC MECHANISMS

Assimilation and disassimilation are so bound up with one another as to form a single feature. The former was a subject of research long before the latter. Knowledge of the existence of certain processes of assimilation goes right back to ancient times, whilst disassimilation is on the other hand a comparatively recent discovery.

Nevertheless greater attention has been paid to the study of disassirailative processes than to acts of assimilation. Both these processes are equally important in forming the metabolic whole. They are of equal effect in the state of nutritional balance; anabolism overrules catabolism during growth, and conversely, disassimilation preponderates in all the aspects of orgainc impoverishment.

Disassimilative changes present nothing which is exclusive and particular to living bodies. As we know, Lavoisier taught that breathing is exactly the same thing as combustion. Hydrogen and carbon are consumed inside living organisms in exactly the same way as in any other combustion process.

Living matter disintegrates and decomposes into smaller particles during disassimilation as it does in death. Disassimilation is like bodily decomposition. In both cases the breaking down operations are exothermic and in disassimilation there resides a store of energy available for the functions. Thus Claude Bernard was able to say: life, that is function, is death. Catabolism, like putrefaction, is an act of decomposition, simplification, a tendency to change back living or organic matter into inorganic matter. Such changes presuppose a positive thermal set-up both inside and outside the organism.

Assimilation, on the contrary, occurs solely in living organisms and is specific to them as we have already seen. Le Dantec in 1904 catalogued the characteristics of the disassimilative processes as compared with the assimilative, putting them into the form of two basic formulae of nutrition. In disassimilation, which in general circumstances hardly differs from any other disintegrative or analytical change, the substances reacting in the first term of the equation are not to be found in the second, a total conversion of substance having taken place both in the nutritive material and in the cell components.

$$N + O = R$$

i.e. Nutriment plus Oxygen equals Residues.

In the assimilation formula, one of the reacting terms, the living matter, is found generally in increased amount in the second term:

$$N + L = nL. + R$$

i.e. Nutriment plus Living matter equals n times the Living matter plus Relsidues.

The object of assimilation is to build up living matter by consumption of foodstuffs. That is why, therefore, foods are divided into respiratory-as they were called in former times-or dynamogenic, and plastic or morphogenic; in other words, foods for heat-generating combustion and foods for the formation of living matter.

These two facets of metabolism would not occur independently. Claude Bernard himself pointed out that prior disassimilation is a factor of assimilation.

To repair a breaking off of portions from the biogenic particle, obviously such amputation must first have taken place. As we have shown, it might be thought there was some contradiction between assimilation and function, but elementary observation proves the contrary, namely, that an organ in operation continues to develop, as for instance a muscle.

Rodrigo Lavin in 1900 emphasized this when he asserted that assimilatory synthesis is a functional phenomenon. In a function, of which the muscle is the most characteristic example, food reserves are consumed mainly as a supply of fuel for the production of energy.

Living matter as such takes part in the chemical changes demanded by the function and that entails an increased biogenic disintegration. The replacement of molecular complexes destroyed by the functions is more than taken care of by reparatory mechanisms, from which we get an increase in the synthetic operations, and an augmentation of the living matter accompanying and distinguishing the function.

## ANOXYBIOSIS AND OXIDATION

Disassimilative disintegration was considered a few years ago to be exclusively of an oxidizing character, according to the ideas expressed by Lavoisier. Our increased knowledge of facts now shows that the chief part in disassimilation is taken by anaerobic operations without the aid of oxygen, these being the chemical reactions labelled by Pasteur as fermentative. Oxygen then acts upon the products of anoxybioses whilst combustion completes the acts of disassimilation. This twofold aspect of the process was brought to light mainly through Gautier's researches in 1898 and has more recently met with repeated confirmation.

Disassimilation takes place through the agency of various diastases leading the living matter or reserves through all stages to excretion. Assimilation is similarly the work of synthetizing diastases, i.e. enzymes, whose exact operation is at the present time even less known than that of the catabolizing diastases.

Each of these two types of diastase regulates its activities in such a way as to ensure exact adjustment of the intensity and quality of nutritional changes to the vital needs of each individual, be it plant or animal.

The result of this is that the intensity of nutritional changes comes about neither through chance nor even by reason of special substances at the disposal of the living matter in its uninterrupted course of metabolism. Living matter automatically regulates the intensity of chemical changes taking place in it, by means of properties which we must perforce call vital, intrinsic in itself, from the very fact of its being living matter.

This impelled Pflüger to say in 1905: "The live body cannot be compared to a fire which burns all the more fiercely the more fuel is thrown on to it; quite the contrary: whatever may be the quantity of material, it oxidizes only the exact amount needed for its functional needs.

If the materials of combustion are excessive in amount, they stay stored up in the form of fat. Consequently, what determines the total of work in our organs is not the amount of matter supplied by the food, but the quantity of material actually used."

To which Bouchard added in 1906: "A portion of body albumin, eight parts in a thousand in young people and four in a thousand in older folk, is probably destroyed every day and replaced by an equal amount of albumin derived from the food consumed.

The amount of living albumin renewed comes out practically constant for each stage in life and for each set of circumstances. It varies according to age, decreasing as time rolls on; it increases during certain illnesses particularly when fever is present and decreases in others, especially in certain diatheses.

It is not affected by muscular effort nor by cold, since the foodstuffs supply — quite apart from the albumin which has to replace that lost in the tissues — sufficient organic matter to liberate the calories required by the effort or by the struggle against cold. Neither is it affected by variations in the intake of oxygen or protein foods, since such foods are ingested in sufficient quantity."Nutritional changes develop therefore in all forms of animal and plant life as the basis of some form of vital manifestation, animal or vegetable.

Disassimilation is particularly manifest in animals, which move and, on getting heated up, require calories which are in fact supplied by the disassimilative, exothermic operations. Apart from such differences in quantity, the metabolic mechanisms in animals and plants — not counting the chlorophyll action—are identical.

## CLAUDE BERNARD

If in regard to inorganic matter it be admitted and rightly so that nothing is lost and nothing is created, nevertheless in the case of living beings it is quite another matter. Everything is created morphologically, becomes organized, and then dies or is destroyed. In the developing embryo, muscles, nerves, bones, etc., appear in the spot corresponding to their place in the organism repeating the form of the individual from which the egg was generated.

The surrounding matter is assimilated to the tissues either as a nutritive principle or as an essential component. Thus is created the organ which can be distinguished by its structure, shape and properties.

On the other hand, the organs are destroyed and disorganized continuously and precisely through their own functioning. Such disorganization constitutes the second phase of the great drama of life. The first of these two orders of phenomenon is individual and special to the living body. This evolutionary synthesis is the very essence of life. I will repeat on this occasion the expression to which I gave voice some time ago, namely: "Life is creation."

The second order of phenomenon, the destruction of life, is of a physico-chemical nature, normally the result of combustion, fermentation, putrefaction or some other action, comparable in short to a large number of chemical acts of decomposition or disintegration.

These are true phenomena of death when they take place in an organic being. And it is worthy of mention that we delude ourselves habitually in such form that when we want to point out phenomena of life, what we really talk about are phenomena of death.

Vital phenomena properly so-called do not attract our attention. Organizatory synthesis works silently within the body, its actual occurrence well hidden away from sight, gathering up secretly the materials which will afterwards be consumed.

We take no account directly of these phenomena of organization and assimilation. Only the histologist or embryologist studying the development of tissues, cell elements or the living body as a whole becomes aware of the changes and phases which reveal this hidden working; in one case it may be the deposit of material, somewhere else it will be the formation of a skin or a nucleus, further on still a multiplication of cells or a renewal of some organ or other. The phenomena of destruction or death and disassimilation are the most obvious to our senses and seem therefore to us to characterize life.

Their signs are clear; a movement takes place because a muscle must have contracted as a result of some corresponding wish; when stimulated a gland secretes. It is the substance of the muscle, nerve, brain or gland which is disorganized, destroyed and consumed. And this is how every living manifestation combines with an organic destruction and that is what I meant when I wrote the paradoxical phrase "life is death."

The existence of all animals and plants is maintained by these two types of necessary and inseparable processes: organization and disorganization. Our science has to fix the terms and conditions which guarantee the correspondence of these two types of phenomenon.

This division of vital phenomena which we have adopted — assimilative and disassimilative — is an exact expression of the truth and comes about simply from proper observation of life phenomena. But life is the combination of both types of manifestation.

## CHARLES BOUCHARD

A common property of all living particles is a particular molecular movement which is not observable in either inorganic or dead matter. The effect of this movement is to introduce into the living particle certain extraneous non-living substances and submit them to chemical metamorphoses which I will call vivifying, since by means of such changes the said substances come to form a constituent part of the live element and participate in its life; there are also other changes which I would call retrograde, whereby living matter ceases to exist and is converted into products of decomposition which have to be expelled from the organism.

The distinctive feature of the activity of any living particle is, therefore, a twofold continuous molecular movement of introduction and expulsion simultaneous with a double operation — also continuous — of chemical transmutation, the one following the introduction of substance and the other preceding its ejection. On the one hand we have translation, and on the other transmutation: a double physical operation simultaneous with a double chemical effect.

If nutritional material conversions are a common characteristic of anything alive, if nutrition is therefore the general characteristic of life, the best definition of life itself will be that given by Aristotle: "Life is the combination of the operations of nutrition, growth and destruction." This idea corresponds with what Blainville set out in other terms: "Life is the double internal motion of composition and decomposition which at the same time is general and continuous."

Nutrition therefore — we may repeat — is life; it is life with its double motion of assimilation and disassimilation, of creation and destruction at the same time.

Assimilation and disassimilation are in general parallel and simultaneous but not invariably so, not even in the physiological state. Growth is the predominance of assimilation over disassimilation. In pathological conditions this predominance is the cause of over-development or hypertrophy, whilst atrophy and cachexia come about through excessive disassimilation.

And we may affirm withal, that even when the physical and chemical actions — each of them twofold in the sense of being integrative and disintegrative — are not found to be indissolubly bound together between themselves at every particular moment, they are jointly and severally equally necessary for the maintenance of life. None of them may be hampered or suppressed without the nutritional molecular motion being held up and life being consequently extinguished.

## ARMAND GAUTIER

Life makes its appearance in the cell and passes from generation to generation thanks to specific substances transmitted from parent to offspring and determining the organic forms and molecular structure whence the elementary functions are derived.

The regular functioning which preserves the individual and perpetuates the species consists of a series of orderly acts in whose fulfilment every cell plays its special part. Each lives, however, more or less independently according to its structure, assimilating nutrition and embodying it in the mould of its own specific constitution.

We do not know the method of molecular differentiation and organization for each species of cell. All we know is that the parts which govern and specialize the working of the cells are the nucleus and the protoplasm, each made up of specific albuminoid substances.

The nucleus rules the activity of the protoplasm and its chief importance is the maintenance of morphological type and the development of the cell. Protoplasm works and assimilates its surrounding medium thanks above all to its plastids or specific granulations, which are differentiated organs proper to each species and frequently varying in nature even within the same cell.

Animal cells cannot from their elements synthetize the albuminoid ingredients of their protoplasm. The proteins arriving within them, or perhaps their more immediate derivatives, are converted into particular alburninoids due to disintegrations, syntheses, and isomerizations which modify, rejoin and assimilate simpler materials absorbed in the digestive tube or arriving from other cells.

In the living cell-protoplasm fundamental protein substances are disassimilated mainly by hydrolysis in a reducing medium and sheltered from any influence of oxygen. From this arises the formation of fresh nitrogenous substances, amides or complex amines, which are definitely converted into urea, carbohydrates and fatty bodies by the effect of simplifying anaerobic reactions.Urea, the definite form in which some fourteen-fifteenths of the total nitrogen in the tissues is eliminated, seems to be produced in the main independently from the oxidation processes. Between this substance and the protoplasmic albuminoids we have a whole range of intermediate nitrogen compounds. The most complex are proteins still full of considerable activity, such as peptones, diastases and toxins. Afterwards we get creatinic substances and ureides which immediately precede the formation of urea itself.

Furthermore, there are xanthine compounds and uric acid arising from nuclear materials. Many of these derivates are eliminated through the urine or bile whilst others serve for transient syntheses or serve to stimulate digestion or even nervous activity.The ternary products, formed by disassimilation of the protein substances in correlation with urea and other nitrogenous bodies, vanish through oxidation controlled by the oxidases.

The sugars are consumed or changed into fats by loss of carbon dioxide. The fatty bodies are saponified and then oxidized by degrees. It is precisely these oxidations of sugars, fats and other ternary compounds which are the main source of energy for the warming-up and work of the organism.The chemical phenomena, developed in the cell, supply the energy necessary for every manifestation of life. The organization determines merely the direction which this energy must take and the order of the phenomena which preserve the cell, the individual and the species.Reproduced by permission of the publisher

## OTTO MEYERHOF

The application of purely chemical methods to the study of the mechanism of muscular contraction has given us the following results:

- Oxidization of the foodstuffs is not the process directly supplying work, but serves solely for the purpose of replacement. Contraction may take place even in the absence of oxygen, at the expense of the energy liberated by anaerobe break-down processes, in which case the decomposition of glycogen makes way for the formation of lactic acid.

- Nor is lactic acid formation the process directly liberating the energy required for contraction; since, as Embden demonstrated, such formation takes place, to a large degree, when contraction has already ceased, and Lundsgaard's experiments showed that through intoxication of the muscle by monolodoacetic acid, the formation of lactic acid comes to a full stop, and the muscle is then still able to carry out a certain amount of anaerobic work. Under such conditions, creatine phosphoric acid, discovered by Eggleton and Fiske, decomposes more intensely than usual and it is this decomposition which supplies the energy necessary for contraction. Whilst in a normal muscle, the formation of lactic acid starting from glycogen, serves partly to resynthetize the creatine phosphoric acid broken down, just as in respiration, oxidization is employed for resynthetizing the lactic acid to glycogen.
- It is quite likely that neither does the decomposition of the creatine phosphoric acid directly supply the energy required for contraction. At least that is what Lohmann's latest experiments lead us to think. Here again, it may be a matter of a process of recovery whilst some other chemical conversion — decomposition of adenylpyrophosphoric acid into adenylic acid and phosphoric acid — is taking place at the same time as the contraction and appears to be reversed during relaxation.

The idea that the breakdown of adenylpyrophosphoric acid might take place prior to that of the phosphocreatine is due to Lohmann's observations verified not only in intact muscles but also in muscular extracts.

We have already been aware for some years that coupled chemical reactions take place in such extracts and must be borne in mind when changes in energy in the muscle are under consideration.

The following facts can be established in connection with healthy muscle and its histological structure. The break-down of creatine phosphoric acid is impossible unless adenylic acid and pyrophosphoric acid is simultaneously synthetized to adenylpyrophosphoric acid. It is therefore necessary that the first reaction to take place after excitation should be the break-down of adenylpyrophosphoric acid into adenylic and phosphoric acids.

Only then will it be possible for the phosphocreatine to split up so that the adenylpyrophosphoric acid previously decomposed now becomes resynthetized. If the decomposition of the adenylpyrophosphoric acid takes place in between the instant of excitation and the breakdown of the creatine phosphoric acid, we may assume, in agreement with Lohmann, that such reaction takes place simultaneously with the mechanical process of contraction.Reproduced by permission of the publisher

## ARCHIBALD VIVIAN HILL

A lecture delivered during the Course in Biochemistry and Physiology of Muscular Contraction and Fermentation at the University of Barcelona (Curs de Bioquómica i de Fisiologia de la Fermentació i la Contracció Muscular — Institut d'Estudis Catalans), 1934.Muscular contraction can be analysed into the four following stages:

*First stage:* In which the response is produced

*Second stage:* Lasting throughout the whole period during which the stimulus remains active

*Third stage:* When the stimulus ceases to act and the muscle relaxes

*Final stage:* When, under the influence of oxygen, the muscle recovers from the changes taking place during the period of activity

An emission of heat is observable during the rising mechanical response, and a continued production of heat, becoming constant in rate during the period in which contraction is maintained. Coincident with relaxation, the production of heat first diminishes, afterwards again increasing, and finally decreasing to zero as relaxation terminates.

The heat of relaxation represents the disappearance of the potential mechanical energy necessarily present in the muscle in isometric. contraction. One day we shall learn what the chemical reactions are which are associated with each one of these stages.

The anaerobic heat curves can be considered as the energy outline of chemical reactions taking place in the first half minute after contraction, as discovered by Embden, Meyerhof and other investigators.

An interesting fact, not yet explained, is that the maximum intensity of the delayed heat comes first and is more intense after a long stimulation than after a short one. Such a difference in the form of the delayed anaerobic heat appears with such regularity and so plainly that there must be some reason for its occurrence. The amount of this heat is not the same in subsequent contractions: as the muscle gets more and more tired the delayed heat diminishes relatively more than the initial heat which it follows.

This delayed heat is liberated also in the presence of oxygen, but to it is added the heat of recovery which has its source in oxidation and is of longer duration. The ratio between the total heat (produced during and after contraction by a muscle to which sufficient oxygen is supplied) and the heat first liberated (that is, during contraction) gives us important information about the chemical mechanism of contraction.

The total heat set free in the quarter-of-an-hour or twenty minutes during and after contraction maintains a ratio to the initial heat of contraction which was originally thought to be absolutely constant.

This is not exactly so. In the case of a relatively long tetanus the ratio is about 2.4; in a short tetanus, or in a series of twitches, the figure falls to about 2.0; and in an isolated twitch, the ratio lies between 1.6 and 1.7.

From such facts, it may perhaps be deduced that recovery is not invariable, but depends on the nature of preceding activity. The efficiency of recovery is greater in the case of a simple isolated twitch than in that of a tetanic contraction. Perhaps, some day, Professor Meyerhof, whom I see here listening to me, and who has contributed so much to the advance of our knowledge of the chemistry of muscular contraction, will be able to tell us why things happen this way!

A curious phenomenon, discovered by my co-worker, Dr. Feng, is that the metabolism of the muscle at rest increases as the load increases. Taking the load away causes the resting metabolism to drop back immediately. The same effect can be seen on measuring the consumption of oxygen by the muscle.

## Chapter 6

# Growth and Reproduction

### CELL DIVISION AND GROWTH

All cells have a certain limit for size, not varying appreciably from one to the other and measurable in thousandths or ten-thousandths of an inch. Of course some cells are larger than others but the differences in size are not really very great.

On the other hand, however, we do find giant vegetables and animals living in the same environment as microscopic and practically invisible creatures and plant organisms, all of them from the least to the greatest containing a huge number of cells.

The fact remains that the size of the cell element is practically constant, and that of the individual depends on the number of cells forming it, this being a characteristic index for any particular species.

Again, in reproduction through germ cells, the whole of the organism derives from progressive division of the originating cell, that is a zygote in the case of sexual reproduction or some other cell in place of the zygote when reproduction is asexual.

From the division of such cell we get two blastomeres which split into four and these again into eight, then into sixteen, thirty-two, and so on to infinity. In this way the embryo keeps on increasing through multiplication of its cells, whilst after the embryonic stage growth still continues until the individual reaches adult size, its number of cells increasing all the time. Growth is therefore in all its aspects a consequence of cellular multiplication developing progressively until the volumetric limit for the species in question has been reached.

Thus we see why growth cannot go on indefinitely; it must cease as soon as the specific limiting size has been reached, and it is at this point generally that the reproductive faculties of the individual make their appearance. Reproduction becomes thus a continuation of the growth process.

There is, indeed, an increase in mass in species because the number of individuals increases, but no single one of the species increases beyond its particular limiting volume, and when that is reached individual growth ceases. The division of its cells then serves only to replace those cells which have died off or, in the case of natural or artificial mutilation, to repair such damages.

Cell division is a distinguishing feature of living matter, a prime factor like nutrition and indeed a consequence of the latter. When assimilation is playing the major part, the cell grows, but there must be some fixed relationship between volume and surface area, since it is through the surface that interchanges of substance and energy with its environment take place.

Increase in size of cell implies mastery of volume over surface and decrease in their ratio, which means that a moment will come when the surface will be insufficient for the continued existence of the cell; this is circumvented by division of the cell, which when it has become sufficiently large breaks down into two or more.

Schleiden and Schwann assumed that the cell was the offspring of the plasmic fluids, the blasteme, brought into existence in much the same way as a crystal appears in a saturated solution, by precipitation of the nucleolus followed by the nucleus which latter is surrounded with protoplasm.

In 1854, Nägeli for the first time established that one plant cell is always born from division of another, whilst later on, in 1858, Virchow confirmed that this takes place in all classes of cell whether plant or animal and epitomized this in his well-known saying "every cell comes from some other cell."

## TYPES OF CELL DIVISION: AMITOSIS AND MITOSIS

Division of the cells can be effected by different processes. The simplest is that of direct splitting to be observed in single-celled organisms such as protophytes and protozoa, and also in some cases metaphytic and metazoic tissues. Ranvier in 1886 verified the direct division in leucocytes and still later the same thing was confirmed in cells of other animal tissues, in the connective for instance.

In direct division or amitosis, the mass of the cell keeps on lengthening at the same time becoming constricted until the nucleus divides into two parts. This division of the nucleus is followed by excision of the protoplasm into two portions each surrounding one of the two halves of the nucleus. This is how the two cells resulting from partition of the mother-cell are formed and the process may be repeated through a long series of generations.

More frequent than this direct or simple division is another form of cellular multiplication, in which changes characteristic of the morphology of the nucleus take place. This is karyokinesis, so called by Schleicher in 1878, or mitosis found by Flemming, also in 1878. Its details were given by Strasburger between 1876 and 1884, and Flemming showed that it also takes place in animal cells.

The nuclear chromatin is arranged in threadlike form, the spireme in the prophase, and this breaks down into a fixed number of sections, the chromosomes, which after curving into hairpin shape form a sort of star in the equatorial plane of the cell, around the achromatic spindle which joins the two protoplasmic centrosomes.

Each of these shapes splits longitudinally into two, and thus the combination thereof, the motherstar or chromatic-aster, divides into two daughter-stars with the same number of chromosomes in the anaphase.

The daughter-stars are attracted as it were each to its respective centrosome and then move away from the equator of the cell along the achromatic spindle, this being the metaphase. Having once reached the centrosomes, the daughter-stars become the seat of an inverse process to that which fostered the individualization of the chromosomes; this completes the telephase. The forks of the daughter-stars lose their inflexion and degrade into straight lines, the various resultant sections becoming joined together in spiremes and very soon in these the morphological signs of reproductive activity disappear and they now become the chromatic formations of the daughter-nuclei.

Around these chromatic formations the respective nucleus is built up again, and around each nucleus the protoplasm is divided, so that, as in the case of amitosis, one cell has become two, its fund of nuclear chromatin being shared symmetrically and equally between the resultant daughternuclei.

Mitosis does not always occur with the regularity just described. It must be pointed out, in the first place, that there are certain multipolar mitoses where, instead of two centrosomes and one achromatic spindle, three, four or more centrosomes make their appearance with their corresponding spindles almost always intertwined and their mother-stars in a plane perpendicular to that of the axis of each one of these spindles.

Understandably in such cases the cell will split into three, four or more according to the section planes dependent on the spindles and this practically always in an irregular manner. However, in some of these mitoses the cellular division will not be completed and the resultant cell may contain several nuclei.

These non-typical or substandard mitoses are particularly to be observed under pathological conditions, for instance in certain cases of poisoning (Hertwig, 1887), in tissues of abnormal growth, cases of irritation, neoplasms (Galeotti, 1893) and others, also in the case of polyspermic fertilization.

Though mitosis is the most typical way in which cell division takes place in the tissues of metaphytcs and metazoa, this does not mean that direct or amitotic division may not occur in such organisms, and in fact this may be so. Direct division however most frequently occurs in the more simple organic forms such as protophytes and protozoa. It is possible, moreover, to observe cases which might be held to be intermediate between amitosis and mitosis, for instance nuclear chromatin counterparts which, though not characteristics of karyokinesis, are more complex than those developed in simple division and recall certain nuclear aspects of mitotic division. There is room for intermediate gradation between the two forms of division of the nucleus and of the cell.

## THE SIGNIFICANCE OF MITOSIS AND AMITOSIS

The significance of the two ways of producing multiplication of the cells has been the subject of much discussion. It is generally assumed that mitosis represents some form of improvement, or a more advanced way of doing things. It takes place in the fertilized female gamete, the zygote, and from this is derived an argument in favour of the superiority of karyokinesis. This opinion does not, however, enjoy the widest acceptance.

Certainly amitosis is found mostly in reproduction of the more simple forms of life, but there do exist cases of amitosis in very active young cells of metazoa and highly differentiated creatures, which allow expression of the opinion — contrary to what is generally assumed by the majority of writers — that direct division may be a manifestation of an advanced degree of specialization, and that by no means can it be held to be a process of the most elementary simplicity.

It seems, however, that within a general framework, amitotic reproduction may be the more easily seen in the rudimentary forms of cellular life than in the more differentiated. Still, retaining such framework, the appearance of one or the other reproductive variety may depend on physiological conditions and special circumstances ruling at particular moments of existence. Within the general biological framework above indicated we should have to debar the idea of any determinism, from causes still unknown, deciding the form of the process of division.

In short, though it be safe to say that amitosis prevails in the lower organisms and karyokinesis in others of higher botanical or zoological lineage, this does not mean that the ruling plan is rigid and inevitable, but that different circumstances might modify it and promote consequently either amitosis or mitosis whether in animal or plant according to immediate circumstances.

We must point out also the possibility of multiple cellular division. Various nuclei are formed by plurality of division of the original nucleus, and each one of these nuclei becomes surrounded by the corresponding part of the protoplasm, the corresponding cells thus being born (conitomy).

This is a process of reproduction observable in certain protozoa. A similar mechanism is that of sporulation. In other cases a species of nuclear gemmation occurs; spots of chromatic material accumulate in the surface of the mother-cell, arising from expansion of the nucleus in such cell, and each one of these spots becomes the centre of a new cell. There are obviously various ways in which cells may divide, but the more usual processes are in any event direct division and karyokinesis.

## EMBRYONIC DEVELOPMENT AND REGENERATION

Such mechanisms of cellular division assure the development of embryo and subsequent growth. Such development consists in multiplication of the number of cells and simultaneous differentiation thereof. Starting with a single cell, a huge number of them are formed in various tissues as parts of the different organs.

Multiplication and differentiation of cells are particularly active during the embryonic stage, but continue also throughout life, though of course with everdecreasing intensity from stage to stage. Differentiations are however still possible even after the embryo has developed.

In certain cases the formative process regains its activity and in mutilations, for instance, there is the possibility of certain organs and even the complete individual being reproduced again. The ability to make good varies from species to species ranging from rebuilding of complete organs and even of a major portion of the whole organism down to the simple cicatrization of wounds.

Division of the cells and differentiation arise from a first assimilative cellular impulse for growth and multiplication, though at the same time the presence of other cells and the influence of the whole being or organism play their part as may be seen in the case of tissue cultures. The cultured cells tend generally to become dedifferentiated (cataplasis), which Champy in 1915 attributed to lack of action on the part of the organism as a whole. The vegetative power of the protoplasm belonging to the cells prevails over the specific differentiating power of the whole with its various parts. Thus it is that Gracianu was able to say in 1931 that cellular proliferation in the test-tube, in cultures, differs entirely from what takes place during the normal evolution of an embryo.

Such influence is exerted by means of various substances, such as trephones, auxins, growth hormones, biotin and others, present in extracts of various tissues and more particularly in young and embryonic tissues. These substances revive the action in cultured cells and are probably the same as those promoting normal development. The agents of such development are provided with a high degree of specificity and profound differentiations govern the formation of particular organs, or "organizers" as Spemann and Mangold, for instance, called them.

Combined action of all the growth factors moulds the ontogeny of an organism and the possibility of repairing mutilations is included in this, more or less in proportion to the phyletic degree at which the creature, or even a plant, may stand. It is natural, therefore, that the capacity for such repair of damage will not be identical at all parts of an organism.

## REPRODUCTION BY REGENERATION

In the vegetable kingdom this capacity for reproduction through regeneration of mutilations is usually unlimited, as for instance in grafts. It is common knowledge that many plants multiply by thrusting into the ground fragments of shoots or branches or in some species even leaves. From the underground portion radicle organs are born and from the top part the breathing system consisting of branches, buds, leaves, and flowers. Plant species can also be propagated by means of roots and likewise by sticking the ends of bent-down branches, and even cuttings and so forth, into the ground. Grafts are of a similar nature to these latter. A shoot from one plant engrafts itself into another in such a way that the cambiums of each of them make contact, thus establishing mutual chemical effects between the cells of the graft and those of the host, whence modifications may occur in the host varying in nature and extent according to the particular case. Such modification impressed on the form and functions of the host can become extremely pronounced.

Regeneration with the formation of new, separate individuals is observable also in particular species of animals. In 1774, Trembley showed that after repeatedly taking sections from a hydra, a new hydra could be formed again from each one of the parts cut off. Up to the eighteenth century it was held that such regeneration was a special feature of the plant world and hence it was argued the hydra must also be a plant. It was not, however, very long before it became possible to show that other creatures, of much greater differentiation than the hydra, were also blessed with the same faculty.

Certain worms, particularly the flatworm Planaria Linnaeus socialis, the earthworm and others, represent remarkable instances of this. Some echinoderms and starfish are able to build themselves up again, starting with the minutest fragments down to a thirtieth part of their body. As the organs become more complicated the regenerative faculty becomes correspondingly less, but some well-known examples are worthy of note, such as the possibility of forming anew the heads of snails, the nippers and eyes of crabs, and the tails of salamanders, tritons and newts, which latter are already among the vertebrata.

Many other examples are forthcoming among various articulated creatures, molluscs, tunicates and the rest. In these processes of regeneration, the first phase is that of dedifferentiation of the cells, i.e. almost as if the cells returned to their embryonic state, this being followed immediately by morphogenous differentiation.

Regeneration is in evidence right from the simplest single-celled creatures and among many protozoa, as shown by Gruber in 1885, Balbiani in 1892 and Nussbaum in 1908 with their researches into merotomy. For the cell to become reconstituted it is essential that the broken-off fragment shall contain a portion of the nucleus, which latter governs the process of regeneration. In polyplastids such regeneration does not depend on the reconstruction of a fractionated cell but on cellular division as we have just seen. In the division of the cells the main part is again played by the substance of the nucleus.

## REPRODUCTION THROUGH SOMATIC OR BODY CELLS

Another process of reproduction by somatic cells is that known as gemmation, which may well be considered a type of genetic or organic growth. Gemmation appears in single-celled beings, and is frequent for instance in the saccharomycete cells and among plants. Undifferentiated groups of cells are able to generate the entire plant, as happens with the buds of the foliate appendages of ferns, and the inflorescent bulbils of Allium, Lilium, Ficus, etc., and the buds or sprouts of numerous aquatic plants and the like.

In the animal kingdom the most prominent case is that of the hydras and hydroids, polyps and jellyfish. In the body surface of the individual there appears a bud or excrescence which gradually assumes the shape of the individual and finally becomes detached, constituting a new hydra by simple growth. Certain worms also can reproduce through gemmation, sometimes through successive regenerations. In certain squirts, ascidia and other tunicates living fixed on to rocks, expansions occur in the basal regions or stolon, at the ends of which buds develop, each forming a fresh individual.

Strobilation or transverse fission is similarly a type of gemmation. The creature places various transverse portions one on top of the other which remain joined together for a time and then become successively separated layer by layer, starting with the first, each layer giving rise to a fresh individual. This happens particularly among the coelenterata of the scyphomedusa group, which divide by this means into different individuals.

In all these cases, reproduction of plants and animals comes about simply by multiplication of the cells forming the tissues, the somatic cells, and no special factors such as germ-cells enter into the question. Reproduction is in these cases nothing more than regeneration or, as we may call it, "growth."

## EDUARD ADOLF STRASBURGER

The most suitable living object for the purpose of studying mitotic or indirect division is the cell, together with its corresponding nucleus, of the spiderwort Tradescantia virginica, though observations should be made at a stage of active development prior to the completed formation whilst the cells are still multiplying in large numbers.

For these investigations flower-buds measuring from a fifth to a quarter of an inch in length without their stalk should be selected, and their anthers removed by means of a fine forceps after opening up. Next, the ovary and filaments under the insertion should be cut transversely with a scalpel and separated from the other flower organs, and then placed in a 3 per cent sugar solution on a slide so that the filaments may be examined under an ordinary microscope.

The resting nucleus of the cells forming the hairs looks as if it were finely stippled, though under a stronger magnification it will be seen that the minute granules forming it are not isolated one from the other but connected closely together in rows of fine coiled threads, which permeate the whole nucleus and form a network or framework entirely surrounded by a delicate membrane. Nucleoli of various sizes can be made out between the coils.

The nucleus is surrounded by the perinuclear cytoplasm connected with the peripheral cytoplasm beneath the cell membrane by means of a number of irregular cytoplasmic links. In addition to the hardly distinguishable microsomes, the cytoplasm includes larger, more strongly refractive grains, some with products of cellular existence, the leucoplasts.

When getting ready for division, the nucleus swells up whilst its chromatic framework shapes into a coarsely granular thread which then commences to lengthen, whilst the coils tend to become arranged roughly parallel to each other. These progressive changes arc easily seen in one and the same cell, but such observation requires a comparatively long time and must be maintained continuously or interrupted only for brief intervals.

The grains gradually assume a homogeneous appearance and at the same time the nuclear membrane disappears. After a while the swollen thread breaks down into segments or chromosomes, which then form into straight rods approximately equal in length, arranged in two groups one above the other with their equatorial ends meeting and their polar ends converging.

It usually takes about an hour to reach this stage from the moment when the threads become coarsely granular. The chromosomes at first sight appear practically homogeneous, but under a higher power, X 1000 for instance, slight indentations can be seen in their surface, indicating that they are in fact formed out of a series of discoid segments.

A few minutes later the two groups, each containing half the chromatic substance of the original nucleus, separate from each other in the lengthwise direction, each one making for the poles of the cell.

After another five minutes they lie quite an appreciable distance apart from each other. The daughter-chromosomes become shorter and thicker, whilst between the two groups the crystal-clear mass of cytoplasm keeps on increasing visibly in quantity. Within this transparent central mass finer threads can be distinguished which assume an arrangement like the staves of a barrel.

Some twenty-five to thirty minutes after separating out has commenced, rows of dark dots make their appearance in the equatorial plane, and form the socalled central cell-plate, by fusion together into a dark line which is the membrane separating the two daughter-cells. A delicate cellulose wall is the next thing to develop. If the central barrel-like formation of cytoplasm has been wide enough to fill the whole breadth of the cell, the partition membrane between the two daughter cells will reach all parts of the side walls of the mother-cell.

If, on the other hand, this formation be not of sufficient volume to stretch right across the cell's cross-section, it will tend to one side of the cell until making contact with a part of the lateral wall, to form the partition membrane of the corresponding portion in its equator. The little barrel then continues its circular motion so as to come successively into contact with the whole internal surface of the cell membrane.

As this movement goes on, the dividing membrane between the two cells continues forming on the spot as previously described. In this manner the membranes are finally completed and the cells divide one from the other. The central body then withdraws slightly from the already existing diaphragm, making up the missing portions thereof out of cell-plate segments formed in the neighbourhood. Thus a cell-plate is first produced equidistant from the two halves of the nucleus, in the middle of the transparent protoplasm and from it the new diaphragm is formed.

During these changes, the daughter-chromosomes bend inwards, sending their equatorial ends towards the axis of the nucleus, ever increasingly until they make contact one with the other and become the meridian lines of a sphere. The nuclear membrane then immediately reappears and clothes each daughter-nucleus, whilst the chromosomes attenuate and lengthen, becoming finely granular and acquiring once more the original appearance of irregular windings of delicate threads on a highly complicated reel.

The coils tighten up more closely together whilst the threads increase in number and stretch out, finally anastomosing or bridging together so that the individuality of the chromosomes is lost, and their substance remains in the form of fine granules, as it was in the mother-cell before mitosis commenced.

The daughter-nuclei swell in volume, most likely at the expense of the neighbouring cytoplasm; they become cloaked in cytoplasm which extends right up to the surface, leaving vacuoles characteristic of the cell, and at the end of about an hour and a half from the start of mitosis, the separation of the daughter-cells is complete and nucleoli are plainly visible again within the nuclei.

## WALTHER. FLEMMING

The various phases of nuclear division are clearly observable in large cells with relatively small nucleus, such as amphibian larva tissues and the embryo sac of plants, where the sphere of evolution of the nuclear elements is sufficiently large for movements to be seen with less vagueness.

We can distinguish several periods in karyokinesis or mitosis:

## THE PREPARATORY PHASE

The nuclear chromatin disperses in the resting state to a distance from any reproductive activity and, when karyokinesis is about to commence, reassembles into a thread formed of groups of beads with channels running through it, as observed by Balbiani in Chironomus.

The chromatin is distributed in stainable granules or small discs, the microsomes, separated by colourless plastin. At the same time, within the cytoplasm, the middle of the cell becomes divided into two centrosomes which develop into star-shaped configurations and act as the directive spheres. Karyokinesis now starts on its course.

## THE NINE PHASES OF KARYOKINESIS

### First Phase

The two directive spheres leave each other and arrange themselves at diametrically opposite ends of the cell, their radii lengthening into two cones coaxial on the same diameter, the apices of which are the corresponding centrosomes whilst the bases of the cones approach one another in order to form the protoplasmic barrel or spindle.

The central radii of these cones penetrate the nuclear membrane and thrust the thread of chromatin towards the equatorial plane. In a very short time the nuclear membrane is reabsorbed into the cytoplasm. This is the "achromatic amphiaster" and "chromatic spireme" phase.

### Second Phase

Development of the equator plane of the cell takes place. The achromatic spindle does not change but the groups of nuclear beads arrange themselves in festoons on the plane of the equator, forming a starlike configuration. This is the "chromatic aster" phase.

### Third Phase

The aster breaks up at its points into as many V- or U-shaped segments bent towards the centre of the cell as there were branches originally. These V-shaped segments are the chromosomes, and this is the "nuclear V" phase.

### Fourth Phase

Each chromosome straightens out lengthways, as each particle of chromatin or microsome splits into two equal grains. This is the "twin chromosome" or "nuclear W" stage.

### Fifth Phase

By double gyration around the equatorial plane, the two elements in each W are arranged symmetrically on one meridian of the achromatic spindle, their respective vertices tending towards one or other of the poles of the spindle. This is the "pole seeking" phase.

### Sixth Phase

Attracted by the respective poles, the two groups of halfchromosomes shift symmetrically on their corresponding meridian. This is known as the stage of "attraction toward the poles." The spindle keeps on producing cones one after the other, each shorter than the preceding one, and between them hardly visible parallel filaments make their appearance as generating lines of a cylinder joining each two semi-chromosomes of the same pair. It has not yet been decided whether we should consider these filaments as originating from the nucleus or from the cytoplasm.

### Seventh Phase

Arriving simultaneously at their respective poles, the groups of half-chromosomes group into two stars, each around the corresponding centrosome. This is the "chromatic amphiaster" stage.

### Eighth Phase

The two asters at the poles change into subdivided groups. This is the "double spireme" stage.

### Ninth Phase

To end up, each of the spireme groups is surrounded by a membrane thus giving place to two daughter-nuclei. This is the "twin nucleus" stage. Simultaneously the rest of the cell splits into two equal daughtercells. In this way the sharing of chromatin from the mother-cell between the two nuclei of the daughter-cells is effected rigorously in precisely equal parts, which is the object of karyokinesis.

## CELL DIVISION THEORIES

Various interesting attempts have been made in an effort to discover the causes of cell division, whether they be mechanical or determinant. With regard to the mechanical causes two forces have been suggested, on the one hand a contraction of the spindle filaments and on the other a chemotactic attraction exerted on the chromosomes by the centrosomes.

The former opinion was propounded by E. van Beneden in 1887, according to whom, the filaments shot out from the spheres of attraction become attached directly to the twin loops and drag them towards the poles whilst contracting. This tractive effect might even be the cause of the longitudinal splitting of the chromosomes.

This idea has been supported by a large number of writers, including Boveri, O. Hertwig, Bergh, Rawitz, C. Schneider, and

Rabl, the last of whom has given a very complete description of the manner in which these phenomena run their course.

However, Flemming in 1891 and Hermann after him raised a most powerful objection against this attractive theory, when they pointed out that the lengthwise splitting of the chromosomes often precedes that of the asters and centrosome. The theory may be partly true, but it is erroneous in certain respects and insufficient in others. Certainly the division of the centrosome and that of the chromosomes are perhaps associated without one being the direct cause or direct effect of the other.

The second alternative was advanced by Strasburger in 1893. This scientist held that the wings slide only on the filaments being attracted by some chernotactic force (or rather should we say biotactic) emanating from the attractional spheres.

Still more theoretical are the forces invoked as determinant causes of cell division. We know that Spencer, quite a long time ago, in 1864 called attention to the fact that when an organic being grows, its surface area increases as the square of its linear dimensions and its volume as the cube.

Assimilation must be proportional to volume and, as it takes place only through the surface, the consequence is that the more a creature grows the more difficult its nutrition becomes. Von Rees in 1887 applying these considerations to the cell thinks that the reasons for division lie here. But, showing that a thing is advantageous does not explain why it happens.

It would be necessary to prove how the difficulty in nutrition became the physiological stimulus for division. Rees sought to show by certain examples that this stimulative action from lack of nourishing power is a fact; in protozoa unfavourable conditions do cause division. Orr strove to determine the source of the movement causing division in the relative asphyxia resulting from diminution of the respiratory surface as compared to volume.

Every cell, whilst living, accumulates within itself spent products which it has to expel, and usually gets rid of them by excretion. Lendl in 1890 thought this could also be achieved by the cell splitting off from itself the portion of its body where these products, which he called ballast, had accumulated. Division in such case would have its origin in excretion of which it would be merely one particular aspect. Conjugation is an inverse process to cell division by which two distinct cells blend into a single one. It results in the formation of one fresh cellular individual formed from the substance of two different cells.

The race thereby gains an afflux of life which is apparent in the cells issuing from divisions following conjugation, and is preserved for a longer or shorter period, gradually becoming exhausted as asexual generations succeed one another until it is renewed by a fresh fusion. Such degeneration of the race after too long a series of divisions was excellently observed in the ciliates by Maupas, who gave it the expressive name of senile decay.

Seeing that conjugation is an essential condition for indefinite reproduction by segmentation among all creatures where it does exist, we might be tempted to generalize on this and believe it to be essential to every creature without exception; but a fair number of algae and the majority of fungi are reproduced exclusively by asexual spores.

There are therefore some beings amongst whom assuredly conjugation does not exist, though in many others we can only say that it has not yet been observed, and that through advances in science this number decreases every day.

# Chapter 7

# Sexual and Asexual Reproduction

## DIFFERENTIATION OF GERM-CELLS

When multiplying, cells may separate out, individual ones breaking away, or perhaps remaining joined together in clusters and increasing continually in volume through the division of individual component cells. In some cases this increase in the total quantity of cells goes on indefinitely. In others, however, the species has a characteristic limit, namely the colony.

Transition is possible from single-celled organism to polyplastid. Such multiplication of cells may occur in one dimension, as in the filamentary systems; in two dimensions, as in membranes; or in three dimensions, i.e. volume. In the more primitive forms of coexistence, no special cells seem specifically earmarked to reproduction, i.e. there are no germ-cells as such.

All the cellular elements composing them may be equally divided, and it is from such individual division that growth and reproduction of the whole is attained. This is observable among bacteria in general, in certain ascomycete fungi or yeasts, and in certain unicellular algae such as the genus Protococcus, etc.

As a contrast, in other algae, some cells become functionally distinct from the remainder, because their protoplasm and nuclear matter are divided into protoplasmic formations constituting spore. When the ripening thereof is completed, the cell membrane is broken and the spores stream abroad. Each spore on finding a suitable environment will give rise to the development of a fresh individual. This is a type of specialization which causes a particular cell to take upon itself the functions of reproduction, and occurs in single-celled algae like Sphaerella or in multicellular filamentary algae like Ulothrix.

## REPRODUCTION BY GAMETES

However, alongside reproduction by spores, a form of sexual reproduction also develops in various species. Certain cells form protoplasts from which issue cellular elements, smaller than spores, namely the gametes. These are similarly released by fracture of the membrane of the cell where they are generated and they have to unite two by two in order to fuse and enable a zygote to be produced. From the zygote, by successive divisions, spores also are obtained. The spores, like the zygote, are capable of latent existence; they multiply only under favourable conditions to the point of producing a fresh individual.

In the same way as among rudimentary plants, among certain protozoa in colonies, there is the twofold possibility of direct reproduction and reproduction through specialized germ-cells. When these make their appearance, the other or somatic cells become more and more incapable of reproduction. Colonies are made up of combinations of single-celled protozoa, of which a good example is the genus Chlamydomonas, whose cells arc ovoid with two flagella, two contractile vacuoles containing a chromatophore with red pigment and another more voluminous with chlorophyll. Because chlorophyll is present and because the cell is surrounded by a thick membrane, this genus was at one time thought to belong to the vegetable kingdom. It reproduces by cellular division with separation of the individuals so formed, or by union of two similar individuals — isogametes — and immediate redivision.

Forms of this class are united in varying quantities according to genus and species. Thus, colonies of Gonium sociale are made up of four individuals, whilst Gonium pectorale contains sixteen, Pandorina morum eight or sixteen,

Eudorina elegans eight, sixteen, thirty-two or sixty-four, Pleodorina illinoisensis thirty-two, Pleodorina californica sixty-four or a hundred and twenty-eight, and so on and on. In volvox types such as Volvox globator, the number of individuals making up a colony may run into thousands, arranged on the surface of a sphere whose interior is found to be full of watery fluid. Each of the component individuals in such a colony is a flagellate cell, with two flagella to the outside, similar to Chlamydomonas.

In the simpler combinations among these organizations, cellular reproduction comes about through simultaneous division of each one of the cells, that is to say symmetric reproduction whereby each colony is divided into two. It is also feasible for individuals to break off from the colony in order to unite again immediately each with an identical mate, like isogametcs, forming a zygote by such union. Divisions of such zygote will generate a fresh colony. Even so, this does not yet establish specialization of germ-cells.

In other cases, in more populous colonies, some of the cells making them up are engaged in the reproduction of the colony whilst the remainder, the somatic cells, lose the reproductive faculty. In Pleodorina illinoisensis,illinoisensis, of the thirty-two individuals making up the colony, twenty-eight are large cells suitable for division to form gametes, whilst the remaining four are small and sterile. In Pleodorina californica with sixty-four or a hundred and twenty-eight cells, approximately half are germ-cells and the remainder somatic.

In Volvox, the greater part of the cells are somatic and a small number becomes differentiated into germ-cells. This is the start of anisogamy, which is the interesting feature of this example. Microgametes, or sperm elements, and macrogametes, or ovular elements, are produced. From the union of the former with the latter, one of each in every case, we get the zygote generating a new colony by progressive division.

These organisms mark the transition between single-celled organisms and polyplastids. They are not yet polyplastids, because there is no morphological and functional differentiation of the somatic cells in various tissues; moreover, physiological correlation to unify and determine individuality is lacking. Among primitive plants, certain fungi, phycomycetes for instance, the formation of gametes is also to be seen in its initial stages.

Hypha cells belonging to various branches of mycelium approach one another before differentiation, eventually making contact, and then each of the touching cells behaves like a corresponding gamete. The two gametes immediately unite and form a single cell, the zygote, whence spring the spores to reproduce the plant.

Sporulation initially, like the formation of gametes, takes place in cells in no way distinguishable from their fellows, which, nevertheless, for reasons still unknown, behave like germ-cells specializing in the function of forming fresh individuals.

## CONJUGATION AND FERTILIZATION

The union of gametes, the addition of two cells one to the other, or amphimyxis betokens fusion of the respective nuclear substances. This is a universal process throughout the world of life, much the most frequent in the production of simple and complex plants and animals. It is a matter of reorganization of the elementary unit of any organism such as a cell, and is the result of nuclear interchange.

The case of Spirogyra in plants is typical. Two filaments lie parallel alongside each other, each emitting an evagination starting through the membrane. These evaginations unite at the tip, perforating each other's end and forming a connecting tube perpendicular to the centreline of the threads, thus joining the two cells.

One of the two penetrates into the other, by means of the tubes, the two nuclei unite and a single cell is thus formed, the zygote, which divides in profusion to reproduce the plant. Comparable with this is the conjugation of certain protozoa. The process is the same as that above described, though here it takes place in rudimentary animals. This was a phenomenon of great interest discovered a lifetime ago by Maupas, in 1889.

In certain Cilia, for instance Paramecia, two individuals approach one another, making contact with their cytostoma, between which a protoplasmic bridge is formed. These protozoa contain two nuclei, a large one or macronucleus which is trophic, and a small one, the micronucleus, which is the reproducer. The large nucleus breaks up and is dissolved in the protoplasm at the start of the process. Meanwhile, the small nucleus divides mitotically twice in succession, so that four micronuclei are thus formed in each of the conjugate individuals.

Three of these then disappear into the cytoplasm and the remaining micronucleus lengthens out, approaching the protoplasm bridge joining the two individuals. Such lengthening-out causes the micronucleus to split into two halves, of which one stays in its own part, whilst the other passes over to the conjugate.

As this passage occurs simultaneously in each of the two individuals, the result is an exchange of half the chromatin from each of the two micronuclei. Once this has taken place, the two paramecia separate, whilst the blended nucleus of each one, being the sum of two half-nuclei, splits into two portions, whereof one hardly grows at all, to become the micronucleus, whilst the other part increases in volume to form a new macronucleus.

It was originally thought that the exchange of nuclear chromatin between two individuals would intrinsically bring about the "rejuvenation" of each, and give them new life, being thus the manifestation of a law according to which germ plasms would enjoy everlasting life. The fashion was to talk of the "immortality" of infusoria, which immortality has, however, not been confirmed by facts. Conjugating infasoria also grow old and die just like anything else.

## SPORES AND GAMETES

Germinal specialization can likewise lead to direct sporulation without previous union of the gametic cells, and to reproduction by cellular conjugation. This can, on the other hand, be brought about by union of like cells or isogamy, or even by different gametes large and small, male and female, which is anisogamy. In the latter case, the microgamete is, as its name implies, small and mobile and seeks out the macrogamete until it can introduce its nuclear substance into the latter. The macrogamete usually contains in its cytoplasm reserves of nutrition to support the metabolic needs for development of the embryo which make themselves felt when cellular divisions of the zygote commence.

Reproduction by gametes appears in very simple forms among living bodies and is usually isogamic under such circumstances, as in the monocellular algae, such as Sphaerella or in the filamentary Ulothrix, or it may be anisogamic as in Oedogonium and other cases, and also in rudimentary animals.

Thus, in the hydroids, the production of macrogametes and microgametes, ovaries and testes, is observable. Temporary conical projections develop in the ectoderm formed by embryonic interstitial cells. Sperm elements sally forth herefrom dropping freely into the fluid and fertilizing the ovule adherent to the ovary of the corresponding individual. The fertilized ovule, the zygote, then loosens off, and becoming progressively divided gives place to a new hydra.

The possibility of forming spores occurs now and then in cells which are indistinguishable from the rest, but shows itself generally in cells forming part of special organs, as in the sporangia. Similarly, gametes originate in gametangia: antheridia, whence proceed the microgametes and archegonia, the formation foci of the macrogametes.

## ALTERNATION OF GENERATIONS

In many cases reproduction through direct sporulation alternates with that through union of gametes. It is to be seen in unicellular animals and plants, various protozoa and algae, and particularly when, at distinct stages in their existence, they inhabit different media.

This occurs in various types of living organisms. Thus, in species of mosses and in the ferns, which exhibit generations of gametophytes interspersed among generations of sporophytes. Among the same higher plants, the phanerogams, we can distinguish an asexual generation followed by a sexuate one in uninterrupted and continuous alternation one after the other. Numerous examples are to be found, particularly among protozoa; one well known case is that of the malarial parasite plasmodium, where we get alternating sexuate, gametic, amphiontic reproduction of haemosporidia in the transmittor, the anopheles mosquito, and sporular, agamic or monontic reproduction thereof in man.

Similar instances are to be found in certain polyplastids and the alternation of generations in hydrozoa is a typical case. The colony is but one phase in the sequence of generations in this creature. From it, at some given moment, through development of a yolk, a free individual breaks loose, the medusa, capable of independent existence on its own in the sea.

The colony is reproduced asexually and the medusa sexually. The hydra forms a similar colony in which the medusa form does not make its appearance. In other coelenterata, the scyphomedusae, the alternating process is likewise seen at work giving asexual generation by strobilation in the fixed forms, and sexuate in the mobile forms of the medusae.

Animal species do exist with germ tissue or gonads which appear temporarily at particular seasons of the year or by reason of other spcial circumstances. In other species, the majority in fact, the sexual organs are permanent.

## DIFFERENTIATION IN GERM-CELLS

The germ cells characterizing these organs come about from a morphologic and functional differentiation like any other, and identical to that forming the source of the various tissues in embryonic development. In some creatures, as for instance worms of the genus Ascaris, or Sphaerium among the molluscs, it is possible to distinguish germcells from the somatics, from which the organs arise, right from the commencement of blastomeric division. In general, however, the distinction between one and the other has to be made at a later stage. Primitive germ-cells form localized areas in the epithelium of the coelum, or on one side or the other of the spine in vertebrates.

The reproductive cells develop there in conjunction with connective tissue, blood vessels and nerves. Diffierentiation may take place much later, when the adult stage has already been reached, even in the higher animal forms, such as the mammifera themselves. Davenport in 1925, Parkes, Fielding and Brambell in 1927, state that in mice, regeneration of the ovary may be achieved after its having been taken out, by formation of fresh germ-cells at the expense of the somatic cells.

Similarly in plants, we find that the meristems, from which branches, leaves, and flowers derive, differentiate germ-cells from somatic, without the former being distinguishable from the others by any special cytological characters even from their first moment. Once differentiation is established, the germ-cells become localized in the corresponding organs and emigration of these germ-cells in the individual then does not appear to be possible.

## GERM PLASM

Following Nussbaum, Weismann in 1885 propounded his theory of the "continuity of the germ plasm." According to this, germ-cells pass from generation to generation, apparently immortal, in contrast to the somatic cells which make up the individuals, and consequently die off. The biological part played by each individual, with its somatic cells, seems to be merely as a mortal support for the germ-cells, in their uninterrupted course through the ages. S. Butler comments: "A hen is apparently nothing more than a means contrived by the egg so that another egg may be hatched."

By considering the germ cells as quite apart from the somatic cells, Weismann came to the conclusion that there was no connection between the one and the other within an organism, but a complete separation, an absolute partition between them, such that the somatic cells would not and could not exercise any influence whatsoever upon the germ-cells, whereby there could be no possibility of hereditary transmission of characters acquired by an individual in the course of its life. This would be worthy of discussion if germ-cells did indeed remain in all cases demarcated in the first division of the zygote.

This is the plan which was mapped out by Walter. It is not however proven that one of the first pair of blastomeres already shows the characters and functions of a germ-cell. There does exist, on the other hand, the possibility, previously pointed out, of reproduction by means of somatic cells, through which differentiation they will form germcells in fresh individuals. This would obviously interrupt the germ line and there would be no continuity in the germ plasm nor immortal cells.

The intrinsic characters of cells charged with the business of reproduction make their appearance at different moments according to the species under consideration. In some of them, the germ-cells are very speedily distinguishable from the somatics; in others this is not possible until later. The germinative quality is the result of one of many differentiations in ontogenesis; cells which acquire special reproductive faculties have to be distinguished through their specialization to carry out genetic functions and in the particular morphology of each one of these classes of cells. Germinal capacity demands only the existence of a genetic inheritance enabling each one of the characters distinguishing a species to make its appearance in due course.

## REPRODUCTION BY SOMATIC CELLS

The said reproductive capacity is not, moreover, exclusive to germcells and it varies even among somatic cells from one class to the other. As far as we know, there is a possibility of reproduction by somatic cells. This faculty shows itself by degrees, being absolute in the zygote and partial in a blastomere existing with one or more of its kind.

If however such blastomere becomes isolated from the others and sets forth from one of the so-called regulation eggs, belonging to certain animal species that partial faculty may become total and the blastomere on dividing will give birth to an embryo, just as would a zygote. This is of course easier the younger the generation to which the blastomere belongs.

The further we progress with embryonic development, the less becomes the range of morphogenic capacity in each cell. The sum total of the ontogenic faculty decreases progressively with differentiation, forming the various tissues. Only the germ-cells then preserve such total formative capacity.

According to what this capacity may be in its actual or its potential state, the possibility of regeneration of injuries will be more or less efficient. Among plants, this faculty is both wide and deep. In certain animals, particularly the lower ones, it is also present. It is to be seen to a more limited degree in other cases whilst in some it becomes reduced to mere cicatrization of a wound.

The greater the differentiation, or what is the same thing, the higher the creature in the scale of life, the less are the possibilities of what we may call automatic regeneration. The differentiatory morphogenic power decreases in exact proportion to the degree of differentiation in the cell. The nerve cell has lost the power of division. Thus differentiation means specialization. For this reason, in the superior forms of living matter, among the animals, the reproductive function remains reserved to the germ-cells, and in plants also reproduction by seeds rather than growth from shoots is generally more certain in its effects.

Reproduction seems to be a specialized function, of use particularly to animate beings of greater morphological and physiological delicacy; it is one improvement, amongst many, of biological progress through differentiation. All of which does not, however, prevent reproduction from being simply one manifestation among others of growth and cellular multiplication, first through somatic cells and afterwards through germ-cells.

## AUGUST WEISMANN

If I may now go so far as to propound a theory of heredity in the manner it has unrolled before me in the course of my own scientific development, I might start off by showing you that the factors, not merely from one single individual but rather from several and indeed often from a great many, are containable in the hereditary substance of the germ-cell of the animal or plant.

I start off with what I believe is the proven fact that the hereditary material resides in the chromatic substance of the nucleus. We have seen that this reaches each type of germ-cell in the form of a definite number of chromosomes, and that this number is previously halved in the germ-cells concerned with fertilization, which is to say in the sexual cells, and in fact as has been demonstrated for a whole range of creatures, this is effected in the last two cell divisions, the so-called ripening divisions.

We know that the whole number is again made up by the process of amphimyxis, since the half-quantities of chromosomes from the male and female germ-cells combine in a single cell, the "fertilized ovum," and in a single nucleus, the so-called fission nucleus.

The hereditary material of the offspring is thus formed half from the paternal and half from the maternal substance of transmission, and we have seen that this remains true during the whole development of the offspring, since in each subsequent cell-division each of the paternal and each of the maternal chromosomes is doubled by partition and the resultant halves are distributed to the two daughter-nuclei.

If now the whole hereditary material of a germ cell prior to the reduction division contains latently the whole of the elements of the body, as is obvious, then after the reduction each germ cell must contain either only half the components from the parents — or the whole number of elements must be contained collectively in the half-quantity of chromosomes. The latter alternative seems to me to be the only logical supposition, as I shall soon show, and by that it is clear that at the very least, factors for two complete individuals must be contained in the chromosomes of the fission nucleus.

That this conclusion is justified follows indeed from the fact that an entire, that is to say a complete individual, develops with all its parts from the egg, but a damaged one does not. Supposing, therefore, that each mature germ-cell contains only half the bodily factors, then it would be impossible for such halves, as they would be distributed by the luck of the division, to complement each other exactly in the two cells uniting in fertilization and it would therefore be much more likely that they were not properly paired, so that from their union an individual might result lacking in certain parts. From the combination of the two sexual cells, we should scarcely ever get an embryo complete with all its parts, for sometimes one and sometimes another set of parts would necessarily be lacking whilst some other might be doubled in size or be formed in duplicate.

Moreover, the facts of heredity teach us that resemblance to female and male parent can occur simultaneously in all or even in particular parts of the offspring, as is especially evident in plant hybrids, whence it inevitably follows that all constituents of the entire organism are in fact present in the half-quantities of chromosomes.

From now on I shall call the inheritable substance of a cell its idioplasm, following the example of Nägeli, who in fact expected it in the body of the cell, not in the nucleus, and theoretically thought it to be different in action, though as we have seen he was the first to conceive and formulate the idea of it as an elemental substance determining the entire structure of the organism in contrast to the ordinary protoplasm.

Every cell contains idioplasm, but since each also contains chromatin in its nucleus, I denote the idioplasm of the germ-cell as germ plasm, the elemental material for the entire organism, whilst the complexes shown to be equally available for the requisite elements of an entire individual I term ids.

These ids may in many cases coincide with the "chromosomes" at least in all those where the chromosomes are of simple composition, that is, not made up of several congruent formations. Thus, in the case of the brine shrimp Artemia salina, with its 168 small granular chromosomes, each of these chromosomes must be considered as an id, since every one of them can under certain circumstances be driven from the egg at the reduction division and combined with a great variety of other chromosomes on fertilization. Each of them must therefore be complete germ plasm in the sense that all parts of an individual are virtually contained therein: each one is a biological unit.

If, however, in many animals we observe sizable interlooped or even rod-like "chromosomes," and if these, as for instance in the oft referred to Ascaris megalocephala, are composed of a series of granules, then each of these granules may be looked on as an id. Indeed, we then discover in other forms of Ascaris a larger number of small globular chromosomes instead of the two or four large rod-shaped chromosomes appearing in Ascaris megalocephala. Combined chromosomes consisting of several ids, as in fact all rod-shaped or loop-shaped elements of nuclear matter are, I term idants.

Because of their smallness, their composition from several single ids does not always clearly stand out, and even in the larger of them only at certain stages. In the salamander sperm-cell the single ids are not visible at the earlier stage, though later on when the loops split a necklace formation becomes evident. We cannot therefore regard each chromosome immediately as corresponding to one only or several ids.

On a more precise delving into the processes of the reduction division, it has been shown that there are multivalent chromosomes made up, that is, of several ids, though their plurivalence is not directly recognizable but can only be deduced from their later development; there are bivalent and quadrivalent chromosomes which we must consider as made up out of two or four ids respectively.

It would lead us too far afield if I were to go further into this matter here, and the plan and purpose of this lecture does not oblige us to go deeper into this most intimate and even now still much debated question.

The germ plasm of each kind of animal and plant is therefore made up of a greater or less number of ids or individual elements, and only through their interaction does the individual developing from the egg become determined. Now at last we have reached the point at which it is most fitting to discuss the organization of living matter in general.

The Viennese physiologist, Ernst Brücke, some forty years ago established the idea that living matter might be not merely a conglomeration of all sorts of chemical particles, but could well be organized, which is to say made up of minute invisible vital units.

If — as we must therefore suppose — the mechanist theory of life is correct, if there is no vital force in the natural philosopher's meaning of the term, then Brücke's hypothesis is incontestable, for a chance mixture of molecules cannot produce the vital phenomena, not even a single molecule, since mere molecules do not live empirically, neither do they assimilate nor do they grow or propagate themselves. Life can therefore only arise through a definite combination of various types of molecule, and all animate matter must consist of such definite groups of molecules.

Herbert Spencer a little while later than Brücke also assumed such vital "units," and still more recently de Vries, Wiesner and I myself have adopted this idea. On the composition of these vitality carriers, or biophors as I call them, we cannot for the time being say anything more precise than that the protein molecules, water, salts and some other substances play the main part in their make-up. This follows from the chemical analysis of dead protoplasm; the form in which these materials are contained in the biophor and the way they act on one another in order to call forth the vital phenomena whilst passing through a continuous cycle from decomposition to reconstitution is, however, still quite hidden from us.

We do not need to bother about that here; it is sufficient to attribute the character of life to the biophors, and to consider all living substances, therefore, cell and nuclear matter, muscle, nerve and gland materials in all their variants, to consist of biophors, which latter must naturally be most varied in their composition.

There must be countless types of biophors, in all the different parts of the millions of forms of life which now exist on our planet; all of them, however, in order to exist must be built up according to some particular basic plan which at the same time determines their remarkable activity; all of them possess the fundamental property of animation, and disassimilate, assimilate, grow and multiply by partition. We must also grant them movement and sensation to some degree and intent.

As to their dimensions, it can only be said that they lie far below the limits of visibility and that all the smallest granules of the protoplasm, as far as they are directly ascertainable by means of our most powerful instruments, cannot be single biophors but masses thereof. On the other hand, however, they must be larger than any chemical molecule whatsoever, since in fact they themselves consist of a group of molecules, among which they are to be found in the most complex formation and must likewise be of corresponding relatively significant size.

It may then be asked whether perchance the previously inferred determinants are not identical with these "biophors" or infinitely small vital particles; this is not generally the case, however. We gave the name of determinants to those particles of the embryonic matter which determine a "hereditary fragment" of the body, i.e. from whose presence in the embryo it depends that a fixed part of the body whether composed of a group of cells, of one cell only, or of a part of a cell, shall be formed and that in a particular manner, and from whose variation furthermore only these particular parts can be induced to vary correspondingly.

We now ask how large and how numerous such hereditary portions are, whether each cell or each part of a cell in the body represents one such fragment, or whether only larger groups of cells play this part. It is now evident that these agglomerations, uniquely definable from the very germ, are quite different in size one from the other according to whether we are dealing with small or large, simple or complex organisms.

The single-celled creatures, for instance infusoria, must indeed possess special determinants for a medley of cellular organs and parts, if we are unable to identify directly even the independent and inheritable variability of their organs; lower multicellular organisms, such as perhaps the chalky sponges, will need only a relatively small number of determinants, but among the higher multiceuulars, e.g. certainly in the majority of articulated animals, their number must already be extremely high and amount indeed to several thousand if not to hundreds of thousands; for here everything in the body is already specialized and must be modified by autonomous variation right from the embryo itself.

## EVOLUTION OF PROTOPLASM AND CHROMATIN

The development of the cell theory after its enunciation in 1838 by Schleiden and Schwann followed first the differentiation of protoplasmic structure in the cellular tissues (histology). Since 1880 it has taken a new direction in investigating the chemical and functional separation of the chromatin.

As protoplasm is now known to be the expression, so chromatin is now known to be the seat of heredity, which Nägli was the first to discuss as having a physico-chemical basis; the "idioplasm" postulated in his theory being realized in the actual structure of the chromatin as developed in the researches of Hertwig, Strasburger, Kölliker, and Weismann, who independently and almost simultaneously were led to the conclusion that the nucleus of the cell contains the physical basis of inheritance and that the chromatin is its essential constituent.

In the development from unicellular (Protozoa) into multicellular (Metazoa) organisms, the chromatin is distributed through the nuclei to all the cells of the body, but Boveri has demonstrated that all the body-cells lose a portion of their chromatin and only the germ-cells retain the entire ancestral heritage.

Chemically, the most characteristic peculiarity of chromatin, as contrasted with protoplasm, is its phosphorus content. It is also distinguished by a strong affinity for certain stains which cause its scattered or collected particles to appear intensely dark. Nuclein, which is probably identical with chromatin, is a complex albuminoid substance rich in phosphorus. The chemical, or molecular and atomic, constitution of chromatin infinitely exceeds in complexity that of any other form of matter or energy known. As intimated above, it not improbably contains undetected chemical elements. Experiments made by Oskar, Gunther, and Paula Hertwig resulted in the conclusion that in cells exposed to radium rays, the seat of injury is chiefly, if not exclusively, in the chromatin; these experiments point also to the separate and distinct chemical constitution of the chromatin.

The principle formulated by Cuvier that the distinctive property of life is the maintenance of the individual specific form throughout the incessant changes of matter which occur in the inflow and outflow of energy, acquires wider scope in the law of the continuity of the germ-plasm (i.e. chromatin) announced by Weismann in 1883, for it is in the heredity-chromatin that the ideal form is not only preserved but, through subdivision, carried into the germ-cells of all the present and succeeding generations.

It would appear, according to this interpretation, that the continuity of life since it first appeared in Archaeozoic time is the continuity of the physico-chemical energies of the chromatin; the development of the individual life is an unfolding of the energies taken within the body under the directing agency of the chromatin; and the evolution of life is essentially the evolution of the chromatin energies.

We are equally ignorant as to how the chromatin responds to the actions, reactions, and interactions of the body cells, of the life environment, and of the physical environment, so as to call forth a new adaptive character, unless it be through some infinitely complex system of chemical messengers and other catalytic agencies. Yet in pursuing the history of the evolution of life upon the earth we may constantly keep before us our fundamental biologic law that the causes of evolution are to be sought within four complexes of energies, which are partly visible and partly invisible.

## SEX BIOLOGY AND SEX PSYCHOLOGY

In the first place, then, we have to consider the evolutionary history of sex. Of its origin, we can say only that it is veiled in complete obscurity. Once present, however, it appears to have a definite function by making possible, through sexual reproduction, all the various combinations of any hereditable variations that may arise in different individuals of a species, and so conferring greater evolutionary plasticity on the species as a whole.

Primarily, sex implies only the fusion of nuclei from two separate individuals; there is no need for sex differences to exist at all. Sex differences, however, are almost universal in sexually-reproducing organisms, and represent a division of labour between the active male cell and the passive female cell, the former taking over the task of uniting the two, the latter storing up nutriment for the new individual that will result from that union.

The subsequent history of sex is, roughly speaking, the history of its invasion of more and more of the organization of its possessors. First the male as a whole, and not merely its reproductive cells, tends to become organized for finding the female. The female's whole type of metabolism is altered to produce the most efficient storage of reserve material in her ova, and later she almost invariably protects and nourishes the young during the first part of their development, either within or without her own body. Appropriate instincts are of course developed in both male and female.

At the outset there is enormous waste incurred in the liberation of sperms and ova into the water, there to unite as best they may. Congress of sexes eliminates the major part of this waste, and is universal above a certain level. This is in itself the basis for other changes. As the mind, or shall we say the psycho-neural organization, becomes more complex, the sexual instinct becomes more interwoven with the general emotional state; and a large number of animals appear not to mate unless their emotional state has been raised to a certain level. The result of this is that special actions, associated generally with bright colours or striking structures, with song or with scent, come into being.

The exact mechanism of the appearance of these courtship-displays is a much-vexed point; but it is undoubted that they only occur in animals with congress of the sexes with minds above a certain level of complexity, and that they are employed in ceremonies between the two sexes at mating-time. There can subsist no reasonable doubt that there exists some causal connection between the associated facts.

An important point, which has been commonly overlooked, is that such characters and actions may be either developed in one sex only, or in both. In a large number of birds, such as egrets, grebes, cranes and many others, the courtship-displays are mutual, and the characters used in them developed to a similar extent in both sexes.

Such characters are therefore often not secondary sexual differences, and we had best use Poulton's term epigamic for them, whether they are developed in one or in both sexes. In vertebrates the gonads form part of what has been called the chemical directorate of the body — the interlocking system of endocrine glands, each of which is exerting an effect upon the rest. It is only in higher groups that these emotion-stimulating sexual characters arise, for only in them has mind reached a sufficient degree of perfection.

When we reach man, however, the whole aspect of the matter changes. The change is most marked, naturally, in his mental organization. Through his powers of rapid and unlimited association, any one part of his experience can be combined with any other; through his powers of generalizing and of giving names to things, his experience is far more highly organized than that of any animal; through speech and writing, he is inheritor of a continuous tradition which enormously enlarges his range of experience. Again, he can frame a purpose and thus put the objective of his actions far further into the future than can lower organisms.

There are, however, also changes of considerable biological importance on the physical side. Man brings with him from his animal ancestors the endocrine secretory mechanism of the reproductive organs: but his life is not subordinated to it in such an iron-bound way. This is not to deny that the sexual life of man is dependent upon the reproductive hormones. It is apparently necessary for proper activation of the sexual centres of the brain that there should occur a continuous liberation of secretion from the reproductive organs into the blood.

Again, the mental activities of man are so much more important than those of other forms, that even the cessation of activity of the reproductive organs, for instance, in the female at the change of life, or even their total removal, need not prevent the continuation, albeit in a modified form, of the sexual life in its varied indirect manifestations.

It is quite clear from these and other facts that in higher vertebrates there are present in every individual of either sex the nervous connections which give the possibility of either male or female behaviour.

## Chapter 8

# Dynamics of Reproduction

### THE FERTILIZATION PROCESS

Whenever a cell divides either directly or by mitosis, the most striking thing is the sequence of events developing in the nuclear substance. The discovery of the various phases of karyokinesis was very soon followed by that of the morphology of fertilization or syngamy. Penetration of the nucleus by the male gamete or pronucleus is the natural way of exciting mitosis in the ovule through the ensuing action upon its nucleus, that is to say on the female pronucleus.

The chromatic substance of both nuclei contributes on an equal footing to karyokinesis, from which division of the fertilized egg or zygote into two blastomeres comes about. This was observed by O. Hertwig as far back as 1875 and confirmed by van Beneden in 1883 and by Boveri in 1886 using material of a different nature and origin.

In 1865 Nägeli had assumed that germ-cells contained directional micelles. Kölliker held that the nucleus itself might be the carrier of hereditable properties, on the assumption that the "idioplasm" resided therein, and it remained only for the protoplasm to fulfil a nutritive function.

Strasburger in 1876 describes the chromosomes, whilst van Beneden in 1883 and Rabl in 1885 indicate their significance. Weismann in 1883 stated that they are the bearers of characters transmissible through heredity. Of late the conviction has grown more and more pronounced that chromosomes play a predominant and often exclusive part in heredity. We may mention, for instance, the voluminous researches of Edmund Beecher Wilson, from 1903 onwards, with regard to chromosomes over an extremely varied range of species and their functions.

In fertilization, chromosomes from each of the two participating gametes draw up alongside each other and form half and half the branches of the mother aster or equatorial plate at the commencement of metaphase. When immediately the hitherto single aster starts to divide into two daughter-asters, it is obvious that these latter will likewise be made up, as to one half, of chromatin of paternal origin, whilst the other half will derive from the maternal source. Similarly, therefore, the nuclei of the two blastomeres are produced from such division, and afterwards likewise all cells generated during successive divisions.

The consequence of this will be that all cells ultimately derivable from the zygote will necessarily possess the same number of chromosomes, as stated by Boveri in 1886, which assertion after being the object of considerable debate has now been repeatedly corroborated. The number of chromosomes in any cell, be it body-cell or germ-cell, is fixed and definite for each particular species.

## CHROMATIC REDUCTION OR MEIOSIS

This constancy in the number of chromosomes necessarily implies nullification by destruction or expulsion of half the chromosomes from each gamete. Thus, when the chromosomes from two gametes unite, the zygote will come out with the same quantity as all the other cells, as will also the blastomeres and their descendants and, consequently, by the same token, all the cells of the new organism. Fol in 1876 noted the reduction in chromosomes on the ripening of germcells, and this was confirmed by Strasburger in 1894, by R. Hertwig in 1906, and by many others subsequently. In 1903, Farmer and Moore suggested the term meiosis to denote this type of process.

The chromosomes in the cells are arranged in pairs of homologous chromosomes or alleles, of which each pair is composed of one chromosome from the male source and a second from the female, resulting in diploid cells containing 2n chromosomes. When the germ-cells ripen, half the chromosomes are lost through meiosis, and cells become haploid with n chromosomes.

The process whereby meiosis takes place is typical in spermatogonium. Spermatogonia derive from an original germ-cell through division, all being diploid cells. On division taking place in the spermatogonia, constituting the primary spermatocytes, chromosomes from each pair adhere to each other lengthwise by synapsis thus: 1-I, 2-II, 3-III and consequent to this, each chromosome of each pair will split also in the lengthwise direction, thus giving as a result a bundle of four chromosomes, two male and two female thus: [1-I, 1'-I'] [2-II, 2'-II'] [3-III, 3'-III'] and so on, forming tetrads.

The primary spermatocyte divides into two secondary spermatocytes. In this mitosis, each tetrad forms a branch of the mother aster, and on these branches separating in the anaphase, the resultant chromosomes will be made up naturally of two elements alongside each other: dyads,perhaps 1-1', II-II', III-III or may be I-I', 2-2', 3-4', and so on.

Thus the secondary spermatocyte will have the same number of chromosomes as the spermatogonium, though they will now be in joined pairs so that their quantity will seem to be only half of what it really is.

Moreover, in the division of the tetrads, each of the resultant dyads will be made up of chromosomes two by two from the same line, that is, each pair will come from chromosomes from the male source or from the female source exclusively, which means that the whole of the chromosomes will be distributed between two secondary spermatocytes in such a way that the male and the female chromatin will complement each other, since the dyad that has a particular class of chromosome originating from the male will not contain any of the same class coming from the female.

That is to say that, just as pairs in the germ cells are formed from one chromosome of male origin and another of female origin, here each pair will be solely male or solely female. From this it comes about that when the secondary spermatocytes split into halves — spermatids — the dyads separate from each other, the nucleus of each of these spermatids attracting the chromosomes which formed the dyads, perhaps 1-II-III and 1'-II'-III' or I-2-3 and I'-2'-3' in the scheme already described. Both spermatids, originating in any particular secondary spermatocyte, will contain half the chromosomes that are in the primary spermatocyte and will be haploid cells, their chromosomes being, as is obvious, those of the corresponding dyads.

Through such a series of operations, there will have been generated from the primary spermatocyte four spermatids, each pair having had its source in one of the two secondary spermatocytes.

Two of these spermatids — A — will be made up of similar chromosomes, and the other two — B — will be offspring from the other secondary spermatocyte and likewise identical in chromosomic content. It is easy to see, therefore, that the spermatid pair A resulting from one of the two spermatocytes may be simultaneously complementary, through their chromosomes, to the two spermatids B descending from the other spermatocyte.

That is, since these spermatids possess half the chromosomes from the spermatogonia, the male and the female chromosomes will be complementary pair by pair — male-female-female and female-male-male as an example — in such manner that the sum of the chromosomes of these two types of spermatid will exactly equal the chromosomic inheritance — malefemale - male-female - male-female — of the original spermatogonium.

The nuclear substance, the sperm head arising from the ripening of the spermatid, will of course have the same chromosomes as the spermatid. Identical phenomena take place during the development of the ovule. When the oogonia divide into primary oocytes, the chromosomes group together and divide into tetrads. The formation of the second oocyte with emission of the primary cell or polar sac takes place when the tetrads split into dyads, one of which goes to the oocyte and the other, not identical, to the polar cell.

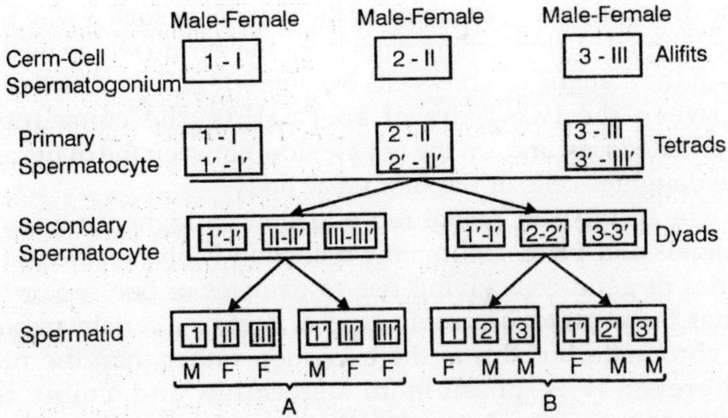


The immediate division of the secondary oocyte into a ripe ovule and second polar body comes about from lengthwise partition of the dyads. One consequence of this is that the ripe ovule and the second polar body will have like chromosomes and these will be complements of those found in the nuclei of the cells resulting from the division of the first polar sac.

The process of chromatic reduction comes about therefore in the same manner in oogenesis as in spermatogenesis. The only distinguishing feature is, that in spermatogenesis the four spermatids and consequently the four derived spermatozoids from successive divisions of the primary spermatocyte are functionally complete and suitable, whilst in the formation of the ovule, one only of the four cells originating from the primary oocyte will ripen and be utilizable. The remainder of the cells, offspring from the first and second polar bodies, will degenerate and vanish.

## CONSTITUTION OF THE ZYGOTE

Fresh pairs of chromosomes arise in the zygote through fertilization and the cell becomes diploid with 2n chromosomes, due to the superposition of the n chromosomes from each of the gametes.

As can be seen, in meiosis there is no question of a simple reduction of chromosomes, but of something deeper and more significant, a "segregation" or "disjunction" of chromosomes, a sharing of chromosomes as definite biological entities from one or other source, the male or the female, in such a way that the original allelomorphic pairs from the germ-cells as well as from the somatic or body cells become irregularly distributed between the two pairs of spermatids, the consequent spermatozoids, and, on the female side, between the ovule and the daughter-cells of the first polar body.

In all types of sexual reproduction, in higher and lower animals and plants, chromatic reduction is observable in the series of germ-cells giving rise to gametes of one sex or the other. Some protozoa simply expel or destroy a certain amount of chromatin but this is the exception rather than the rule; wherever it is possible to determine and count the chromosomes, it can be verified that the reduction comes about more or less in accordance with the scheme previously outlined.

It never comes about by mere chance. The ripe gametes have their definite heritage of chromatin containing the portion belonging to them from the bearers of hereditary factors from one or the other line of the parent stem. Thus it is that in the make-up of the zygote, the paternal or maternal chromosomes reproduce part of the chromosomes of the four grandparents and further up the ancestral fine, throughout all generations.

These combinations of chromosomes will be the more plentiful the greater the number of chromosomes from each species. This may be calculated by the mathematical laws of permutations and combinations, with however, one limitation — each gamete must receive a derivate from one of the members of each pair in the germ-cell, so that the gamete will contain one complete haploid group.

It will be very difficult for such group to be formed exclusively by paternal or by maternal chromosomes alone; chromosomes from each of the two main stocks alternate one with the other in each group. In the human race, with forty-eight chromosomes in every cell, forming therefore twenty-four pairs, we can get 224, or pretty nearly seventeen million (16,777,396) types of gametes for each sex, which, through union of the gametes in fertilization means getting on well towards three hundred billion (281, 481, 016, 540, 816) chromosomic combinations possible for each pair. It is not to be wondered at, then, that different children of the same parents are not necessarily alike in every respect!

Fertilization reunites the chromosomes from two haploid gametes into a diploid zygote. It can be understood, therefore, that in the zygote, chromosomes of the same type will be duplicated from father and mother and will group up in pairs. As the chromosomes divide along their length, when the daughter-asters are generated by division of the matrix, during successive mitoses, not only during development of the embryo but throughout the whole existence of the individual, the cells of the body without exception, that is to say diploids, body cells and germ-cells alike, will always have the same number of pairs of chromosomes.

## CHROMOSOMES AND GENES

The behaviour of the chromosomes in cellular division and the relation this has to the transmission of hereditable characters from one line or the other, paternal or maternal, has resulted in the recurrence of ancient theories which attributed heredity to the action of certain material particles contained in the nuclei of the germ-cells.

Nägeli's "micellar" theory, Spencer's "physiological units", Darwin's "gemmules" and de Vries's "pangenes", finally culminated in Weismann's theory, according to which inherited characteristics were held to be due to the intervention of material elements combined in the idioplasm forming part of the nuclear substance.

Such idioplasm or germ plasm comprises echoes from every part of the organism, or ids. In certain cases, it will be possible to identify the ids with small chromosomes; in others, the chromosomes will be idants or combinations of various ids. At the same time, such ids may be complex systems made up of numerous determinants or organizations of minute, invisible, biological units, the living molecules or biophors.

Originally, Weismann considered that the material unit or id might be the chromosome itself, or small chromosomes, especially when such chromosomes occur in great number in the corresponding cells. Later on, however, on taking into consideration that the chromosomes are often small in number whilst the characters which have to be accounted for are on the other hand multitudinous, he assumed that each chromosome consisted of many ids. Every id must, moreover, contain, in addition to the potential source of the character or characters it represents, the similar possibility of a source of the whole combination of characters, i.e. of the totality of characters defining the product. Each id will therefore possess a special representative element and complementary thereto a complete representation of all the characters of the individual or of the species.

In guiding the course of development, the ids play their part, each one on its own account, and, on a higher level, organization of the ids will also come into the picture. Each id will therefore be a sort of microcosm, a little world in itself. Individual variations will be due to combinations of ids from the male line combining with those coming from the female source.

By progressive division of cells constituting the development of the embryo, the ids keep on dividing in proportion to the multiplication of the cells, until each type of cell forming the body organs receives a determinant, the final result of such divisions of the ids. On the other hand, in the generation of germ-cells, there will be no disintegration of ids.

Wilhelm Ludwig Johannsen in 1909 asssumed that chromosomes are merely systems of genuine hereditary elements, and suggested the name genes as a label for these latter. Thomas Hunt Morgan in 1926 published his celebrated work The Theory of the Gene, which forms a conspectus of results obtained by the noted school of genetics which revolved around him under his direction.

The genes, distributed in exact location in the chromosome, thus facilitating the preparation of the gene charts which we now possess thanks to Mackensen's work in 1935 and to Bridges, are present in large numbers and each one of them is responsible for the blossoming of some particular character or set of characters. Alteration of a gene is bound up with some modification of the corresponding character; the disappearance of a gene is accompanied by the disappearance of the corresponding character.

On the other hand, there is withal an interrelation or mutual influence between different genes in various special rigid relationships. From the properties of each gene on its own account and from those arising from its location and proximity to other genes, we get the embryonic production of living forms and the full development and functions thereof.

## CYTOPLASM AND REPRODUCTION

All these facts which we have just been reviewing corroborate the part played by the nucleus in reproduction. The question still arises, however, whether nuclear intervention is the sole factor, or whether the cytoplasm does not have some share in transmission of characters from cell to cell.

To such queries the results of experiments in merogony, or the fertilization of fragments of ovule by sperm penetration, give an adequate answer. Such fertilization was achieved by Boveri in 1889, when he broke up sea-urchin egg-cells by shaking. Break-up of an egg-cell is nowadays effected with precision under the compound dissection microscope. These researches have reached their present extent and importance through the work of Yves Delage in the first place, and the later investigations by Edmund Beecher Wilson, von Ubisch, S. Hörstadius, Baltzer, Curry and many more.

If fertilization is effected on a fragment of egg-cell containing its nucleus, the zygote will be diploid; but if it take place in a piece of cytoplasm denuded of nuclear matter, the resultant cell will be haploid and the maternal nuclear matter will not be able to exert any influence on the characters of the product, and indeed the ovular cytoplasm only will act as the mother-cell.

It is generally deduced from experiments of this nature, that the nucleus is not an essential factor in heredity, even though it may predominate to a fairly wide extent. We must, in fact, bear in mind the arrangement of the cytoplasm, given the anisotropy of the egg; the germs of the characters show a precise layout, just as does the nucleus in the cytoplasm. It is the cell as a whole which determines the transmission of characters, even when for motive power it is inevitable that the nucleus must play a most important part.

## PARTHENOGENESIS

In merogony without ovular nucleus we come up against the problem of chromosomic multiplication in the haploid zygote. The same occurs in certain cases of parthenogenesis. On the other hand, natural fertilization results from the activation brought about by penetration of the nuclear substance of the male gamete, which immediately dives in and simultaneously sets the division of the female nucleus in motion, the chromosomes of the two nuclei pulling up alongside each other as we have already observed.

Nevertheless in certain species and in particular circumstances, division of the egg occurs without the help of the sperm, i.e. parthenogenesis, and furthermore such parthenogenesis can be achieved by artificial stimulation of the nuclear substance of the egg-cell, in what we call artificial or experimental parthenogenesis.

These facts have an important bearing from a theoretical standpoint.Parthenogenesis was demonstrated in aphides by Bonnet in 1845, when ten generations of banana fly were produced without fertilization. Strasburger confirmed the occurrence of parthenogenesis among plants. There are various types of parthenogenesis. Weismann in 1887 thought that the parthenogenetic egg-cell emitted only one polocyte, and that this occurred moreover without meiosis, so that such cell would be diploid. This is not, however, the general rule.There are several types of parthenogenesis:

- The so-called optional parthenogenesis of certain hymenoptera, xylocopa, bombyx, osmia, apis, hemiptera, the aleurodyde family, acarus, tetranychus, etc., wherein the egg is separated from its two generated polar bodies and remains haploid. From such parthenogenetic eggs we get males exclusively — arrhenotoky — and this generation alternates with sexuate, that is with fertilized offspring, and therefore obviously with diploid zygotes which generate females -thelyotoky. This case is comparatively rare.

- Accidental parthenogenesis is exceptional and of a primitive type, occurring in some echinoderms, asterias and lepidoptera such as the silkworm bômbyx mori. Only certain of the eggs develop and the remainder degenerate forming two polar bodies with meiosis, though, according to Boveri, Hertwig, and others, it may be possible for the second polocyte to fuse with the egg in a sort of self-fertilization. The egg is diploid.
- Cyclic parthenogenesis, so called because it alternates with sexual reproduction and takes place seasonally only at certain periods in the year. It is the most common type of parthenogenesis. In it, the egg ejects a single polar body and meiosis does not occur; this is the type of parthenogenesis studied by Weismann.
- Polyploid parthenogenesis: the egg commences its chromosomic divisions in the absence of fertilization and division of the cells does not invariably accompany that of the nucleus, so that the resulting cells may well be polyploid. This is to be seen in nematodes and some crustacea. Artemia furnishes a typical example: the sexual generation produces specimens with diploid cells of forty-two chromosomes, whilst the parthenogenetic gives polyploid cells with eighty-four chromosomes.

All these classes of parthenogenesis are alike in so far as at a given instant, a female germ-cell starts to divide, thus commencing the embryonic development without participation of a gamete of the opposite sex.

Siebold in 1871 compared this process to an internal gemmation, which is furthermore similar to the apogamy to be observed in certain plants. The difference depends on the fact that in parthenogenesis the dividing cell, although the reason therefor is not known, is a germ-cell or ovule, whereas in apogamy it is a somatic cell which, through successive division, gives birth to a new individual.

## ARTIFICIAL OR EXPERIMENTAL PARTHENOGENESIS

Nothing is known about the mechanism which sets off the multiplication of the ovule. In 1785, Spallanzani made a mixture of lemon juice and vinegar with frog spawn, with the object of stimulating development of the latter. A very important discovery was that of J. Loeb in 1899, namely experimental parthenogenesis, or the production of embryos starting with the ovule and without the intervention of a male gamete.

The procedure followed originally by Loeb consisted in raising the osmotic pressure of the sea water in which eggs of the arctic urchin Strongylocentrotus were immersed. Afterwards, he added fatty acids to the water and placed the ovules straight into the hightensioned sea water. Delage in 1901 also obtained parthenogenetic development by alternate action of dilute bases and acids, whilst Lefevre in 1907 used dilute organic acids, and other researchers had recourse to physical agents such as heat, X-rays, and the like, or mechanical effects like stirring, shaking and pricking.

Positive results have been obtained under such tests for experimental parthenogenesis in amny species — echinoderms, nemertinea, gephyrea and other annelids, certain fish and batrachia. It has also been possible to activate the egg-cells of certain mammals. Every species is specially sensitive to its particular type of stimulus. Normally, according to Bataillon parthenogenetic development, especially in frogs, comes about after ejection of the polar bodies and the cells come out obviously haploid, which, says Wilson in 1925, serves to explain why many of the larvae obtained by experimental parthenogenesis never succeed in reaching the adult stage.

The presence of diploid nuclei in certain cases of blastomeric development does not however appear to be exceptional. Adult forms are sometimes obtained, as Loeb proved even in the frog itself, and it has been stated that frogs can thus be born having a female parent but no male. The biological significance of all these results is very evident.

## MATERIAL FACTORS IN DEVELOPMENT

We have just seen that various substances can foster parthenogenetic development of the egg. Equally material agents are those which ensure the attraction of both gametes. Pfeffer in 1897 examined the action of malic acid, whilst Lillie in 1914 showed that the ovule secretes a substance, which he called "fertilisin," the effect of which is to impel the sperm to the ovule and to act as the agent causing penetration by such sperm. Evidently, from this, material factors of a distinct type promote and govern development of the embryo — hormones, harmosones and parahormones — in much the same way as the growth and regeneration of mutilations is developed under the influence of agents of this category.

These are the "auxins" or "auxetics" of Rous and Drew, Centanni's "blastins," Fischer's "atraxins," Haberland's "wound hormones," or Caspari's "necrohormones," Fischer's "desmons, Carrel's "trephones," the older "erguse" and "archuse," biotin, the effects of which have been carefully studied. There are, on another level of vital integration, Spemann's "organizers," also Needham's and Waddington's "evocators."

The genes too in their turn may well be material factors. Morgan in 1928 questioned whethe the gene might reach the size of some of the large organic molecules, e.g. polypeptids, proteases. He asked himself if the organizers might be cyclic hydrocarbons, phenanthrenes, like certain of the cancer-producing substances, sterols, like some morphogenic hormones, carbohydrates, and so on. It seems more probable that the nature of such agents approaches more to a plasmic supramolecular structure than to a collection of isolated molecules.

Nevertheless, chemical substances of very simple molecular composition such as indolacetic acid, or paradichlorophenoxyacetic acid, amongst others, show intense harmosonic properties on the development of plants. A number of very distinct bodies, of which a typical example is represented by certain amino-acids, such as lysin, tryptophan, para-aminobenzoic acid, some hormones and specific vitamins, also influence growth in animals and the development of the organs.

Chemical effects in the most diverse classes are responsible for ontogenic development. Indeed starting from the first divisions, this is accurately proven by the study of experimental parthenogenesis. Through the results thereof these general biological principles approach those of physiology and are the commencement of the scientific study of the determinant conditions governing them.

It is necessary to get rid of the preconception that development of the embryo is something specially mysterious or magically different from other organic functions. Growth through assimilation is but one of many functions, and furthermore, the growth and evolution throughout life of an individual, with its characteristic cytological features at any given instant, is in its essence exactly the same thing as embryonic development. .

This development does not cease when the individual reaches its adult stages; it commences when division of the egg-cell is started and is arrested only by death. Existence is an uninterrupted morphogenic track through time, in the same way that it is also a continuous functional track. Nowadays there is no ground for separating function from form.

## EDUARD VAN BENEDEN

It was only after the appearance of the remarkable works of Baer, Purkinje, R. Wagner, Coste, Prévost, Dumas and Rusconi on the vertebrates, and of Rathke, Hérold, Siebold and P. J. van Beneden on the lower animals, that the foundations of comparative embryogeny and ovology were definitely established. The nature of the egg of the higher animals and of a certain number of lower creatures became known, and the notion gradually arose that in all cases the egg was made up of the same essential parts, namely: a membrane, a yolk, and a germinal vesicle containing one or more refringent corpuscles in suspension. Furthermore, the segmentation in batrachia detected by Prévost and Dumas had just been discovered in fish by Rusconi and Baer, whilst Siebold had recently confirmed it in nematoda, and Dumortier, Beneden and Windischmann in gastropoda.

What a mystery this segmentation is, occurring with similar characters in all cases! What relationship could it have with the formation of the embryo, and what was its purpose? This was the seemingly unfathomable puzzle and nobody knew the reason for the yolk segmenting any more than they knew why the egg should contain a vesicle fated to fade away.

The real nature of the egg was still beyond our ken. In 1839, however, Schwann's treatise was published, and the discovery that all animal tissues originate in cells worked a radical change in scientific thought. Schwann's cellular theory threw, perforce, an entirely new light on anatomy and physiology, just as much as it did on embryogeny and was enough to make its author's name famous for evermore.

The cellular theory was now set on a firm foundation, and simultaneously the deep mist hitherto surrounding the question of the nature of the egg and the purpose of segmentation vanished. From his study of its composition, Schwann was the first to proclaim that the egg is a cell; and as all tissues are made up only of cells, it became clear that the purpose of segmentation was to multiply the egg-cell. Bergmann, Reichert and Remak made their particular contributions by showing the part played by segmentation in the production of cells.

Although a great number of physiologists, following Schwann's lead, regarded the egg as a simple cell, others, like Henle, looked on it as an agglomeration of cells, and saw in the germinal vesicle a cell complete in itself. Among those of this opinion we may mention Bischoff, Steinlein, Stein, and others. It is essential to keep a middle course between these two opinions, and to consider certain eggs as simple cells whilst regarding others as complex. Is it possible, for instance, to consider the threadworm. (trematoda) and tapeworm (cestoidea) eggs as single cells, when they are seen to be formed by the union inside a common shell of a protoplasmic cell with others formed by clearly different glands, erroneously bearing the name of yolk-formers (vitellogenes)?

Our own study of the method of formation of the egg and primary embryonic manifestations gives a plain answer to this question. In every egg, whether it be from mammal or bird, crustacean or trematode, we find a protoplasmic cell whose germinal vesicle is its nucleus, the touch or Wagner's corpuscle, the nucleolus. This cell, which we have termed the germ or egg-cell, and which may be considered as being the original cell of the embryo, is in all cases formed the same way; it always exhibits the same characteristics and gives birth, by division, to the original cells of the embryo.

The egg yolk is, however, made up of two elements, one protoplasmic representing the main substance of the egg-cell, and the other, serving as food, forming what we have called the egg's deutoplasm.This deutoplasm is the portion subsidiary to the yolk, and sometimes it is missing, or is produced in a different manner, exhibiting very irregular relation to the protoplasm, and behaving with extreme diversity in the course of the first appearance of the embryo. Often it comes to life right inside the protoplasm and is fashioned by the eggcell itself; at others, it is formed by special cells either in an exclusively deutoplasm-forming gland, or in the same gland where the germs are formed though in a particular portion of such organ.

This deutoplasm may be formed from specific cells and see the light in a special organ; and though it forms part of the egg, it cannot be regarded as an integral part of the egg-cell. It follows that the generally accepted thesis "every egg is a cell" does not bear that stamp of precision which should be a feature of every scientific principle. But in every egg there exists one EGG-CELL, one germ which is the original cell of the embryo.

Besides this cell, there resides in the egg a supply of food material, which can be found mixed in with the protoplasm of the egg-cell, and formed in its interior, as may be observed in many vertebrate creatures. In this case they may be included in the cell, and it may be said in agreement with Schwann, who had the vertebrates particularly in view, that the egg is a cell.

If, however, the deutoplasm is discovered outside the egg-cell, it cannot be considered as forming an integral part of the germ, and may itself be composed of cells, as can be seen from examples in many of the lower animals which are distinguished by their extreme fecundity. In such case evidently the egg is not one cell but an agglomeration of cells.

## THEODOR BOVERI

In conjugation the chromatin content of two similar nuclei is united. Not only does one of the nuclei consist of the same number of chromosomes as the other, but it can be shown in many cases, where observation has been possible, that the chromosomes from the sperm have the same size, shape and identical properties as those from the ovule nucleus. We may therefore state that each sperm chromosome meets its mate in the female ovular nucleus.

If we denote the chromosomes from one of these nuclei by the letters a, b, c, d, . . ., the other nucleus will also contain a similar series a, b, c, d, . . . . These two "haploid" nuclei, on uniting in conjugation, will form a "diploid" nucleus containing 2a, 2b, 2c, 2d, . . . chromosomes, which we will call the nucleus of the zygote. The double diploid series will develop into identical chromosomes when the zygote divides into two blastomeres and likewise in every subsequent division, since, as is well known, each of the chromosomes forming the matrix splits lengthwise so as to form two daughter matrices in mitosis.

When the nucleus of each cell comes out from the resting stage, the chromosomes become visible as individualized elements always equal in quantity. It is therefore probable that, even during the resting or reproductive period of the nucleus, the chromosomes are at least potentially in existence. The number of chromosomes is characteristic of the particular animal or plant species. Furthermore, as we have just seen, in mitotic division the chromosomes are equally divided and become paired off during fertilization, so that we may feel justified in contending that the chromosomes possess "individuality."

In blastomeric division after fertilization the two series, male and female, are transmitted in exact balance, making up mated pairs of elements from each source. This process is repeated indefinitely throughout all the cell divisions which take place during development and subsequently throughout the whole course of existence.

This is made possible precisely because the chromatic reduction, taking place during the ripening period of the sex cells by reason of the haploid character acquired by these cells, brings in its train the natural result that the fertilized ovule or zygote is in its turn diploid. Such homogeneous distribution of chromosomes to all the cells in the body is only feasible on the condition that the mitotic forms are normal, regular and bipolar. Mitosis with three or more irregular poles is accompanied, as may well be understood, by a heterogeneous arrangement of the chromosomes, unsuitable for normal development.

## EDMUND BEECHER WILSON

In fixed material, especially as viewed under relatively low magnification, the nuclei commonly appear deeply stained after treatment by certain dyes, such as carmine, haematoxylin, methyl-green, or gentian violet, while the cytoplasm remains relatively pale. Such dyes, accordingly, are often designated as "nuclear dyes," in contradistinction to the "plasma-dyes" which stain especially the cytoplasmic substance.

The cytoplasm often contains various formed elements (granules, fibrillae, etc.) that may likewise be deeply stained by the "nuclear" dyes. The term "plasma-dyes," therefore, only denotes their predominant effect on the cytoplasm considered as a whole; examples of the latter are offered by eosin, acid fuchsin, orange G or light green.

It was shown by Ehrlich and his successors that the nuclear dyes in general and, in particular, the aniline dyes or coal-tar colours, are "basic," the plasma-dyes "acidic," and it is convenient, accordingly, to designate the various cell-components as basophilic and oxyphilic according to their tendency to take up the basic or the acidic dyes. On what this tendency depends — whether on chemical affinity, on physical processes of adsorption, or on both — need not here be considered.

The earlier cytologists, employing for the study of the nucleus mainly the basic or nuclear dyes (especially carmine, haematoxylin, and later safranin and gentian violet) observed that in fixed material, and after certain technical manipulation, only certain components of the nucleus were stained by these dyes.

To the substance thus stained Flemming gave the name of chromatin, to that which stains slightly or retains the colour feebly upon extraction (by acids, etc.) achromatin. "Chromatin," as thus defined, was considered by Flemming to be composed wholly or in part of the chemical substance "nuclein" and to form the more conspicuous part of the nuclear framework and also certain types of nucleoli.

Under the conception of "achromatin" Flemming included all the remaining nuclear substance except the enchylema. Strasburger and Carnoy recognized that the framework itself appears to consist of two constituents, namely, a continuous "achromatin" basis, and of more or less discontinuous granules or clumps of "chromatin" suspended in it. The first of these was found to be oxyphilic and was accordingly designated by Strasburger as nucleo-hyaloplasm, by Carnoy as plasmatic network (composed of "plastin") and later by Schwartz (1887) as linin, a term still in common use. To the foregoing differences may be added the fact that "chromatin," as thus defined, shows a high degree of resistance to hydrochloric digestion, while the oxyphilic "linin" is less resistant in varying degree.

The nuclear division is of two widely different types, which came to be known as direct and indirect. In the direct and simpler type the nucleus, like the cell-body, undergoes a simple mass-division into two parts. In the indirect and more complex type, the nucleus is not destroyed, but is spun out into long threads which split lengthwise so that every portion is exactly divided between the daughternuclei. Before the separation of their longitudinal halves these threads shorten, and thicken to form more condensed bodies known as chromosomes (so named by Waldeyer because of their intense stainingcapacity). The products of their fission (daughter-chromosomes) separate, and pass to opposite poles; and from the two groups of daughter-chromosomes are rebuilt two corresponding daughternuclei which by reason of the preceding processes are exact duplicates of each other and of the mother-nucleus. In this operation we now recognize one of the most fundamental mechanisms of heredity.

By Schleicher this process was called karyokinesis, a term still widely employed for cell-division of this type. Flemming proposed the more appropriate term mitosis, in allusion to the characteristic thread-formation, while the direct mode of division was called amitosis; and this usage gradually became firmly established.

Other terms are karyodieresis, cytodieresis (Henneguy), kinesis and akinesis (Fol, Carnoy), but these are less generally used. Strictly speaking, all these terms refer to division of the nucleus, but by an extension of meaning they are often applied to cell-division as a whole. It is often convenient to employ the term cytokinesis to designate the associated changes taking place in the cytoplasmic cell-body, though in practice it is sometimes difficult to draw any definite line of distinction between the nuclear and the cytoplasmic activities, e.g. in the formation of the spindle.

Cytokinesis includes not only the division of the cytosome as a whole but also the orderly distribution of smaller elements within it; such as the chondriosomes of the Golgibodies; and these processes have received corresponding names.

## EDMUND BEECHER WILSON

The splitting of the chromosomes is, in Boveri's words "an independent vital manifestation, an act of reproduction on the part of the chromosomes." All of the recent researches in this field point to the conclusion that this act of division must be referred to the fission of the chromatingranules or chromomeres of which the chromatin-thread is built. These granules were first clearly described by Balbiani in the chromatin-network of epithelial cells in the insect ovary, and he found that the spireme-thread arose by the linear arrangement of these granules in a single row like a chain of bacteria. Six years later Pfitzner added the interesting discovery, that during the mitosis of various tissue-cells of the salamander, the granules of the spiremethread divide by fission and thus determine the longitudinal splitting of the entire chromosome.

This discovery was confirmed by Flemming, and a similar result has been reached by many other observers. The division of the chromatin-granules may take place at a very early period. Flemming observed as long ago as 1881 that the chromatinthread might split in the spireme stage (epithelial cells of the salamander), and this has since been shown to occur in many other cases; for instance, by Guignard in the mother-cells of the pollen in the lily.

Brauer's recent work on the spermatogenesis of Ascaris shows that the fission of the chromatin-granules here takes place even before the spireme stage, when the chromatin is still in the form of a reticulum, and long before the division of the centrosome. He therefore concludes: "With Boveri I regard the splitting as an independent reproductive act of the chromatin.

The reconstruction of the nucleus, and in particular the breaking up of the chromosomes after division into small granules and their uniform distribution through the nuclear cavity, is, in the first place, for the purpose of allowing uniform growth to take place; and in the second place, after the granules have grown to their normal size, to admit of their precisely equal quantitative and qualitative division.

I hold that all the succeeding phenomena, such as the grouping of the granules in threads, their union to form larger granules, the division of the thread into segments and finally into chromosomes, are of secondary importance; all these are only for the purpose of bringing about in the simplest and most certain manner, the transmission of the daughter-granules to the daughter-cells." "In my opinion the chromosomes are not independent individuals, but only groups of numberless minute chromatin-granules, which alone have the value of individuals."

These observations certainly lend strong support to the view that the chromatin is to be regarded as a morphological aggregate — as a congeries or colony of self-propagating elementary organisms capable of assimilation, growth and division. They prove, moreover, that mitosis involves two distinct though closely related factors, one of which is the fission of the chromatic nuclear substance, while the other is the distribution of that substance to the daughter-cells. In the first of these it is the chromatin that takes the active part; in the second it would seem that the main role is played by the archoplasm, or in the last analysis, the centrosome.

## THE CHROMOSOMES AS BEARERS OF HEREDITARY MATERIAL

The evidence in favour of the view that the chromosomes are the bearers of hereditary factors comes from several sources and has continually grown stronger, while a number of alleged facts, that seemed opposed to this evidence, have either been disproven, or else their value has been seriously questioned.

It has been argued that since the sperm transmits equally with the egg, and since only the sperm head, consisting of the nucleus, enters the egg, inheritance is only through the nucleus.

But it must be admitted that around the entering sperm nucleus there may be a thin enveloping protoplasm, which, however scanty, might suffice to transmit certain cytoplasm factors. Moreover, while the tail of the sperm appears in some cases to be left outside the egg, in other cases it appears to enter and to be absorbed.

Behind the head of the spermatozoön, and at the base of the tail, there is a middle piece which contains a derivative of the old centriole or division centre. Since the centrosome carried by the sperm has been found in some forms to give rise to the new centrosomes that occupy the poles of the first cleavage spindle of the egg, it may appear that a paternal contribution can come about in this way. It is true that the continuity of the centrosome of the sperm with that of the dividing egg has been disputed in some forms; but it is difficult to prove that the sperm centrosome is lost, even though it may disappear owing to loss of staining power.

The nucleus contains a sap which is probably of cytoplasmic origin. The presence of this sap may again be appealed to by those who do not accept the chromosomes as the bearers of heredity, as a weak link in the evidence. It is true that the nuclear sap appears to be squeezed out of the nucleus of the sperm head, leaving a compact and apparently solid mass of chromatin, yet its complete elimination can not be proved. Hence, while those who favour chromosomal transmission find in the facts of normal fertilization strong indications favourable to that view, yet it is also true that those who are inclined to dispute this view find several loopholes in the arguments of their opponents.

The importance of the nucleus in heredity has further been shown by experiments of Bierens de Haans, Herbst and Boveri on giant eggs of sea-urchins fertilized by sperm of another species. The hybrid larvae produced when normal eggs of one species are fertilized by sperm of the other species are intermediate in character between the two parental types of larvae; while those from giant eggs of the same species fertilized by sperm of the other, also intermediate, incline more to the maternal side.

The nucleus of the giant egg is double the size of that of the normal egg, and according to Bierens de Haans, the chromosomes are also double in number. Consequently the amount of maternal chromatin should be double that introduced by the sperm, and might produce a corresponding influence on the hybrid character. But since in these giant eggs the cytoplasm is also doubled it is not evident that the results are due to the chromosomes rather than to the cytoplasm.

Boveri's studies upon dispermic fertilization of the egg of the seaurchin bear directly upon the question at issue. He found that when two sperms simultaneously enter the same egg, each brings in a centrosome so that a tetra- or tri-polar spindle is formed for the first division. Instead of a double set of chromosomes, as in normal fertilization, there are three sets. At the first division, the chromosomes are irregularly distributed upon the multipolar spindles. In consequence, some cells may get one of each kind of chromosome, while other cells may get less than a full complement. These dispermic eggs almost always give rise to abnormal embryos, as several observers have recorded.

The result can best be attributed to the irregular distribution of qualitatively different chromosomes; only those embryos in which each cell has a full complement developing normally. The view that the chromosomes are persistent as individual structures in the cell has steadily gained ground during the last twenty years.

The process of karyokinetic division by means of which at each cell division the halves derived from a lengthwise split of each chromosome are carried to opposite poles, so that a genetic continuity is maintained between corresponding chromosomes (and parts of chromosomes) in mother- and daughter-cells, has been found to be almost universal in both plants and animals.

During the resting state the chromosomes spin out in such a way that they appear to form a continuous network in the nucleus. They cannot be identified individually during this period. When the chromosomes again become visible, preparatory to the next division, it has been found by Boveri in Ascaris, which is particularly well suited for the study of this point, that in sister-cells the configuration of the groups of chromosomes is the same.

The similarity of the sistercells would be expected had the chromosomes retained during the resting stage the same shape and size and relative location that they had at the end of the last division. In many animals and in some plants the chromosomes are of very different sizes and shapes, and many, or even all of them, can be identified at each division. It is found that these size relations hold throughout all divisions of the cells. While this evidence appears at first sight to show that the chromosomes are structures that perpetuate themselves, preserving their identity, yet it might be maintained, in fact it has been maintained, that each species has its own peculiar protoplasm from which chromosomes of a particular kind and number are, as it were, crystallized out anew before each cell division. This point of view cannot however be reconciled with the evidence that follows.

In general, it may be said that even an abnormal set of chromosomes, once established in a cell, tends to persist through all succeeding cell generations. This evidence indicates that the chromosomes are not mere products of the rest of the cell but are self-perpetuating structures.

## Chapter 9

# Heredity

### HEREDITARY TRANSMISSION

Children resemble their parents, and the characters of the species are transmitted from generation to generation. Darwin in 1859 stated that heredity must be subject to natural laws at that time as yet unknown. He noted that if individuals, of different races be crossed, their descendants will exhibit combined and intermediate characters which in the course of succeeding generations tend to revert to their original forms, that is, revert to type.

It is common knowledge indeed that cross-breeding gives rise to offspring exhibiting characters from the male line in combination with others from the female. The distribution of these characters from different sources is not constant in all cases, though the offspring will always exhibit morphological and functional elements, which have descended to it from its two parent strains.

Sometimes the influence of one parent will predominate greatly over the other, giving what we call alternating heredity, whilst in others and indeed the greater number of cases the form of the child is intermediate or halfway between that of the two parents, either in mosaic where the interpolation of the paternal and maternal characters is visible, or in blending or fusion.

Naudin in 1863 crossed individuals of different races and even of different species and showed that offspring of the first generation were usually intermediate between their two progenitors, bearing greater resemblance to the male parent in some cases and to the female in others, though in general there seemed to be a balanced distribution of characters. Naudin also pointed out a reversion to original type of one of the parental lines releasing the offspring from the influence of the least dominant progenitor after a series of generations.

## QUANTITATIVE LAWS OF HEREDITY

Francis Galton, who in fact was a cousin of Darwin, studied the phenomena of heredity from the statistical angle in accordance with methods which had been proposed by Quetelet in 1845. In 1889, Galton was in a position to formulate the quantitative laws of heredity, according to which the parents contribute in equal shares a half of the characters exhibited by the child; the four grandparents contribute jointly a quarter, i.e. a sixteenth part each; the eight greatgrandparents are responsible for an eighth altogether, and so on and on.

Pearson, Davenport and others confirmed and elaborated these findings, proposing various algebraic formulae to express them.Such participation by each individual in the ancestral chain, which, as we see it, keeps on diluting with the succession of generations, explains the tendency to revert to specific type, as expressed by Galton's second law.

The offspring inherits more from its immediate ancestors than from the more remote, but heredity tends to erase individual differences and return any individual to the common mass of the population or to a merely statistical mediocrity.

This in itself implies the existence of the species which, precisely because it is a species, is evidence of a certain constancy.The science of the laws of heredity underwent great progress thanks to the labours of Gregor Johann Mendel. In his paper on cross-breeding experiments Versuche über Pflanzenhybriden in 1865, which at the time went practically unnoticed, Mendel gives a summary of his researches, and his main conclusions were:

- Characters transmitted by heredity do not mingle, but maintain their particular quality without blending with the characters imparted by the other parent; this is disjunction of characters, i.e. characters from the male parent and from the female parent are not added to one another or blended into a compound.

- As the paternal and maternal influences coexist and one only of them is evident in the pairs of characters or allelomorphs from each of the lines, male and female, there must be a dominant character giving visible effects and a recessive. The characters are arranged in pairs, such as the chromosomes, or the genes, but from the pair for each one of the characters, only one element appears active; the other lies dormant and will be effective only when there is no longer a dominant coexisting with it to overshadow it and prevent it from making itself evident.

Mendel effected crosses between races of the same species, thus obtaining infinitely fertile series of specimens which allowed him to study a great number of generations. He called these crossings hybridisms, and not as would have been correct cross-breeds, and the offspring obtained by the crossing of two different progenitors were called hybrids, whether from the same species or from different species.

Mendel got easily evaluable results from his researches, since he chose a few well-defined, readily observable, differentiable characters. His first observations were made with pea vines Pisum sativum.

For instance, he crossed tall standards with short ones, and, in other experiments, plants giving smooth round peas with those giving wrinkled peas, green peas with yellow peas, sweet with tasteless, plump pods with shrivelled ones, and so on.

Having selected the two kinds he wished to cross, he prevented their self-fertilization in the flower, and on the contrary fertilized the blossoms of the one variety with pollen from the other, denoting the generation by the letter P (parents). Afterwards, he sowed the seeds obtained and observed the characters of the first generation of descendants (F1). He then allowed the individuals of this generation to fertilize themselves and sowed the seeds thus obtained, thereby getting the second generation (F2), and so on.

Crossing an individual from the tall variety with one from the short, all the F1 individuals were tall without any noticeable difference. Thus the character "tall" must be dominant and "short" recessive. In the second generation of the progeny, which are hybrids, F2, we get groups of four offspring, three tall and one short, or 75 per cent dominant individuals and 25 per cent recessives. Of course, working with a small number of specimens, the ratio will be only approximate, but mathematically and statistically it will become more and more exact as the number of individuals increases in total.

Self-fertilization of recessive standards constantly gave short, recessive offspring. On the other hand, self-fertilization of the tall plants gave different results – one-third of them produced dominants exclusively which on self-fertilization again conserved their dominant characteristic without exception; the other two-thirds were hybrids repeating the F1 experience, namely groups of four with two pures, one dominant and one recessive, and two hybrids.

This quantitative relationship was repeated on trial with other characters and also using other species of plants, such as kidney beans, wheat, Indian corn, barley, oats, and so on, as well as with animals, e.g. mice, rabbits, guinea-pigs, etc., so that it must therefore be a general law. Since self-fertilization of hybrids gives birth to other hybrids, though these include also a pure paternal and a pure maternal form for every two hybrids, it may be deduced that the factors determining the characters are not fused but are disjoint or separate in the germ-cells or gametes.

Accepting, for the purposes of simplifying the argument, that a determinate factor, a gene perhaps, defines a certain character and that the genes are arranged in pairs, one gene in each pair being respectively from the male and one from the female parent, a very simple combination of genetic elements in fertilization win give one zygote in which will be found solely elements from one of the progenitors, and another in which occur those from the other parent making up the homozygote offspring. In the other two specimens we shall find the paternal element jointly with the maternal giving heterozygote offspring or half-castes exhibiting characters imposed on them by the dominant element.

| Pollen \ Ovules | *A* | *a* |
|---|---|---|
| *A* | *AA* | *Aa* |
| *a* | *aA* | *aa* |

Sixteen combinations of gametes are possible through fertilization, since there are four possible distinct classes of sperm as well as four of ovules. Thus the generation F2 will give four phenotypes YS, Ys, yS and ys in the proportion 9 YS, 3 Ys, 3 yS and 1 ys. This 9: 3: 3: 1 proportion is the same as the 3: 1 ratio found in the monohybrids and multiplied by 3 in the first two places.

The same occurs with trihybrids. In the F3 generation, there will be eight phenotypes and the proportion of the number of individuals will be as 27: 9: 9: 9: 3: 3: 3: 1. The same mathematical laws hold, however intricate the combinations become.

## STATISTICS AND CYTOLOGY

The importance of the facts laid bare by Mendel was realized only many years afterwards. At first, they seemed to be merely a collection of simple numbers, but later on cytological science was seen to show that these figures tallied with the make-up of the components of nuclear substance. For some time consequent to 1865, Mendel's investigations remained little less than unknown, but at the beginning of the twentieth century, de Vries in Amsterdam, Correns at Tübingen, and Tschermack in Vienna, all simultaneously arrived at the same conclusions as Mendel when investigating the laws of hybridization.

From that time forward, researches connected with the Mendelian laws of heredity have progressed by leaps and bounds, especially in the United States, starting with Bateson in 1903 and proceeding particularly under the leadership of T. H. Morgan from 1910 onwards.

Sutton brought to light the relation between transmission of Mendelian characters and the data furnished by cytology, showing how the quantities observed tally with the behaviour of the chromosomes with their genes, constituting allelomorphs in the cell, which become separated in meiosis by the lengthwise splitting of such chromosomes.

## ABSOLUTE AND RELATIVE DOMINANTS

Countless observations which have been made in the present century up to now, with regard to Mendelian transmission, have fully confirmed the two basic laws:

- Disjunction,
- Dominance and recession.

It is, however, becoming clear that the influence of the dominant is The zygotes will be 1 AA, 2 Aa, 1 aa. But since A (tall) is dominant over a (short), the offspring will be three talls and one short. The three talls are not, however, identical, they correspond to the same phenotype but differ in their germinal constitution, a third of them — one specimen in the case of a single group — being homozygote or pure AA, and two-thirds being made up of hybrids Aa, i.e. heterozygotic, giving tall plants since that character is dominant and overclouds the recessive character, though the gene of such recessive is also present. From this it is obvious that the tall phenotype will be made up of two different genotypes, the pure and the hybrid, which can be distinguished only through the offspring to which they give birth.

The numerical proportions are maintained when, instead of taking into account a single set of characters or alleles, say tall-short, smoothwrinkled, green-yellow, and so on, we study what happens on crossing varieties with two sets of alleles, for instance yellow, smooth peas with green, wrinkled ones, obtaining thus two dihybrids. Trihybrids can also be obtained by crossing tall specimens having smooth, yellow seeds with short ones having rough, green seeds. And in this manner we can go on progressively complicating the combinations.

In the dihybrids the genetic elements make their appearance independently as usual, disjunct and separate as shown by the combinations YS and ys, the dominants yellow and smooth, and the recessives green and wrinkled.

Sixteen combinations of gametes are possible through fertilization, since there are four possible distinct classes of sperm as well as four of ovules. Thus the generation F2 will give four phenotypes YS, Ys, yS and ys in the proportion 9 YS, 3 Ys, 3 yS and 1 ys. This 9: 3: 3: 1 proportion is the same as the 3: 1 ratio found in the monohybrids and multiplied by 3 in the first two places.

The same occurs with trihybrids. In the F3 generation, there will be eight phenotypes and the proportion of the number of individuals will be as 27: 9: 9: 9: 3: 3: 3: 1. The same mathematical laws hold, however intricate the combinations become.

## STATISTICS AND CYTOLOGY

The importance of the facts laid bare by Mendel was realized only many years afterwards. At first, they seemed to be merely a collection of simple numbers, but later on cytological science was seen to show that these figures tallied with the make-up of the components of nuclear substance. For some time consequent to 1865, Mendel's investigations remained little less than unknown, but at the beginning of the twentieth century, de Vries in Amsterdam, Correns at Tübingen, and Tschermack in Vienna, all simultaneously arrived at the same conclusions as Mendel when investigating the laws of hybridization.

From that time forward, researches connected with the Mendelian laws of heredity have progressed by leaps and bounds, especially in the United States, starting with Bateson in 1903 and proceeding particularly under the leadership of T. H. Morgan from 1910 onwards. Sutton brought to light the relation between transmission of Mendelian characters and the data furnished by cytology, showing how the quantities observed tally with the behaviour of the chromosomes with their genes, constituting allelomorphs in the cell, which become separated in meiosis by the lengthwise splitting of such chromosomes.

## ABSOLUTE AND RELATIVE DOMINANTS

Countless observations which have been made in the present century up to now, with regard to Mendelian transmission, have fully confirmed the two basic laws:

- Disjunction,
- Dominance and recession.

It is, however, becoming clear that the influence of the dominant is not complete in every case and therefore, in certain specimens, the hybrid forms a phenotype differing from the dominant, since the corresponding character is added to that of the recessive. Such is the famous case of Mirabilis jalapa (the "Four o'clock") among others, where on crossing red-flowered specimens of this species with whiteflowered ones, the F1 hybrids give pink petals, that is to say, petals of a colour halfway between those of the two parents.

Thus in F2 we get one individual with red flowers, another with white and two heterozygotic hybrids with pink. Such blending of characters in the hybrids does not signify lack of disjunction, since self-fertilization of the F2 hybrids results in pure male and pure female homozygotic offspring in accordance with the standard law.

This case, which is by no means unique, only goes to show that sometimes the dominant element is not sufficiently in the ascendant as to overcome entirely the influence of the recessive, which state of affairs causes a certain interference in the powers of the dominant and must be taken into account when seeking to interpret certain cases of hereditary transmission.

There arises also the possibility that the recessive character, not remaining latent, may unite with the dominant without forming a more or less uniform blend, but giving rise to blended intermediate characters wherein both dominant and recessive characters alternate in effect to give a mosaic. This is to be observed in shorthorns, where the crossing of white stock with red gives roan offspring.

The appearance of this character concurs with Mendelian statistical laws, the roans being heterozygotes and their intercrossing giving offspring in the ratio 1 red, 2 roan, 1 white, which is clear evidence of disjunction of the characters. Another example is that of the so-called Andalusian Blue poultry, to obtain which it is necessary to cross blue-speckled whites with blacks. The F1 offspring is the blue hybrid and crossing two specimens of this generation gives I speckled white, 2 blues and 1 black.

## TRANSMISSION OF COMPLEX CHARACTERS

The action of multiple genes may mount up in such a manner that it is difficult to retrace the characters of the original homozygotes, which can be explained by reference to the formulae of the calculus of probabilities applicable to the case. This is particularly noticeable in crossings of the human species. The crossing of white with negro gives a more or less dark mulatto, according to the predominance of the white or the negro characters in one progenitor or the other. The offspring of the two mulattoes will not, however, revert to the pure white or pure negro. In other half-breeds the possible effect of "lethal" germs must be taken into consideration, which by unfailingly bringing about the death of some of the offspring apparently modifies the normal quantitative relationships.

Such examples, among others which we could adduce, show that apparent exceptions to Mendelian statistical forecasts in no wise invalidate the general rule, even though it may be necessary to bear in mind the concatenation of special conditions in such exceptional cases. These problems are, however, practically always of such complexity that their analysis is frequently most difficult. Our knowledge of heredity is today rather more extensive than the ideas Mendel had available to take as a basis for his famous experiments, and the tremendous wealth of material now at our disposal only serves still more to corroborate the accuracy of his original deductions.

## LINKAGE AND CROSS-OVER

The dependence of a character with regard to the corresponding genes contained in the chromosome — the intrinsic material basis upon which Mendelian statistics subsist — is shown by the fact of character linkage. The characters are independent and referable to particular genes, preserving always their biological individuality. They do not, however, appear or act separately. Some characters carry others in. their train.

Goethe in 1780, on discussing the Metamorphoses of Plants, even that long ago, mentioned the correlation or subordinacy of various characters. Bernard and Jussieu in 1798 likewise demonstrated the "subordinacy" of characters, namely, that a main character implies the appearance of certain secondary characters.

Bateson and Punnet in 1905 rediscovered such subordinacy, i.e. the mutual dependence of certain characters, which nowadays is termed linkage, arising from the presence and more or less close vicinity of corresponding genes in one and the same chromosome. An interesting example of linkage is that of the relations set up between the chromosomes which determine sex and certain cognate characteristics.

The same system of charting which helps to explain linkage allows us also in many cases to interpret the cross-over of certain characters. "The presence of associated genes," says Woodruff in 1946, "in the same chromosome by no means indicates that these genes must always be distributed together.

Thus during synapsis, homologous genes (alleles) often reciprocally cross over from one synaptic mate to the other, and so become separated from their former gene associates in the same chromosome." Consequently, when chromosomes separate through division of the equatorial plate or mother aster, the descendant chromosomes may show alternation of genes and the characters of the offspring will modify the results accordingly.

The cross between two homologous chromosomes may be unique or variable and in this way recombinations of genes may come about, to which will correspond recombinations of characters.

This obviously grants a certain flexibility to the genetic mechanisms. The study of so important a phenomenon has again confirmed the justice of attributing characters to their chromosomic elements, and has consolidated the basis of the Mendelian theory, imparting to it the necessary flexibility, and facilitating the elaboration of admirable gene-maps which show accurately in each chromosome the spot where, theoretically, the various genes must lie.

Theory is moreover justified by practical results. In few other domains of science has it been possible to find so perfect a correlation between theory and actual practical results. Facts demonstrated in cytology help us to interpret the Mendelian laws, though all this does not mean to say that the gene hypothesis alone can explain the development of living organisms, any more than the assumption that a phenomenon may be due to the effect of a particular factor, or of several factors more or less hypothetical, will explain such particular phenomenon in any of the realms of biology.

With these reservations, we can however assert that knowledge of the conditions leading to transmission of hereditary characters has made extraordinary progess, and so great has been the advance that in many cases it is possible to foretell the forms that offspring from certain given progenitors will take, as well as the arithmetical proportions of different types of such progeny.

Genetics is rapidly becoming an exact science and is profiting from the facts we have been able to analyse statistically, even when we do not know — as indeed no science really can know — exactly what those facts are in their intrinsic significance.

## GREGOR JOHANN MENDEL

Numerous experiments have shown that if two plants differing constantly in one or more characters be crossed with each other, the common characters will be transmitted without change to their hybrid progeny, but that each pair of different characters will on the other hand combine together to form a new and usually variable character in the offspring. The object of my experiments was to observe such variations in connection with pairs of different characters and to deduce the laws governing their appearance in succeeding generations. Various forms of peas were selected for crossing, showing differences in:

- Length and colour of stem.
- Size and shape of leaves.
- Position, colour and size of flower.
- Length of flower stalk.
- Colour, shape and size of pod.
- Shape and size of seed.
- Colour of seed-coat and cotyledons.

There are some characters which do not allow of a sharp and clear distinction, since their differences are merely a matter of degree often difficult to define, and they cannot therefore be used in these particular researches which of their very nature had to be confined to features allowing of no confusion. The particular characters selected for experiment were accordingly:

- *Differences in Shape of the Ripe Seed:* These are either round or roundish with shallow wrinkles on the surface, or irregularly angular with deep wrinkles.
- *Differences in Endosperm Colour:* The ripe seed endosperm ranges from pale yellow or bright yellow to orange, or else possesses a more or less intense green tint. This variation is easily visible in the seeds since their coats are transparent.
- *Differences in Seed-coat Colour:* This is either white, when it is invariably in correlation with white flowers, or grey, grey-brown or leather-brown with or without mauve spots, in which latter case the colour of the

standards will be mauve, that of the wings purple, and the stem of the leaf-axils reddish. The grey seed-coats turn dark brown in boiling water.

- *Differences in Shape of Ripe Pods:* These are either smooth and plump or more or less wrinkled and deeply constricted between the seeds.
- *Differences in Colour of Unripe Pods:* These are either in the range from light to dark green or else of a bright yellow, the stalks, leaf-veins and calyx partaking of the same colour.
- *Differences in Position of the Flowers:* The flowers are distributed either axially along the main stem or bunched together at the top, in which latter case the upper end of the stem is more or less wide in section.
- *Differences in Length of Stem:* The length of stem varies considerably in certain plants, but is nevertheless a constant character for any particular species in so far that healthy plants grown in the same soil are subject to merely unimportant variations in this character. In experiments dealing with this character, long stems of 6 to 7 ft were crossed always with short ones of 9 in. to 1 ft 6 in. so as to allow certainty in discrimination.

Each pair of the different characters listed above was united by crossfertilization, in accordance with the following table: From a quantity of plants of the same variety only the most vigorous were selected for fertilization.

| *1st* | *Test,* | *60* | *Fertilizations* | *on* | *15* | *Plants* |
|---|---|---|---|---|---|---|
| 2nd | " | 58 | " | " | 10 | " |
| 3rd | " | 35 | " | " | 10 | " |
| 4th | " | 40 | " | " | 10 | " |
| 5th | " | 23 | " | " | 5 | " |
| 6th | " | 34 | " | " | 10 | " |
| 7th | " | 37 | " | " | 10 | " |

Ailing plants will give only uncertain results since, even in the first generation of hybrids, and still more so in subsequent ones, many of the offspring will either fail entirely to flower or will form only a few seeds of inferior quality.

## FORMS OF HYBRIDS (F1)

Hybrids are not, as a rule, exactly midway between their parental species.

This certainly holds for pea hybrids. In each of the seven crosses indicated above, the hybrid-character is so near to that of one of the parental forms that the other escapes observation entirely or at any rate cannot be detected with certainty.

Those characters which are transmitted entire or with hardly any change on fertilization, and are consequently in themselves the main characters of the hybrid are termed dominant, while those which become latent are called recessive.

This latter expression has been chosen because the characters thus designated withdraw or disappear to all intents and purposes in the hybrids, even though they make their appearance again unchanged in the progeny of such hybrids, as will be shown further on.

These experiments showed, furthermore, that it is absolutely immaterial whether a dominant character belongs to the seed-bearer or to the pollen-parent; the form of the hybrid remains identical in either case.

This interesting fact was also confirmed by Gärtner, who stated that even the most practised expert cannot determine in a hybrid which of its two parent species was male or which female.Among the differentiating characters employed in these experiments, the following are dominant:

- Round or roundish seed with or without shallow wrinkles.
- Yellow colour of endosperm.
- Grey, grey-brown or leather-brown colour of seed-coat, in conjunction with purple blossoms and reddish spots in the leaf-axils.
- Smooth, plump pod.
- Green colour of pod, with same colour in stems, leaf-veins and calyx.
- Flowers distributed along the length of the stem.
- Long stem.

## THE FIRST GENERATION BRED FROM THE HYBRIDS (F2)

Recessive characters with all their peculiarities reappear in this generation alongside the dominant characters, in the definite average proportion of 1: 3, so that in every four plants of this generation, three show the dominant character and one the recessive. This applies without exception to all the characters covered by the experiments; in no case were any transitional forms observed.The relative quantities of each pair of differentiating characters obtained were as follows:

*Test 1:* Shape of Seed: During the test year under consideration, 7,324 seeds were obtained from 253 hybrids, giving 5,474 round or roundish seeds and 1,850 angular, wrinkled ones, whence we get the ratio 2.96: 1.

*Test 2:* Endosperm Colour: 258 plants yielded 8,023 seeds of which 6,022 yellow and 2,001 green, giving the ratio of 3.01: 1.

*Test 3:* Seed-coat Colour: Of 929 plants, 705 bore purple flowers and grey-brown seed-coats, whilst 224 had white flowers and white seed-coats, giving the ratio 3.15: 1.

*Test 4:* Pod Shape: Out of 1,181 plants, 882 were plump and 299 constricted, whence the ratio 2.95: 1.

*Test 5:* Pod Colour: 580 plants were submitted to test, of which 428 had green pods and 152 yellow ones, or a ratio of 2.82: 1.

*Test 6:* Position of Blossoms: Among 858 specimens, 651 blossoms were axial and 207 terminal, giving the ratio 3.14: 1.

*Test 7:* Length of Stem: Out of 1,064 plants, the stem was long in 787 cases and short in 277, whence the ratio of 2.84: 1.

The dominant character may here possess a double significance representing a pure parental character or a hybrid character. In which particular aspect it appears in any particular case can only be determined in the succeeding generation. As a parental character it will pass over unmodified to all offspring, whereas if on the other hand it belongs to the hybrid, it will follow the same behaviour as in the first generation.

## THE SECOND GENERATION BRED FROM THE HYBRIDS (F3)

Those forms which maintain the recessive character in the first generation do not vary in regard to it in the second generation but remain constant in their offspring.

Those which possess the dominant character in the first generation from the hybrids behave otherwise, for from them two-thirds yield offspring displaying dominant and recessive characters in the proportion of three to one, and showing thereby the same ratio as the hybrid forms, whilst one-third only behaves as a pure dominant transmitting the dominant character as a constant without modification.

Some of the experimental results are given below:

**Experiment 1.**

Among 565 plants raised from round seeds of the first generation, 193 gave round seeds only, remaining accordingly constant as far as this character is concerned, whereas 372 gave both round and angular seeds in the proportion of 3: 1. The number of hybrids therefore as compared with the pure constants is 1.93: 1.

**Experiment 2.**

From 519 plants raised from seeds with yellow endoderm in the first generation, 166 yielded yellow exclusively, whilst 353 gave yellow and green seeds in the proportion of 3: 1. The division into hybrid and pure forms comes out therefore at 2.13: 1.

In every experiment a certain number of plants always came out with the dominant character. The above two experiments are of special value in determining the proportion in which the forms with the constantly persistent character occurs. The average of the ratios 1.93: 1 and 2.13: 1 taken together gives practically the exact ratio 2: 1.

The three-to-one proportion governing the distribution between dominant and recessive characters in the first generation, resolves itself therefore over the whole series of experiments into the ratio of 2: 1: 1 if the dominant character be differentiated in accordance with whether it be a hybrid character or a pure parental one.

Since members of the first generation spring directly from seeds of the hybrids, it now becomes clear that the hybrids produce seeds containing one or other of the two differentiating characters, of which one-half again give fresh hybrids, whilst half yield offspring which remain constant in their genetic properties and receive the dominant or recessive in equal quantities.

## MARIE GOLDSMITH

The fundamental question in any theory of heredity is that referring to the physiological process whereby an organism in course of development acquires similarity to its forebears. The problem may, however, be posed another way, namely, by leaving aside the innermost phenomena occurring within the fertilized egg and the various tissues which make their appearance and become differentiated, at the same time taking for granted the resemblance between parents and offspring, we may consider merely similarities of features, gradations therein and their variations over the course of time. It was along these lines that Galton and Mendel carried out their researches and formulated the laws which we will now proceed to expound.

In the first place, we should mention analysis of hereditary effects in the light of statistical methods, for the purpose of deducing certain general laws from observation of a large number of facts. The idea of applying statistics to biological questions and particularly to variational phenomena is due to Francis Galton, who laid the foundations of a new science "Biometry" by his two famous books Hereditary Genius and Natural Inheritance.

He was soon followed by many scientists such as his direct successor, Karl Pearson, and others like Weldon, Bateson,

Darbishire, whose labours were for the most part published in the journal Biometrika devoted especially to this subject.

The first general principle Galton was able to establish, on the basis of abundant statistics covering variations relating to 150 families and embracing a huge number of distinguishing physical and mental factors, was that if variations of a particular character or ability be taken into account, it seems certain that a constant average exists for each generation and that differences as compared with such average mutually balance one another.

If, for instance, a parent be much taller or shorter than the average, his children have a leaning in the opposite direction. There is a tendency to revert to the average type, since such norm represents the most stable equilibrium.

Heritage does not come solely from the immediate parents, but from all previous ancestors, and their contribution is the greater the closer they are to the particular generation under observation. Galton combined all these considerations into one scheme which he called the "Law of Ancestral Inheritance." Each generation dating back from the immediate parents contributes on a decreasing scale to the make-up of the final offspring.

This law of ancestral heritage tells us to what degree the progressive thinning-out or exhaustion of heritable effects is unavoidable and explains the gradual disappearance of variations left to go their own gait. This can be adduced as an argument against the idea of definite fixation or the permanence of chance variations operating independently from the direct influence of the environment; which argument has been used in controversy raging around natural selection by those opponents who do not accept its exclusive nature for gospel.

In contrast to the Galtonian theory, Mendel's results are invoked to show that, at least in certain cases, hereditary characters do not become erased, and do not proceed to disperse into thin air in the course of a number of generations until vanishing entirely. This is doubtless so in some instances and in connection with particular characters, but not in all by any means. Characters arising from scarcely perceptible variations, which in accordance with Darwinian principles would necessarily continue to show themselves progressively, are not suitable for genetic research and hybridization studies. In such investigations, we must employ progenitors distinguishable by well-marked characters, and this is one reason for certain reservations when seeking to generalize Mendelian conclusions.

For reasons given elsewhere, we cannot accept the principle of representative particles. The very idea of a concrete particle in place of an abstract concept is illogical, and for that reason proponents of the theory have never been able to offer any evidence on the basis of experiment or actual observation, nor are they likely ever to do so.

The explanation for the independent appearance of characters and of all other facts of inheritance which at first sight appear so easily accounted for by the theory of particles must therefore be sought elsewhere.

Agreement is, nevertheless, general in saying that the form of an offspring will resemble that of its direct ancestors, omitting now to include stability of characters from generation to generation. Heredity consists of an always orderly transmission of chemical, structural, anatomical, physiological and mental features of a distinct nature in the biological sense, from which the characters of the individual and the sum total of individuals making up a species all have their origin.

## THOMAS HUNT MORGAN

We are now in a position to formulate the theory of the gene. The theory states that the characters of the individual are referable to paired elements (genes) in the germinal material that are held together in a definite number of linkage groups; it states that the members of each pair of genes separate when the germ-cells mature in accordance with Mendel's first law, and in consequence each germ-cell comes to contain one set only; it states that the members belonging to different linkage groups assort independently in accordance with Mendel's second law; it states that an orderly interchange

-crossing-over — also takes place, at times, between the elements in corresponding linkage groups; and it states that the frequency of crossing-over furnishes evidence of the linear order of the elements in each linkage group and of the relative position of the elements with respect to each other.

These principles, which, taken together, I have ventured to call "The Theory of the Gene", enable us to handle problems of genetics on a strictly numerical basis, and allow us to predict, with a great deal of precision, what will occur in any given situation. In these respects the theory fulfils the requirements of a scientific theory in the fullest sense.

In the light of what has just been said we can give a reasonable explanation of the differences that follow when a mutant change involves a whole chromosome (or part of one) and when only a single gene is involved.

The former change adds nothing intrinsically new to the situation. More or less of what is already present is involved in the change, and the effects are small in degree but involve a large number of characters. The latter change — mutation in a single gene — may also produce widespread and slight effects, but, in addition, it often happens that one part of the body is changed to a striking degree along with other changes less striking.

This latter kind of change, as I have said, supplies materials favourable for genetic study; these have been widely utilized. Now it is these mutational changes that have occupied the forefront of genetic publication, and have given rise to a popular illusion that each such mutant character is the effect of only one gene, and by implication, to the fallacy, more insidious still, that each unit character has a single representative in the germ material. On the contrary, the study of embryology shows that every organ of the body is the end-result, the culmination of a long series of processes.

A change that affects any step in the process may be expected often to effect a change in the end-result. It is the final visible effect that we see, not the point at which the effect was brought about. If, as we may readily suppose, very many steps are involved in the development of a single organ, and if each of these steps is affected by the action of a host of genes, there can be no single representative in the germ-plasm for any organ of the body, however small or trivial that organ may be. Suppose, for instance, to take perhaps an extreme case, all the genes are instrumental in producing each organ of the body.

This may only mean that they all produce chemical substances essential for the normal course of development. If now one gene is changed so that it produces some substance different from that which it produced before, the endresult may be affected, and if the change affects one organ predominatingly it may appear that one gene alone has produced this effect. In a strictly causal sense this is true, but the effect is produced only in conjunction with all the other genes. In other words, they are all still contributing, as before, to the end-result, which is different in so far as one of them is different.

In this sense, then, each gene has a specific effect on one particular organ but this gene is by no means the sole representative of that organ, and it has also equally specific effects on other organs, and, in extreme cases, perhaps on all the organs or characters of the body.

Continuing our comparison, the effect of a change in a gene (which, if recessive, means a pair of like genes) frequently produces a more localized effect than a doubling or trebling of the genes already present, because a change in one gene is more likely to upset the established relation between all the genes than is an increase in the number of genes already present.

By extension, this argument seems to mean that each gene has a specific effect on the course of development, and this is not inconsistent with the point of view urged above, that all the genes or many of them work together towards a definite and complicated end-product . . .

Evidence points to the correctness of the theory that the sex-linked genes are carried by the chromosomes. Our present understanding of the mechanism of sex-determination has come from two sources. Students of the cell have discovered the role played by certain chromosomes and students of genetics have gone further and have discovered important facts as to the role of the genes.

Two principal types of mechanism for sex-determination are known. They both involve the same principle, although they may seem, at first, to be the converse of each other. The first type may be called the insect-type, because in insects we have the best cytological and genetic evidence for this kind of sexdetermining mechanism.

The second type may be called the avian type, because in birds we now have both cytological and genetic evidence for this alternative mechanism. It is also present in moths. In the insect type the female has two sex-chromosomes that are called X-chromosomes. When the eggs of the female ripen (that is, after each has given off its two polar bodies), the number of the chromosomes is reduced to one-half. Each ripe egg, then, contains one X and, in addition, one set of ordinary chromosomes.

The male has one X-chromosome only. In some species this X has no mate, but in other species it has a mate that is called the Y-chromosome. At one of the maturation divisions the X and the Y pass to opposite poles. One daughter-cell gets the X and the other the Y. At the other maturation division each splits into daughter chromosomes. The outcome is four cells that later become spermatozoa; two contain an X-chromosome, two contain a Y-chromosome.

Any egg fertilized by an X-sperm gives rise to a female that has two X's. Any egg that is fertilized by a Y-sperm gives rise to a male. In the other type of sex-mechanism, the avian-moth type, the male has two like sex-chromosomes that may be called ZZ. These separate at one of the two maturation divisions and each ripe sperm-cell comes to contain one Z. The female has one Z-chromosome and a Wchromosome. When the eggs mature, each egg is left with one or the other of these chromosomes. Half the eggs contain a Z- and half contain a W-chromosome. Any W-egg fertilized by a Z-sperm produces a female (WZ). Any Z-egg fertilized by a Z-sperm produces a male (ZZ).

One of the surprises of the year 1923 was the simultaneous announcement by several independent workers (Santos, Kihara and Ono, Winge, Correns, and Blackburn) that in some of the flowering plants with separate sexes a mechanism is present that follows the XX-XY type.

Several years before these observations in flowering plants had been made, it had been shown by the Marchals that when the spores are formed in dioecious mosses — mosses that have separate male and female gametophytes (or sexual prothallia) — two of the spores derived from the same sporophyte mother-cell produce female gametophytes and the other two male gametophytes.

These experiments are interesting in showing how artificial hermaphroditic individuals may be made from plants that normally have separate sexes by combining the two sets of elements. The results also show that the sequence in which the sexual organs develop is determined by the age of the plant. More important is the actual reversal of this time relation by changing the genetic composition in the opposite direction . . .

If the same number of genes is present in a white blood corpuscle as in all the other cells of the body that constitutes a mammal, and if the former makes only an amoeba-like cell and the rest collectively a man, it scarcely seems necessary to postulate fewer genes for an amoeba or more for a man. The only practical interest that a discussion of the question as to whether genes are organic molecules might have would relate to the nature of their stability. By stability, we might mean only that the gene tends to vary about a definite mode, or we might mean that the gene is stable in the sense that an organic molecule is stable. The genetic problem would be simplified if we could establish the latter interpretation.

If, on the other hand, the gene is regarded as merely a quantity of so much material, we can give no satisfactory answer as to why it remains so constant through all the vicissitudes of outcrossing, unless we appeal to the mysterious powers of organization outside the genes that keep them constant. There is little hope at present of settling the question.

A few years ago I attempted to make a calculation as to the size of the gene in the hope that it might throw a little light on the problem, but at present we lack sufficiently exact measurements to make such a calculation more than a speculation.

It seemed to show that the order of magnitude of the gene is near that of the larger-sized organic molecules. If any weight can be attached to the result, it indicates, perhaps, that the gene is not too large for it to be considered as a chemical molecule, but further than this we are not justified in going. The gene might even then not be a molecule but only a collection of organic matter not held together in chemical combination.

When all this is given due weight it nevertheless is difficult to resist the fascinating assumption that the gene is constant because it represents an organic chemical entity. This is the simplest assumption that one can make at present, and since this view is consistent with all that is known about the stability of the gene it seems, at least, a good working hypothesis.

## THE CHROMOSOME AND THE GENE

Chromosomes as such have never been chemically analysed; they are too small for present methods. The nearest material which can be collected in quantities large enough for ordinary chemical investigation is sperm, particularly fish sperm. The head part of the sperm consists almost entirely of nuclear material, and analyses of this show that the two main constituents are thymonucleic acid (about 60 per cent) and simple proteins of the kind known as protamines (35 per cent) — Caspersson and Hammarsten, 1935. Nucleic acid combines very easily with proteins to form complex nucleoproteins, and it probably occurs in this combined form in the nucleus.

The distribution of nucleic acid in the nucleus can be investigated by means of ultra-violet spectroscopy, since it has a characteristic strong absorption at wave-lengths near 2,600 Ångströms. During division stages, when the chromosomes are contracted and can be seen as separate bodies, almost all the nucleic acid is attached to the chromosomes; it may also be in the chromosomes in the resting stages, but the fully extended chromonemata are not separately distinguishable from the nuclear sap, and the evidence is therefore not clear. The metaphase chromosomes can also be shown to contain protein, since they are attacked by proteolytic enzymes. In the salivary gland nuclei, the chromosomes contain the same two constituents, which perhaps makes it likely that the resting stage chromosomes are built up in the same way, and that the chromosome constitution is constant throughout the division cycle.

The chemical make-up of protein is not yet fully understood. It is known that some of the chemically rather inert proteins such as hair and silk are formed from fibrous elements consisting of chains of "polypeptide links," each link having the constitution-CO-CHR-NH-, where R is a group (the side-chain) which may be a simple hydrocarbon, an alcohol, or a base such as arginine. In the fibrous proteins we have mentioned, the links are arranged in linear chains, the chains being connected together by means of the side-chains.

There are, however, other types of proteins in which the molecules seem to be spherical rather than elongated; these are known as the globular proteins, and there are others intermediate between the fibrous and globular types. We do not know to which of these types the chromosome proteins belong, since the protein isolated from sperm (clupein) has not been examined from this point of view.

The thread-like appearance of the extended chromosome, and the two-ended nature of the chromomeres suggests that the chromosome consists of protein fibres arranged more or less parallel to its length. But this does not by any means necessitate the assumption that the chromosome protein is itself fibrous, since the orders of magnitude are quite different. The polypeptide links in a fibrous protein chain are about 0.35 mì long by 0.45 mì thick by I m ì wide in the direction of the side-chains.

The visible chromonema is a few hundred millimicrons thick, while Muller's estimate, based on its length in salivary gland chromosomes, gives it a width of about 20 mì. Fibres as large as this can just as well be formed from globular as from strictly fibrous proteins, since the units (molecules or repeat cells in crystals) of the former are about 6 mì in diameter, and cases are known, for instance in some of the virus proteins, in which these units unite to form fibres a few tens of millimicrons thick.

Studies on the extensibility, and particularly the reversible extensibility, of the chromosomes give some, and could probably give much more, information about globular or fibrous nature of the chromosome proteins. In completely fibrous proteins the polypeptide chains lie fairly parallel and are more or less unfolded; they can only be stretched by actual straining of the chemical bonds. It is probable, however, that in globular proteins, the same or very similar polypeptide chains exist in a folded configuration, so that extension of a fibre constructed of globular protein involves only the unfolding of the chains and can proceed much farther before the fibre is ruptured.

Duryee has shown that the lampbrush chromosomes of amphibian oocytes can be reversibly extended to about 3½ times their normal length, at least under favourable conditions (in absence of calcium or other heavy metallic ions). Salivary gland chromosomes easily stretch to at least twice their length, and probably can be stretched farther when special efforts are made to do so by micro-dissection methods.

Thus even in chromosomes in which the chromosome thread or chromonema is apparently uncoiled, the thread itself has considerable elasticity, and may perhaps be constructed of globular proteins in which the polypeptide chains are folded on a molecular scale. Much further study is required, however, before this can be taken as more than a suggestion.

When we turn to consider the other main constituent of chromosomes, the nucleic acid, a series of facts emerge which are extremely suggestive of an essential connection between nucleic acid and proteins, but whose exact significance cannot yet be stated. Nucleic acid itself easily forms fibres, and X-ray studies have shown that these consist of a chain of phosphoric acid residues to the side of which are attached a series of flat, plate-shaped groups each of which contains a purine base attached to a sugar. The first remarkable fact is that the repeat distance along the chain, i.e. the distance between neighbouring phosphoric acid residues, is almost exactly the same as the repeat distance in a polypeptide chain: 0.336 mì for the nucleic acid, 0.334 mì for the polypeptide. The difference, which may not be significant, is at least so small that it is easy to imagine that the polypeptide and nucleic acid chains might unite parallel to one another to give protein-nucleate chains.

This can in fact actually be observed; Astbury has prepared the nucleate of clupein, the protein isolated from fish sperm, and shown that it is a fibrous material. Further confirmation comes from a study of the double refraction. Protein fibres have a somewhat weak double refraction which is positive in the direction of the fibre, while nucleic acid, in which there are large flat plate-like groups sticking out at right angles to the length of the fibres, has a much stronger double refraction which is negative in the direction of the fibre axis. The clupein-nucleate shows a double refraction negative in the fibre direction due to the nucleic acid.

So do fully uncoiled chromosomes such as those of salivary glands and zygotene stages; when the chromonema is presumably coiled in a single coil (e.g. mitotic metaphase), so that it runs perpendicularly to the length of the chromosome, the sign of the double refraction changes and becomes positive in the direction of the chromosome axis, while in meiotic metaphase, where the minor spiral is coiled again in a major spiral, the double refraction reverses again and becomes once more negative in the direction of the chromosome axis. All these data fit in very well with the idea that the protein and nucleic acid have combined to form composite fibres in which the two constituent fibres lie parallel to each other.

The cytological evidence makes it quite clear that the chromosomes are not homogeneous structures. In the first place, there is a differentiation in salivary gland chromosomes between the darkly staining bands and the non-staining inter-band regions. The property of stainability depends on the content of nucleic acid, and the concentration of this substance in the bands can be demonstrated directly by studies of ultra-violet absorption.

One must suppose that the proteins in the band regions have a particular affinity for nucleic acid, and Wrinch has suggested that this may be due to a higher concentration of basic groups, particularly arginine, which is known to be present in remarkably high amounts in clupein. The difference between the bands and inter-bands appears, however, rather larger and more sharply defined than would be expected if it were due to a merely quantitative difference; but this appearance may turn out to be illusory when actual measurements of nucleic acid content become available.

The extensibility of the bands seems to differ sharply from that of the inter-bands, the former being much the more rigid. There are two factors to be taken into consideration here. Firstly, the nucleic acid fibre itself appears to be inelastic, and the rigidity of the bands may be due simply to their nucleic acid content.

Secondly, while it is easy to see how nucleic acid may combine with fully extended polypeptides, it is not so clear how it can fit on to a globular protein; it is possible then that the proteins of the bands, when combined with nucleic acid, are in the extended form, and thus have themselves lost much of their extensibility; but whether this should be regarded as a result or as a contributory cause of their affinity for nucleic acid is as yet quite unknown.

The differential staining behaviour of the heterochromatic regions presumably depends on a chemical composition or physical state different to that of the euchromatin; but in general very little is known about this. In salivary glands of Drosophila, the inert regions show a structure of longitudinal striations and transverse bands which is somewhat similar to that of the euchromatic: regions, except that the bands are more feebly staining and the whole structure less clear-cut. Although the inert region of the X chromosome, for instance, is fairly short in the salivary gland chromosomes and has only a small number of bands, it occupies a large proportion of the whole chromosome at metaphase of mitosis. This may perhaps be partly due to a lesser degree of spiralization in mitosis, although, since the region at that time is definitely shorter than it is in salivary chromosomes, some spiralization must occur. It is probable that the large relative volume of the inert region in mitosis is at least partly due to an abnormally large concentration of nucleic acid on to it at this stage. Muller (1938) claims that the greater bulk of the mitotic region is produced under the influence of only two loci in the region, and it is conceivable that the region contains loci specially concerned with the synthesis of nucleic acid.

The physical and chemical basis of this structure is unknown, but it is remarkable to find that the conditions underlying it appear to be transmissible; when inert regions are brought, by translocation, into contact with euchromatic parts of the chromosomes, there is a tendency for the latter to be modified in their appearance in salivary glands, so as to assume more nearly the inert structure. This suggests that the inert regions differ from the euchromatic regions only in some general condition which overlies the same basic differentiation into band and inter-band.

The centromeres are probably quite differently constituted from the rest of the chromosome. They seem to be unable to transmit torsional stresses, since the directions of coiling at metaphase are apparently independent of one another in the two arms of a chromosome with a central centromere; and similarly interference in crossing-over does not extend across a centromere. It has also been shown that when centromeres divide at metaphase they do not always split along a plane parallel to the length of the chromosome, but may occasionally be divided transversely or at any angle. All these facts tend to suggest that the centromeres, unlike the rest of the chromosome, are not fibrous structures.

## THE NATURE OF THE GENE

Before we can discuss the chemical nature of the gene, we must reconsider the definition of the word (Darlington, 1938).

The Mendelian factor was originally defined simply as an entity which obeyed Mendel's laws and had an action in determining the characters of the adult organism.

This definition could apply to whole chromosomes, or large sections of chromosomes. With the discovery of linkage and crossing-over, the definition of the factor, or, as it was now called, the gene, became narrowed down by adding the property that genes act as units in crossing-over, which occurs between them but not through them. At the present time, this definition has become unsatisfactory because there are cases to which it is inapplicable.

We know that genes may be guarded from crossing-over, for instance in the parts of the Y chromosome which have no homologue or in the complexes in a ring-forming heterozygote such as Oenothera. Moreover, we know of inert sections of the chromosomes, which appear to consist of particles much like those in the other regions of the chromosome, sharing with them the properties of multiplication and attraction, but lacking any effect on development; we may wish to stretch our definition to cover inert genes.

Finally, the position effect shows that genes which may cross-over independently of one another may not be independent in their developmental actions; in this case the two parts of the definition do not tally.

It is clear that the old picture of the chromosome, as a linear array of individual indivisible particles, each of which is a gene, is too simple. In attempting to work out a more adequate picture, one can start from the fundamental fact that the chromosome is an elongated structure which, whenever we can analyse it, has differences arranged in a linear order along it; these differences can be detected by linkage studies, chromosome structures, etc.

The units, between which differences are noted, may be of different sizes according to the different methods of investigation; there are, in roughly descending order, inert or precociously condensing regions, large chromomeres, ultimate chromomeres or salivary gland chromomeres, and the units of cross-over and X-ray breakage. One might symbolically represent the chromosome thus: abcd' e'f'g'hijklMNOPQ RSTU'V'W', where there are differences on three scales, between the capitals and the lower-case letters; normal, underlined and dashed letters; and finally the letters themselves.

The smallest units of this scheme, symbolized by the individual letters, are the units of crossing-over and X-ray breakage, and probably measure, as we have seen, about 100 mì in length.

If we view the chromosome as it were through the other end of the telescope, attempting to build it up from chemical units, we arrive at a somewhat similar scheme of a linear order of units of different orders of magnitude. The ultimate units now are the links in a polypeptide chain, with a length of only 0.334 mì.

Exactly what the larger units are is more doubtful, but we have a range of possibilities; there are the periodicities along the chains, the repeat units out of which protein crystals are built, the protein molecules such as they exist in solution, and finally virus particles, all of which may be considered as providing suggestions as to the kinds of units which may be involved.

These units range in size nearly up to the 100 mì which we took as an estimate of the smallest units to be considered when we approached the chromosome structure from the other end. It is, then, possible to conceive of the chromosome as a linear array of units, the units themselves forming a hierarchy all the way from heterochromatic and euchromatic regions, some tens of thousands of millimicrons long, to polypeptide links only a few tenths of a millimicron long.

This apparent homogeneity in the type of formal order exemplified by the chromosome on different scales should not tempt one to suppose that other properties may be just as easily conceived of in any of these scales.

For instance, it is sometimes suggested that because the nature of one link in a polypeptide chain may chemically affect the properties of a neighbouring link, the same type mechanism may explain the phenomenon of position effect.

But in the latter case, the influence is between neighbouring genes (i.e. breakage units) and extends over distances about a hundred times as great as in the former case. No direct analogy between mechanisms of two phenomena is possible; and in fact no example of a direct chemical influence extending throughout such a distance appears to be known in protein chemistry.

## TABLE OF SIZES

The sizes are given in millimicrons (1 mì = 10 Å = 10-6 mm). Where only one dimension is given, it is the diameter of a spherical unit.

| | |
|---|---|
| Vaccinia virus | 175 |
| Rous sarcoma virus | 100 |
| Tobacco mosaic virus | 430 × 12.3 × 12.3 |
| Bushy stunt virus | 28 |
| Haemocyanin molecule | 59 × 13.2 × 13.2 |
| $S_{13}$ Bacteriophage | 10 (?shape) |
| Repeat unit of virus crystal | 15 × 15 × 7 |
| Haemoglobin molecule | 2.8 × 0.6 × 0.6 |
| Protein fibre (repeat unit) | 0.334 × 0.45 × 1.0 |
| Nucleic acid (repeat along fibre) | 0.336 |
| Gene (estimated maximum dimensions) | 100 × 20 × 20 |
| Sensitive volumes: | |
| Gene mutations (Timofeeff-Ressovsky) | c.1 |
| Gene (somatic, Haskins and Enzmann) | 15 |
| Cytological effects (Marschak) | 5 |

Certain of the properties of the genes give kind of units which may fill the gap between the 0'334 mì polypeptide links and the 100 mì genes. The most important is the property of identical reproduction. Between two cell divisions, each gene causes the formation of another gene exactly like it; if the gene mutates into an abnormal form, it is the mutated gene which is reduplicated.

The gene, then, must in some way act as a model on which the new gene is formed. This can only occur if chemical forces originating in the radicals in the gene can extend far enough to influence the nature of radicals formed in the equivalent places in the new gene. The thickness which we can postulate for the gene is therefore limited by the distance through which we can imagine such chemical forces extending.

Probably the maximum estimate which is chemically reasonable is about 10 mì, which is the order of magnitude of the thickness of the repeat units out of which protein crystals are built. This is of the same order of magnitude as the estimate given above for the maximum thickness of the chromosome thread. It is therefore impossible to reject, from consideration of gene reduplication, the idea that the gene is a single unit.

On the other hand, a further difficulty arises in this connection, namely the necessity to find some mechanism which accounts for the fact that only two genes, the old one and the new, are present at the end of each intermitotic period. The reduplication occurs only once. No plausible hypothesis to account for this has been put forward.

Alternatively, we may assume that the gene is compound, consisting of a number of identical sub-units. Such a supposition probably simplifies the task of accounting for gene reproduction. The chemical forces on which the identity of the new and old gene depend would not have to extend so far from the radicals to which they were due, since the thickness of the sub-units would be less than that estimated for the whole chromosome thread.

Similarly the reproduction might continue gradually, and the gene grow until it eventually split into two by reason of some instability which increased with increasing size, such as that which causes a drop to break up when it passes a certain size limit.

The difficulty of this hypothesis, as was pointed out before, is the fact that some genes (though only a few) show more or less equal rates of back and forward mutation.

It appears not unlikely that nucleic acid plays some important role in the process of gene reduplication. For instance, the most rapid synthesis of nucleic acid occurs just before the prophase of mitosis, at the time when the chromosome appears to split or reduplicate.

Again, it is remarkable that the virus proteins, which share with the genes the property of identical reproduction in living systems, and of mutation also, contain large quantities of nucleic acid.

Conceivably there is some connection here with the remarkable fact recently revealed by Schultz and Caspersson, that nucleic acid is in some way connected with the stability of the gene; when parts of the inert region in Drosophila are translocated into the euchromatic regions, they frequently cause the neighbouring loci to become unstable and undergo somatic mutations which give rise to phenotypic spotting such as that found with other mutable genes; and this instability appears to be correlated with an increase in the nucleic acid content of the corresponding bands in the salivary gland chromosomes.

All the above considerations apply to genes considered as units of crossing-over and X-ray breakage. It is quite possible that only a small part of the gene defined in this way is actually active in the control of development. We cannot rule out the possibility that this activity is due to some particular group within the large protein-nucleic acid complex we have been discussing.

In fact, the small size of the "sensitive volumes found for particular steps of mutation might suggest that only quite restricted regions are concerned in producing the phenotypic differences between two allelomorphs; but we have pointed out the many uncertainties in the interpretation of the sensitive volume measurements.

On the other hand, it is quite possible that all primary gene products are enzymes and therefore probably proteins, which may be similar in composition to the genes themselves. It would then be in order to suggest a connection between gene activity, in which enzymes were produced and liberated into the cytoplasm, and gene reproduction, in which similar bodies were formed but retained in the neighbourhood to form a new chromosome.

It will be apparent from the above discussion that the exact knowledge at our disposal is so meagre that very many alternative hypotheses are still possible as to the nature of the chromosome, and the gene in its different senses.

However, the enormously important effects of the genes on development, their capacity for identical reproduction, and the fact that they, rather than the cells of an earlier time, seem to be the most ultimate units into which we can analyse living organisms, make the problem of their constitution one of the most fundamental questions of biochemistry, well worthy of discussion even long before it can be fully answered.

# Chapter 10

# Individual and Species

An individual can be an animal or a plant. Common observation shows that the definition of the word individual should be simple, e.g. dog, shark, rose-bush, beech-tree, etc. Nevertheless, sometimes insuperable doubts arise when we try to define what individuality is. Since all living organisms are formed from cells, since there are some plants and animals in existence composed of one cell only, whilst others are combinations of several different cells generally great in number, it is at times difficult if not impossible to distinguish a colony of several single-celled individuals from a polyplastid. We have already alluded to this point.

Hertwig in 1894 defined the individual organism as a living unit capable of preserving its shape, endowed with the general functions of existence, and able to retain its stability or constancy despite variations in the surrounding medium. Individuality may then be defined by its power of making, preserving, rebuilding and reproducing a particular shape or form.

It is all the same whether we are dealing with a single-celled organism or one of the more outstanding plants or creatures in this world. A part of an animal or of a plant which goes on living for a period after it has been detached from the main body will not become a fresh individual unless it is able to reconstruct a complete animal or plant identical to the one from which it had its origin.

The individual will therefore exist and produce fresh individuals by reproduction, whatever may be the process by which such reproduction is effected, and will follow a course in time, which is life in its various succeeding stages all exactly mapped out until the end comes in death. Individuality may be said therefore, from the morphological and functional standpoint, to be a complete whole in itself, which in the case of mankind is subjectively translated into consciousness, feeling, and knowledge of the ego.

It is through anatomy, physiology and psychology that the individual becomes apart from the world, and opposes himself to his surroundings, forming thus an independent entity. Withal, existence of an individual without a surrounding medium would be impossible.

## SPECIES

An aggregate of individuals, animals or plants, which present broad similarities without being absolutely identical, forms the species. The source of such resemblances lies in the fact that individuals of the same species reproduce and give birth to others like their progenitors. Cuvier in 1817 defined species as a corection of individuals descending one from the other having their source in some common parentage and resembling therefore each other as much as they resemble such original parents.

A purely morphological criterion cannot be applied in all cases, since, even within a single species, differences of form are to be found, for instance: according to sex; to the period of existence -larva or imago; to conditions in the environment, and so forth.

It would be more precise to attribute the characteristics of individuality to the chemical composition and physico-chemical structure of the living plasma. Individuals of the same species will be therefore those whose cells and humours present similar chemical and physicochemical features, from which the functional and morphological characteristics are precisely derivable.

## RACE OR VARIETY

A species includes races and varieties. Racial differences may perhaps lie so deep that it is not always easy to distinguish a specific character from a racial one; in certain cases two races may from their form appear to be more remote from each other than two species. Here too the morphological criterion is not sufficient; only the properties of plasm and tissue will allow us to define a species and distinguish it from a variety. As a consequence of this the safest character for differentiating is the capacity for reproduction and fertility. In general half-breeds are distinguished from hybrids. The increasing affinity of living plasma as relationship becomes closer is one explanation of the facts observed when we try to reproduce different breeds.

There is no method of crossing two sufficiently remote species, though, on the contrary, we certainly can and do get hybrids of species fairly close to each other. Such hybrids are usually sterile, either in the immediate offspring or petering out after a few generations.

The crossing of animals and plants of the same species, and of different race, gives rise to half-breeds which are just as fruitful as pure stock. These differences of reproductive capacity are due to the greater or lesser identity of cell plasm in the various cases and constitute a further important argument in favour of the theory that the chemical and physico-chemical quality of the living elements is the essential characteristic of a species.

## BALANCE AND SPECIES

Species represents a biological balance, and an individual is also a balance or system in equilibrium. Within a species itself, individuals are similar but not identical. The differences are called variations, and when numbers of individuals exhibit the same variation they form a variety.

A variety is distinguished from a race because the latter has greater constancy, whereas the former fluctuates in general and tends to disappear, whilst individuals showing such variation engender a line which continues over a larger or smaller number of generations and becomes approximated to the specific type. This is what Galton in 1889 termed "filial regression."

Specific characters do exist representing a middle term, and on one side or the other of this average, we find variations which arithmetically follow Gauss's Law, or the Law of Mean Errors of the Calculus of Probabilities. This can be confirmed by taking the more distinctive characters. Thus height, weight, colour of eyes, basic intensity of metabolism, pulse, blood pressure, and so on.

In species other than the human we can take for instance the weight of certain seeds, the height of a plant, the number of vertebrae in certain fish, and so forth. The graph showing such variations quantitatively is called a "Galton's curve," and demonstrates that individuals showing typical average characters of the species are the most numerous, whilst divergencies above and below such average diminish in proportion to the degree of variation of the character from the average type.

Thus, for instance, the height of individuals: there is an average height, a distinctive feature for each species and this will be the height of the majority of specimens. There are taller individuals and shorter ones, and the numbers of these outsiders decreases as their height differs more and more from the standard.

There is furthermore, as we have repeatedly mentioned, a tendency to revert to the specific standard. Short parents and tall usually have offspring whose height is different from their own and more or less in accordance with the variance of the parents from the average standard. In all these cases the tendency for "reversion to type," to "balance" in the species, is evident.

## ELEMENTAL SPECIES

There may well exist primitive species with a certain degree of stability, fairly constant in form, such as those distinguished by Jordan in 1848 as between the plants and the midpoint of the limits of a Linnaean species. Thus, for instance, in Viola tricolor, Jordan distinguished several dozen various species, independent and stable as regards the transmission of their characters through the seeds: with large, small and middling blossoms respectively identifiable through particular features of the component parts of the blossoms, seeds, and so on.

The same thing occurs in many other plants, and animals too, though the latter usually have greater stability. It is consequently often difficult to make a distinction between this idea of elemental species and that of variety and race. The most rigorous test is founded on the possibility of hereditary transmission previously pointed out. A variety tends to disappear in the course of a few generations; race is more constant, whilst species is still more so.

The researches of Ewing on greenfly, of Johannsen on kidney beans, of Zeleny, Mattoon and Nay on the fruit-fly Drosophila, of MacDowell, Payen, de Hoge and others, using in some cases parthenogenetic stock and in others taking observations on the effects of sexual reproduction, all lead to the same conclusion, namely that the genes are invariable and heredity is homogeneous.

If individuals be submitted, through a series of generations, to outside influences which provoke variations in them, then in so far as such influences cease to act or lose their power, the species will after a few generations revert to its normal type. Daphnia, for instance, kept at a high temperature alter the shape of their head, and such variation may pass on to the next generation, but when the modifying agent ceases to apply, the very next subsequent generation will come out with heads of normal shape and size. There are many other examples of a similar nature.

## THE EVOLUTION OF SPECIES

Variations in the species which populated the earth during past geological periods do not seem to corroborate such stability. Palaeontology shows that in ancient times species existed which have since disappeared, whilst conversely we now have species which were not formerly in existence, if we accept as evidence the absence of their remains among fossils. Fauna and flora have changed in the course of time, generally for the better.

This led to the idea that species might have evolved by orthogeny, with variations accumulating from generation to generation and diverging thus ever increasingly from the primal type by a kind of differentiation which continuously made itself evident by setting in motion a "principle of perfection," or some "internal force" or "law of progress," that is to say a phylogenetic differentiation comparable to the differentiation which comes about during the development of the embryo, and which is similarly subject to internal forces intrinsic in the properties of organic matter, though itself lying totally outside our present knowledge.

It is not to be wondered at that phylogenetic development has many a time and oft been compared to that of the embryo. As far back as 1821, Meckel asserted that there was some similarity between the formation of an embryo and the various forms of species in the animal kingdom.

Von Baer in 1828 noted that in ontogeny the higher vertebrates pass through stages at which the lower ones are arrested: "The embryos of mammals, birds, lizards and serpents are extraordinarily alike during the early stages of their development, in both the general and the particular, whence it is difficult to distinguish one from the other." Thus it is that Häckel in 1866 came to propound his fundamental law of biogenesis: "The organized individual repeats during the rapid course of its own development the more important phases through which its ancestors passed in the course of their long palaeontological evolution." In other words, "Ontogeny may be said to be a conspectus of phylogeny."

From the instant that orthogenetic evolution is not ascertainable and there is no further assurance that the species may tend to become differentiated by divergence, it is no longer feasible to make a comparison between phylogeny and ontogeny, where indeed there is a progressive multiplicative and differentiable evolution starting from the zygote.

It must, however, be conceded that fossil plant and animal forms certainly show greater and greater complexity and a more evident differentiation in their organs the closer the period of their existence approaches the present time.

## ADAPTATION TO ENVIRONMENT

The theory of orthogeny is opposed by that of adaptation to environment. Lamarck in 1809 founded his philosophy of the animal kingdom mainly on the idea of adaptation. Living bodies assume their distinguishing forms and characters through the influence of the medium in which they exist; and the characters thus acquired are transmitted by inheritance, whereby such forms continue evolving in accordance with the needs of existence. The species is thus the result of a conflict between the living body and its surroundings and changes in accordance with such surroundings.

Lamarck's ideas have been bitterly assailed by those biologists who do not agree that acquired characters can be transmitted by inheritance; but, in spite of all objections, these notions still continue to exercise wide influence on our present framework of ideas, which cannot by any means rest satisfied within the narrow limits of the Weismann system.

It is not in fact reasonable to doubt that some correlation between the medium and the living creature does exist and that the state of the latter is conditional upon that of its environment, for unless this were so life would become impossible. We shall deal with this point in greater detail later on. The influence of environment is shown with special effect by the appearance and development of suitable forms and shapes in animals and plants, thus enabling the continuance of existence in a particular medium.

## MUTATION

Another way of changing the forms of living creatures, which could give us some explanation of how new species came to swell the number of those already in existence, is mutation or sudden variation. From time immemorial; practical botanists, zoologists and naturalists have been aware of what Darwin eventually labelled "sports." Waagen in 1849 suggested the name "mutation," and Kölliker in 1864 and Dall in 1867 made certain researches in this direction. It was not however until 1901 that de Vries gave an exact description and noted the great significance of this phenomenon. He describes the case of the Oenothera Lamarckiana which produces monster shrubs Oenothera gigas, apparently spontaneously and without any valid reason, alongside normal plants. There are also other types recognizable from the shape of their leaves, and others from different characters. The mutate form can be transmitted through inheritance.

At first it was thought, on the basis of de Vries's discovery, that mutation was the exception rather than the rule and that at a particular moment a form might arise different from the specific, but this only occasionally. It was seen later that Oenothera is by no means a unique case, quite the contrary. Thus de Vries writes: "Species are not continuous, they exhibit sudden changes and leaps, to new types representing a spccics distinct from the one whence they proceed. Such new species appear without warning and without any intermediate transition stage."

After the discovery of mutations, it was possible to prove that the matter was not one of exceptional fact. It is indeed a normal and very frequent process of modification. In Drosophila alone, among two and a half million fruit-flies studied by Morgan and his co-workers, practically a thousand forms of mutation were counted.

It is understandable then that as fresh characters are transmitted by inheritance, mutation may well constitute a starting-point for variations of species perhaps of considerable importance. Such variations will sometimes be temporary, sometimes permanent.

The main interest in mutations rests in the fact that it has been possible to establish the relation thereof with modifications in the chromosomes or genes. In this way, the study of mutations has been of inestimable value in the field of genetics, leading for instance to knowledge of the multiple alleles which is of far-reaching importance in human pathology, and to the experimental production of artificial mutates. These experimental cases which at first were thought to be always spontaneous in origin and without appreciable cause can nowadays be directed in such a manner as to produce fresh forms of species.

Müller in 1927 produced mutations for the first time by means of radium beta-rays. The flow of electrons increased the mutation process by as much as 150 to 1. Mutations can also be obtained by means of X-rays, also through changes of temperature.

The metabolic changes which they entail also foster mutation. Indeed much influence must be attributed to the metabolic condition of the subject and to chemical factors which have yet to be determined, as well as modifications within the body and in the medium.

Blaringhem obtained mutations by causing violent mutilations in Indian corn stalks. Mutations are more frequent among plants than among animals and still rarer the higher the creature's place may be in the animal kingdom. We must not forget, however, that even among the mammals there are typical cases of mutation. For instance, the short-legged sheep of Panama first bred in 1791, horn-less Herefords in Kansas in 1889, tail-less or fur-free mousers and ratters both canine and feline, and sharp modifications of character in horses, rats and other animals.

Mutation cases correspond essentially to what Lloyd Morgan in 1928 termed "emergences," the appearance of fresh properties thanks to fresh combinations of factors. It is said, for instance, that if a person had no knowledge of water but did on the other hand know oxygen and hydrogen, he would never be able to foretell that the combination of the two gases would result in a compound like water with properties so different from those of the original two simple elements, nor would it be possible to foresee the vital properties of living matter resulting from complex chemical combinations. Changes in composition and structure of molecules might give rise to the formation of live matter and to the special characteristics thereof.

## NATURAL SELECTION

Natural selection operates consequently on differences of species occurring through orthogenesis, adaptation or mutation. By its effects the unfit disappear and the fit survive, so that selection determines the species which inhabit each of the various regions on this earth.

Natural selection is a term applicable to two distinct ideas, namely the Darwinian "struggle for life," or the mere "effort to exist" which was Darwin's original thought; or alternatively a violent conflict according to the interpretation put on it by so many, which in plain words becomes the result of a battle of the living creature against its physical and biological surroundings.

Nobody can deny that environment determines the beings which will survive in it. If, in reality, mechanisms exist which cause the appearance of fresh species in the course of time, some of these species will disappear because they cannot withstand the surrounding conditions, whilst others will persist, being of such nature that the medium will be favourable to them. This is an incontestable fact, whatever one's personal opinion may be with regard to the origin of species.

Natural selection has since Darwin's time come to be considered as the certain means of originating specific forms. Weismann, and with him many others, considered that natural selection would resolve all difficulties when seeking to explain the origin of species. The historic controversy between Weismann and Spencer on this question. will long be remembered. Selection is not a metaphysical principle or a driving force, it is merely a biological fact to which a disproportionate theoretical significance has been given.

Species, as we have seen, represents a position of balance, an equilibrium between internal factors among themselves and between them and outside influences. By this balance, which is one among the many which make life possible, certain living forms preserve, for a longer-or shorter period, characters which they transmit through heredity and which therefore endure from generation to generation. Huxley in 1880 stated that species is an abstraction, a somewhat strong statement assuredly, since species is an aggregation or total of real beings and individuals, and any aggregate contains the reality of its components.

Species has its history, its life course, just as an individual has. A species may make its appearance, wax strong, decay and finally become extinct, because each individual varies in vitality, since the fruitfulness of such individuals changes, or because the power to withstand the medium alters, or by reason of various combinations of these factors. These are what cause a species to progress or to deteriorate. The problems relating to species, its origin and its conditions of existence are the most basic applying to biological science. As much attention must be devoted to consideration of the biology of species as to that of the individual. However, these serious problems must be denuded of anything which may not be of pure scientific value and interest, so as to avoid unjustifiable and irksome disputes.

## CAROLUS LINNAEUS

A Man of Science contemplating the World, a Workroom of the Almighty, embellished where'er one looks with Miracles produced by that Supreme All-knowing Power, a place of Delight when one has attained a true Knowledge of Our Lord. Though but a lowly guest among such great Endowments, I still know how to appreciate the Benefits thus conferred.

To render ourselves worthy of such Liberality, we must study with all due Reverence these Works of the Creator of all things, in so far as they may be understood by application of our Reason and our Human Capacities.

Knowledge implies Peculiar Ideas and suitable Designations, some of which allow us in ourselves to imitate certain things, and others which make possible an Intercourse with the rest of Mankind by means of some comprehensible Manner of Parlance. The Natural Sciences surrounding Man include Physics and Chemistry dealing with the elements, whilst Zoology, Botany and Mineralogy cover the Study of the Natural Entities.

Botany, the Science of Plant Life, in its early Beginnings knew but few plants. Now, it constitutes a vast Science thanks to the Labours of countless Learned Men throughout the Ages. Despite this, the names given to Plants were determined by

Caprice, for which reason its study was rendered difficult and the results obtained from it were but doubtful. Learned Men investigated the Orders of Nature and built up this Science upon solid Foundations by establishing a proper System. Let us mention Gessner, Cesalpino, Bauhin, Hermann, Tournefort and Vaillant among countless others.

The Open Sesame to Method commences with Genus, divided into Species, a logical concept differentiating all things with Certainty. To distinguish the intrinsic Characters of plants, in order thus to find ourselves in a position to give them Specific Names is by no means an easy Task. It necessitates abundant and carefully gathered collections, painstaking examination of the separate parts, selection by differences, and a correct application of conclusions. Let us be on Guard against those who desecrate the Good Name of Botany, when they propound trifling denominations, lacking in specific Discrimination, which are liable to lead us back toward a primitive Crudeness in Science.

## GEORGES LÉOPOLD CUVIER

Any investigation relating to natural beings needs some way of distinguishing each one of them in such a manner as to avoid confusion with the countless others which Dame Nature brings to our notice. Natural History ought therefore to possess as a basic tool some "Natural System" or comprehensive list containing an agreed name for every single being, thus enabling it to be identified by reference to its distinctive features.

As no living organism is entirely simple in its character, it is insufficient to try to distinguish it merely by one single feature of its make-up. We must therefore, nearly always, rely on a combination of several distinct features if we wish to differentiate any particular individual from its fellows, which latter may in part, if not entirely, through their very similarity show traits pertaining to the first specimen.

In a like manner, such original sample may present characters common to itself as well as to all the others and will then be distinguishable only through those features not possessed by the remainder. Some beings are very closely related to one another in the biological sense and possess many characters in common, whilst others show far less a degree of similarity in detail.

The larger the quantity of beings requiring identification, the greater the number of characters will it be necessary to take into consideration, even perhaps down to the point of making a complete and exhaustive specification of every single detail of the living form.

This state of affairs necessitates the setting-up of divisions and subdivisions. As mentioned above, some individuals will show a large number of features in common, and may then be grouped together as a genus, wherein differences in detail are slight. The genera will then have to be classified and, repeating the operation, closely related genera will be combined to form an order. Similarly, orders related in rank will form a class.

All this implies some framework and method, and nothing could be more suitable than the "Natural Method," that is to say, an arrangement showing us how individual members of one particular genus are more like each other than to members of other genera, and how the genera of a particular order are more alike to each other than to those of other orders, and so on.

This is the system towards which Natural History must necessarily tend, since obviously if an exact system of classification can be attained, the complete portrayal of Nature will have been achieved. Every single living body can be distinguished from the rest by means of its similarities and differences, and if a classification like that mentioned above be possible, then the relationships of living bodies among themselves will become clearly demonstrable.

The Natural Method is the proper scientific way of doing things, and every step made in improving it will be a step forward towards the objectives of our Sciences. Since existence itself is the most important of the qualities and the ultimate supreme characteristic of Nature, it is not surprising that throughout the ages, the most outstanding principle of distinction has been that leading to the classification of natural organisms into two huge divisions: the living and the inorganic.

## CHARLES DARWIN

How will the struggle for existence, briefly discussed in the last chapter, act in regard to variation? Can the principle of selection, which we have seen is so potent in the hands of man, apply under nature? I think we shall see that it can act most efficiently. Let the endless number of slight variations and individual differences occurring in our domestic productions, and, in a lesser degree, in those under nature, be borne in mind; as well as the strength of the hereditary tendency. Under domestication, it may truly be said that the whole organization becomes in some degree plastic.

Let it also be borne in mind how infinitely complex and close-fitting are the mutual relations of all organic beings to each other and to their physical conditions of life; and consequently what infinitely varied diversities of structure might be of use to each being under changing conditions of life. Can it, then, be thought improbable, seeing that variations useful to man have undoubtedly occurred, that other variations useful in some way to each being in the great and complex battle of life, should occur in the course of many successive generations? If such do occur, can we doubt (remembering that many more individuals are born than can possibly survive) that individuals having any advantage, however slight, over others, would have the best chance of surviving and of procreating their kind?

On the other hand, we may feel sure that any variation in the least degree injurious would be rigidly destroyed. This preservation of favourable individual differences and variations, and the destruction of those which are injurious, I have called Natural Selection, or the Survival of the Fittest. Variations neither useful nor injurious would not be affected by natural selection, and would be left either a fluctuating element, as perhaps we see in certain polymorphic species, or would ultimately become fixed, owing to the nature of the organism and the nature of the conditions.

We shall best understand the probable course of natural selection by taking the case of a country undergoing some slight physical change, for instance, of climate. The proportional numbers of its inhabitants will almost immediately undergo a change, and some species will probably become extinct.

We may conclude from what we have seen of the intimate and complex manner in which the inhabitants of each country are bound together, that any change in the numerical proportions of the inhabitants, independently of the change of climate itself, would seriously affect the others. If the country were open on its borders, new forms would certainly immigrate and this would likewise seriously disturb the relations of some of the former inhabitants.

Let it be remembered how powerful the influence of a single introduced tree or mammal has been shown to be. But in the case of an island, or of a country partly surrounded by barriers, into which new and better adapted forms could not freely enter, we should then have place in the economy of nature which would assuredly be better filled up, if some of the original inhabitants were in some manner modified; for, had the area been open to immigration, these same places would have been seized on by intruders. In such cases, slight modifications, which in any way favoured the individuals of any species, by better adapting them to their altered conditions, would tend to be preserved; and natural selection would have free scope for the work of improvement.

We have good reason to believe that changes in the conditions of life give a tendency to increased variability; and in the foregoing cases the conditions have changed, and this would manifestly be favourable to natural selection, by affording a better chance of the occurrence of profitable variations. Unless such occur, natural selection can do nothing. Under the term of "variations," it must never be forgotten that mere individual differences are included. As man can produce a great result with his domestic animals and plants by adding up in any given direction individual differences, so could natural selection, but far more easily, from having incomparably longer time for action. Nor do I believe that any great physical change, as of climate, or any unusual degree of isolation to check immigration, is necessary in order that new and unoccupied places should be left, for natural selection to fill up by improving some of the varying inhabitants.

For as all the inhabitants of each country are struggling together with nicely balanced forces, extremely slight modifications in the structure or habits of one species would often give it an advantage over others; and still further modifications of the same kind would often still further increase the advantage so long as the species continued under the same conditions of life and profited by similar means of subsistence and defence.

No country can be named in which all the native inhabitants are now so perfectly adapted to each other and to the physical conditions under which they live, that none of them could be still better adapted or improved; for in all countries, the natives have been so far conquered by naturalized productions, that they have allowed some foreigners to take firm possession of the land. And as foreigners have thus in every country beaten some of the natives, we may safely conclude that the natives might have been modified with advantage, so as to have better resisted the intruders.

As man can produce, and certainly has produced, a great result by his methodical and unconscious means of selection, what may not natural selection effect? Man can act only on external and visible characters: Nature, if I may be allowed to personify the natural preservation or survival of the fittest, cares nothing for appearances, except in so far as they are useful to any being. She can act on every internal organ, on every shade of constitutional difference, on the whole machinery of life. Man selects only for his own good: Nature only for that of the being which she tends.

It may metaphorically be said that natural selection is daily and hourly scrutinizing, throughout the world, the slightest variations; rejecting those that are bad, preserving and adding up all that are good; silently and insensibly working, whenever and wherever opportunity offers, at the improvement of each organic being in relation to its organic and inorganic conditions of life. We see nothing of these slow changes in progress, until the hand of time has marked the lapse of ages, and then so imperfect is our view into long-past geological ages that we see only that the forms of life are now different from what they formerly were.

In order that any great amount of modification should be effected in a species, a variety when once formed must again, perhaps after a long interval of time, vary or present individual differences of the same favourable nature as before, and these must be again preserved, and so onwards step by step. Seeing that individual differences of the same kind perpetually recur, this can hardly be considered as an unwarrantable assumption. But whether it is true, we can judge only by seeing how far the hypothesis accords with and explains the general phenomena of nature. On the other hand, the ordinary belief that the amount of possible variation is a strictly limited quantity is likewise a simple assumption.

## THOMAS HENRY HUXLEY

All the individual crayfish referred to thus far, therefore, have been sorted out, first into the groups termed species; and then these species have been further sorted into two divisions termed genera. Each genus is an abstraction, formed by summing up the common characters of the species which it includes, just as each species is an abstraction composed of the common characters of the individuals which belong to it; and the one has no more existence in nature than the other. The definition of the genus is simply a statement of the plan of structure which is common to all the species included under that genus; just as the definition of the species is a statement of the common plan of structure which runs throughout the individuals which compose the species.

The southern crayfishes, like those of the northern hemisphere, are divisible into many species; and these species are susceptible of being grouped into six genera: Astacoides, Astacopsis, Chaeraps, Parastacus, Engaeus and Paranephrops forming the Southern family Parastacidae, on the same principle as that which led to the grouping of the Northern forms into two genera Astacus and Cambarus of the family Potamobiidae, the two families forming the general tribe of Prostacidae.

The same convenience which has led to the association of groups of similar species into genera, has given rise to the combination of allied genera into higher groups, which are termed families. It is obvious that the definition of a family, as a statement of the characters in which certain number of genera agree, is another morphological abstraction, which stands in the same relation to generic, as generic do to specific abstractions. Moreover, the distinction of the family is a statement of the plan of all the genera comprised in that family, and carrying out the metaphorical nomenclature of the zoologist a stage further, we may say the families form a tribe — the definition of which describes the plan which is common to several families.

It may conduce to intelligibility if these results are put into a graphic form, the plan of an animal in which all the externally visible parts which are found, more or less modified, in the natural objects which we call individual crayfishes. A represents the plan of the tribe. B exhibits such a modification of A as converts it into the plan common to the whole family of Parastacidae (Southern crayfishes). C stands in the same relation to the Potamobiidae (Northern crayfishes). All these figures would represent abstractions — mental images which have no existence outside the mind.

The truths of anatomy and embryology are generalized statements of facts of experience; the question whether an animal is more or less like another in its structure and its development, or not, is capable of being tested by observation; the doctrine of the unity of organization of plants and animals is simply a mode of stating the conclusions drawn from experience. But, if it is a just mode of stating these conclusions, then it is undoubtedly conceivable that all plants and all animals may have been evolved from a common physical basis of life, by processes similar to those which we every day see at work in the evolution of individual animals and plants from that foundation.

## HUGO DE VRIES

From all these statements and a good many others which can be found in horticultural and botanical literature, it may be inferred that mutations are not so very rare in nature. Moreover we may conclude that it is a general rule that they are neither preceded nor accompanied by intermediate steps, and that they are probably constant from seed from the first.

Why then are they not met with more often? In my opinion it is the struggle for life which is the cause of this apparent rarity; which is nothing else than the premature death of all the individuals that so vary from the common type of their species as to be incapable of development under prevailing circumstances. It is obviously without consequence whether these deviations are of a fluctuating or of a mutating nature. Hence we may conclude that useless mutations will soon die out and will disappear without leaving any progeny. Even if they are produced again and again by the same strain, but under the same unfavourable conditions, there will be no appreciable result.

Thousands of mutations may take place yearly among the plants of our immediate vicinity without any chance of being discovered. We are trained to the appreciation of the differentiating marks of systematic species. When we have succeeded in discerning these as given by our local flora lists, we must rest content.

Meeting them again we are in the habit of greeting them with their proper names. Such is the satisfaction ensuing from this knowledge that we do not feel any inclination for further inquiry. Striking deviations, such as many varietal characters, may be remarked, but then they are considered as being of only secondary interest. Our minds are turned from the delicately shaded features which differentiate elementary species.

Even in the native field of the evening-primroses, no botanist would have discovered the rosettes with smaller or paler leaves, constituting the first signs of the new species. Only by the guidance of a distinct theoretical idea were they discovered, and having once been pointed out a closer inspection soon disclosed their number.

Variability seems to us to be very general, but very limited. The limits, however, are distinctly drawn by the struggle for existence. Of course the chance for useful mutations is a very small one. We have seen that the same mutations are as a rule repeated from time to time by the same species. Now, if a useful mutation, or even a wholly indifferent one, might easily be produced, it would have been so, long ago, and would at the present time simply exist as a systematic variety. If produced anew somewhere the botanist would take it for the old variety and would omit to make any inquiry as to its local origin.

Thousands of seeds with perhaps wide circles of variability are ripened each year, but only those that belong to the existing old narrow circles survive. How different would Nature appear to us if she were free to evolve all her potentialities! Darwin himself was struck with this lack of harmony between common observations and the probable real state of things. He discussed it in connection with the cranesbill of the Pyrenees (Geranium pyrenaicum). He described how this fine little plant, which has never been extensively cultivated, had escaped from a garden in Staffordshire and had succeeded in multiplying itself so as to occupy a large area.

In doing so it had evidently found place for an uncommonly large number of plantlets from its seeds and correspondingly it had commenced to vary in almost all organs and qualities and nearly in all imaginable directions. It displayed under these exceptional circumstances a capacity which never had been exceeded and which of course would have remained concealed if its multiplication had been checked in the ordinary way.

From this discussion we may infer that the chances of discovering new mutating species are great enough to justify the utmost efforts to secure them. It is only necessary to observe large numbers of plants, grown under circumstances which allow the best opportunities for all the seeds. And as nature affords such opportunities only at rare intervals, we should make use of artificial methods.

Large quantities of seed should be gathered from wild plants and sowed under very favourable conditions, giving all the nourishment and space required to the young seedlings. It is recommended that they be sown under glass, either in a glass-house or protected against cold and rain by glassframes. The same lot of seed will be seen to yield twice or thrice as many seedlings if thus protected, compared with what it would have produced when sown in the field or in the garden. I have nearly wholly given up sowing seeds in my garden, as circumstances can be controlled and determined with greater exactitude when the sowing is done in a glass-house.

The best proof perhaps, of the unfavourable influence of external conditions for slightly deteriorated deviations is afforded by variegated leaves. Many beautiful varieties are seen in our gardens and parks, and even corn has a variety with striped leaves. They are easily reproduced, both by buds and by seeds, and they are the most ordinary of all varietal deviations. They may be expected to occur wild also. But no real variegated species, nor even good varieties with this attribute, occurs in nature.

On the other hand occasional specimens with a single variegated leaf, or with some few of them, are actually met with, and if attention is once drawn to this question, perhaps a dozen or so instances might be brought together in a summer. But they never seem to be capable of further evolution, or of reproducing themselves sufficiently and of repeating their peculiarity in their progeny. They make their appearance, are seen during a season, and then disappear. Even this slight incompleteness of some spots on one or two leaves may be enough to be their doom.

It is a common belief that new varieties owe their origin to the direct action of external conditions and moreover it is often assumed that similar deviations must have similar causes, and that these causes may act repeatedly in the same species, or in allied, or even systematically distant, genera. No doubt in the end all things must have their causes, and the same causes will lead under the same circumstances to the same results. But we are not justified in deducing a direct relation between the external conditions and the internal changes of plants.

These relations may be of so remote a nature that they cannot as yet be guessed at. Therefore only direct experience may be our guide. Summing up the result of our facts and discussions we may state that wild new elementary species and varieties are recorded to have appeared from time to time. Invariably this happened by sudden leaps and without intermediates.

The mutants are constant when propagated by seed, and at once constitute a new race. In rare instances this may be of sufficient superiority to win a place for itself in nature, but more often it has qualities which have led to its introduction into gardens as an ornamental plant or into botanical gardens by reason of the interest afforded by their novelty, or by their anomaly.

Many more mutations may be supposed to be taking place all around us, but artificial sowings on a large scale, combined with a close examination of the seedlings and a keen appreciation of the slightest indications of deviation seem required to bring them to light.

## GENETICS AND EVOLUTION

The modern approach to the critical analysis of significant variations was opened by the work of two botanists: de Vries and Johannsen. De Vries laid stress on the importance of discontinuous variations which he called mutations; while Johannsen made clear that in a homozygous germ complex, or pure line, selection is ineffective.

Some of the problems of selection will be clear from an example. Take, say, a quart of beans and sort them into groups according to the weight of each bean. Then put each group into a separate cylinder and arrange the cylinders in a series according to the weight of the enclosed beans. Now if we imagine a line connecting the tops of the bean piles in the cylinders, it takes the form of a normal curve of probability, or variability curve. A similar figure would be obtained by the statistical treatment of nearly all fluctuating characters among the members of any large group of organisms, or of the size of the grains in a handful of sand, or the deviations of spots from the bull's-eye in a shooting match. Therefore the variations with respect to a given character very closely approximate the expectation from the mathematical theory of probability, or chance, and the reasonable conclusion is that such finely-graded fluctuating variations are a resultant of a large number of factors, each of which contributes its slight and variable quota to the expression in a given individual.

The question is, what results are obtained by breeding from individuals which exhibit such a fluctuating variation to, let us say, a greater degree than that of the mean of a mixed population? With Galton's principle of filial regression in mind, one will naturally expect, and rightly, that the offspring usually will exhibit the character to a less degree than the parents but to a greater degree than the population. The top (mode) of the curve will have moved, so to speak, slightly in the direction of selection.

Now, by continuing generation after generation to select as parents the extreme individuals, is it possible, with due allowance for some regression, to take one step after another indefinitely, or until the character in question is expressed to a degree which did not exist previously? The experience of practical breeders gives a partial answer, since the continual selection of the best animals for mating and the best plants for seed has been a profitable procedure. But it has long been known that after a certain amount of selection has been practised it may cease to be effective, and thenceforth serves chiefly to keep the character at the higher level attained.

The crux of the matter is in regard to exactly what the variations are. Both modifications and recombinations are usually included, and this mixture of non-heritable and heritable variations is what makes confusion. If we rule out recombinations by inbreeding or by selffertilization of homozygous individuals, soon we establish pure lines. Then the variations are all modifications and selection is ineffectual with characters which are not inherited.

The importance of this point was discovered by Johannsen in careful experiments on the inheritance of characters in single pure lines of a brown variety of the common garden bean. For example, by keeping the progeny of each individual bean separate from that of all the rest, he was able to isolate a number of pure lines which differed in regard to the average weight of the beans. Thus selection resolved the bean population with which he began into its constituent "weight types," or lines, each of which exhibited a characteristic variability curve of its own with a mode departing more or less from that of the population.

But when Johannsen selected within a pure line (ruled out recombinations) nothing at all resulted; he was unable to shift the mode because he was dealing with non-heritable characters. In other words, selection sorts out pre-existing pure lines (lines with homogeneous germinal constitution) from a population and then stops. Thereafter mutation must occur in the pure line for selection to be effective — but by the mutation the single pure line becomes two.

The pure line concept has served to clarify our ideas in regard to selection by focusing attention on the actual nature of the variations being dealt with. It discriminates sharply between modifications, which are a result of environmental influences, often recurrent in each generation and so seemingly inherited, and variations which are heritable because they are the result of changes in the germ plasm.

## Chapter 11

# Organisation of the Germ-cells

The problem of the transmission of form in animal and plant species from generation to generation, and consequently of the evolution of the embryo, has puzzled observers throughout the ages. In the fourth century B.C. Aristotle had already contrived experiments relating to the development of the hen's egg.

Until the discovery of the sex elements, however, such experiments could not be said to have been of a scientific nature, that is to say until 1667 when Leeuwenhoek discovered spermatozoa.

Spallanzini in 1754 and Prevost and Dumas some time later in 1824 showed the necessity of having spermatozoa in the fertilization process, since on filtering the sperm secretion of a frog prior to mixing it with the egg-cells, the latter failed to develop.

Not until 1875 did Hermann Fol and Oscar Hertwig, working independently of each other, describe the penetration of sperm into the egg-cell and the formation of the zygote nucleus, the egg, by combination of nuclear chromatin from such sperm with that of the egg-cell nucleus.

The male and female sex cells must possess, therefore, organogenetic properties, since their progressive differentiation and increase gives place to the evolution of the embryo. The most direct theory could not help but be the assumption that in such cells there must be material or potential representations of every part of the organism. Hartsoeker in 1694, for instance, gave a sketch of a large-headed mannikin dwelling in the sperm chromatin, growing up to become the foetus and feeding on the egg substance at the beginning and then on sustenance directly supplied by the mother.

Other writers of the eighteenth century held that a microscopic ready-formed being was to be found in the egg-cell whilst the sperm must be some sort of parasite absolutely superfluous as far as reproduction went.

The theory of preformation has had its partisans in all ages: Swammerdam, Bonnet, Haller, Malpighi and Buffon in earlier times and Baer, Remak, Kölliker, Haeckel and others in more recent centuries, and still closer to our own times Wilhelm Roux with his mosaic theory, not to mention many pure philosophers, particularly Malebranche and Leibnitz.All writers who sought to explain the formation of the organs as the work of representative particles existing in germ-cells must be classed as preformationists.

Such representative particles are presumably contained in the primary germ-cells and become liberated successively in the course of generations until one day or other the stock becomes effete, as mentioned by Bonnet for instance. Or they might be reproduced indefinitely by division in the same way as the cells containing them, as are also the genes and living molecules.

## PHYSIOLOGY OF EMBRYONIC DEVELOPMENT

Opposed to such morphological ideas, of what we might call the anatomical type, are those of other biologists who consider development to be but one among many other functions.

Forms, they say, appear through epigenesis, because in the embryo stage the circumstances necessary to the purpose arise and determinant causes come into play at the right moment, so that inevitably the organs naturally follow one after the other differentiating successively at each stage of development.

In the seventeenth and eighteenth centuries particularly, disputes were constantly taking place between the preformationists and the epigeneticists.

Wolff in 1759, at his inaugural lecture at Königsberg, maintained that embryogeny was merely a function and, referring particularly to the growth of domestic fowl, stated that it is by no means possible to agree that the germ will contain, even invisibly, all the elements of the mature creature which is distinguishable by the great amount of differentiation and tremendous complexity of its individual organs; embryogeny, he said, is something more than a mere growth from a number of preformed outlines.

He agreed that organ-forming regions exist in the egg, but at any instant during the production of the embryo the first buds of the organs also start to make their appearance, each of them developing on its own account, whilst influencing and being influenced by the remainder. This is the first reference to ideas which are nowadays commonly accepted, commencing particularly with the work of W. Roux since 1883. Differentiation takes place.

- Through properties intrinsic in the germ-cells (auto-differentiation from the outlines),
- Through correlative differentiation, mutual influences between the various parts of the embryo, or dependent differentiation.

These two aspects correspond, as is evident, respectively to preformative and epigenetic mechanisms.

Driesch in 1895 gave preformation the credit of being of prospective significance: each part of the germ-cell or of its descendants will in normal conditions give rise to a particular portion of the organism. He also named as prospective potential, the epigenetic possibility that each morphological element, from the zygote or from the blastomeric aggregate, might generate organs different from those which they ordinarily produce.

It is easy to show how during embryogeny, as well as throughout the whole course of life, essential properties from the organized living matter help in morphogeny and function, and how, moreover, they influence conditions which follow one another as time passes on, outlining the vital course of a creature's existence.

## ANISOTROPY IN THE EGG

Pflüger's error, for instance, in assuming that the egg was isotropic, that is to say, homogeneous in its formative determinates, has been demonstrated many a time and oft. It is incontestable that germ localizers, in His's sense, reside within the egg. There is no exception, even though the egg may present the most remarkable example, for the same thing occurs in all reproductive cells and is proportionate to the activity of reproduction.

The polarity of the egg, resulting from the polarity of the egg-cell, is easily verified. The ejection of the polar bodies, resulting from double division on ripening, takes place through a spot on the surface of the egg-cell thus determining the polarity of this latter. The point at which this takes place becomes the animal pole, whilst at the diametrically opposite end we get the vegetative antipole. In vertebrates, the pole is that part of the egg-cell adherent to the ovary; in invertebrates this becomes the antipole.

From the blastomeres issuing from the polar hemisphere the ectoderm is formed, this being the source of the organs of animal existence, whilst those deriving from the antipolar half give us the endoderm or source of the vegetative organs. In the spot corresponding to the antipole, the blastopore develops, making way for the primitive digestive apparatus, the archenteron which ultimately becomes the gut cavity.

In a frog's egg-cell, polarity is indicated by pigmental arrangement; the antipolar region is white, whilst the remainder of the egg-cell is dark, and transition from one zone to the other is observable through parallel strata located beneath its equator. There are other similar examples. Anisotropy of the egg-cell becomes more evident at the instant of fertilization. The entry of the sperm causes profound alterations in the cytoplasm, and most certainly, in the nucleus, the structure thus obtained acting on the segmentation.

It is the rule that only one sperm can enter the egg-cell, whereupon the latter, after penetration has been achieved, acquires a genuine immunity and repels any subsequent sperm, in some cases even filming over adventitiously with an impermeable cover. Polyspermy is something out of the ordinary; it may possibly happen, for instance in damaged egg-cells subjected to the action of anaesthetics. Eggs very rich in yolk, especially telocytes, may become polyspermic, but only one single sperm, one single male pronucleus, will actually combine with the ovular nucleus. Other sperms will disappear and only under abnormal circumstances can polyspermic fertilization take place.

In the frog's egg-cell, a grey area appears immediately after fertilization below the equator and above the white antipolar cap. Its shade is halfway between the dark of the pole and the pale of the antipole, whilst its appearance is that of the moon's last quarter, or a "grey crescent" as Morgan termed it. Now, the symmetry of this crescent is determined by the point at which the sperm makes its entry, and its upper part corresponds to the longitude at which such entry took place.

The plane of section of the sphere, i.e. of the egg, containing the pole, antipole and entry meridian, which is the axis of symmetry of the crescent on which the entry-point is located, will in seventy cases out of a hundred, according to Brachet, be the primary plane of cleavage for the egg when it divides into its two blastomeres, i.e. the sagittal plane; in 10 per cent of cases it will be the frontal plane, i.e. the perpendicular to the one last mentioned, and in the remaining 20 per cent it will take up an oblique position varying according to circumstances.

In the first case, the crescent is shared in halves between the two first blastomeres; in the second, it is contained fully within one of the blastomeres, the other having no part of it; whilst in the third eventuality the division of the crescent will be unequal.

The primary plane determines the remainder. The successive planes of division are always perpendicular among themselves. If the division of the zygotes into two blastomeres takes place in the sagittal plane, the following division into four will occur on the frontal plane, the next into eight on the horizontal and so on. Roux's observations tell us the importance of the bearing of the planes in blastomeric division.

If the cleavage plane in the zygote is sagittal, and one of the two blastomeres resulting from such segmentation is afterwards destroyed, only half an embryo is obtained; if the segmentation plane is frontal, the surviving blastomere will give either a whole embryo or an undifferentiated conglomeration of cells according to whether it comes from the part of the egg with the crescent or without it.

Herlitzka's experiments in 1893 have supplied confirmatory data on this point, as did Spemann's ligature experiments in 1901, and Winterberger's more recent researches in 1929 on killing one of the blastomeres with X-rays. All this tells us how, before cleavage, the march of morphological events to develop in the genesis of the embryo has already been predetermined as if obeying some already well-laid master-plan.

What we have just described is not exclusive to the frog. Other batrachians and amphibians, anura and urodela, show analogous phenomena, even though the modification of the cytoplasm following on fertilization may not be so evident to the sight, since the grey crescent does not show itself in these species.

## THE SIGNIFICANCE OF BLASTOMERIC DIVISION

Boveri found, on the other hand, that in the sea-urchin, after fertilization, a pigmented ring made its appearance above the equator of the egg, and had a definite connection with the division into blastomeres. At the "eight" stage, that is the third division, the ring appears in four antipolar blastomeres and at the bottom of the four polar ones. At the next or "sixteen" stage, when the blastomeres are no longer equal in size, the eight upper polar ones of medium size, or mesomeres, have the lower portion pigmented and the main part of the ring surrounds the four blastomeres situated beneath them, i.e. the larger blastomeres or macromeres; whilst the four antipolar blastomeres which are the smallest, or micromeres, are quite colourless.

In the sea-urchin, the invagination which will form the gastrula commences with a flattening of the antipole, and then the cells deriving from the micromeres drop into the blastocele or segmentation cavity where the mesenchyme or primary mesoblast will develop. The portion which is invaginated and forms the source of the archenteron comes from pigmented macromeres. In this fourth cleavage it is still possible to tell the origin of the cells forming the essential parts of the gastrula and the primitive streaks, thanks to the localization of pigment. It is found, also, that blastomeres continuing to multiply, like the organs which will later develop from them, are already roughly outlined in the egg, even before any cleavage whatsoever.

In certain Ascidians, Styela partita, Conklin was able in 1905 to describe similar occurrences. Insemination of the ovule similarly provoked alterations in the morphogenic structure thereof. In a particular portion of the surface a yellowish crescent rich in lipoids develops, corresponding to the region which later becomes the rear region of the embryo, and it is easy to relate the cells deriving from this region, as far as to the gastrula and still farther even, since they are distinguishable by their pigmentation.

Likewise in Amphioxus, Conklin showed in 1932 that already in the first stages of cleavage, up to the fourth (with sixteen subdivisions), it is possible to define the cells from which the different parts of the gastrula proceed. Bilateral symmetry is determined in the first cleavage, and the dorsal and ventral faces in the second. As we have seen, morphogeny develops with precision, starting at the first instant of the blastomeric division.

Artificial marks, similar to these natural markings, which are found in certain species, can be imprinted artificially by experiment, and this will help us likewise to follow the route and destination of strings of cells during cmbryogeny. This was the idea W. Vogt had in 1925, when he suggested the use of stains. Various dyes, which of course must be indelible, can be used, for instance Nile blue, neutral red, Bismarck brown, bright blue, etc.

Ansell and Winterberger used electrolytic markings. Such staining allows us to study the fate of various portions of the egg, of different blastomeres and so on right up to a fairly advanced stage in development.

Vogt made use of various kinds of batrachians; von Ubisch and Hörstadius worked with sea-urchins' eggs, and Teissier with hydra. It has been possible to establish the origin of vestiges which do not show themselves until later, thus proving that each species, has a definitely fixed characteristic scheme of development. It has been proved, moreover, that relative movements of the cells during increase, spiral cleavage and cell migrations do take place, in other words that there are such things as morphogenetic motions which are quite contrary to Oscar Hertwig's original idea.

## PREDETERMINATION IN THE EGG

It is quite evident that there is predetermination in the egg, arising from the ovule and from the sperm, though it is neither inviolable nor decisive. Countless researches have proved that if during the first instants of embryo formation a blastomere or group of blastomeres be destroyed, there are sufficient others to take their place in the general development.

Investigations in this direction have been numerous, and contradictory results have at times been obtained during their course; W. Roux in 1886 used a red-hot needle to destroy one of the first-stage blastomeres from a frog, and got half-embryos both right and left.

These anomalous embryos, nevertheless, went on to full development later with more or less conformity to standard according to circumstances. Chabry in 1887 likewise killed some of the initial blastomeres by pricking, working with Ascidians, and as in Roux's case, the result was a semi-embryo. Conklin in 1901, a considerable time after this, was able to corroborate the same thing experimenting with the ascidian Styella.

On the other hand, Driesch reached important conclusions to directly the opposite effect. Isolating the two primary blastomeres of a sea-urchin, he obtained two complete larvae, though they were much smaller than normal. He observed, however, that as the development of the embryo proceeded, the blastomeres went on gradually losing their total power, until this vanished entirely in the fourth or fifth cleavage. From then on the blastomeres ceased to be equivalent among themselves and to possess the same faculties. This has been confirmed by various writers — Wilson, Zoja, Yatsu, Zeleny, Herlitzka, Conklin, Morgan and others — using various procedures for isolation, such as destruction of one of the blastomeres, ligature (Herlitzka, Spemann), dissection under the stereoscopic microscope (Driesch, Hörstadius), pressure with a thin glass rod (Mangold), gentle stirring in water (Driesch, Boveri, Wilson, Conklin), contact with calcium-free brine or sea water (Herbst), and so forth.

Merogony, cleavage of the blastomeres, gives different results not only according to the moment of development at which it takes place but also according to species. Thus, in Ctenophora, Driesch and Morgan and Fischel and Yatsu noticed that blastomeres isolated in the first, second and third cleavages gave respectively half, quarter and one-eighth of a normal larva. Crampton in 1896 mentioned that the same happens in certain gastropod molluscs, Wilson found the same in Scaphopoda (elephant's tusk shells), and Conklin in another mollusc the Crepidula. Chabry demonstrates similar results in Annelids and Ascidians.

## MOSAIC EGGS AND REGULATION EGGS

There are, therefore, some eggs which behave as if the functional and morphogenic traces were working in a definite way, whilst others, on the contrary, have blastomeres giving place to organs which they would not form under normal conditions. There are some eggs in whose development self-differentiation prevails, and others wherein correlative differentiation, in Roux's sense, is predominant. The former are called eggs of real potential in mosaic, and the latter regulation eggs of full power.

Blastomeres proceeding from the first cleavage of an egg in mosaic give birth to incomplete embryos, such as we have seen occurring in the case of certain Ascidia and Annelidae (Chabry), in frogs (W. Roux), in Ctenophora (Driesch and

Morgan). In regulation eggs, each blastomere produces a complete embryo, as in the sea-urchin (Driesch), in Amphibia (Herlitzka), and in Amphyoxus (Wilson). This shows that a local influence such as position in the germ of those organo-genetic traces of which W. His spoke, may be possible and even coincide with epigenetic factors which make themselves evident during development.

These are two aspects of embryogeny which are not evidenced with equal effectiveness in each species — since both mosaic and regulation eggs exist — nor at every instant during ontogeny. Generally total power prevails at the start of development and weakens progressively as the contours of the organs become plainer. The egg is not a rigid prototype of the embryo since parts are also determined by interaction. This in no way means that the moments of morphogenetic evolution are not genetically imposed right from the germcells. It does not seem difficult nowadays to reconcile these two apparently opposed points of view, which divided embryologists so long ago into two camps.

Recent advances in experimental embryology have, however, led to results of the greatest importance. We must specially mention researches based on the union of blastomeres proceeding from different zygotes, either of the same species or each from different species. Hörstadius in 1928, working on blastomeres, fragmented from seaurchin embryos, obtained pluteus larvae of different form according to graft. He was able to "animalize" or "vegetalize" at will any particular part of the embryo, by adding suitable complementary portions thereto.

These results were confirmed by fresh researches in 1936 using ribbon-worm eggs. Von Ubisch completed similar experiments but with the main object of obtaining chimeras or graft hybrids by uniting two distinct species in various combinations. In some cases the grafted elements are eliminated; in others they develop in more or less incomplete and isolated groups within a single embryo, derived from each one of the original blastomeric groups; in yet others, however, regulation phenomena are in fact observed tending to accommodate the whole to make up an individualized system which will indeed be a true crossbreed.

On a like foundation with the foregoing are based the researches of Spemann and his school, who implanted parts of one embryo in another host embryo. This method consists in alternating fragments a of embryo A with portions b of the second embryo B; thus a is inset in B and b in A, changing their anatomic disposition. It will be seen whether the grafts a and b in their new position develop in the same way as they might have done in their original location, i.e. in accordance with their origin and by virtue of their inherent properties "preformational self-differentiation" or if they develop under the influence of their new location "correlative differentiation."

## EPIGENESIS AND PREFORMATION VARYING WITH TIME

These researches have led to a confirmation that in the initial moments of development, the total power of the blastomeres is greater than the actual potentiality, in other words that epigenesis is at that stage predominant. The fate of each one of the parts of the embryo is not at that moment irrevocably determined and extensive regulation generally can set in. As embryogeny proceeds this capacity for regulation diminishes more and more.

The work of Spemann and his co-workers was carried out on Urodela and various species of tritons. The original test consisted in changing grafts between two young gastrulas of like age, namely fragments of epidermis, one of which, a, was taken from the region in front of the blastopore which later becomes the neural plate, whilst the other, b, came from the latero-ventral region where skin develops naturally without any special physiological significance. Graft b will therefore give birth to nerve elements, whilst a on the other hand will give skin cells.

This shows that the fates of the a and b fragments were not definitely determined at source, but that morphogenic fixation springs from the spot where the grafts begin to develop locally.

Later, Spemann carried out further, more convincing tests, still using two different species, namely Triton taeniatus, whose eggs are highly pigmented, and Triton cristatus with light-coloured eggs, the eggs giving similarly coloured embryos, dark and light respectively. In this way plainly visible chimeras are obtained. The pigmented graft stands out as a dark stain on the light embryo, and conversely. Depth of section is also easily defined by the presence or absence of pigmentary granulations in the cells of the grafted tissues.

This technique has enabled us to make some very interesting deductions. At the start of gastrulation, the antipolar hemisphere can be replaced, through implantation, by the polar, provided that the section be made above the zone corresponding to the crescent.

The meridians of the cap and the remainder of the blastula may be allowed to coincide, or the cap may be given a twist of one or two right angles at the time of positioning for implantation. A normal embryo is obtained; the neural plate in this case develops before the blastopore, just as if there had been neither substitution nor rotation, and this occurs at the expense of cells which would normally have generated skin and not nerve elements.

If the horizontal section be made lower down, so as to affect the crescent zone, and a rotation of 180° be given, a double embryo will be obtained, one of whose elements differentiates in its normal position, i.e. in the dorsal surface of the egg, and the other at 180° in the ventral surface. Under such conditions, working only with comparatively fresh gastrulas, we get proof that the fate of the polar hemisphere is not determinate up to this point.

On the other hand, in the neurula stage, the outlines have already been definitely differentiated. Grafts therefore do not become differentiated according to location, but in accordance with their source. The moment of change from epigenesis to determinancy occurs in this species between the end of gastrulation and the formation of the neural plate.

The existence of germ localization, anisotropy, is consistent with the extremely extensive regulational capacity of certain eggs. Spemann's own work proved that even in the first cleavage, there is an inductor centre conditioning the formation of the blastopore. This blastopore later becomes a gastrular localization endowed with high inductional powers. Isolation of blastomeres in the second and third cleavages causes only those to gastrulate which contain the necessary substance for forming a blastopore.

## INDUCTORS AND ORGANIZERS

Grafting, as in Hilde Mangold's experiments, a fragment from the anterior lip of the blastopore of one species of Triton, say the pigmented type, to the latero-ventral region of the clear embryo from another species, will determine in the host the development of a secondary embryo characterized by axial organs, spinal chord, nervous system, etc., and incite the "neural organization of tissues" which, without its influence, would have given rise to other organs more usual to that region. The only active factor is the anterior lip of the blastopore, which not only brings about differentiation of suitable tissues but also that of the tissues of the host. The graft may be said to induce the formation of organs in the host.

On the other hand, if we implant other tissues not from the anterior blastopore lip, they are induced by the neighbouring host tissues and contribute to formation of the organs which should be differentiated in the locality under consideration.

The effects of the graft of an organizer are definite evidence of epigenic influences in the course of development, the organizer seems to be some one or several chemical substances and shows no zoological specificity. Inductive action can be transmitted by contact.

There are other inductions of a different category; for instance, primary inductions which are the most intense and come from the blastopore lip in the first instance, secondary and remoter ones which are more and more localized and subordinate. The case of development of the crystalline lens of the eye, studied by Spemann in 1901, Lewis in 1904, King in 1905, and still later by Filatov in 1925, shows yet again that organs become differentiated thanks to a double hereditary influence, preformational and epigenically inductive, a common twofold instrument called a "double safety device" by Spemann.

We could mention other experimental results to corroborate the foregoing. Functional and morphological characters are certainly already foreshadowed in the egg by the genetic factors contained therein, without which inheritance would become impossible. This implies determinancy in the ovule, and thus there is no doubt that organism results from properties in the egg itself. The parts, however, which go on to production are determined by mutual action and this allows regulations which will be more or less far-reaching according to species.

Influences are exercised, as we previously saw, by material agents -inductors, organizers, hormones, metabolites, and so on — and perhaps also by as yet undetermined dynamic factors acting nevertheless in fields of force, in a well-defined spatial framework, and evidenced by dynamic gradients as mentioned by Child in 1921, the importance of which latter has been dwelt upon by Huxley and de Beer.

It can be deduced from all this, that in embryonic development, the same processes of correlation and organic unification are at work as in other biological and physiological phenomena in general: elementary properties of the cell in the first place and chemical and energetic integrated mechanisms in the second.

A perfect adjustment between the elementary properties and the functional processes of differentiated parts to form an integral whole or organism makes possible the development of life in its most diverse aspects, the most remarkable of which at present seem to be instinct on the one hand and on the other embryological or reproductive morphogeny and the auto-regeneration of morphological characters. Constantly, even in the most distant reaches of biology, we find properties of cell structure and correlation between cells all in synergetic activity. In development, the former is preformation and the latter epigenesis.

It is therefore unjustifiable in these days to carry on ancient feuds. Recent advances have shown us that in ontogenic evolution, preformation and epigenesis exist side by side. Furthermore, they have been shown to be one and the same. Everything has its own moment in life. In the formation of a new body, everything lies naturally implicit in the germ-cell. The offspring unfailingly follows its set course, existence, and this course will be one of many biological properties distinguishing the individual and the species. Preformation does not only assume certain forms, but assumes that certain instants also are of importance.

When the time is due, a particular phenomenon will make its appearance, and this may be the creation of a bodily part just as much as of a function. Also, among so many functions, we have the production of one particular substance or the liberation of a force, and therefore, a relation among the cells and their mutual influence; in short — epigenesis. Epigenesis is therefore just one more piece of evidence, amongst so many, of genetic preformation, which in its turn is not mere anatomy.

## JULIAN SORELL HUXLEY

It is now possible to give a brief summary of the chief points which have emerged from our study of development, during which attention was focused on differentiation and its origin as the central problem. In the first place, animal development is truly epigenetic, in that it involves a real creation of complex organization. It is also predetermined, but only in the sense that an egg cannot give rise to an organism of a species different from its parent. The development of each individual is unique. It is the result of the interaction of a specific hereditary constitution with its environment. Alterations in either of these will produce alterations in the end result.

Determination is progressive. In the earliest stages, the egg acquires a unitary organization of the gradient-field type in which quantitative differentials of one or more kinds extend across the substance of the egg in one or more directions. The constitution of the egg predetermines it to be able to produce a gradient-field of a particular type; however, the localization of the gradients is not predetermined, but is brought about by agencies external to the egg. The respective roles of internal predetermination and external epigenetic determination are clearly seen in regard to the bilateral symmetry of the egg. The amphibian egg is predetermined to be able to give rise to a gradientfield system of bilateral type through the establishment of the grey crescent at a particular latitude of one meridian.

The particular meridian is not predetermined, but is normally decided by the point of sperm-entry; the precise latitude is determined as a result of the primary axial gradient of the egg, impressed upon it by factors in the ovary. On the other hand, the egg of a radially symmetrical animal like a Hydroid is incapable of developing bilateral symmetry; the predetermined capacity to react to stimuli localized in one meridian is not given in its constitution.

The agencies which determine the position of the various axes involved in the gradient-field system may be of very various nature; they may be factors in the maternal environment (ovarian conditions), biological factors (point of sperm-entry), or external physical factors (as in the determination of the polarity of the egg of Fucus). In any case they are external to the egg. They may also operate at very different times relatively to fertilization.

A number of chemical processes are set going by fertilization. These will proceed differently in the quantitatively different environments provided in different parts of the gradient-field system, until qualitative differences are set up. In most cases, these differences are at first not visible, and are presumably of chemical nature; this step in differentiation is therefore spoken of as chemo-differentiation.

These chemical differences appear at first to be reversible (e.g. labile determination of the presumptive neural tube region in the Urodele before gastrulation) but after a certain point to become irreversible. From this moment onwards, the organism consists of a mosaic of chemo-differentiated regions, each determined to give rise only to one or a limited number of kinds of structure. These are what we have called partial fields. The attainment of the mosaic stage often takes place under the influence of a dominant region or organizer. This may determine the extent and form of the whole gradient-field within which chemodifferentiation occurs, as in Planarian regeneration, or may interact with a previously established gradient-field orientation in another direction, as in amphibian organizer grafts.

The organizer may exert its effects at a distance, as does the regenerated head on a cut piece of a Planarian, or may supplement such distance effects by more powerful contact effects, as happens when the amphibian organizer comes to underlie a certain portion of the animal hemisphere, and at once determines it irrevocably as a nervous system.

Modifications of the gradients by external agencies will entail alterations in the structures produced. These alterations may consist in changed proportions, or in the total absence of certain regions (temperature-gradient experiments with frogs' eggs, cyclopia in fish, modification of regeneration in Planarians).

Here again, there is a predetermined capacity to produce a certain type of structure in certain conditions; but the precise localization of the structures produced depends upon the form of the gradients in the field-system.

Once the mosaic stage has set in, further differentiation may be brought about by the influence of one point on its neighbours. The classical example of this is the induction of a lens from epidermis by the optic cup. During the period when the organization of the developing animal consists of a single field-system, far-reaching regulation is possible; after-irreversible chemo-differentiation has occurred, it is not. The precise time at which irreversible chemo-differentiation sets in varies markedly in different groups. In Amphibia it occurs during gastrulation; in Ascidians at fertilization.

After the establishment of a mosaic of partial fields, it does not follow that all the cells of any given partial field necessarily give rise to the organ characteristic of the field. Thus, more cells are capable of giving rise to the amphibian fore-limb than do in fact give rise to it in normal development. Further, the boundaries of the partial fields overlap: a given group of cells in the limb-rudiment of the chick may contribute to the formation of either a thigh or a shank, according as to whether it is allowed to remain attached to or is isolated from one partial field or the other. Gradients may exist in such fields: the capacity of cells within the fore-limb field to give rise to a limb decreases with their distance from a subcentral portion of the field: the same is true for many other organ-fields.

Up to a certain time, regulation is still possible within each of the partial fields; but as development proceeds, each of these becomes split up into progressively smaller fields, each with its own determined fate: for instance, the fields for leg, shank, and foot, within the originally single hind-limb field. Each area in the mosaic passes from the state of invisible chemodifferentiation by the process of histo-differentiation to full visible differentiation, and so reaches the functional stage. After the organism as a whole has reached the functional stage, many new morphogenetic agencies come into play. The organism also, through acquiring the power of regeneration, re-acquires much of the regulative capacity which it lost in its passage through the mosaic stage.

The type of organization characteristic of one stage appears to persist, in whole or in part, throughout subsequent stages. Thus, the main gradient-system of the embryo permeates the partial fields of the limb, neural folds, ears, gills and heart, and determines their axis; and the growth of the lateral line along a particular level of the flank can best be interpreted in terms of a persistent total gradient-field.

Again, a total field-system certainly exists in adult Planarians and appears to reveal its presence in late stages of other groups through the presence of growth-gradients permeating the whole organism.

The persistence into adult life of the partial field-systems of the mosaic stage is shown by the phenomena of regeneration, by the existence of localized growth-gradients within single areas, and notably by phenomena such as those found in newts, where, for instance, indifferent regeneration-buds produced by an amputated limb will produce legs when grafted into a certain area round the leg, while if grafted near the base of the tail they will produce tails.

With this, of course, only a start has been made with the scientific analysis of development. It remains for the future to discover such fundamentals as the physiological bases of the field-systems, and the elaborate physico-chemical processes which must be operative at the time when the quantitative differences of the early gradient-field system are converted into the qualitative differences of the chemodifferentiated mosaic stage.

It is, however, already a good deal to have arrived at this first outline of development on the biological level. To have established the fact that organizations of quite different type succeed one another during development is important.

The recognition of the gradient-field system, with its purely quantitative differentials, as the basis of early organization, is a great step forward, since it provides an adequate formal explanation of many phenomena of regulation which have been considered by various authors, notably by Driesch, as affording proof of vitalistic theories of development.

Further, the epigenetic analysis of development is pointing the way to a large extension of the field of heredity, in the shape of physiological genetics. It is only through a study of development that it will be possible to understand what the term "genetic characters" really stands for — in other words, what are the basic processes involved in the action of a particular Mendelian gene.

Experimental embryology as a separate branch of science was initiated by Roux; in its next phase, in which Driesch, Boveri, Wilson, Herbst, Morgan, Brachet and Jenkinson are outstanding names, a large body of facts was amassed, and the experimental proof of epigenesis provided; in the third phase, Spemann and Harrison are the outstanding figures within the sub-science, while the theories of Child have not only linked the facts of regeneration with those of embryonic differentiation, but have provided a scientific basis to early development, thus filling a large gap in the theoretical aspect of the subject.

Meanwhile, experimental embryology has been making fruitful contacts with physiology, notably in the field of hormone action, with genetics, and with growth studies. The fourth stage is now beginning, in which this framework of general principle will be filled in through intensive research, and the whole science deepened by a search for the physicochemical bases of the empirical biological principles which have been discovered in its earlier stages.

## MAURICE CAULLERY

The question of differentiation in the embryo lies in the field of chemistry and perhaps more particularly in the province of physicochemistry when we come to the study of general problems relating to nutrition and growth. Such a basis will also furnish a useful insight by means of the investigation of reproduction in the lower single-celled organisms, in which, though it cannot be said there are actually any embryological problems, we do discover cell reactions of great value. At the present time, for instance, several investigators are attempting the culture of single-celled species, particularly of flagellata, in accurately determined conditions of the nutritional environment.

By such means we shall be able to decide what may be the most satisfactory substances for fostering the multiplication of such organisms, by "factors of growth" as they were named by Lwoff in 1932. Furthermore, data obtained from the examination of tissue cultures are also of major importance.

Experimental researches along independent lines are likely to converge and coincide, thus setting in relief phenomena of a general order which are of fundamental importance in the existence of cells and therefore in embryology. Here again we may observe the mutual dependence of essential problems which when they make their appearance are at first glance quite unrelated, though upon going into them more deeply, we find that they are distinct manifestations of the same essential properties.

Morphological epigenesis, of which the extent of its influence has been shown by experimental embryology to have direct natural consequences in physiological epigenesis, has been rightly stressed by Wintrebert during recent years. The one is complementary to the other. Development leads to a properly constituted functional organism. The embryonic period has generally been held to be antithetic to the rest of the individual's existence. Functional existence commences indeed in the egg.

The functions of the embryo correspond to morphology in its various stages; there is an exact relation between form and function at any given instant, and things cannot be otherwise.

Each stage in embryogeny conditions the succeeding one, and that is the reason why temporary organs may appear quite different from those occurring in the adult with its particular activities.

This is remarkably evident in those species which pass through larval stages and metamorphosis. I have tried to stress the idea that physiological epigenesis is demonstrated by experimental embryology.

It is precisely because of its functional reactions that the embryo continues to evolve during its earlier stages of development as well as in the more advanced periods of its existence.

Physiological research, on the other hand, is not concerned with form, except as a sideline to function, and in the primary stages where the functions are not so developed, the general factors governing cell life become manifest.

This lends interest to the problems of experimental embryology, in which we can analyse the most interesting factors in the processes of vital activity, those of the widest effect, namely the cell processes in their most profound aspects. In every biological manifestation, the intrinsic properties of the cells take first place. After these come the phenomena resulting from relationships among the cells, and this is confirmed yet again in the development of the individual organism.

## CONRAD HAL WADDINGTON

A complementarity to the ideas of physics and chemistry is often alleged not only of life, but of the notion of biological organization. Indeed, in recent years, there has been a tendency either to regard organization as one of the irreducible fundamental bases of all biology, or to invoke it, as though it were a well-defined concept, to fill up any awkward gaps in a theoretical structure.

The latter use is perhaps the more reasonable; but it is obviously dangerous without a clear idea of exactly what is meant by the term, and when an attempt is made to reach such clarity, I think it will be found that the scope of the notion, as an explanatory principle, is not so great as has sometimes been suggested. On the other hand, it does provide the key to an extremely valuable method of thought. But before it can be accepted as a guide, it requires more precise formulation than it has yet received, at least in scientific contexts.

In the first place, it is clear that the fundamental notion is organization, which is capable of quantitative variation, rather than organism, which is not. The degree of organization of an entity is usually considered to depend on the degree to which the parts of the entity are dependent on the whole. Now it is immediately clear, but is rarely pointed out, that the degree of dependence will be different in respect of different properties of the parts.

For instance, if we consider a certain mechanism, such as a motor-car engine, the parts such as the cylinders, pistons, etc., are not in any way existentially dependent on the whole; they can exist perfectly well in isolation from it. They are, however, highly dependent on it in a certain context, namely with reference to the functioning of an internal combustion engine.

When we speak of the dependence of the parts on the whole we must always have in mind some particular context; thus the parts of an entity can be said to be dependent on the whole, in a particular context, if, in order to express the properties of the parts in that context, some reference to the whole is necessary. For example, the eye may be considered either with reference to the function of vision, or as a developing entity; in the first context, the relations of the retina and lens are those of two bodies which have to be adjusted to one another in order to focus light, etc., while in the second their relations are those of an inductive reaction and the mutual adjustments of growth rates.

There is obviously no reason why the internal co-ordination of the eye, or, to put it in the way previously used, the dependence of the parts on the whole, should be the same in the two cases. The eye may, and at certain periods of its existence undoubtedly does, have very different degrees of organization in these two contexts.

The purpose for which the concept of organization is usually invoked is to form part of the theoretical system for dealing with phenomena which seem to involve the subjection of otherwise self-sufficient parts to some overriding whole. For instance, one might be tempted to advance "the organization of the eye-forming region" as an explanation of the formation of a normal eye after some experimental disturbance of its rudiment.

Instead of the individual cells behaving independently, they are subordinated to a general influence which causes them, even after disarrangement, to form a single normal organ. This is often expressed by saying that the tissue can no longer adequately be regarded as a mere mass of cells, but has attained a higher level of organization in which the relevance of the whole organ to the constituent cells can no longer be disregarded.

In some ways this type of expression suffers from the same difficulties as we noted in discussing fields. Where does the organization come from? And what is meant by a higher level of organization? The second of these questions is easier to answer. The statement made earlier that organization must be defined with reference to some context provides the clue. A new level of organization is in fact nothing more than a new relevant context. When it is said that an organ rudiment has a higher level of organization, as a developing entity, than a mere mass of cells, what is meant is that some organ is relevant to the former, while the latter has nothing to do with an organ, either because it is not yet competent or because it has passed the regulative stage and reached a point at which its development is completely mosaic, each fragment differentiating on its own without any reference to the whole.

The question of whence a new level of organization is derived is more delicate. In discussing the analogous question about fields, we stated that the field must be regarded as a product of the interaction of its parts.

If we applied this directly to the problem of organization we might seem to arrive at the conclusion that all levels of organization are implicit in the levels beneath them, which constitute their parts; and that would lead to denial of any true arising of new contexts, and thus to a denial of levels of organization. In fact, we should find ourselves in a vicious circle.

We do, however, seem to be confronted with the fact, which is very difficult to deny, of new contexts and new levels of organization. It does seem that we are thinking in terms which are in some ways at different levels when we think of genes and their immediate products, and of the mirror symmetries of reduplicated legs.

Or, to take a more extreme case outside the present field, psychology is hardly relevant in connection with the chemistry of sugars.

We must then accept the existence of different levels of organization as a fact of nature. On the other hand, we cannot easily suppose that the arising of a new level involves the appearance out of the blue of completely new properties of the elements of which the organized unit is composed. But the way out of the dilemma is clear.

When elements of a certain degree of complexity become organized into an entity belonging to a higher level of organization, we must suppose that the coherence of the higher level depends on properties which the isolated elements indeed possessed but which could not be exhibited until the elements entered into certain relations with one another. For instance, we have suggested that it may be possible to explain the organization of regions of embryonic tissue into organ rudiments by supposing that they contain some orientated protein microstructure similar to that of a liquid crystal.

Such a microstructure is on a different level of organization to that of ordinary molecules in solutions. It depends on the mutual attractions and repulsions of fibre-like particles, which cannot be exhibited unless the molecules first form elongated fibres which then come together in large numbers.

But the existence of the fibre level of organization is not accounted for in terms of the elementary molecules plus some entity of a higher level, such as an overriding field. Instead we account for the formation of the fibre field by enlarging our ideas about the elementary molecules to include the fact that they can polymerize into fibres, which will then attract and repel one another in certain ways. That is to say, a new level of organization cannot be accounted for in terms of the properties of its elementary units as they behave in isolation but is accounted for if we add to these certain other properties which the units only exhibit when in combination with one another.

According to this view, one cannot explain any peculiar behaviour by postulating the existence of a new level of organization. The advantages of the concept are not explanatory, but simply that it provides a terminology in which it is easy to admit the recognition that phenomena do not present themselves as being all of the same kind, and this is certainly an important advantage. In its absence we are practically forced to argue that the phenomena of sociology and chemistry are not significantly different, a point of view which cannot easily be made very plausible. The admission that different levels of organization exist frees us from such preconceptions.

We feel no conviction that, for instance, the behaviour of a mass of tissue must be explicable in terms of the properties of its isolated cells. Instead, we hope that* investigation of the tissue will reveal new data about the mutual interactions of the cells when aggregated in a mass. Our aim is not merely to explain the complex by the simple, but also to discover more about the simple by studying the complex.

## CHROMOSOMES AND DETERMINATION

In what sense can the chromosomes be considered as agents of determination? By many writers they have been treated as the actual and even as the exclusive "bearers of heredity"; numerous citations from the literature of the subject might be offered to show how often they have been treated as central, governing factors of heredity and development, to which all else is subsidiary.

The most complete example of this conception, perhaps, is embodied in the theory of the germ-plasm as developed by Weismann; but in one form or another it has persisted almost to the present day. Many writers, while avoiding this particular usage, have referred to the chromosomes, or their components as "determiners" of corresponding characters; but this term, too, is becoming obsolete save as a convenient descriptive device.

The whole tendency of modern investigation has been towards a different and more rational conception which recognizes the fact that the egg is a reaction-system and that (to cite an earlier statement) "the whole germinal complex is directly involved in the production of every character."

Genetic research is constantly bringing to light new cases of the co-operation of several or many factors in the production of single characters (e.g. in that of sex); and it is possible that all the chromosomes, or even all of the units which they contain, may be concerned in the production of every character.

Beyond this it is evident that every character is produced during development by an activity in which the cytoplasm, and what we call the "organism as a whole" plays a most important part. When, for example, we speak of the X-chromosome as a "determiner" of sex, we mean only that it is a differential factor or modifier the relative quantity of which in relation to the autosomal material conditions a particular reaction by the developing germ. The value of the chromosome-theory of heredity does not lie in our identification of this or that "determiner" or "bearer of heredity," but in its practical importance as a means of experimental analysis. In this respect, in the writer's opinion, the theory has the same kind of value as the molecular and atomic constructions of physico-chemical science; and the "mystical" and "unscientific" character ascribed to it by some writers is purely imaginary.

## THE UNKNOWN FACTOR IN DEVELOPMENT

We are now arrived at the furthest outposts of cell-research and here we find ourselves confronted with the same unsolved problems before which the investigators halted. The first question raises once more the old puzzle of preformation or epigenesis. The pangen hypothesis of de Vries and Weismann recognizes the fact that development is epigenetic in its external features; but like Darwin's hypothesis of pangenesis, it is at bottom a theory of preformation, and Weismann expresses the conviction that an epigenetic development is an impossibility.

He thus explicitly adopts the view, long since suggested by Huxley, that "the process which in its superficial aspect is epigenesis appears in essence to be evolution in the modified sense adopted in Bonnet's later writings; and development is merely the expansion of a potential organism or 'original preformation' according to fixed laws."

Hertwig, while accepting the pangen hypothesis, endeavours to take a middle ground between preformation and epigenesis, by assuming that the pangens (idioblasts) represent only cellcharacters, the traits of the multicellular body arising epigenetically by permutations and combinations of these characters.

This conception certainly tends to simplify our ideas of development in its outward features, but it does not explain why cells of different characters should be combined in a definite manner, and hence does not reach the ultimate problem of inheritance.

What lies beyond our reach at present, as Driesch has very ably urged, is to explain the orderly rhythm of development — the coordinating power that guides development to its predestined end. We are logically compelled to refer this power to the inherent organization of the germ, but we neither know nor can we even conceive what this organization is.

The theory of Roux and Weismann demands for the orderly distribution of the elements of the germ-plasm a prearranged system of forces of absolutely inconceivable complexity.

Hertwig's and de Vries's theory, though apparently simpler, makes no less a demand; for how are we to conceive the power which guides the countless hosts of migrating pangens throughout all the long and complex events of development?

The same difficulty confronts us under any theory we can frame. If with Herbert Spencer we assume the germ-plasm to be an aggregation of like units, molecular or supramolecular, endowed with predetermined polarities which lead to their grouping in specific forms, we but throw the problem one stage further back, and, as Weismann himself has pointed out, substitute for one difficulty another of exactly the same kind.

The truth is that an explanation of development is at present beyond our reach. The controversy between preformation and epigenesis has now arrived at a stage where it has little meaning apart from the general problem of physical causality.

What we know is that a specific kind of living substance, derived from the parent, tends to run through a specific cycle of changes during which it transforms itself into a body like that of which it formed a part; and we are able to study with greater or less precision the mechanism by which that transformation is effected and the conditions under which it takes place.

But despite all our theories we no more know how the properties of the idioplasm involve the properties of the adult body than we know how the properties of hydrogen and oxygen involve those of water. So long as the chemist and physicist are unable to solve so simple a problem of physical causality as this, the embryologist may well be content to reserve his judgment on a problem a hundredfold more complex.

The second question, regarding the historical origin of the idioplasm, brings us to the side of the evolutionists. The idioplasm of every species has been derived, as we must believe, by the modification of a pre-existing idioplasm through variation, and the survival of the fittest. Whether these variations first arise in the idioplasm of the germ-cells, as Weismann maintains, or whether they may arise in the body-cells and then be reflected back upon the idioplasm, is a question on which, as far as I can see, the study of the cell has not thus far thrown a ray of light.

Whatever position we take on this question, the same difficulty is encountered; namely, the origin of that co-ordinated fitness, that power of active adjustment between internal and external relations, which, as so many eminent biological thinkers have insisted, overshadows every manifestation of life. The nature and origin of this power is the fundamental problem of biology.

When, after removing the lens of the eye in the larval salamander, we see it restored in perfect and typical form by regeneration from the posterior layer of the iris, we behold an adaptive response to changed conditions of which the organism can have had no antecedent experience either ontogenetic or phylogenetic, and one of so marvellous a character.

That we are made to realize, as by a flash of light, how far we still are from a solution of this problem. It may be true, as Schwann himself urged, that the adaptive power of living beings differs in degree only, not in kind, from that of unorganized bodies.

It is true that we may trace in organic nature long and finely graduated series leading upward from the lower to the higher forms, and we must believe that the wonderful adaptive manifestations of the more complex forms have been derived from simpler conditions through the progressive operation of natural causes.

But when all these admissions are made, and when the conserving action of natural selection is in the fullest degree recognized, we cannot close our eyes to two facts: first, that we are utterly ignorant of the manner in which the idioplasm of the germ-cell can so respond to the play of physical forces upon it as to call forth an adaptive variation; and second, that the study of the cell has on the whole seemed to widen rather than to narrow the enormous gap that separates even the lowest forms of life from the inorganic world.

## Chapter 12

# Life on Earth

### FLORA, FAUNA AND ENVIRONMENT

Every period and every locality has its particular flora and fauna, and when the environmental conditions change in any way, the forms of life also change. This is a fact which for anyone not ridden by prejudice is practically impossible to deny. Furthermore every living species exhibits those characters which are the most propitious for its continued existence in its particular environment.

Aquatic animals assume certain forms and develop certain functions which are quite different from those in the land animals; plants which live in damp surroundings are of a particular pattern more or less, whilst those of the and deserts are quite different.

Temperature, nourishment, warmth from the sun and degree of exposure, the concurrence of other living creatures and many other factors play their part in determining the growth of form and functions in everything that is alive.

It is inevitable that this should be so. If it be impossible to conceive of life without environment, if it be impossible for a creature to live in isolation, independent from all others, if life be an incessant replacement of substance and energy from the environment, the living creature takes from its environment what it needs and gives up thereto what it has a surfeit of; if life is the interrelation between the individual with everything that surrounds it, then it becomes easy to understand such environment as a decisive factor in the development of life. "Organism would be impossible of explanation without an environment" according to Child. Each of the characters of an organism has its relation to external factors. This becomes still more evident when one considers the organism as a whole, which is what it really is. "Unity, order, physiological differences, relationships and the just adaptation of parts among themselves have no significance unless it be in relation to an outside world."

Correlation between form and function in a living creature and the conditions affecting it has always been a cause for wonder in those who study such things. Scientists may differ from one another in opinion in many respects, but the facts, outstandingly clear as they are, admit of no discussion, and though certainly many different explanations may be propounded, reality itself remains indisputable. Even the most radical fixationist theorists have had to admit that each living form is without exception the most suitable for fostering the continued existence of its particular species. According to these latter opinions, this comes about through metaphysical causes implicit in the end of all things.

## ADAPTION TO ENVIRONMENT

Jean-Baptiste Pierre Antoine de Monet de Lamarck published his work on Zoological Philosophy in 1809. He attributed the origin of various species in existence to a process of evolution, as had also been done in ancient times by Thales, Herachtus, Empedocles, Anaximander, Theophrastus, Lucretius, and others.

Species continue acquiring their characteristic forms because they become adapted to their surroundings and the modifications which the environment impresses on individuals are then passed on by inheritance. Climate, manner of life, relations with other species are all environmental factors determining the characters of living creatures.

Such fresh acquired characters keep on accumulating in influence throughout countless generations under the same environmental conditions, thereby converting the species progressively in accordance with its vital needs. This explains how such characters happen always to be the most expedient to allow a species to live in its corresponding medium. Function, answering to an objective, creates the necessary organ, and disuse on the other hand makes for atrophy. The fresh organ passes along to descendants, in the same way that an organ which degenerates through uselessness will disappear in the course of generations.

According to Lamarckian theory, the individual is active, reacting to its environment, withstanding it or bending before it, and in the last instance becoming adapted thereto by its own labours. Hence it happens, when an individual has eventually acquired the suitable characters needed, that such adaptation becomes naturally a cause for survival.

Species, therefore, become adapted, change their form and functions, for the purposes of continuing in existence. Active evolution creates suitable forms and thus preserves life on this planet, despite the changing conditions of our environment in time and in space.

## NATURAL SELECTION

Charles Darwin in 1859 explained evolution after a different fashion. Species do not survive because of adaptation, but evolve and change. From impartial and dispassionate observation it can be shown that they vary with time and according to the place and the influence of various circumstances, or perchance through instability in the species, and such variations become permanent or vanish through natural selection.

Fit offspring resist and the unfit become extinct. The same thing occurs in the wild state as may be observed in the breeding of domestic creatures, so carefully studied by Darwin. It amounts to selecting suitable individuals for the purposes of transmission and thus to fix the most suitable characters for getting races which become step by step stronger and more efficient.

Natural selection is merely the possibility, or impossibility, that a species has of living in a particular manner. By evolution and variation of the species, only those will survive which find the conditions necessary for their existence in their more or less immediate world where they are in the course of development.

An individual is always affected by surrounding circumstances. It may be that it win disappear in the course of its life through incompatibility of the environment or that it may withstand conditions; obviously if all the individuals disappear then so also does the species, or the line of the species may be so influenced at the end of a few generations that it gradually becomes weaker and weaker to the point of extinction; or, on the other hand, if the environment be favourable, the species will develop and expand in profusion. The environment, the factor of natural selection, and equally the properties of the species will determine the latter's fate. This same natural selection will explain also the origin of those species now extant.

According to the Darwinian theory, environment has no influence on the production of forms. These make their appearance independently from the medium and eventually, according to whether the forms are suitable or not, selection dictates their destiny. It is obvious that the Darwinian idea differs entirely from that of Lamarck. According to Lamarck, the whale takes the form of a fish because it lives in the sea, whilst according to Darwin it lives in the sea because it has the form of a fish.

Darwin does not deny that the surroundings may exert an influence on living creatures, but he does not consider such effect when explaining the production of shapes nor of any other of the biological characteristics of the various species. Genetic factors alone are decisive, and natural selection works on the differences resulting therefrom.

## GERM PLASM

After Darwin, opinions of the greatest radicalism came to be expressed, and from this point we get the modern tendency to underestimate the influence of environment on the development of vital phenomena, starting with those of formation. Weismann stated that Nägeli's idioplasm or germ plasm is found only in germ-cells and that these bear no relation whatever to the somatic cells formed of food-plasm. Through complete isolation of the one type of cell from the others, germ plasm continues from generation to generation, whlist the individual with its body cells plays merely the part of a temporary carrier for the never-dying germ cells. This being So, it becomes plain that even if the individual during its short existence can stand the effect of environment, that is of no account so far as the species is concerned, since the germ cells remain unaffected. This theory is indeed a total negation of the Lamarckian idea of the transmissibility of acquired characters.

Weismann's ideas have gained more and more ground with biologists over the years, including the geneticists themselves, who for the greater part have vigorously championed the theory that the environment is lacking in influence on the individual and consequently upon the species.

Variation, according to Weismann, comes about from amphimixis, that is, fertilization or the fusion of two different germ plasms, whereby distinct ids and determinants are blended, and realize between themselves a type of natural selection, say germ selection, which makes the opportunity for certain factors to endure and for others to perish.

From these highly complex unions, new forms issue which in their turn are preserved or destroyed by the action of natural selection. Weismann is responsible also for the idea that inert determinants, or what we now call genes in the latent state, coexist alongside the active ones. The inerts may come into action through the presence of definite internal or external stimuli. In this way there may come about a struggle between determinants, whereby certain features would prevail and become manifest because their determinants were stronger or more numerous. This would explain how in certain cases there might appear in the progeny, characters which were not evident in previous generations.

## SEGREGATION

Some idea of the difficulty attending the separation of the idea of purely genetic action from that of environmental action with its living factors, as if these did not also form part of the environment, can be gained from disquisitions regarding the effects of topographical isolation. Say segregation or isolation fosters the fixation of fresh specific characters, becoming thereby one more means of natural selection, thus preventing crossing with individuals of the original species who do not possess the distinctive characters, then in this way, the newly developed form becomes easily stabilized. This would explain the genesis of odd species through migration or geographical isolation, for instance after some upheaval of nature, or after some remarkable engineering feat which might restrict the availability of large expanses of water or land.

Moritz Wagner maintained that any new and stable form starts with the separation of forms becoming isolated from their originals. He accepts the effects of adaptation and the immediate and primary increment of the modified characters thanks to the crossing of individuals in both sexes which possess them, and the exclusion of those progenitors lacking in them.

Romanes in 1886 and Jordan in 1911 also went into this important question, which however contributes but little to the problem of the effect of environment upon the characters of species. We can see, nevertheless, whether any of those propositions deducible from such observations favours more the physiogenic and morphogenic influence of environment through, of course, gene mechanism as suggested, for instance, by le Dantec in 1907.

## ADAPTATION AND TRANSMISSIBILITY OF ACQUIRED CHARACTERS

We must now make a distinction between the two related questions: Is there any adaptation of living creatures to their surrounding conditions, or not? and, Are the characters ordained by environment transmitted through heritage?

By reason of Weismann's efforts, many biologists nowadays tend to give a negative answer, though examination of actual cases allows no doubt that life embraces the whole world, since, as Rabaud said in 1922, life does not exist in the abstract, for it consists of living organisms which have to develop within an environment. "Adaptation is no more than the possibility of existence," he goes on to say, "and selection does not actually preserve the best but simply eliminates the most unfit." The living mechanism, wrote Cuénot in 1925, "has a past which has left its mark, and the aim of every vital activity of a being is to exist and to endure, well or ill, but certainly to survive."

Without arriving at the same conclusion as Jennings, who in 1934 asserted that the same character can be just as much occasioned by genes as by environment, there are some independent naturalists who do not give way before the general current of opinion, and who grant a due morphogenetic and functional value to outside influences. Woodruff in 1941, under the heading "Nature and Nurture" in his Foundations of Biology, has this to say, for instance: "Since the life of an organism is one continuous series of reactions with its surroundings, it follows that nurture plays an immensely important part in moulding the individual on the basis of its heritage.

Indeed we are apt to overlook the fact, already mentioned, that every character is a product both of factors of the heritage and of the environment and can be reproduced only when both are present. Those characters that appear regularly in successive generations are those whose development depends upon factors always present in normal surroundings.

Other characters, potentially present, do not become realized unless the unusual environmental conditions necessary for their development happen to be met. Witness the examples of the supernumerary legs in Drosophila, and sun-red in Indian corn.

In Drosophila, the abnormal condition of extra legs is inherited in typical Mendelian manner when the flies are reared at low temperatures; whereas supernumerary legs do not appear in flies with the same gene heritage when bred at a higher temperature. Or again, the so-called sun-red colour of the kernels of Indian corn is a heritable character that develops only when the kernels themselves are exposed to light by the removal of the husk. In short, the environment, in certain cases at least, may act as a differential intensifying or diminishing gene action, and thus influence the realization of the potentialities of the heritage."

There are strong reasons in favour of this. Surrounding temperatures, barometric pressures, different types of radiation, meteorological and climatic conditions, the metabolic condition, in viviparous creatures, of the mother during the first periods of development of the embryo, all exhibit an influence upon such development, affecting the weight of the offspring, its height or length, its sex and its ability to lead a separate life.

This had already been pointed out ages ago by Hippocrates: "Of course the breed may vary even though the seed remain the same, according to whether it be winter or summer, a wet season or a period of drought," and this was again confirmed in 1942 by Petersen and Mayne's lengthy and accurate statistical investigations.

Researches on twins, such as those carried out by Newman, Freeman and Holzinger in 1937, likewise show that environment is the cause of differences even between individuals endowed with the same hereditary potentialities. Not everyone, of course, would be willing to subscribe to the thesis that "what heritage can achieve can also be achieved by environment," though, on the other hand, everybody must admit that the action of environment does get added to the influence of the genes.

If it were only the gene factors which were effective in the production of functional qualities or forms, then the only worry of eugenics would be when seeking improvements in human social groupings or in animal and plant stocks. It is easy to show that this in itself is not enough; of equal importance is euthenics, that is instruction in betterment in its most varied aspects.

On the basis of heredity we build up the framework of the triangle of life, whose shape and height as well as its area can change considerably, even though its base, heredity, remain constant in all cases. The other two sides of the triangle represent respectively the environment in general and the special case of environmental intervention which is training or education. Though enjoying the same hereditary birthright, the area of the triangle, the capacity of direction in life, may vary considerably.

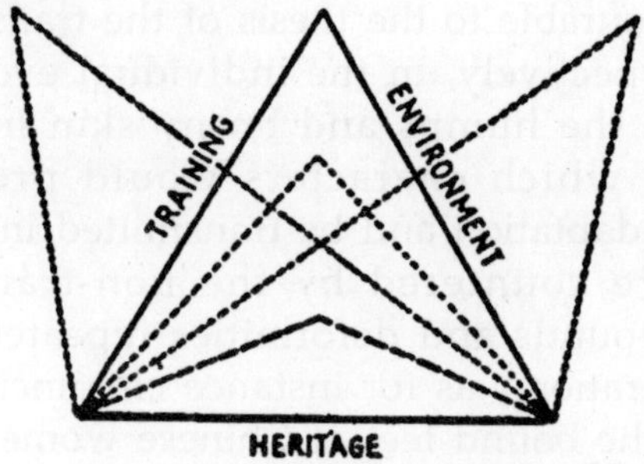


There is no room therefore to forget the circumstances which contribute throughout life, and especially in its earlier periods when the future and the fate of the individual is decided, to the way in which such life shall develop. Conditions in the environment will take part in the production of forms and in the development of functions.

If an individual during its lifetime becomes adapted to environmental circumstances, is it necessary then to discuss the adaptation of species: We find ourselves up against the so well-known question of the transmissibility of acquired characters. We have seen that Lamarck made such transmissibility the keystone of his theory, and that latterly it has been denied practically unanimously.

The material nowadays at our disposal is profuse. We may remember among many others the well-known experiments of Kellog and Bell in 1907 on the silkworm larva Bombyx mori. By varying nutrition, functional and morphological modifications were obtained which were transmitted by heredity to a certain number of generations.

Tower in 1906 carried out almost simultaneously researches making use of beetles' larva, which he subjected to environments of various degrees of humidity, whilst Picted in 1925 went into the same question, using larva from different sorts of butterflies and moths, which he also caused to vary through feeding and humidity. Variations transmitted from parent to progeny were observed in general even though they always maintained a tendency to revert to their original type.

Cunningham and Cattaneo thought they had found arguments favourable to the thesis of the transmissibility of characters, respectively, in the individual evolution of the flounder, and the humps and horny skin in camels and dromedaries, which characters would presumably be produced by adaptation and be transmitted in series. These arguments are countered by the non-transmission of mutilations, wounds and deformities repeated throughout countless generations, as for instance circumcision amongst the Jews and the bound feet of Chinese women as formerly practised.

The docking of the tails of dogs, cats or rats, as effected by Weismann, does not cause the birth of tail-less animals. Against these tests it might be objected that mutilation is not carried out for a sufficient number of generations and that casual excision of an organ of so little biological interest as a tail could not be considered as an acquired character. Everything rests therefore on determining what exactly is an "acquired character," and when does a particular modification become such an acquired character.

Obviously, prior to a character becoming permanent there is no possibility of its being transmitted through inheritance. Normally also such fixation on the genetic potential must be difficult and the influence of environment must be extended through quite a long time for fixation to succeed. Hence it is that investigations in this direction are usually attended by negative results.

It is therefore impossible to state that acquired characters are indisputably transmitted by heredity, even if everything leads to the assumption that such transmissibility may be true.

Neither the one side nor the other of this extremely important problem has been definitely proven, and it still remains therefore without a definite solution one way or the other.

## MODIFICATIONS, RECOMBINATIONS AND MUTATIONS

We shall say, in short, that individuals may vary through so-called modifications, through "recombinations" and through "mutations." The first are those changes which supervene by reason of the conditions under which the embryonic or adult life of the individual is developing, these being influences on the body or soma and not on the germ-cells.

The environment with all its perquisites, to wit, temperature, humidity, foodstuffs, friends, enemies, etc., as well as education, work, general reaction to the environment, become translated into functional and morphological characteristics such as the blacksmith's brawny arm, the supple fingers of the violinist, the reasoning powers of the mathematician, and so on. These are "acquired characters," and they do not appear to be transmissible.

Recombinations consist in changes in the form of offspring as compared with that of their progenitors, these being exactly proportional to the genetic qualities of such progenitors, and germ characters transmitted definitely in accordance with the laws of heredity. Mutations give rise to sudden jumps and changes. Normal parents bear children with different forms and functions from their normal selves, occurring by reason of some disturbance of the genetic elements. Mutations constitute an incontestable reason for variation, because the ancestral line will tend to have fresh characters starting with the mutate.

From which we may deduce that an individual can be "modified" merely by the effect of environment, but that fresh characters will not be transmitted by heredity or will be transmitted only with difficulty and perhaps even then only because of some sustained environmental pressure repeatedly exerted upon a large number of generations.

Breeding can be another reason for variation in form or recombination. The genetic heritages from male and female parent, which are different when dealing with cross-breeding, will give rise, through Mendelian segregation of characters, to different combinations and the offspring will be perhaps quite different from either or each of its parents. The new characters thus attained will naturally be transmissible. Natural cross-breeds, so effectively studied by Darwin, have had a major part in the generation of fresh specific forms. Crossing, however, which is possible through the cohabitation of two strains then becomes a circumstance arising out of biological environment.

Mutations were formerly considered to be a matter of chance, something mysterious which was produced by unknown means. The transmissibility of mutate characters is now, as we have seen, no longer a matter for discussion. Fresh strains of races and even of species arise from mutation.

We are now beginning to glimpse the mechanism of mutation and it is constantly being shown how influences disturbing the balance of the genes and deficiencies of the most varied nature, physical, various radiations, temperatures, different chemicals, and other actions, are capable of causing mutations by acting on chromosomes and their genes. Mutations are not produced, therefore, spontaneously, but come about from a concatenation of external conditions as part of the environment.

Thus we have an acquired character; not necessarily a visible character exhibited by an individual who may have suffered from environmental factors, but it certainly is a fresh character in its germ-cells. Moreover, such character which can cause changes of species is transmitted by inheritance. It is a striking case of hereditary transmission of an acquired character.

Mutation may be of varying degrees of depth and intensity, from the change of an insignificant minor character to a total change of form and function. A slight mutation might resemble an acquired character and this would be transmitted indefinitely. The adaptation might excite the appearance of fresh specific forms through systematic mutation processes consisting of more or less progressive minor mutations. This is the Lamarckian method of variation. If the abyss between body cells and germ-cells postulated by Weismann does not exist, and if the former are affected by environment and are able to exert their action upon the latter which so plainly do contain genetic powers, and if the germ-cells themselves are also subject to environmental conditions, the problem of transmissibility of acquired characters has lost much of its importance and questions seemingly of the greatest seriousness to us years ago now become mere pastimes.

Firm and incontrovertible remains the fact, which nobody can deny, of the fitting of our living world to the qualities of the outer universe in which it exists, both in the spatial as in the temporal sense. Life has therefore to surrender itself without reserve to the circumstances in which it develops.

## JEAN BAPTISTE PIERRE

Extensive changes in the conditions surrounding animals will lead to wide modifications in their requirements and perforce to corresponding changes in their activities. Now, if the new requirements become permanent, the creatures then acquire fresh habits which also become fixed like the requirements evoking them.

It is obvious, therefore, that a deep-reaching change in surrounding conditions on making itself felt by any particular race of animals will induce in them fresh habits which will, in turn, entail the use of certain parts of the body in preference to those previously in use and in a similar manner will lead to the disuse of any part no longer necessary.

All this is known fact and not mere theory. Now certain well-known facts show that fresh requirements which make any particular part essential will by reason of repeated effort bring about the formation of such part, and consequently it will gradually increase in strength and development and end up by reaching an appreciable size. Moreover, when fresh conditions and requirements render a part entirely useless, the creature will cease to use it and because of such lack of exercise the part in question will gradually cease to share in the development enjoyed by the other portions of the animal, whilst if such conditions of disuse persist over a long period, it will finish up by completely disappearing.

Among plants, where there are no active movements and consequently no habits properly so-called, extraordinary changes in conditions nevertheless also produce wide differences in the development of parts, since in these living organisms everything happens by reason of changes occurring in nutrition, absorption and transpiration, in the quantities of heat and light, air and moisture which they normally obtain, and finally by reason of the ascendance which certain vital motions are able to acquire over others. If such conditions endure, the state of those individuals which are ill-fed, sickly or weak becomes habitual and permanent, whilst their internal organization becomes modified to such a degree that their offspring also preserve the acquired changes and end up by producing a race different from that of such individuals lucky enough always to live in conditions favourable to their development.

What Nature achieves over a long period can be done by man in a day by suddenly altering the conditions in which a plant and all the individuals of a species are situated. Botanists know that plants transplanted from their native heaths to nurseries undergo such profound changes that they may become quite unrecognizable. Cultivated plants become in this way so different from the wild specimens of the same species, that botanists are loth to identify garden plants.

Where can we find wild cabbage or lettuce identical with those grown in our market-gardens? How many new strains of domestic pigeon and fowl which it would have been impossible to find in the wild state, have been obtained by breeding them under varying conditions?

However, even discounting changes of site and conditions, we know that living creatures, although they may never stir from where they always have been, are also exposed to outside influences, though the consequent changes may only become apparent very, very gradually. Places change in exposure, climate, nature and local conditions, though with such slowness in comparison with our length of life that we are liable to credit such conditions with being perfectly stable and unchangeable. Now, when conditions change, they exert fresh influences on living creatures, which latter consequently change their habits and become modified, as previously indicated, so as to comply with such alterations of environment. They adapt themselves gradually to the altered circumstances.

## TRUE FUNCTIONAL ADAPTATION

All other cases of morphological adaptation among animals, and several in the vegetable kingdom as Well, belong to our second group of phenomena which, in our analytical discussion, we called adaptations to functional changes that result from the very nature of functioning, and which we shall now call by their ordinary name of "functional adaptations."

It was Roux who first saw the importance of this kind of organic regulation and thought it well to give it a distinctive name. "Byfunctioning, the organization of organic tissues becomes better adapted for functioning." These words describe better than any others what happens.

It is well known that the muscles get stronger and stronger the more they are used, and that the same holds for glands, for connective tissues, etc. But in these cases, only quantitative changes come into account. We meet with functional adaptations of a much more complicated and important kind, when for instance, as shown by Babak, the intestine of tadpoles changes enormously in length and thickness according to whether they receive animal or vegetable food, attaining nearly twice the length in the latter case. Besides this, the socalled mechanical adaptations are of the greatest interest.

It has long been known, especially from the discoveries of Schwendener, Julius Wolff, and Roux, that all tissues whose function it is to resist mechanical pressure or mechanical tension possess a minute histological structure specially suitable to their requirements. This is most markedly exhibited in the stems of plants, in the dolphin's tail, and in the lime lamellae in the bones of vertebrates. All these structures, indeed, are such as an engineer might have made who knew the kind of mechanical conditions they would be called upon to meet.

Of course all these types of mechanically adapted structure are far from being "mechanically explained," as the verbal expression Might perhaps be taken to indicate, and as has indeed sometimes been the opinion of non-discriminating writers.

These structures exist for mechanics, not by it. On the other hand, all these structures which we have called mechanically "adapted" ones, are far from being mechanical "adaptations," in our meaning of the word, simply because they are "adapted." Many of them, indeed, exist prior to any functioning; they are for the most part truly inherited if for once we may be permitted to use that ambiguous word.

However, the merely descriptive facts of mechanical adaptedness having been ascertained, there have now also been discovered real processes of mechanical adaptation. They occur among the static tissues of plants, though not in that extremely high degree which has sometimes been assumed to exist; they occur also in a very high degree of perfection in the connective tissues, in the muscles, and in the bone tissues of vertebrates.

Here indeed it has proved possible to change the specific structure of the tissue by changing the mechanical conditions which were to be withstood, and it is in cases of healing broken bones that these phenomena have acquired a great deal of importance both theoretically and practically; new joints also, which may arise through force of circumstances, correspond mechanically to their newly created mechanical functions.

We now have to ask ourselves, of course, if any more intimate analysis of these facts is possible, and indeed we easily find out that here also, as in the first of our groups of morphological adaptations, there are always individual agents of the environment which might be called "causes" or "means" of adaptive effects, the word "environment" being taken to embrace everything that is external to the reacting cells.

Of course here also the proof that there are single formative agents does not in the least detract from the adaptive character of the reaction itself. We may perhaps say then that localized pressure is the formative stimulus for the secretion of skeletal material at a particular point of the bone tissue, or of the fibres of the connective tissue, whilst merely quantitative adaptations of the muscles might even allow of a still simpler explanation. Adaptations, nevertheless, remain adaptations in spite of that, even if they deserve only the name of "primary" regulations.

## LORANDE LOSS WOODRUFF

To epitomize — these facts from genetics, taken in connection with the wealth of data from geographical distribution, the succession of types in the geologic past, and so on, give us the modern background for attempting to form an opinion of the method of evolution. The opinion of most biologists is that natural selection in general is a guiding principle underlying the establishment of the adaptive complexes of organisms. Evolution is the result of mutations, germinal variations largely, though not entirely, independent of environing conditions.

Many of these variations give rise to characters which neither increase nor decrease the adaptation of the organism, and consequently are neutral from the standpoint of its survival. With regard to such characters natural selection is essentially inoperative. Other germinal changes occur, some of which produce adaptive and others unadaptive characters, and here natural selection is effective.

It may eliminate the unadaptive and leave the adaptive variations and so make possible the survival value of the latter in the struggle for existence. The germ plasm never ceases to experiment or natural selection to discover. Variability affording opportunity for adaptability is expressed in "evolvability," a profoundly significant characteristic of life.

So, it will be noted, this is essentially a clarified view of Darwin's idea of natural selection that has been made possible by recent intensive studies of the intrinsic nature and the origin of variations. Natural selection still affords the most satisfactory explanation of that coordinated adaptation which pervades every form of life: it shows how nature can be self-regulating in establishing adaptations. But it is probable —

indeed, positive — that there are more factors involved than are dreamt of in our biology.

In the words of Thomson: "The process of evolution from invisible animalcules has a magnificence that cannot be exaggerated. It has been a process in which the time required has been, as it were, of no consideration, in which for many millions of years there has been neither rest nor haste, in which broad foundations have been laid so that a splendid superstructure has been secured, in which, in spite of the disappearance of many masterpieces, there has been a conservation of great gains. It has its outcome in personalities who have discerned its magnificent sweep, who are seeking to understand its factors, who are learning some of its lessons, who cannot cease trying to interpret it.

It looks as if Nature were Nature for a purpose." Indeed, Aristotle, in effect, emphasized that the essence of a living being is not protoplasm but purpose. However, this thought carries us beyond the accepted sphere of science into the great fields of philosophy and theology.

## Chapter 13

# Living Creatures on the Earth Crust

The forms assumed by living matter are countless, and vary, as we have seen, in both space and time. They are not identical at different places on the planet nor in different geological periods. We have seen that each region of the earth, in the seas or on the land, has its own characteristic fauna and flora. The species living in any particular region are determined by a multitude of circumstances, for instance, humidity, temperature, particular foodstuffs available, and other co-existing and competing types of life, resulting in a balance between the various creatures which live contemporaneously or succeed one another. Thus it is that each country is characterized by its flora and fauna.

The areas of diffusion of the various species are also defined according to the circumstances obtaining in them. Plants generally show a more limited distribution, because it is impossible for them to move about. On the other hand, animals, being possessed of the power of movement, are able to wander to greater or less distances and spread.

Even so, there are differences among the animals also, particularly observable in fossils, where we find deep-sea sedentary types or benthos living on the ocean bed more than a thousand fathoms down and usually showing themselves only in circumscribed areas which sometimes help us to define a particular geological facies or period of the earth's history. Wandering creatures, as for instance nekton and plankton, have a wider habitat.

Fortuitous chances of geography may impede or, on the contrary, foster the spreading of a species. Such particularly is the effect of mountains, streams and oceans. A species at a particular place tends to propagate itself, and the success or failure of its efforts will depend on its own special qualities, which include, for instance, greater or less incentive to live, adaptability, strength, reproductive powers, and so on, whilst on the other hand environmental conditions, be they avourable or adverse to the species, will also play their part. Among environmental conditions we must place foremost the possible existence therein of other living species likely to compete with the particular one in question.

Plants and animals differ according to whether they live in water or on land. Among the former we have marine, lake-dwelling and river forms, whilst on land existing forms vary according to altitude, climate, humidity, and similar factors, and according to whether they live underground or can travel through the air.

The temperature of the sea, its depth, and in connection with this the degree of bathymetric pressure and light, saltiness, osmotic pressure, and chemical composition, nearness to or distance from the coast, ocean currents, presence of certain plants, and particularly of animals, etc., will be conditions which make possible the establishment and survival of any particular species at a particular spot.

Among sea and river creatures, certain species remain fixed and others migrate, which latter at certain seasons of the year undertake long journeys in which practically all the individuals of the species move away. Such emigrations are made of course under the influence of those very conditions which, as we have just seen, distinguish the waters where they five, namely temperature, salinity, other plants and animals, foods, currents, and so forth.

Plants and animals in lakes and rivers differ from those to be found in the sea and they vary also according to natural or artificial features of their surroundings. The shutting-off of stretches of water, salt or soft, through natural events or even because of some engineering feat of man, cutting a reservoir of water off from the rest of the region, have as their consequence more or less remarkable variations in the species populating a neighbourhood. The same occurs on land. Take, for instance, the poor vegetation in deserts or moorlands, the richer pastures and highlands, and the luxuriance of the jungles, all of which lend their stamp to their particular territories. Climate, richness or poverty of the Soil, humidity, neighbourhood of particular species, all will cause the flora to exhibit special characters at any particular point, forming spots in which certain species proper to that region will prevail.

There are also areas widely distant from each other where a particular plant lends itself so well to identical climates, that from that fact a common origin is deducible. Such is the case, for instance, of the walnut family Juglans which grow in confined districts well separated from one another in both the eastern and the western hemispheres, in Europe, Asia, and America, whilst in past geological ages from the cretacean to the phocene, plantations of such trees extended over vast zones more than covering their present habitats.

In the distribution of animal life, identical facts are observable, though, as we have said, animals have greater powers of movement. There are, withal, numerous species special to certain limited localities, such as the Arctic and Antarctic fauna, and that to be found in the temperate, warm and tropical regions of the equatorial zones, all of which exhibit their own peculiarities.

We have tigers in India and jaguars in America, lions in Africa as well as camels, whilst in South America we find pumas, llamas, and vicugna, kangaroos in Australia, and so on and so forth. In numerous other cases, similar species populate different continents, where in the distant past they ranged over extensive territories, which is the reason why their present-day descendants form isolated groups widely separated in certain instances.

The opposite situation is however more frequent — segregation. Isolation will give different characters to animals of the same origin. Darwin's observations in this respect are of historic value. During his famous three-year journey round the world in the Beagle, he made some findings of the greatest importance in the Galapagos Islands at some six hundred miles distance from the Pacific coast of South America. He describes how the fauna, which as a whole resembles that of the mainland, embraces nevertheless some species not only different from those of the continent but also morphologically distinct from the same species inhabiting other islands in the same archipelago.

He mentions the case of certain starlings, which differ according to whether they are found on Charles Island, Albemarle, James or Chatham. At another place, he writes: "It seems remarkable to me to have discovered thirty-eight plants exclusive to these Galapagos Islands, and not to be found anywhere else on earth."

There are over-ground animals which are able to fly and spread abroad, capable of immensely distant migrations in obedience to definite temporal rhythms; and there are other underground species buried deep down in the bowels of the earth, protozoa, worms, arthropods, fish, batrachians, reptiles, mammals, sthenothermic or sluggish in the main, clinging to damp surroundings and many of them blind. Spelaeological exploration constantly reveals the existence of special fauna and flora for each particular system of caverns.

The seas are variously populated according to their situation and depth, marine climate, latitude, and the in-sea distance from a coastline. Everywhere, one can remark a littoral fauna and flora up to a furlong along the coast, whilst deep-sea fauna and abyssal fauna and flora can also be distinguished.

The species constituting these groups depend on the depth, the nature of the sea-bottom, the salt content and temperature of the water around. Microscopic creatures from plankton give us animal plankton, zooplankton, and plant or vegetable plankton, phytoplankton. Then we get phaeoplankton down to 100 ft below the surface, knephoplankton from about 100 ft down to somewhere about a third of a mile deep, and scotoplankton from there downwards to the uttermost depths. And every one of these groups has its particular and characteristic plant and animal species.

The deep sweet waters also differ from the depths of the sea by the nature of their denizens. In some few lakes, such as Baikal, depths down to a mile can be plumbed. In such waters there are no echinodeans, very few coelenterata and seaworms, plenty of articulates, crustaceans, insects and water-mites or their larvae, some fish and reptiles. In lakewaters in general, life is more sparse than in the sea, and still more so in rivers. Every soft-water region, much more than the limitless sea, displays its plant and animal denizens, from the surface and from the depths.

Understandably then, the study of biological geography is fraught with enthralling interest. Apart from the possibility of direct practical applications, its theoretical importance must also be considered. It provides irrefutable proofs of the adaptation of living forms to their environment.

## LIVING CREATURES THROUGHOUT THE AGES

In the same way that species vary in space, that is by geographical distribution, according to their circumstances of existence, obviously they will have varied also during the course of time, since in millenium after millenium the conditions in which they have developed have changed time and again.

The knowledge that each geological stratum is identifiable through the fossil remains therein dates from slightly more than a century ago. The majority of substances forming the earth's crust are ancient sediments deposited in ocean depths or in their land-locked connections such as lakes or seas. In this way the various geological strata have been built up, and thus it is possible to determine with greater or less exactitude the period to which they belong. This is done in the first place by the fossils to be found in each stratum, as William Smith first suggested in 1813 and Lyell subsequently confirmed.

The geological age is recognized by the so-called facies of the terrain, which is a combination of mineralogical and chemical lithologic features of the ground with the palaeontological characters of the fossils. These give valuable data regarding the biological conditions of the age and environment, height, climate, depth, temperature, saltness, currents, nature of the surroundings, nutritional capacity, and so forth. From all these details we can ascertain the distribution of creatures in existence at a particular time according to altitude and range, which enables us to make up palaeo-geographical charts, as suggested for instance by Lapparent. The facies and formations which are distinct facies though with certain properties in common, since they originated under analogous conditions whether marine, lacustrine, littoral, continental or inland, etc., allow the chronological determination in scale and knowledge of the changes therein and the consequences of adaptation. The same applies to flora as to fauna, so that alongside palaeozoology we must study also the science of palacobotany.

Together with the influences of physical environment, we repeat, it is important to bear in mind also the biological environment. The composition of a flora and of a fauna in a particular region affects the conditions of existence of each of the beings constituting the one and the other, so that biological reactions between living creatures in competition take first rank in our considerations.

Each geological era is distinguished by the plants and animals which were in existence at that time. There are difficulties in the way of exactly determining the duration of each of such ages, but we do have certain appreciable data which allow at least an adequate perspective of relative durations, even though it may still not be possible to define precisely the length of each period or the time of any particular occurrence. The figures suggested by different writers vary considerably one from the other.

It may be possible perhaps to calculate the total duration of the period included in our geology, that is the formation of the various layers of the earth, as some two billion years. As Woodruff writes: "Ten thousand years are but a yesterday in the history of our earth." From this assumed two billions, more than half would be taken up by the period of no life.

In the long stretch of time reaching from the Precambrian, starting with the Laurentine period, to the Archaeozoic as far as the Algonquin, we can take it that there must have been some bacteria and fungi as well as primitive algae, and most certainly protozoa in existence, all of them single-celled organisms. Polyplastid organization naturally must have come later. The mosses and giant ferns led up to the conifers and these to the spermatophytes. The plants first were aquatic, from sea and swamp, in huge quantities at certain periods; afterwards came the land plants.

In the lower palaeozoic the smaller crustaceans, worms and sea creatures abounded, after which came the sea scorpions. In the Devonian and Silurian epochs came numerous fish species. Life was still more marine in nature. In the Carboniferous period it extended to dry land; creeping creatures issued from the waters, amphibians descendant from the ancient sea scorpions, giant crabs and huge insects. The Carboniferous is indeed a period of great fertility, and in the strata corresponding to it we find a huge number of fossil remains. Afterwards came a period of drought, when existence became difficult.

Temperature underwent considerable changes and long glacial epochs followed. The abounding vegetation of the swamp-lands of the previous age disappeared at certain points, covered by fresh deposits, compressed and mineralized, and carbon, becoming isolated from the organic principles, formed the coal measures embodying hydrocarbons in a gaseous state.

At the start of the mesozoic period, when climatic conditions improved, life on earth renewed its strength and proliferated, amphibians and, even more so, reptiles taking first place. Huge reptiles made their appearance, some of which remained on land whilst others returned to the sea. It is thought that the mesozoic period may have lasted some two hundred million years; the reptilian era may have extended another eight million decades. It was a period of exuberant vitality and many and vigorous species. In the aftermath, however, the generative impulse seems to have become spent, a period of decadence in life started and totally fresh forms of existence made their appearance. A new era of glacial cold kills the mesozoic fauna and flora, the gigantic reptiles and their kin, leaving only those animal and plant species which are able to resist cold.

Serious disturbances within the earth mark the end of the mesozoic era and glacial periods alternate with cycles of tremendous heat until we reach the camozoic and the commencement of the tertiary period, when temperature becomes stabilized once more and life is renewed with fresh vigour. In the tertiary period birds appeared, descending from the reptiles, whilst mammals also started to develop. The realm of life on earth becomes consolidated and takes upon itself from that time forth the qualities still ruling generally in our present-day existence. In the primitive quaternary epoch the first men come to life.

## THE PLASTICITY OF EXISTENCE

Climatic characters and other conditions of existence in each epoch determine the flora and fauna of that period. Environment moulds the species which become adapted thereto or perish; some endure and others become extinct. Fresh forms are constantly being produced in an uninterrupted and incessant burst of generative activity, either by progressive variation, i.e., orthogeny or by mutation. It is likely that in such geological eras the instability of species may have been greater than is shown in the present with all its changes and present frequency of mutations.

Life runs through a course higher than the individual, through groups and collectively; likewise each organism follows its own course through time. This is due to the action of the genetic principle, to which becomes added the unceasing influence of environment. From the union of these two factors we get not only the individual organism, but also, and this is the more important, the species.

The distribution of living organisms in space and in time is thus proved. If life is impossible without environment, if there be no living organism which can subsist independently from the world surrounding it, there is no miracle in the environment giving evidence of its effects on the course of existence, and on the most varied manifestations of life. Genetic agents alone are insufficient to decide the appearance of shapes and forms and other vital peculiarities.

Variations of germ, orthogeny, mutations, cross-breeding, hybridization and segregation bring about fresh characters which are transmitted by heredity and thus become fixed and permanent. Environment also plays its part in the production of distinctive physiological and morphological characters. The coincidence of internal influences with those from the outside ends up in the existential birthright of the world with all its immense wealth. It is that, moreover, which causes individuals and species to be more or less robust and therefore capable of survival or condemned to premature extinction.

A species may become extinct through lack of vigour, through lability, or through its inability to reproduce, and similarly, because the environment in which it occurs is adverse. All these various influences add together and determine whether a particular species survives or not. If it is weak, however slight the adversity of the environment, the species will fail, and conversely if it be strong it will have greater chances of survival however antagonistic the environment. The destiny of a species resides in itself, but also in the surrounding world. Fate simultaneously marks both the being and the circumstance.

## ALEXANDER VON HUMBOLDT

The physical sciences are interconnected by those same links which unite all the phenomena of nature. The classification of species, which must be taken as the basis of botany, and the study whereof has been made easier and more attractive through the introduction of logical methods, is to plant geography what descriptive mineralogy is to the survey of the rocks making up the outer crust of our globe. In order to grasp the laws governing these rocks in their strata, to determine the ages of their successive formations and their similarity in the most diverse regions, the geologist must first of all know the simple fossils making up the mountain masses and of which the study of mineralogy teaches their distinguishing signs and nomenclature.

The same applies to that branch of natural history dealing with connections between plants, whether amongst themselves or with the soil in which they exist, or with the air they breathe and transform. Advances in plant geography depend to a large extent on those made in descriptive botany, and it would be detrimental to progress in the sciences to raise them to the level of general ideas whilst neglecting the value of individual facts.

I have endeavoured to bring together within a single framework the sum total of the physical phenomena offered by that part of the New World lying within the torrid zone, from the Pacific coastline up to the topmost peak of the Andes, that is to say, to combine all the facts about plant life, animals, geological relationships, soil cultivation, air temperatures, the limits of the eternal snows, the chemical composition of the atmosphere, its electric potential, barometric pressure, decline in gravitational pull, depth of blue in the sky, fading of light during its journey through the various layers of the upper atmosphere, the horizontal bending of light, and the heat of boiling water at various altitudes.

Fourteen scales arranged alongside a cross-section representing the Andes show the variations sustained by these phenomena through the effect of the height of the ground above sea level. Each group of plants is shown at the height assigned to it by Nature, and the prodigious variety of their forms can be traced from regions of palm and fern up to those of the fever-repellent Joannesia or Chuquiragua, JUSS., graminaceae and lichen. These regions form the natural divisions of the vegetable kingdom and just as the everlasting snows are found at a definite height in any particular climate, so likewise do the febrifuge species such as Cinchona have fixed limits which I have shown on the botanical map attached to this essay on plant geography.

During the course of my tour, I made quite a number of personal notes on the races of man in South America, on the Orinoco settlements, on the hindrances which the climate and vigour of the jungle set in the way of the progress of civilization in the torrid zone, on the character of the country in the High Andes as compared with that of the Swiss Alps, on the analogies observable between the rocks of the two hemispheres, on the physical composition of the air in the Tropics, and so forth.

I had left Europe with the firm resolve not to write what is usually called a traveller's tale, but to publish the fruit of my investigations in a purely descriptive form. I set my facts not in the order of their happening, but according to the relationship they bear to each other. Amid the awe-inspiring majesty of Nature, keenly occupied with the stupendous sights she reveals at every step, the traveller is hardly in the mood to jot down in his diaries that which concerns himself alone and the trivial details of his daily life.

Thus the unity of composition which distinguishes good books from those of faulty design cannot always be faithfully respected, in so far as man cannot help but describe in a more lively manner what he has seen with his own eyes, provided the main part of his attention has been fixed less on scientific observation than on the habits of the peoples he meets and the mighty works of Mother Nature.

Now, the most faithful portrayal of customs is that which best reveals the affinities between the members of the human race. The character of nature in the raw or of civilized existence is depicted either by the hindrances which rise up before the traveller or by the feelings evoked in him. What we want to see is he himself ceaselessly in contact with the objects surrounding him, and his story becomes the more interesting the more a touch of local colour is diffused over his description of the countryside and of those who dwell there.

## SUCCESSIVE DEVELOPMENT OF ORGANIC LIFE

In preceding chapters I have considered many of the most popular grounds of opposition to the doctrine that all former changes of the organic and inorganic creation are referable to one uninterrupted succession of physical events governed by the laws of Nature now in operation. Against this doctrine some popular arguments have been derived from the great vicissitudes of the organic creation in times past; I shall therefore proceed to a discussion of such objections, especially as they have been formally advanced in these words by a late distinguished philosopher, Sir Humphry Davy.

"It is impossible," he affirms, "to defend the proposition that the present order of things is the ancient and constant order of nature, only modified by existing laws: in those strata which are deepest, and which must, consequently, be supposed to be the earliest, deposited forms even of vegetable life are rare; shells and vegetable remains are found in the next order; the bones of fishes and oviparous reptiles exist in the following class; the remains of birds, with those of the same genera mentioned before, in the next order; those of quadrupeds of extinct species in a still more recent class; and it is only in the loose' and slightly consolidated strata of gravel and sand, and which are usually called diluvial formations, that the remains of animals such as now people the globe are found, with others belonging to extinct species.

But, in none of these formations, whether called secondary, tertiary or diluvial have the remains of man, or any of his works, been discovered; and whoever dwells upon this subject must be convinced, that the present order of things, and the comparatively recent existence of man as the master of the globe, is as certain as the destruction of a former and a different order and the extinction of a number of living forms which have no types in being.

In the oldest secondary strata there are no remains of such animals as now belong to the surface; and in the rocks, which may be regarded as more recently deposited, these remains occur but rarely, and with abundance of extinct species; there seems, as it were, a gradual approach to the present system of things and a succession of destructions and creations preparatory to the existence of man."

In the above passages, the author deduces two important conclusions from geological data: first, that in successive groups of strata from the oldest to the most recent, there is a progressive development of organic life, from the simplest to the most complicated forms; secondly, that man is of comparatively recent origin, and these conclusions he regards as inconsistent with the doctrine "that the present order of things is the ancient and constant order of nature only modified by existing laws."

It is on other grounds that we are entitled to infer that man is, comparatively speaking, of modern origin; and if this be assumed, we may then ask whether his introduction can be considered as a step in a progressive system by which some suppose the organic world advanced slowly from a more simple to a more perfect state.

The principle of adaptation to which I allude is the same as that in force at the present time and which causes genera and species to be distributed heterogeneously through the various regions of the globe, so that they or the races may exhibit contemporaneously characters differing according to climate, be it arctic, temperate, or tropical, which is not merely a question of temperature. From local conditions we get the fauna and flora appropriate to each territory. And this is evident in geography, which is space, as it is in geology, which is time.

## CHARLES DARWIN

The inhabitants of the world at each successive period in its history have beaten their predecessors in the race for life, and arc, therefore, higher in the scale, and their structure has generally become more specialized; and this may account for the common belief held by so many palaeontologists that organization on the whole has progressed. Extinct and ancient animals resemble to a certain extent the embryos of the more recent animals belonging to the same classes, and this wonderful fact receives a simple explanation according to our views. The succession of the same types of structure within the same areas during the later geological periods ceases to be mysterious, and is intelligible on the principle of inheritance.

All the chief laws of palaeontology plainly proclaim, as it seems to me, that species have been produced by ordinary generation; old forms having been supplanted by new and improved forms of life, the products of Variation and the Survival of the Fittest. As the late Edward Forbes often insisted, there is a striking parallelism in the laws of life throughout time and space; the laws governing the succession of forms in past times being nearly the same with those governing at the present time the differences in different areas. We see this in many facts.

The endurance of each species and group of species is continuous in time; for the apparent exceptions to the rule are so few, that they may fairly be attributed to our not having as yet discovered in an intermediate deposit certain forms which are absent in it, but which occur both above and below: so in space, it certainly is the general rule that the area inhabited by a single species, or by a group of species, is continuous, and the exceptions, which are not rare, may, as I have attempted to show, be accounted for by former migrations under different circumstances, or through occasional means of transport, or by the species having become extinct in the intermediate tracts.

Both in time and space, species and groups of species have their points of maximum development. Groups of species, living during the same period of time, or living within the same area, are often characterized by trifling features in common, as of sculpture or colour.

In looking to the long succession of past ages, as in now looking to distant provinces throughout the world, we find that species in certain classes differ little from each other, whilst those in another class, or only in a different section of the same order, differ greatly from each other. In both time and space, the lowly organized members of each class generally change less than the highly organized; but there are in both cases marked exceptions to the rule.

According to our theory, these several relations throughout time and space are intelligible; for whether we look to the allied forms of life, which have changed during successive ages, or to those which have changed after having migrated into distant quarters, in both cases they are connected by the same bond of ordinary generation; in both cases the laws of variation have been the same, and modifications have been accumulated by the same means of natural selection.

## ALFRED RUSSEL WALLACE

An accurate knowledge of any groups of birds or of insects and of their geographical distribution, may enable us to map out the islands and continents of a former epoch — the amount of difference that exists between the animals of adjacent districts being closely related to preceding geological changes. By the collection of such minute facts, alone, can we hope to fill up a great gap in the past history of the earth as revealed by geology, and obtain some indications of the existence of those ancient lands which now he buried beneath the ocean, and have left us nothing but these living records of their former existence . . .

The preceding remarks are all I now venture to offer, on the distinguishing features of the various groups of land-animals as regards their distribution and migrations. They are at best but indications of the various lines of research opened up to us by the study of animals from the geographical point of view, and by looking upon their range in space and time as an important portion of the earth's history. Much work has yet to be done before the materials will exist for a complete treatment of the subject in all its branches; and it is the author's hope that his volumes may lead to a more systematic collection and arrangement of the necessary facts. At present all public museums and private collections are arranged zoologically.

All treatises, monographs and catalogues, also follow, more or less completely, the zoological arrangement; and the greatest difficulty the student of geographical distribution has to contend against, is the total absence of geographical collections, and almost total want of complete and comparable local catalogues.

Till every well-marked district, every archipelago, and every important island, has all its known species of the more important groups of animals catalogued on a uniform plan, and with a uniform nomenclature, a thoroughly satisfactory account of the Geographical Distribution of Animals will not be possible. But more than this is wanted. Many of the most curious relations between animal forms and their habitats, are entirely unnoticed, owing to the productions of the same locality never being associated in our museums and collections. A few such relations have been brought to light by modern scientific travellers, but many more remain to be discovered; and there is probably no fresher and more productive field still unexplored in Natural History.

Most of these curious and suggestive. relations are to be found in the productions of islands, as compared with each other, or with the continents of which they form appendages; but these can never be properly studied, or even discovered, unless they are visibly grouped together.

When the birds, the more conspicuous families of insects, and the land-shells of islands, are kept together so as to be readily compared with similar associations from the adjacent continents or other islands, it is believed that in almost every case there will be found to be peculiarities of form or colour running through widely different groups, and strictly indicative of local or geographical influences.

Some of these coincident variations have been alluded to in various parts of this work, but they have never been systematically investigated. They constitute an unworked mine of wealth for the enterprising explorer; and they may not improbably lead to the discovery of some of the hidden laws (supplementary to Natural Selection), which seem to be required, in order to account for many of the external characteristics of animals.

In concluding his task, the author ventures to suggest, that naturalists who are disposed to turn aside from the beaten track of research, may find in the line of study here suggested a new and interesting pursuit, not inferior in attractions to the lofty heights of transcendental anatomy, or the bewildering mazes of modern classification. And it is a study which will surely lead them to an increased appreciation of the beauty and harmony of nature, and to a fuller comprehension of the complex relations and mutual interdependence, which link together every animal and vegetable form, with the ever-changing earth which supports them, into one grand organic whole.

## CAUSATION AND DESIGN

Even the most primitive view of the world impresses on man a coviction of intent. Since man acts from his own knowledge of objects within his understanding, phenomena are attributed to the intervention of "entities," "entelechies," as conceived from the standpoints of human consciousness and will. The primitive, in time and space, populates things with mystical spirits through whose agency the activity of the whole is maintained.

Such anthropomorphic myths observe events and foresee the consequences whereby their eventual effect will come to pass, finally, in a manner suitable for achieving a particular object in mind. Happenings develop teleologically "in order" to arrive at a particular result, in physical nature and still more so in the more complex sphere of living things, thus causing Aristotle (to mention but one instance) to say "life is not form but objective."

This proves to be no obstacle against even Aristotle himself, among so many other active minds, to assert that knowledge must be based upon the study of tangible events, on a basis of observation and experience, in order to establish by inductive reasoning the causes of such events. The profound ideological revolution which such a proposition signifies should be noted; namely, the endeavour to explain the world by the world itself without the help of myths. Hence we cast aside the symbol of Prometheus who wrested the secret of fire from the gods, attempting to eliminate imaginary deities in favour of an actual explanation of things as they enter our perceptions.

If when we achieve some act or do some deed, it comes about through our own will, arbitrarily, and we can likewise not do it, or omit to do it, and if in connection with such act we observe some particular effect which does not occur unless that act be done, we cannot fail to conclude that such deed or act is indeed the cause of the effect, that one occurrence conditions and decides a second, and hence that such causal relationship occurs in acts depending upon ourselves just the same as in those which are outside us.

"Show the cause, and you show the effect." "Do away with the cause and the effect is absent." "Vary the cause and the effect is varied." These are Bacon's laws of determinism. An occurrence makes its appearance and develops because it has been preceded, or is accompanied, by other events which are its cause.

And Nature thus becomes a continuate and vast concatenation of causes and effects stretching to infinity. Mechanical and physical analysis in all its simplicity demonstrates the relations between causes and effects and hence we arrive at the conception of practical science. Once indeed the conditions under which events develop are known, the recurrence of such events may be secured by reproducing those same conditions.

This conclusion is of great importance for mankind and represents an instrument of incommensurable value in action. The mysterious "beings" are not to be handled by the will of man, though this latter can indeed on the other hand reproduce the causal circumstances and bring about the effects desired. Empirical knowledge, founded on observation or experience, is positive knowledge and there is no need in fact to run to empyrean forces for an explanation of events taking place in this world.

Proceeding on these lines, however, we renounce the idea of a high intellectual hierarchy. Thought "in the mechanical plane" restricts the world to immediate reality, every action proceeds from given antecedents in the same world of such reality. Causes and effects succeed one another or take place simultaneously without end and in this manner events continue to evolve and develop.

Clearly, a determinist framework gives man a knowledge of when and how a particular event will take place and consequently endows him also with the option of repeating it. This is positive science. It is not then a long step to apply experimental standards to the study of the sciences relating to existence and living, which thus naturally become determinist.

This has brought inestimable progress in its train. Positive observation and experimentation are possible in biology and its related and derived sciences and, thanks to the application of such a positive criterion, advances in these branches of knowledge — as in many others — have been tremendously valuable. Observation and experiment — "experience" to put it in a single word — are a sign of science in all its phases, and it would in any event be impractical to deny ourselves the use of experimental standards.

Vital phenomena under their extremely diverse guises are studied through analysis of the conditions determining them. In this way, biological knowledge is built up and thus, likewise, does man continue to acquire means of ever wider scope, and more perfect in application, for taking the measure of vital reality. The standard of causality has enabled its proofs to be definite in the domains of the biological sciences just as in all other branches of positive scientific knowledge.

## CONTINGENT REMAINDERS OF PHENOMENA

What we have just said is by no means the whole story. There remains in every science a residue of phenomena, and in a science so advanced as physics we must reckon with Heisenberg's so-called "Uncertainty Principle," the result of which is that some authorities argue that the law of cause and effect is at a turning-point in science. From this, it will be understood that there may be in biology numerous facts which seem to elude the concatenation of cause and effect.

Throughout the ages vitalistic doctrines of various shades have arisen and endured. We may leave on one side those vitalists who might be said to be so "from impetuosity," because they endeavour to explain all vital phenomena by mechanical means and finding this impossible under present-day conditions of the sciences, thunder forth on the inefficiency and breakdown of causal explanations.

There are more serious-minded people who, recognizing that reality is perhaps wider than the frameworks set up by determinism, admit that this does not in any way mean that such frameworks are not efficient and necessary, even though they may be insufficient. Claude Bernard, himself the great protagonist of determinism, wrote: "The great problem dominating everything else is to find out whether logical relationships occurring in events of the mind do indeed tally inevitably and in all cases with external reality, and whether the links we naturally establish between events, for the purpose of satisfying our logical requirements, do indeed really exist.

There might be a series of evolutive events following one another in time, which need not necessarily be determinative; a chain in which each link has no cause-and-effect connection either with the one preceding it or with the one coming after it. We shall never reach absolute determinism for everything. There will always be indeterminism in the sciences. But the intellectual conquest of man consists in lessening and rejecting indeterminism as, with the aid of experimental method, determinism gains ground. Only by this can he satisfy his ambition because it is thus that he extends and will extend step and step his power over nature."

## TELEOLOGICAL ARGUMENTS

The opinions of the majority of vitalists were always based upon the fact that activities of living organisms respond to possible future conditions. Everything in life seems to respond to some aim of value to that life itself, and this teleology is difficult to square with a purely causal determinism. This is a matter of those "final causes" which so infuriated Francis Bacon. Assumption that an event may answer to a cause arousing it, pushing it from behind, as to a motive conditioning it by attraction, is what motivates it with a determinate object. This is precisely the working method of consciousness and will to which we have previously referred; human deeds take place in order to attain a foreseen result.

The study of biology is impossible without taking into account the teleological sense of the phenomena observable. This is an argument dear to vitalists of every class. We may quote, among so many of them, von Uexküll's dictum repeated even in 1936 by Noltenius: "Nature, the world of things, has three dimensions, namely 'substance,' 'soul' and 'spirit,' so that life may thus be 'material,' 'vital' and 'organismic' or 'systematic.' In the sphere of the material, the dominating motive of events is the 'causal' connection, whilst in the domains of soul and spirit the prevailing action is finality, the determinate end-in-view or purpose, a means of connection of a special order characterized by the active influence of the present upon the as yet unfulfaled future."

This extremely simple framework does not seem acceptable just as it stands, but it is as well to point it out as a typical example of the way of thought adopted by numerous biologists, many of them quite eminent, at the present time. Bertalanffy, on the other hand, wrote in 1932: "If the maintenance and renewal of an organism, reproduction, shows that such organism is a functioning whole with a purpose, consideration of the purpose will necessarily entail consideration of the complete whole."

## FITNESS OF FUNCTIONS

It is constantly to be observed that if we remain satisfied with purely teleological explanations, physiology of the organs would without exception give us every ground for satisfaction. If we thought that by saying "such function is fulfilled in this manner for such purpose," we had arrived at some fresh knowledge, study would become very easy for us, for it is certainly true in fact that the functions, both individually and in combination as a whole, do seem to obey an intended object.

Of course there is no "archeus" dwelling in the organs to ensure that the working thereof shall be effected in the best possible manner, but on the other hand there is indeed a combination of functional inter-relations from which the appropriate working of the organism is brought about, or, we might perhaps say, an archeus implicit in the event itself.

Completely automatic functions of the viscera develop as if in answer to some definite call, nervous functions still within the realm of physiology, the unconscious workings of the mind

— every single one of them acts as if moved by teleological beings. This happens not only among a few more highly differentiated animals, man for instance, but is repeated throughout the length and breadth of the animal kingdom. The more primitive forms of life, the more complex, and those higher in the scale, all function as if in their actions they were responding to some final cause. This unquestionable fact lends yeoman support to those who argue that fatalism, the standard of causality, is not always applicable when the phenomena of existence have to be explained.

This biological adequatio rei, this making of things fit their purpose, was a discovery of physiologists and biologists. Everything in each and every organism comes about in the best possible way for the fulfilment of its functions, as if a conscient intelligence were presiding over the development of events from which life results. A sense of use rules in the world of the conscious as well as in that of physiology. A particular thing, done for a purpose because it is useful, is the teleological criterion.

## CONSCIOUS AND UNCONSCIOUS INTFNTIONALITY

Modern philosophy is generally characterized by its being founded upon the notion of intentionality, especially in Scheler and Husserl when reference is made to vital intentionality prior to cognition. The primordial emotive impulse — without cognition, feeling or image — is a "leaning towards," or a "deviation from," an adaptation in fact according to need. In animal life initial impulses to action arise from existence itself being directed and governed from that source. "In the higher animals, Scheler writes, "the stimuli exerted on the brain by the endocrine glands represent the more primitive feelings and sensations constituting the basis of organic sensations as well as of sensations proceeding from external processes."

This is the same idea which was sustained many years before him by the Catalan philosopher Turró, starting from a basis of purely biological speculation. And it is fitting to refer to it here since the general "urge" or "thrust" of growth and reproduction, the "affective impulse," is to be found at the bottom of all vital activities and of all phases of their development, being the direct motive or cause of their expression, which meaning was glimpsed at by Darwin as the primordial phenomenon of existence. Hence is derived the instinct and antecedently the successive strata of psychic existence. The conditioned reflex is a teleological act, just as is the whole field of the mental processes.

Scheler opposed "mind" to "existence," making a distinction between process or performance and the "centre of activities" in the person, distinguishing the mind from the functional centres of existence, the mind being, according to him, the focus of the forces combating the pressure of the organic and everything pertaining to life as such. Such life, the psychophysiological condition of an animal, is the starting-point of behaviour in relation to environment.

Behaviour, then, is adjusted to vital needs so that life may become possible. The mind directs man's deeds. Behaviour governed by the mind may be opposed to his conduct in life, but even when the results of one act or another happen to be opposed one to the other, in each case they nevertheless obey some standard of utility. Divergence would occur from the different nature of these two types of existence.

An animal cannot carry out this peculiar removal to a distance and substantivation which turns an "environment" into a "universe," any more than it can emotively and impulsively convert the definite centres of resistance into an "objective." It is encrusted and immersed in the vital reality corresponding to its organic conditions, without ever objectively learning anything about the latter. The mind shrinks into itself, becoming conscious of itself, converting resistances into "objects" and environment into a "world." Our impulses, our impulsive life, our original vital impulse precedes all cognition.

Awareness of reality is not, therefore, subsequent to but prior to every portrayal of the world. That, however, always implies a movement of desire or of repulsion, a directed activity in all cases with utilitarian ends and purposes, an "intentionality," which not only must be unconscious but which must also in the majority of cases develop properly outside or beneath the consciousness.

"The living body is a system of opportunities seeking fulfilment. The end of every process is its goal. Youthful zest for life and for the world, to fulfil remote wishes and attain distant aims, that is the vital pursuance of an objective," as Jung might still well say. Philosophers, however, generally ascribe intentionality exclusively to the subjective sphere. Things whose object is known are done on purpose. "Between the plane of action," says Bergson, for instance, "and the plane of pure memory, in which the mind preserves in full detail the picture of past existence, a thousand different planes of consciousness are to be seen within the aggregate of lived-through experience. Portrayal of an idea consists in a controlled movement between the image and the action. The psychic mind will become a neutralized consciousness and will therefore become latent, ending up in an individual consciousness.

The advance of living matter consists in a differentiation leading to formation first and to a gradual complication afterwards of a nervous system able to lead excitations through set channels, and to organize actions. The greater the development of the nerve centres the more numerous will be the motive routes among which a single excitation will be able to set in motion various alternative possibilities at choice. Such internal indeterminacy will have, however, to pass through the mesh of necessity. Freedom thrusts her roots deep in necessity and is organized in close intimacy with it."

The life of the consciousness is intentionality and nothing else. Everything done with knowledge and volition is done with a purpose. Brentano thought that the difference between physical and psychic phenomena was the intentionality of the latter. "Intentionality shows itself only in and is characteristic of the acts of consciousness." Again, according to Husserl, intentionality is the essence of consciousness and consciousness is the objective of phenomenology. The subject is not something which may first exist and then subsequently become directed on to the object. Intentionality is not an internal phenomenon. By its agency, consciousness stands in direct contact with its "world," and resolves itself into the being existent for the purpose of subjectivity in the concrete fulness of life.

Heidegger when stating that "heedfulness" is the essence of living, being the essence of existence, identifying heedfulness with existences exhibiting intent, did not however imagine that such intentionality was a mere characteristic of acts of consciousness but that it would explain existence in its entirety, thus agreeing with the conclusions of the biologists previously reviewed. Over the stream of twenty-three centuries Aristotle still has his say on biological standards of judgment. In our times, Thomson says in 1925: "It looks as if Nature were Nature for a purpose." We may bring Aristotle up to date by stating that the essence of every living being is not protoplasm but purpose.

## LIFE AND INTENT

There are nowadays a fair number of biologists who do not admit the validity of the thesis that causality is the giver of shape to vital phenomena, and it is curious to see how physiologists for the most part are mechanists, whilst zoologists, botanists, and naturalists tend more frequently to teleological explanations.

The explanation of this is nevertheless quite simple: the physiologists built up their science by means of experiment, biologists by observation. To experiment is to divide reality into small sections of space and time, and in such limited fields to investigate the circumstances and conditions of phenomena and bring about the repetition or reappearance of such occurrences, all of which is impossible without some causal import. Naturalists, on the contrary, freely observe the course of life on the big scale from an entirely different viewpoint and in another perspective presenting a distinctly different picture.

It is a question now of finding out whether these two views of our world are incompatible; whether determinism and teleology are mutually exclusive or not. Thus it was frequently assumed, though doubtless quite fittingly, that there is some synthetical theory embodying both these contrary opinions.

Determinism assumes order and mutual dependence of events — nothing takes place in the world unless conditions are right. That is the last word! Intent or purpose is not possible without some order. How could anything happen unless there were an intermediate line of chances along which one must run until the end is attained? This, in the long run, is the same road covered in the same direction, onwards, along certain stages, whose destination in the first case (causality) is unknown to us and of which in the second case (intentionality) the goal is but imperfectly conjectured.

Who can doubt at the present time that in order to know the development of an event, it is necessary to investigate its determinant conditions? Not even the most hidebound teleologist would dare deny this! And conversely, who can dispute that events in life are developed consciously or unconsciously in such a way as if they tended to a particular end? An elementary observation of reality suffices to prove in all cases the exact functional fitness of certain vital manifestations.

Everything appears to pursue one definite object: to exist! Every live creature seems animated by a single purpose: to preserve itself! Life appears effectively to show one purpose in the end: to be perpetuated! The living world runs along certain tracks leading it to a terminus, whilst the engine pulling it on its way is intentionality: a hidden and transcendental purpose whose final designs escape our grasp, and therefore, when the mind of man discovers itself, reflecting upon itself, imbibing knowledge of the universe in its own world of existence, at that magic moment when the mind becomes conscious, universal purpose is reflected in the life of the individual, and like some diaphanous aura from it the voluntary deed is seen as in a glass darkly — "done because it is known to what such act must lead." It is not nature which intentionally works in a similar way to man.

It is, on the contrary, man who works with purpose because the events of the Universe are developed on purpose — and independently from man, whose existence is not really essential. Man, a sheer accident, is drawn by an insuperable and transcendental stream towards his fate!

## THE MENTAL FACULTIES

Of the mental faculties previously mentioned, some creatures possess all, some only a few of them, and others one alone. These faculties to which we have referred are those relating to growth and sustenance, perception, movement and thought. Plants possess only the former, growth and sustenance, whilst other living creatures possess perception as well.

Any creature endowed with perception must also have needs, for need implies desire, passion and volition. Now all creatures possess at least one sense, that of touch, and anything that can feel may experience pleasure and pain, and where these are to be found there must be desire, that is to say a craving for something pleasurable. Moreover, all creatures experience the need for food, for the sense of touch is a searching for satisfaction.

All living beings are satisfied by food either solid or liquid, and warm or cold, which qualities are discovered by the sense of touch. Other qualities are found out only indirectly by contact. Sounds, colours, or smells in no way satisfy hunger; whilst as regards taste that comes within the realm of the tangible. Hunger and thirst are types of need, hunger for what is solid or warm, thirst for what is liquid or cooling, whilst taste is an adjunct to both.

These points must be decided later, but for the time being, it may suffice to say that creatures possessing the sense of touch are also subject to needs. As to the imagination, we cannot clearly define it and must discuss it separately. Some types of creatures possess, moreover, the power of movement; others, such as Man or beings like him or perhaps even superior to him, possess also mind and the power of thought.

Obviously, therefore, we may define the idea of mind only in a like manner to shape, since as there is no shape which is not dependent on the triangle and its compounds, neither can there be mind other than in the forms above described. As a matter of fact, with regard to shape we may frame a broad notion applicable to all shapes, though not specifically identical with any particular one. The same applies to the forms of the mind. Thus it is unreasonable to expect a blanket definition either here or elsewhere to cover everything in existence, or on the contrary, in the lack of such overall definition, to try and define each particular species.

Some similarity does in fact subsist between the principles governing shape and mind, since in each case, the outcome implies the origin; thus, for instance, a square implicitly contains triangles, and perception implies needs requiring satisfaction.

## THE FACULTY OF GROWTH

First of all, therefore, we have to consider the questions of sustenance and development, since the growing mind is an element of all other forms of the mind and is indeed the most embracing faculty by virtue whereof any entity may be said to possess life, evidencing itself by the consumption of food and the process of reproduction. The most natural act of any undamaged and fully developed living creature, not spontaneously generated, is the production of other beings like itself; thus an animal will reproduce itself in an animal, and a plant in a plant, and in this way as far as it is possible they stand their share in the everlasting and supernatural groping towards this end by the very essence of their nature.

These words, "this end," are however double in meaning, one interpretation being "the end to be attained," and the other "the end attainable by reason of the act." Now no one living creature can share without interruption in the everlasting and supernatural, for nothing mortal can remain for ever exactly identical with itself throughout the whole course of its existence, whence each plays its part as far as is possible to it with varying proficiency according to its powers. Thus any particular being continues its course not identically in its original form but in some modified form still similar thereto, not in its own original single body, but in other bodies produced from it.

The mind is the origin and mainspring of the living creature, which terms are susceptible to many interpretations. The mind is however the origin for three definite reasons, namely:

- It is the source of movement,
- It' is "this end,"
- It is the ruling factor of the living creature.

Obviously it is the ruling factor, for in everything the ruling factor is its reason for being, and in living beings that reason is their existence its very self, and the origin and mainspring of such existence is the mind. Moreover, such reality as there is in the latent flows from the conception thereof. Clearly the mind is also "this end," since nature acts in the same way to a purpose in view, and that purpose in view is its final objective or end. In living creatures the mind is of this nature, for all natural bodies are agents of the mind, and whether they be plants or animals it is the mind which governs their existence. It is well to remember here the two senses of "this end," namely:

- "The end to be attained"
- "The end attainable by reason of the act."

The mind is also the mainspring of movement, though this faculty is not intrinsic in every living thing. Changes in nature and size are also due to the mind, whilst perception, too, may be said to be a type of change which cannot be effected unless a mind is present.

This also applies to growth and disintegration, since nothing grows nor decays in the physical sense unless it takes in sustenance, and nothing acquires sustenance unless it be subject to some facet of a mind. Empedocles, however, is not correct is stating in this connection that growth reaches downwards in plants where they take root because the earth naturally has this bent, and upwards because of the similar natural tendency of fire to strive upwards.

## RENÉ DESCARTES

Just as a profusion of laws often furnishes some pretext for wrongdoing, so that a State is the better regulated when having only a few they are the more strictly respected, so also did I think that, instead of that huge number of precepts whereof logic is composed, the four following would be quite enough, provided that I took the firm and steady resolve never once to fail in keeping to them.

The first was never to take anything for granted that I did not know by evidence to be true; that is, scrupulously to avoid headlong thoughtlessness and prejudice, and not to embrace in my judgment anything more than what was so clearly and distinctly presented to my mind, that I should have not the slightest reason to be in doubt about it.

The second, to break down every problem that I considered into as many separate simple elements as might be possible and requisite, the better to solve them. The third, to organize my thoughts, starting with the simplest objects most easy of discernment, in order to rise gradually step by step to an understanding of the most complex, assuming some arrangement even amongst those which did not naturally fall into sequence one after the other. And finally, in every case, to make such detailed classifications and such general summaries as to assure myself that nothing relevant had been omitted.

Those long series of quite easy and simple axioms customarily employed by mathematicians in order to arrive at their most difficult proofs, had given me some reason to suppose that everything coming within mankind's range of cognition was similarly interwoven, and if only we refrain from accepting as true anything about which there may be the slightest doubt, and keep also to the necessary sequence in deducing each step from its predecessors, there cannot be any conclusion so remote that it will not eventually be reached nor anything so recondite that it cannot be revealed.

I did not have much difficulty therefore in deciding how I ought to commence, since I already knew that it was by way of the simplest and most easily discernible facts; and, considering that among all those who have formerly sought for truth in the sciences, only the mathematicians themselves had been the ones to find some proofs, i.e. a few clear and obvious reasons, I had not the slightest doubt that what I ought to do was to start with the same things that they had put under examination, although I expected no other benefit than that this would accustom my mind to feed upon truths alone and never allow itself to be fobbed off with false premises.

## CLAUDE BERNARD

A fact means nothing by itself alone. Its value lies solely in the idea intrinsically borne within it or in the evidence it may serve to form. We have said elsewhere, that when a fresh fact is labelled as a "discovery," the discovery is not the fact itself but the new idea which that fact leads to. Similarly, when a fact demonstrates something, the demonstration is not the fact itself, but the logical relation which it helps to establish between the effect and its cause.

The truths of mathematics are absolute and belong to the mind, because the ideal or non-substantial conditions of their existence are also mental and perceived by us in an absolute sense. Experimental truths, on the other hand, are relative and not mental, because the real substantial conditions of their existence are also non-mental and cannot be known to us except in their relation to other things in the state that science stands at present.

But though the experimental truths which serve as the basis for our reasoning be lavishly enwrapped in the complex reality of natural phenomena, from which it follows that they cannot appear to be anything more than isolated fragments, that does not mean that such truths of experiment are not founded upon principles which are as absolute as those of mathematics, and that like these they are accessible to the mind and to logic.

The absolute principle of experimental science is a rigorous determinism of the conditions of the subject, so that, given a natural occurrence, no variation whatsoever shall occur in it without changes having been simultaneously produced in the conditions in which such occurrence comes to pass. It shall not be valid to admit any variation observed not answering to a particular cause. Having now the antecedent a priori assurance that each variation is determined by one particular cause, it follows that rigorous and logical relationships between cause and effect may exist in the evolution of the occurrence.

Experience teaches us the form of the phenomena, but the relation of an occurrence to a determinate cause is necessary and independent of experience — it is mathematical and absolute. Thus we reach the conclusion that the criterion of experimental science is at rock bottom essentially identical to that of the mathematical sciences, since in each case the principle becomes evident through a similar necessary and absolute relation.

What apparently distinguishes the one case from the other is the complexity of the phenomena in the natural sciences as against the inexorability of mathematical reasoning. Natural phenomena are to be found involved in and clouded by other numerous, complex and varied factors which accompany them, and render difficult an inquiry into the causal conditions of the phenomenon under consideration. By trial and error we analyse and disassociate the various phenomena, for the purpose of bringing them down to ever more simple terms and relations, and we hope thus to arrive at the real form of scientific truth, or, in other words, discover the law which allows us to interpret changes in the phenomena.

Such analysis by trial and error, or what we prefer to call experimental analysis, is the only means of finding out truth in the natural sciences, and the absolute determinism of its occurrences of which we have the a priori knowledge has to be the criterion directing and sustaining us in our investigations. We find ourselves still very far from an absolute truth, however much the effort we may have made, and it is probable, particularly in the biological sciences, that such truth may never be reached.

This must not discourage us by any manner of means, since, despite everything, each step forward in science represents an approximation and because; through experimental research, fresh and closer relationships are becoming established one by one among the natural phenomena and facts, which relationships, partial and relative though they be, allow us to expand our mastery over nature on an ever increasing scale.

From the foregoing, it could be deduced that if an experiment gives us an event apparently at contradictory variance at any point with what is already known, and which does not tally logically with the determined conditions of existence, reason would have to "repudiate the fact" as being non-scientific. It would, however, be necessary first to seek by experiment to find out if some cause of error might perhaps have disturbed or upset our factual observations. Most assuredly, there is bound to be some mistake or insufficiency in the observation. To admit the appearance of an occurrence without valid reason, and without determinative conditions, is the same as denying science itself.

Confronted with a case of this nature, the scientist cannot hesitate for an instant. He must give science his full confidence and throw doubt on the other hand upon his qualifications as an investigator. He will have to perfect his methods, and use every effort to find the reason for any apparent contradiction, but it must never enter his head to deny the determinism of the phenomena, since it is precisely the feeling and conviction of such determinism which characterizes the true scientist.

Facts are frequently adduced in the practice of medicine, which have been faultily observed and are therefore apparently indeterminate, constituting by the same token veritable stumbling blocks in the way of science. It is said: "That is a fact — it must be admitted." Rational science based on a stringent determinism can under no circumstances evade a properly observed and exact fact.

However, for that very reason, it must never be cluttered up with facts gathered inaccurately without discernment and lacking therefore any real meaning. Science rejects the indeterminate, and when, for instance, in medicine, somebody attempts to found opinions upon the "magic touch," the "sixth sense," inspiration, or a more or less vague intuition, he will fall unpardonably into a fanciful, arbitrary, anti-scientific and ever dangerous type of practice, since he thus delivers the patient to the whims and imaginings of a fanatical ignoramus. True science knows how to doubt and to refrain from ignorance.

If the physicist and the physiologist are distinguished by the fact that one studies things which occur in inanimate matter whilst the other is concerned with those which happen in living matter, there is no difference between them, on the other hand, as to the object each has in view. Both indeed have the common objective of investigating the proximate cause of events. Now what we call the proximate cause of an occurrence is the material or physical condition of the thing in being or the manifestation thereof.

The object of the experimental method or the end of any scientific investigation is exactly the same when studying events in inert objects as when dealing with live creatures, and consists in recognizing the relations uniting an event with its proximate cause, in determining the conditions necessary for the event to be produced.

When the experimenter has succeeded in learning the conditions surrounding an event, he becomes, so to speak, master of such event. He can cause it, prevent it, help it, hinder it, predict its course and its forms of appearance. The experimenter will then have gained his object and in this manner, by means of science, have extended his sway over a natural occurrence.

We shall therefore define physiology as "that science the object whereof is the study of events in living creatures and the 'determination' of material conditions for the appearance of such events."

Only by the analytical or experimental method shall we be able to arrive at the determination of such conditions, in either lifeless or in living things, for in all sciences logical reasoning remains the same and consequently the idea of experiment is basically and in all fields essentially identical.

## AUGUSTINE OF HIPPO

In what way, however, can the as yet non-existent future shrink or be consumed, and how can the now no longer existent past grow, unless three processes take place in the mind regarding them? For, the functions of the mind are to consider, to experience, and to remember, and what it considers and experiences passes into what becomes memory.

Can it therefore be that what is to happen in the future is not yet in existence, even though there may already be in the mind an anticipation of things to come? Can one deny also that the past is now no longer existent, even though there remain in the mind a memory of things past? Furthermore, can it be denied that the present lacks duration, since it passes instantaneously into the past?

Moreover, experience lingers, so that what formerly was to come to pass, becomes something which has happened. It is not, therefore, the as yet nonexistent future itself but the anticipation of a distant future which is a "long time," any more than it is the now no longer existing past which is long, but the back-reaching memory thereof.

Say I have a mind to recite a piece I have learnt. Before starting, my consideration extends over the whole of it, but when I have begun, every part of it as I bring it out drops away successively into the past and enters the domain of my memory, and the existence of this action of mine becomes divided into a memory of what I have said and a reflection of what I still have to repeat.

My experience remains, however, in the present, through which what is future travels into the past. The further I go with my repetition, the shorter becomes my foreknowledge and the longer becomes my recollection, until the moment when the whole of my reflection is consumed and the completed action passed into memory. What is true of the piece as a whole is so for every single part of it, and for every syllable, and similarly for any longer action of which the piece in question may be but a small portion.

This applies also to the whole course of man's life, which is made up of all his separate actions, and to the whole history of mankind through the ages, the parts whereof are the complete lives of all men put together.

## MAX SCHELER

We have climbed a bit into the sublime. Let us return to the problem of human nature lying nearer to our experience. For the present age, the classical theory of mankind found its most effective form in Descartes' doctrine, which doctrine we have in fact only quite recently begun to shake off absolutely and entirely. By his classification of all matter into either "conceptual" or "protensive," Descartes introduced into Occidental thought a whole host of most weighty errors regarding human nature.

By such division of the entire universe, even he himself had to adopt the absurdity of denying mental qualities to all plants and animals, and interpret the "semblance" of mind in animals and plants — which in all ages prior to him had been taken for granted — as an anthropopathic "groping" of our vital feelings into the external images of organic nature, as well as to declare purely "mechanical" everything which is not human consciousness and thought.

The result of all this was not only to overestimate in a most absurd manner the particular status of man, emancipating him from the very arms of Mother Nature, but also by a stroke of the pen to do away with the basic classification of life and its primary manifestations as far as this world was concerned.

The universe, according to Descartes, is composed consequently of nothing more than points of contemplation and a powerful mechanism which must be explored purely from the mathematical standpoint. One thing only in this doctrine is of any real value, and that is the novel independence and sovereignty of the mind, and recognition of its superiority over and above everything organic and merely animate. All the rest is utter nonsense.

Nowadays we are entitled to say that the problem of body and soul that has occupied mankind for so many centuries has, so far as we are concerned, lost it metaphysical significance. The philosophers, doctors and scientists concerned with this question, concur more and more harmoniously on a basic first principle. That there is no such thing as a locally determined incorporeal matter — as postulated by Descartes — is therefore quite self-evident, since neither in the brainbox nor anywhere else in the human body is there a central point where all the sensitive nerve fibres can meet together and all nervous processes combine.

Another radical fault in the Cartesian theory is his premise that the mind consists only in "consciousness" and resides exclusively in the cerebral cortex. Exhaustive research on the part of psychiatrists has proved to us that the decisive mental functions constituting the foundation of the human "character," particularly everything belonging to instinctive activity and the passions, which we have indeed acknowledged as the basic and primal form of the mind, does not in general have its parallel physiological processes in the upper cavity of the skull but in the neighbourhood of the brain-stem, partly in the grey matter of the central cavity of the third ventricle, partly in the thalamus, which acts as a central clearing-house between sensations and impulses.

Moreover the ductless gland system (thyroid, genital, pituitary, adrenal glands, and so forth) whose operation determines human passions and emotive life, as well as abnormalities of height and thicksetness, giantism and dwarfism, and probably also racial characteristics, has been shown to be the exclusive seat of regulation between the whole organism including its morphological development and that small appendage to mental activity which we call the waking consciousness.

It is nowadays by no means the brain by itself but the whole body which once more becomes the field of physiological parallelism to the mental processes. There can no longer be any serious support for such an external linkage between the mental substance and the corporeal, as assumed by Descartes.

It is one and the same life that possesses mental form in its "subjective being" and bodily shape in its objective relationships. We may not adduce against this, that the "ego" is therefore simple and unique, whilst the body on the other hand is a complex "community of cells." Present-day physiology has completely destroyed the idea of a cellular community, just as it has also broken with the basic idea that the workings of the nervous system combine merely by summation of parts, that is to say not in a ready-made integrated whole, and at times are rigorously determined in location and morphologically in their points of departure.

If, indeed, like Descartes we deem the physical organism to be a type of machine, and this in the strict sense of the old mechanist philosophy of the age of Galileo and Newton but now quite ousted by physical and chemical theory and cast on the scrap heap; if in addition, like Descartes and his followers, we disregard the independence and the certainly established predominance of the combined emotive instinctive life over all conscious imagery.

If we also restrict all mental activity to the waking consciousness, discounting the enormous cleavages between entire co-ordinated groups of functions of mental occurrence and the conscious ego.

If, furthermore, we deny any repression of emotions and disregard any conceivable recalling to mind of whole phases of life, such as the well-known disassociation manifestations of the conscious ego itself in schizophrenia — then, to be sure, we arrive at the following false antithesis: whilst on the one hand there is unity and simplicity of a primitive type, on the other we have a multiplicity of bodily parts combined in a secondary manner as well as a plurality of processes primarily founded thereon. Such a picture of the mind is quite as erroneous as the notion of the spontaneous generation of life presented by a more ancient philosophy.

In resolute opposition to all such theories, we may therefore say that the physiological and mental processes of life are rigorously identical from the ontological point of view, as indeed Kant had already surmised. They differ only in appearance, but are apparently rigorously identical in their laws of structure and in the rhythm of their progress through time: both processes are non-mechanical, the physiological just as much as the mental; both are directed to the final aim, teleoclinic and based upon totality. The physiological processes are all the more of this nature the lower (not necessarily the higher) the position of the parts of the nervous system in which they run their course; the mental processes likewise are more complete and directed to their target, the more primitive they happen to be.

These two processes are merely two aspects of the single process of existence, which is higher than purely mechanist because of its own configuration and the interplay of its functions. What we thus call "physiological" and "mental" are only two facets from which to view one and the same vital process. There is a "biology from within" and a "biology from without."

The latter starts with a perception from the structural form of the organism and progresses to the vital processes properly so-called, but it must never be forgotten that every living form from the ultimate differentiable cell elements through the gamut of cells, tissues, and organs up to the organism as a whole, is at every turn and instant dynamically sustained and renewed by such vital process, and that in development it is the "formative functions," sharply distinct from the operational activities of the organs, which produce the static forms of organic matter thanks only to the assistance of the chemico-physical "state of affairs."

The late Heidelberg anatomical specialist Braus and E. Tschermack on physiological grounds, justly took these ideas as the focus for their joint research. It may be said that this same concept nowadays permeates all sciences concerned with this famous problem. The old "psychomechanical parallelism" now verily belongs as much to the scrapheap as the "theory of reciprocal action" as resuscitated by Lotze, or the mediaeval scholastic hypothesis of the soul as a corporeal shape.

The physiological "function" is in its basic conception an independent regulated succession of mobile form, a figure moving through time and from its very beginning in no way rigidly tied down to a particular location but able with much greater freedom to choose and indeed build its field of operation in the available cellular substrates. A cumulative organic reaction of a determinate and rigorous nature does not exist either among those physiological functions which possess no sort of correlation in the conscious; in fact they do not exist, as has recently been proved, even for so simple a reflex as the knee jerk.

Apparently, moreover, the physiological behaviour of the organism is quite as "clever" as the conscious processes, and by the same token these can often be just as "silly" as the organic processes.

In my opinion, research should straightway be set the methodical aim of testing in the widest measure how far the same behaviour methods of the organism can be induced and varied in the first place by physico-chemical stimuli from without, and again through mental stimulation — suggestion, hypnosis, all kinds of psychotherapy, changes in social environment — on which illnesses are dependent much more than is suspected.

Let us, therefore, definitely guard ourselves against setting too high a false value on exclusively "psychological" interpretations. As we know from experience, a gastric ulcer may be conditioned and governed just as much through the mind as through any particular physico-chemical process; and not only nervous illnesses but also organic indispositions each have entirely definite mental correlatives.

Also we can quantitatively evaluate both types of our influence on the intrinsic unitary vital process, to wit through the channel of the consciousness and through that of external bodily stimulation, in such a manner as to husband the one stimulus to the same extent as we squander the others. All these are merely different methods of approach, which we retain in our experience and conduct for one and the same ontologically centralized vital process.

Even the highest mental functions, such as so-called "connected thought," do not repudiate a strong physiological parallelism. And finally, in accordance with our theory, the intellectual acts must also possess a physiological and parallel linkage, since they derive their entire active energy from the living sphere of impulse, and cannot without some sort of "energy" manifest themselves for the purposes of our experiments even to the individual experimenter himself.

## INTENTIONALITY: VOLITION OR PURPOSE

Since the world of reality depends on cognition, and is established on that as a foundation, then in order to discover the ultimate basis to which we strive, it becomes essential to determine the intrinsic structure of consciousness in all its aspects and limits.

For such purpose, it is vital to commence with the concept of intentionality or design, which, in line with Brentano's terminology, Husserl ascribes to the consciousness as its ruling characteristic.

As we saw in the brief survey previously outlined, this fact of design is not intention in the sense of finite purpose or practical directivity of existence. It means simply, that every perception is perception of an object perceived, every thought an idea regarding an object thought about, every wish a want regarding an object desired, every affection a feeling for an object held in esteem, every notion and idea of an object is a notion and idea of an object conceived and meditated over.

From all the foregoing, we gather that Husserl's concept is something decidedly distinct from Brentano's, which is its immediate antecedent, and from the Aristotelean and Scholastic, which for the first time defines consciousness as a mental intention. For Scholastic philosophy and even for Brentano, with certain profound differences which it is needless to enter into at this point, the relation of intent comes to pass between two previously formed and antecedent points of reference.

A connection becomes established between subject and object, by means of which reality is presented to the subject against which it sets itself as something externally opposed or objective in the etymological sense of the word, and such objective amounts to an understanding and possession of the mind. Intention is a relation between a psychological act and its content, between a real subject and the content of a real objectivity.

It is thus a property of the consciousness characterizing it and even defining it, but only on the assumption of its existence and the existence of the object to which it refers. Among all the substances, in the physiological meaning of the term, one only — consciousness or cognition — possesses the faculty of placing itself in a position of direct contiguity with the remainder of such substances which latter are, by definition, transcendental.

Husserl's intentionality has a much deeper function. A substance, anything whatever from the world of things or what we call reality, has its own structure and consistency. The whole existence of things is based upon their mutual and reciprocal exclusion and they enter into relation only when certain private limits are overstepped. Relationships between them are the consequence and are to a certain degree external to them.

Cognition has no reality outside the relationship it sets up; its whole existence is exhausted in the mere act of setting itself up in relation to something other than itself. Intentionality is not a relation added to the previous existence of subject and object; it is the primary and governing fact of cognition, wherein and whereby subjectivity and objectivity are produced and brought face to face. The existence of consciousness subsists in intent. Thus intentionality is not a property of cognition, a characteristic following on its existence. It is something integral in its very existence itself. Furthermore, it is not directed towards something "outside" cognition — as in the realist theory — nor towards some concrete "content" of the conscious — as in the idealist theory.

The object on which it is trained is not a "real" element of the conscious, nor a "reality" beyond it, but an ideal objectivity. By the act of intent animating it, existence passes beyond itself, becomes transcendental, and acquires a specific form of objectivity.

Intentionality does not signify the road whereby a subject previously existent alongside the object confronts the latter and makes contact with it. It is the very essence and definition of subjectivity.

The subject is not something which first exists and consequently thereafter becomes directed on the object. Intentionality is not an "internal" phenomenon. By its means, cognition makes direct contact with the "world," and the world becomes reduced to an entity existing for subjectivity, in the concrete fullness of life. The life of the consciousness is life in the presence of "being."

## Chapter 14

# Organic Functions

### TROPISM AND TACTISM

External agents of various types, in addition to their own internal impulses, bring about in living organisms a variety of acts which, in general, promote the survival of the individual or species. Such reactions are in their most diverse applications favourable to the development of life.

This is the case with tropisms in particular among very many possible examples. Plants shoot forth their foliage into the air — negative geotropism — and thrust their roots into the ground — positive geotropism — under the influence of gravity. The climbing plants, through the action of thigmotaxis, wind around tree-trunks or other solid objects which serve them as support as they stretch forwards. Another frequent occurrence is phototropism or bending towards the light by plants, and we could mention countless other instances.

Such automatic positioning comes about through the agency of certain factors particularly sensitive to definite impulses, located in the cytoplasmic extensions of the epidermal cells of climbing species, or in the hairs or tactile protuberances of Drosera (Sundew), Dionaea (Venus's Fly-trap), Mimosa (Sensitive Plant), and others among the coleoptiles in general. Material factors of metabolic or hormonic nature may also come into play.

In even the most primitive independent forms in the animal world, and even among plants as well, motor reactions to various stimuli are observable, acts of taxis aroused by light, heat, osmotic pressure, electric energy of various kinds, or chemical substances. Movements are steered in one direction or another, approaching or fleeing from the stimulus, according to the nature of the latter and the condition of the organism itself. There is a huge variety of tactism, and study thereof has permitted the discovery that movements, even in the rudimentary forms of life, evolve in accordance with a standard of usefulness, as if answering a particular purpose, namely sheer survival.

Phenomena develop in the best possible manner, which caused Binet and Jennings to say that, in the protozoa and infusoria, movements come about as if governed by intelligence and will. The most effective stimulant is made up of nutritional matter. Jennings and Wallengreen showed that in paramecia, slipperanimalculae, activity increases and sensitivity to excitants, particularly chemical excitants, becomes livelier when the creatures are in a state of hunger, lacking food. Nutritional chemotaxis in amoebae is common knowledge.

The behaviour of protozoa answers to the conditions in which the life of the creature develops. Mast investigated, for instance, the influence of light on Euglena. Jennings published his famous work on the reactions of the lower animals, studying especially the case of controlled movements in slipper-animalculae. These ciliates evince movements in defence, of trying to get away from an unfavourable stimulus, flight, avoiding reaction, or taxis, or, on the contrary, approaching the point of origin of a favourable stimulus, i.e. positive taction. Motions may be organized in such a manner that the creature weaves tentatively around the object which is the cause of the excitation process, in accordance with a system of trial and error, terminating in the adjustment of the living organism to the circumstances of the environment. In this way, by suitable action on the part of the creature, environment contributes to its survival.

The vital, motor, responses of such rudimentary, unicellular organisms, before the incidence of external conditions, are fulfilled as if there were some governing purpose, some intelligence or some will set in action by such intelligence. Facts of this nature have been repeatedly observed in various species of protozoa and protophytes. When the structure of the living organism becomes complex, which entails intrinsically a parallel increase of complexity in reactions, a like congruence obtains. Behaviour becomes more marked in creatures endowed with movement.

In the spongiaria obvious reactions to definite stimuli are seen, consisting in occlusion of the pores and osculum and contractions of the cells which might obliterate the small cavities of the canalicular system. These reactions may be local or may even extend throughout the sponge, and are followed by a slow return to the original state of rest. The movements are due to specialized contractile cells, without intervention of nerve or sensory cells, which such creatures do not possess.

These contractile cells are "independent effectors" responding to exterior stimuli through the irritability of their protoplasm and answering similarly to the chemical conditions of the cells which in combination make up the individual. This transmission of impulses from cell to cell, which is still not yet a nerve transmission, is called "neuroid transmission," promoted by excitations proceeding from the environment and through changes in state of the actual cell plasma of the creature. The movements, in this case also, respond mainly to metabolic conditions.

## THE NERVOUS SYSTEM

In hydrozoa, among the coelenterata, nerve elements are to be found, as well as differentiated effectors. Setting aside the cnidoblasts, extremely simple in function, we must count as instruments of motion in the hydra, the muscular arrangements, both round and long, in the body, hypostome, base and tentacles. The contraction of these muscular fibres comes under the control of nerve elements.

The ectoderm contains a nerve-net in which at least four classes of sensitive cells are to be found. There are further nerve cells associating among themselves and sending out extensions to the muscular fibres. In the endoderm there reside only a small number of isolated nerve cells and a few others which appear to be sensitive. The ectodermal nerve elements are abundantly in evidence in the hypostome around the stoma, in the base of the tentacles and in the foot. They form a continuous syncytium, the extensions continuing from cell to cell, without break, in contrast to what happens in the case of the nerve centres of more highly differentiated creatures, in which each neurone — as revealed by Cajal -maintains its cytological unity.

By means of the nervous system, not yet centralized, the effects of environmental stimuli of distinct type are developed and, furthermore, also by the state, mainly metabolic, of the individual itself. The hydra reacts to contacts, to light, heat and electrical stimuli or excitations and to various substances in solution. Such stimuli combine their effects, and through them the creature behaves in accordance with the state in which it finds itself.

In green hydra a very marked phototropism is observable, and individuals congregate in the better illuminated spots, wending their way slowly as a rule towards likely foodstuffs and if these latter happen to be microbes or small creatures they first paralyse them by means of nematocysts protruding through the cnidoblasts. Immediately these creatures, like other prey, are carried to the stoma by the tentacles and devoured through movements of the hypostome.

In the digestive cavity peristaltic movements take place and the corresponding digestive juices are secreted. Digestion is completed in a way comparable to what takes place in the stomach and gut of a carnivorous animal. Reactional complexes determined by the behaviour of the coelenterata are better exhibited in more differentiated and larger species, such as certain medusae, the Gonionemus.

These respond moderately to simple contact and chemical agents such as acids, but the reactions are more intense when such stimuli are added to each other, and particularly when the stimulus approaches the natural excitor, for instance, contact with a pipette containing meat extract, if a small amount of the food be allowed to reach the creature. A flexion is then produced in the tentacles towards the stoma, whilst the hypostome turns towards the contacted tentacle and the nematocysts are touched off. This combination of experimental stimuli resembles those experienced by the medusa under natural conditions in the presence of a small crustacean, for instance, which may serve it for food, and thus the response to stimulus will be identical in each case.

## EXTERNAL AND INTERNAL STIMULI

Activity in an animal does not come about solely from some outside stimulus. Gonionemus alters its mode of swimming when it gets hungry, as mentioned by Wagner. Loeb described the inversion of the oesophageal ciliary motion in the actinia Metridium (which normally takes place from inside towards the outside) when the creature comes in contact with feeding matter. Such inversion, from outside inwards, to facilitate food ingestion, does not occur in individuals already sated. Allabach showed, also, that in these creatures hunger increases sensitivity just as among the protozoa.

The tenacles of Stoichactus, according to Jennings, give no signs of motions anticipatory of food in well-fed specimens. Allabach noted the same in Metridium. Gee in 1913 arrested movements of the tentacles in certain Cribrinos by squirting food juices into the digestive cavity. Also, Pearl in 1903 observed variations in motion of planarians according to their state of satiety. Similarly, we could quote countless other instances of a like nature.

From the simplest structures upwards, animals react to life in accordance with their environmental conditions and the conditions obtaining between their own internal organs, particularly those handling food. This becomes all the more evident the more advanced the degree of morphological differentiation and functional specialization.

Taking worms as our point of departure, the nervous system makes an important leap forward whilst response to outside stimuli, as well as to internal ones, comes about by means of nerve elements answering to changes outside the cells affecting them and likewise to the state of the cells themselves. Thus every action of the nerves depends on excitation and the conditions in which the corresponding structures of neural make-up are placed. These last conditions are in the main of a metabolic and chemical nature.

Thus it is that throughout the whole scale of the animal kingdom, the same thing is repeated and the behaviour of every creature shows itself equally fit as regards responses to external excitants as to stimuli of internal origin; the behaviour of any creature with respect to its environment and under the force of functional or metabolic conditions in the organs of the creature itself is evidence of this.

## FITNESS OF ORGANIC FUNCTIONS

For many years now we have laid stress on the adaptability of the organic functions of digestion, circulation, respiration and so on, to the requirements of the individual. "The iron physiological bond linking together the secretory operation of the whole digestive tract, bringing about the secretion of digestive juices," we wrote, as far back as 1908, "becomes the most important factor for stimulating what comes after it, and this together with the natural effect of food on the amount and properties of the various digestive secretions — or to put it in a single word, 'adaptation' of the secretory machinery to the variable digestive circumstances and requirements — are to be seen equally in regard to the dynamics of the various gastro-intestinal departments.

The sensitivity of each section of the digestion and the consequent motor responses are adjusted to functional purposes so as to ensure fulfilment of their physiological task in the best possible manner. Motor coordination obeys the laws of functional economy, which likewise govern vegetative activities as well as conscious acts, so that each class of reaction may be exactly superimposed one on the other."

We have repeatedly stated that all functions — especially the vegetative — develop as if governed by some intelligent power. In these involuntary acts the same thing is to be seen as we have just said takes place in animals, in all their forms, right from the very simplest in. which there can be not the slightest glimmer of intelligence. Life in the internal organs is, in this respect, superposable on the life of the lower animals. Organic activity evolves quite unconsciously and automatically, but, like other involuntary actions, it corresponds perfectly to some physiological purpose. This explains why one's thoughts always turn towards the idea of the intervention of archei, supposedly projected on to vegetative activity. Everything, in vital manifestations, is accomplished in the best way possible, and if that were not so, life would become extinct, or indeed might never have become established. The living organism — from all departments of life — does its utmost to survive, to be preserved as an individual and to preserve its species.

This is most evidenced in conscious acts. An act is performed for a particular purpose, "with intent." And yet even the higher innervations themselves which, in normal circumstances, become manifest in states of consciousness, are founded upon an unconscious pyschophysiological substructure. It can be understood, therefore, that many of such acts may, in particular cases, be developed in an entirely automatic manner outside the consciousness. Higher nervous activity is no exception to the general harmony of existence, which comprises the subjection of all vital activities to actual needs. This is the reason why the objective study of psychology is of such tremendous interest.

## OBJECTIVE PSYCHOLOGY

Psychology, the science of neuro-psychic existence, is not concerned only with manifestations of the consciousness. Bergson himself admits, for instance, that psychology may never be referred exclusively to acts of consciousness since the domain of the extra-conscious is enormously wider than that of the conscious.

The arrival of a nervous current at the nerve centres or even a change of state in such centres, mainly through metabolic conditions, will promote a reaction. In the first instance it will be a reflex, or act as a spur on the centres. These evolve the more frequently in the outer reaches of the consciousness. The nervous function here again shows itself to be according to plan, directed towards the accomplishment of its particular purpose. In every neural organization, that is to say between a greater or less number of neurones, according to the nature and rank of the acts, relationships are established between the functions from which arises the due fulfilment of the operation.

In the same way that sensations are perceived and integrated by means of images, neural organizations prior to the conscious act, so also are physiological effects of a motor or other type produced along the corresponding nexi, in the form of motor images which may be identified with sensory images, and thus reflexes and automatic motions develop as if they were voluntary acts. There are various categories of reflex, from the simplest which do not cross the threshold of the physiological, by way of the instinctive reflexes as classified by Bekhterev and the mimic and symbolic reflexes due to nervous concentration, until in the end we reach the most complex types of personal reflex.

Results of the most diverse nature, obtained by physiological investigation of the functions of the nervous system, serve to prove time and again, that in the higher, as well as the lower, nervous activities, as in so many other existential phenomena, the function bows to necessity and obeys its fate, behaving as if for the precise purpose of attaining a particular end, just in the same way as a known and voluntary act may be performed by action of the conscious mind.

This subjection of the whole nervous life to necessity, including therein the self-same voluntary acts of which we have just spoken, arises from the most intimate recesses of existence. Not only, as we have already said, does it come from the excitation of receptors, but is due also to the physiological state of the neurones. Let us not overlook the fact that a neurone is a cell, subject like any other cell to the conditions which make its existence possible. The first condition of its activity is its metabolic state, the way in which nutritional changes from hunger to repletion follow one another, the effects of various substances, metabolites or hormones. An excellent example is to be found in the effect of the respiratory gases according to their respective concentrations on the breathing centres.

Variations in the chemical composition of the blood which flows to irrigate the centres, and the neurones themselves in such centres, will be factors to be reckoned with in the various functional acts. And we may repeat again that conscious acts also are included.

To these local influences, derived from the state of the cells, we may add the impulses arriving at such centres by way of the nerves, which routes tally not solely with the somatic nervous system, but also with the vegetative. In this last case also, the currents spring from internal excitations originated in the organism itself. The interposition of internal material or nerve factors is now universally admitted, and the credit for having brought this fact into prominence is due in an extraordinary degree to Turró.

Knowledge and will are not indispensable factors in purpose. Intentionality precedes them by a long stretch. For, everything in life, and for all we know in the physical world as well, is accomplished as if governed by the appearance and course of the phenomena, knowledge, will and intent.

## INNERVATIONS AND CONSCIOUSNESS

However, things being thus, it is also certain that the highest innervations are accompanied by consciousness. A subjective world evolves side by side with the objective. An afferent current causes a reflex, but at the same time it also arouses a sensation. Such a process is evidenced simultaneously by subjective phenomena and by objective variations.

The mystery, scientifically insoluble, consists in the fact that certain physiological mechanisms residing principally in the cerebral cortex are attended by phenomena which are accessible only to what we have called internal observation. There is a knowledge of the ego — consciousness — which bears with it also knowledge of something which is not the ego, and consequently knowledge of the universe, and of complete reality starting with the existential reality of the subject itself. From the certainty such subject has of the reality which is it itself — "I think, hence I exist" in Descartes' farrious phrase — it deduces the certainty of existence of an exterior reality or world of things, which becomes known precisely through mechanisms identical to those which bring us to a knowledge of subjective existence.

As moreover the subject may bring about or not, consciously or at will, certain determinate acts and in this way cause the appearance of certain foreseeable sensations, the subjective criterion becomes naturally converted into an objective existential criterion, as proclaimed by all shades of philosophy throughout the ages.

It is then to be understood how the spectacle of the mind itself, of the consciousness and of the will of the active subject, able to set things in motion and to inhibit, may have constantly permeated its understanding of the universe, and that explanations founded on the subjective standard will have attained so much greater authority the more difficult were the problems and the more complex the questions under consideration.

In so far as any awkward phenomenon comes into evidence, the anthropomorphic argument comes to the aid of hesitating reason. From the most primitive savage to the most cultured of men, from the basest and most ignorant to the most learned and stately, it is the road of least resistance which leads us to the explanation that everything occurs in our universe through agents referable to the consciousness and the will resulting from human actions of each member of mankind. Consciousness allows the planning of existence, the projection in series of the acts constituting such existence, movement to a purpose and development of will.

This is the foundation of the subjective world, the most profound faculty within the powers of the soul, and the most decisive of rational behaviour! And if this indeed be true, then it is equally probable that the fate of the universe is determined in very fact by an all-pervading Purpose and Will; an Intent and Will beyond mankind, extrahuman and superhuman, supernatural and above -infinitely above — our own individual consciousness and will.

## CONSCIOUSNESS AND INTENT

Biology has proven that every vital manifestation is fulfilled with a view towards the future, for the accomplishment of a purpose which makes life possible. If this be not so, there would be no such thing as life. It has always been held that opportunity, adaptation of reactions, might be an exclusive characteristic of conscious acts; and in mechanisms similar to those of the conscious it was naturally claimed that they were of the same order as that observed in the vital phenomena.

Facts now are viewed from the opposite end of our perspective. It is not consciousness any more than it is will, or any other powers similar to them, which determine the rules of life; but the knowledge of the ego, the will implied by cognition are merely special manifestations in the individual of universal principles, from which emanates the power of the human mind to turn and twist and regard itself from within, thus acquiring knowledge essential to, for, and of its existence and power!

## SANTIAGO RAMON Y CAJAL

As we have already repeatedly explained, the internal prolongations of the spinal ganglia wind their way backwards and emerge from the hind end of the ganglion, forming the posterior roots of the spinal nerves, and, finally, after entering the spinal cord through the postero-lateral groove, fork into two in the substance of the white matter of the posterior cord, to form two branches, one going upwards and the other downwards. From the radicular stem as well as from both the rising and descending branches there sprout out at right angles an infinite number of collateral threads or fibrils which branch out into terminations reaching into various zones of the grey matter.

The nervous impulse arriving from the ganglion through the dorsal root reaches the spinal cord, where it splits into two currents, upward and downward, of equal or maybe unequal intensity. However, as both the stem and the branches of the sensitive root system start off side by side, the nervous wave will be propagated along these latter in order that it may penetrate the grey matter for transmission to the motor and funicular cells.

The terminal stems or branches will constitute the path of conscious impression as well as of widespreading or remote reflexes. This theory, formulated with more or less clarity by ourselves and adopted by various writers, notably Marquez, who used it in order to explain the physiological make-up of the spinal cord, is in agreement with physiological experiments which show us that the excitational energy necessary to arouse a dispersed and distant reflex is much greater than that needed for the production of a unilateral and limited reflex.

There are grounds for supposing that the bonding substance between the terminal branches of the collateral processes and the body or dendrites of the nerve cells is not so perfect a conductor as the nerve protoplasm, but offers a certain resistance evidenced by delay in the passage of the wave. This explains why the delay in travel of the reflex current becomes greater the larger the number of neurones entering into the chain of transmission.

"Reflex actions" is the name given by physiologists to involuntary movements caused by stimulus of the sensitive nerve-endings. In this field of physiology, as in many others, the new theory of structure of the spinal cord is singularly illuminating, allowing an easy understanding of the course of the current in every type of reflex and of the laws governing it, which laws to a certain extent portray mere consequences of the principles of morphology and connections of the medullo-gangliar neurones.

As far back as 1890, we expounded, subject to these new ideas, the first outlines relative to the course of nerve currents from sensory to motor root, without or with the participation of the commissural and funicular cells. Kölliker, van Gehuchten, Waldeyer and Lenhossek followed along these same lines, amplifying our ideas and submitting others in which all the paths which could be taken by a sensory impression in the production of both reflex and voluntary movements were shown in graphic form.

It is moreover an experimental fact that every sensory impulse reaching the outer surface of the brain at the motor-sensory region, through the specific action of the pyramidal cells, produces a sensation.

There are paths leading the impulses up to the cerebral cortex, and of all the paths along the cerebrospinal axis followed by tactile excitations, warmth and pain, from the muscular and tendinous sense, our only exact knowledge at present is of the path pertaining to the sense of touch. There are also paths carrying impulses up to the cerebellum. All these transmissions towards the cortex, in the cerebrum or cerebellum, are made up of various neurones connected in series.

Clearly, if a sensory nerve current travels along collaterals to a higher or lower point of the spinal cord bulb, it will give rise to reflexes. If on the contrary, it reaches the cerebral cortex, the same sensory current will produce a sensation.

## IVAN PETROVICH PAVLOV

We are now better able to appreciate the marvellous precision of the working of the digestive glands, which at any given moment supply exactly what is demanded of them — neither more nor less, neither too much nor too little! We can, moreover, rest assured of one fact unique to glandular activity, namely, that these glands are able to prepare secretions of different composition according to requirements, with greater or less amounts of diastases and different proportions of each when — as in the case of the pancreatic juices for instance — several diastases appear in the same secretion. Other properties of the digestive juices also, in addition to their enzyme content, vary according to the requirements of the functions.

If note be taken of the quantity secreted in a given period, say an hour, and also of the composition of the corresponding juice, it will be seen that the higher or lower concentration thereof does not depend exclusively upon the rate of secretion.

The most diverse proportions are to be observed in ferments between their fluid content and the richness of the juice; a strong digestive capacity is observable whether the secretion be plentiful or sparse.

In one and the same secretion, considerable variation can take place independently in the amounts of the various ferments, which shows that glands, like the pancreas, developing complex chemical actions, can supply, according to circumstances and at different times, one product at one instant and another different one later. What is here stated in regard to the diastases is equally true of the amount of fluid, salts and acids present in the secretions.

One important physiological question is to ascertain if these variations in composition of the secretions take place according to digestive requirements for the better development thereof. A great many facts tell us that the glands are adapted in their working to the various and successive stages of treatment of the food taken in. In the first place, different foodstuffs arouse different intensities of secretion in the organs traversed by them along the digestive tube, for foods are complex and extremely varied in composition. A suitable quantity of each digestive juice will be secreted in succession according to what the meal was made up of.

The working of the glands seems to obey some objective which has to be fulfilled in the best possible way. For instance, vegetable proteins and bread require a great deal of ferment for their digestion. This requirement is met less by an increase in the amount of secreted fluid than by an extraordinary concentration of diastase. It may thus be deduced that what is needed here is an increased amount of ferment, whilst a proportionate concentration of hydrochloric acid might be useless or even harmful. During the digestion of bread, in fact, the presence of an excess hydrochloric concentration in the stomach is usually avoided. The total amount of gastric juice secreted for digesting bread is only slightly greater than that produced for the digestion of milk.

The secretion is distributed, nevertheless, over a fairly long period, so that the hourly amount secreted, in the case of bread, comes out at one-and-a-half times less than that observed when dealing with milk or meat. The consequence of this is the low hydrochloric content of the juice for dealing with bread. All of which is in accordance with what biochemistry has to tell us, namely, that the digestion of starch, so plentiful in bread, would be hindered by an excess of acid. A succession of hospital cases shows us quite clearly that with excess gastric acidity, a large portion of the starch fails to undergo digestion and appears therefore in undigested form in the faeces, whilst meat is digested without difficulty.

All these facts and many more show that the working of the digestive glands is extremely complex and flexible, taking place with surprising precision and with the designed purpose of fulfilling a definite physiological need.

## SIR CHARLES SCOTT SHERRINGTON

If we regard the nervous system of any higher organism from the broad point of view, a salient feature in its scheme of construction is the following. At the commencement of every reflex-arc is a receptive neurone extending from the receptive surface to the central nervous organ. This neurone forms the sole avenue which impulses generated at its receptive point can use whithersoever be their destination.

This neurone is therefore a path exclusive to the impulses generated at its own receptive point, and other receptive points than its own cannot employ it. A single receptive point may play reflexly upon quite a number of different effector organs. It may be connected through its reflex path with many muscles and glands in many different regions. Yet all its reflex-arcs spring from the one single shank or stem, i.e. from the one afferent neurone which conducts from the receptive point at the periphery into the central nervous organ.

But at the termination of every reflex-arc we find a final neurone, the ultimate conductive link to an effector organ (muscle or gland). This last link in the chain, e.g. the motor neurone, differs obviously in one important respect from the first link of the chain. It does not subserve exclusively impulses generated at one single receptive source, but receives impulses from many receptive sources situate in many and various regions of the body. It is the sole path which all impulses, no matter whence they come, must travel if they are to act on the musclefibres to which it leads.

Therefore, while the receptive neurones form a private path exclusively serving impulses of one source only, the final or efferent neurone is, so to say, a public path, common to impulses arising at any of many sources of reception. A receptive field, e.g. an area of skin, is analysable into receptive points. One and the same effector organ stands in reflex connection not only with many individual receptive points but even with many various receptive fields. Reflexes generated in manifold sense-organs can pour their influence into one and the same muscle. Thus a limb-muscle is the terminus ad quem of many reflex-arcs arising in many various parts of the body. Its motornerve is a path common to all the reflex-arcs which reach that muscle.

Reflex-arcs show, therefore, the general features that the initial neurone of each is a private path exclusively belonging to a single receptive point (or small group of points); and that finally the arcs embouch into a path leading to an effector organ; and that their final path is common to all receptive points wheresoever they may lie in the body, so long as they have connection with the effector organ in question. Before finally converging upon the motor neurone the arcs converge to some degree. Their private paths embouch upon internuncial paths common in various degree to groups of private paths. The terminal path may, to distinguish it from internuncial common paths, be called the final common path. The motor-nerve to a muscle is a collection of final common paths.

Certain consequences result from this arrangement. One of these seems the preclusion of essential qualitative difference between nerveimpulses arising in different afferent nerves. If two conductors have a tract in common, there can hardly be essential qualitative difference between their modes of conduction; and the final common path must be capable of responding with different rhythms which different conductors impress upon it. It must be to a certain degree aperiodic. If its discharge be a rhythmic process, as from many considerations it appears to be, the frequency of its own rhythm must be capable of being at least as high as that of the highest frequency of any of the afferent arcs that play upon it; and it must be able also to reproduce the characters of the slowest. Baglioni's results support this inference.

A second consequence is that each receptor being dependent for final communication with its effector organ upon a path not exclusively its own but common to it with certain other receptors, such nexus necessitates successive and not simultaneous use of the common path by various receptors using it to different or opposed effect. When two receptors are stimulated simultaneously, each of the receptors tending to evoke reflex action that for its end-effect employs the same final common path but employs it in a different way from the other, one reflex appears without the other.

The result is this reflex or that reflex, but not the two together. Excitation of the central end of the afferent root of the eighth or seventh cervical nerve of the monkey evokes reflexly in the same individual animal sometimes flexion at elbow, sometimes extension. If the excitation be preceded by excitation of the first thoracic root the result is usually extension; if preceded by excitation of the sixth cervical root it is usually flexion.

Yet though the same root may thus be made to evoke reflex contraction of the flexors or of the extensors, it does not, in my experience, evoke contraction in both flexors and extensors in the same reflex-response. Of the two reflexes on extensors and flexors respectively, either the one or the other results, but not the two together. Thus, in my experience, excitation of the seventh or eight root never causes simultaneously with reflex contraction of the flexors of elbow a contraction of that part of the triceps which extends the elbow. The flexor-reflex when it occurs seems therefore to exclude the extensor-reflex, and vice versa.

If there resulted a compromise between the two reflexes, so that each reflex had a share in the resultant, the compound would be an action which was neither the appropriate flexion nor the appropriate extension. Were there to occur at the final common path algebraical summation of the influence exerted on it by two opposed receptive arcs, there would result in the effector organ an action adapted to neither and useless for the purposes of either. In the Coelenterate, Carmarina, a mechanical stimulus applied to the sub-umbrella causes, as in another Geryonid, Tiaropsis indicans, a reflex movement that brings the free end of the manubrium to the spot touched.

Bethe reports that if two stimuli are applied simultaneously to opposite points of the discoid sub-umbrella, the points chosen being such that the manubrium is midway between them, the manubrium is moved toward the point at which the stimulus applied was the stronger. He adds that if both stimuli are of exactly equal strength the manubrium remains unmoved and uncontracted.

To obtain such a result as this last with antagonistic spinal reflexes in the vertebrate would obviously be more difficult, because the more complex the preparation and the nervous system involved, the more difficult it will be at any moment to balance exactly the two reflexes. But, apart from that, the observation on Carmarina is an analogue of that in the monkey's arm.

This dilemma between reflexes would seem to be a problem of frequent recurrence in reflex co-ordination. We note an orderly sequence of actions in the movement of animals, even in cases where every observer admits that the co-ordination is merely reflex. We see one act succeed another without confusion. Yet, tracing this sequence to its external causes, we recognize that the usual thing in nature is not for one exciting stimulus to begin immediately after another ceases, but for an array of environmental agents acting concurrently on the animal at any moment to exhibit correlative change in regard to it, so that one or other group of them becomes — generally by increase in intensity - temporarily prepotent.

Thus there dominates now this group, now that group in turn. It may happen that one stimulus ceases coincidently as another begins, but as a rule one stimulus overlaps another in regard to time. Thus each reflex breaks in upon a condition of relative equilibrium, which latter is itself reflex. In the simultaneous correlation of reflexes some reflexes combine harmoniously, being reactions that mutually reinforce. These may be termed allied reflexes, and the neural arcs which they employ allied arcs.

On the other hand, some reflexes, as mentioned above, are antagonistic one to another and incompatible. These do not mutually reinforce, but stand to each other in inhibitory relation. One of them inhibits the other, or a whole group of others. These reflexes may in regard to one another be termed antagotiistic; and the reflex or group of reflexes which succeeds in inhibiting its opponents may be termed "prepotent" for the time being.

The nervous system functions as a whole. Physiological and histological analysis finds it connected throughout its whole extent. Donaldson opens his description of it with the remark: "A group of nervecells disconnected from other nerve-tissues of the body, as muscles and glands are disconnected from each other, would be without physiological significance." A reflex action, even in a "spinal animal" where the solidarity of the nervous system has been so trenchantly mutilated, is always in fact a reaction conditioned not by one reflex-arc but by many. A reflex detached from the general nervous condition is hardly realizable. The compounding together of reflexes is therefore a main problem in nervous co-ordination. For this problem it is important to recognize a feature in the architecture of the grey-centred (synaptic) nervous system which may be termed the principle of the common path.

## THE FOUNDATIONS OF PSYCHOREFLEXOLOGY

The aspect of psychology with which we propose to deal in the present book has little in common with what has hitherto invariably been the target of research. In objective psychology or, more accurately, psychoreflexology which is the study of mental reflexes to which we intend to devote the present work, the problems of subjective or conscious processes play no part.

As we already know, up to the present, psychological manifestations have been held mainly to embrace conscious phenomena. The interpretation of such phenomena depends of course on our comprehension of the causes and conditions governing their formation and effects, in so far as these may generally be ascertainable." The material of the study of psychology hitherto, and as it still stands at present, constitutes the inner or subjective world which, being accessible only through individual observation, implies necessarily that the only basic method of procedure open to present-day psychology is that of personal observation.

Some writers do indeed introduce into psychology the notion of unconscious states, but they clothe even these with the properties of conscious conditions and treat them to all intents and purposes as latent conscious phenomena. The whole question of unconscious mental processes in modern psychology is a matter of controversy.

According to Ziehen, for instance, the yardstick of the mental is that which is presented to our consciousness and nothing more than that. The mental and the conscious, as far as we are concerned, according to this writer, run in double harness; we cannot by any manner of means imagine what an unconscious sensation, notion, or the like really is. We are aware of sensations and notions only in so far as we become conscious of them. So-called unconscious activities are nothing more than past and gone material processes which merely excite a mental or conscious process by delayed action. As a concept the unconscious mental process has no meaning.

Other writers besides Ziehen take this attitude. According to them, we cannot postulate an unconscious mental existence in its literal sense. "Though we oft-times speak of unconscious mental existence, this expression is in general either absolutely lacking in meaning or it is not sufficiently precise." Experience shows, however, that personal observation does not in the least suffice for proper perception of one's own mental life. One opposite example of how misleading it is to allow oneself to be guided by the subjective processes, even in such manifestations as memory and recollection, is given by the memory researches of H. Ebbinghaus who, experimentally, through his own investigation into mechanical learning by rote, became convinced that mental processes which had once taken place, and then apparently vanished from the unconscious, had in fact actually been in existence the whole time.

On the other hand, it is clear that to subjective psychology the whole field of exploration of the conscious processes in individuals other than oneself is barred, since there is absolutely no suitable and accurate means of investigation at its disposal.

In this case the method of scientific investigation clearly constitutes the relation of the individual ego with the extraneous. That this method does not, however, avail is indeed quite plain. Similarity or identification in this connection relates, furthermore, to manifestations of two kinds of consciousness of self, which in a different sense are not comparable, and are perceptible only through internal self-observation where no accurate standards are available. Richet is quite right when he says that "inner self-observation, however powerful a method it be, can be applied only to one field — the field of selfknowledge.

Outside this it is barren and full of pitfalls." "The ego knows itself, perceives itself; it considers and observes itself, and we must not therefore overstep the boundaries of such ego, a domain at once so vast that much still remains to be created from it, and yet at the same time so narrow that our unassuaged thirst for knowledge strives ever to go still further on." Beyond these limits, however, only science with its rigid methods and accurate instruments and measurements, and its slow but sure intellectual development has the power to go. In short, inner observation can rely only on perception of conscious phenomena. The general properties of living matter remain unexplored; their place is in the field of physics, chemistry and physiology.

From our point of view, the usual definition of psychology as a study of the states or manifestations of the consciousness is quite wrong. Psychology must not in reality be restricted to the consideration of conscious phenomena, but must also investigate so-called unconscious mental phenomena and also even the external vital activities of the organism, so far as these are the expression of its mental activities. In brief, the biological basis of mental activity must also be an object of its study.

Does not this imply then, that our movements, whether they be voluntary or involuntary, intelligent or instinctive, must also form an object of investigation from the point of view of subjective psychology. Do not changes in respiration and heartbeat, evoked by mental processes, also then become an object of psychological study, especially if we reflect that knowledge thereof shows us the correlation between mental phenomena and our bodily functions?

A whole series of investigations on the influence of mental processes on the internal organs and bodily workings in general, as well as the effect of intellectual activity on the functioning of the internal organs, must on that account definitely be considered as one of the tasks of psychology, since knowledge of these facts will give us a clue to the explanation of mental phenomena as such, and allow us to understand the essential conditions of the outward manifestations of those activities of the nerve centres which are known under the name of mental processes. Our knowledge of the neuropsychical processes in other creatures similar to ourselves depends also on such conditions.

The intrinsic notion of psychology needs in general to be considerably extended. From our point of view, psychology is the study of mental life in general, and not merely of its conscious manifestations. The business of psychology consists therefore in the diagnosis of all mental processes in the widest sense of the term, that is to say, of conscious as well as unconscious impressions of mental activity, and in appreciating the external expressions of mental activity so far as these serve to determine the properties and characters of such mental activity; furthermore the biological processes, standing in close and direct relation with the mental processes, must also form an object of psychological study. We must not forget that it is a question not only of knowledge of the individual activities of the mind, but also of the mental life of whole groups (such as crowds, communities, nations, and so forth), as well as of the animal kingdom. Hence the division of psychology into individual, social, national, and comparative racial psychology and so-called animal psychology.

Psychoreflexology, which is our subject here, recognizes the material side of the so-called conscious as well as of the unconscious processes, but looks on the mental processes only in their objective manifestations and does not in the least enter into a consideration of the subjective side of the mental world. It has also to investigate and consider the processes in the brain which are accessible to objective research through the use of the finest and most delicate of physical apparatus.

Psychoreflexology does not enquire if a process runs its course consciously or unconsciously, leaving that as a matter for subjective psychology on the ground that no objective signs exist entitling us to call certain processes conscious. As we are considering the objective side exclusively, we cannot answer the question whether a process has taken its course within or without the domain of the consciousness. In any event, all researches in this direction lack any scientific significance, since they are little more than slightly confirmed conjectures.

Thus, as is well known, Auerbach discovered that if a frog deprived of its forebrain had its back irritated by acid, it would try to wipe the acid off with the leg corresponding to the irritated side. Is such action conscious or unconscious? This question has not been answered even yet. Ziehen does indeed say that there is no reason to assume the existence of mental (that is, conscious) analogous processes just because of this performance. On what, however, is such assumption based? We cannot actually penetrate into the mind of the brainless frog, and if we allow ourselves to be guided by personal observation of complicated reflexes within ourselves, it then becomes clear that we cannot deny a certain consciousness in such reflexes — at least, we cannot exclude it.

Ziehen denies a parallel course of mental, i.e. conscious processes even in the case of the complex reflexional movements of a frog with its cerebral hemispheres removed as far back as the optic thalami, when, as has been shown, in its hopping it avoids obstacles casting a dark shadow. We have absolutely no need to go into the grounds for this opinion, since there are but few, unless we may accept the writer's adduced analogy of these movements to the automatic motions of a pianist playing his piano or of a man walking mechanically downstairs.

Here, however, the fact is overlooked that the pianist and the man going downstairs can carry out these same movements not only automatically but also entirely consciously, whilst we cannot differentiate conscious movements from mechanical or automatic or, in other words, from unconscious movements of this type.

## SIGMUND FREUD

Psychoanalysis has taught us that the essence of the process of repression lies, not in abrogating or annihilating the ideational presentation of an instinct, but in withholding it from becoming conscious. We then say of the idea that it is in a state of "unconsciousness," of being not apprehended by the conscious mind, and we can produce convincing proofs to show that unconsciously it can also produce effects, even of a kind that finally penetrate to consciousness. Everything that is repressed must remain unconscious but at the very outset let us state that the repressed does not comprise the whole unconscious. The unconscious has the greater compass: the repressed is a part of the unconscious.

How are we to arrive at a knowledge of the unconscious? It is of course only as something conscious that we know anything of it, after it has undergone transformation or translation into something conscious. The possibility of such translation is a matter of everyday experience in psychoanalytic work. In order to achieve this, it is necessary that the person analysed should overcome certain resistances, the very same as those which at some earlier time placed the material in question under repression by rejecting it from consciousness.

In many quarters our justification is disputed for assuming the existence of an unconscious system in the mind and for employing such an assumption for purposes of scientific work. To this we can reply that our assumption of the existence of the unconscious is necessary and legitimate and that we possess manifold proofs of the existence of the unconscious. It is necessary because the data of consciousness are exceedingly defective; both in healthy and in sick persons mental acts are often in process which can be explained only by presupposing other acts, of which consciousness yields no evidence. These include not only the parapraxes and dreams of healthy persons and everything designated a mental symptom or an obsession in the sick; our most intimate daily experience introduces us to sudden ideas of the source of which we are ignorant and the results of mentation arrived at we know not how.

All these conscious acts remain disconnected and unintelligible if we are determined to hold fast to the claim that every single mental act performed within us must be consciously experienced; on the other hand, they fall into a demonstrable connection if we interpolate the unconscious acts that we infer. A gain in meaning and connection, however, is a perfectly justifiable motive, one which may well carry us beyond the limitations of direct experience.

When, after this, it appears that the assumption of the unconscious helps us to construct a highly successful practical method, by which we are enabled to exert a useful influence upon the course of conscious processes, this success will have won us an incontrovertible proof of the existence of that which we assumed. We become obliged then to take up the position that it is both untenable and presumptuous to claim that whatever goes on in the mind must be known to consciousness.

We can go further and in support of an unconscious mental state allege that only a small content is embraced by consciousness at any given moment, so that the greater part of what we call conscious knowledge must in any case exist for very considerable periods of time in a condition of latency, that is to say, of unconsciousness, of not being apprehended by the mind. When all our latent memories are taken into consideration it becomes totally incomprehensible how the existence of the unconscious can be gainsaid. We then encounter the objection that these latent recollections can no longer be ascribed as mental processes, but that they correspond to residues of somatic processes from which something mental can once more proceed.

The obvious answer to this should be that a latent memory is, on the contrary, indubitably a residuum of a mental process. But it is more important to make clear to our own minds that this objection is based on the identification — not, it is true, explicitly stated but regarded as axiomatic — of conscious and mental. This identification is either a petitio principii and begs the question whether all that is mental is also necessarily conscious, or else it is a matter of convention of nomenclature. In this latter case it is of course no more open to refutation than any other convention.

The only question that remains is whether it proves so useful that we must needs adopt it. To this we may reply that the conventional identification of the mental with the conscious is thoroughly impractical. It breaks up all mental continuity, plunges us into the insoluble difficulties of psycho-physical parallelism, is open to the reproach that without any manifest grounds it overestimates the part played by consciousness, and finally it forces us prematurely to retire from the territory of psychological research without being able to offer us any compensation elsewhere. At any rate it is clear that the question -whether the latent states of mental life, whose existence is undeniable, are to be conceived of as unconscious mental states or as physical ones -threatens to resolve itself into a war of words.

We shall therefore be better advised to give prominence to what we know with certainty of the nature of these debatable states. Now as far as their physical characteristics are concerned they are totally inaccessible to us: no physiological conception nor chemical process can give us any notion of their nature. On the other hand, we know for certain that they have abundant points of contact with conscious mental processes; on being submitted to a certain method of operation they may be transformed into or replaced by conscious processes, and all the categories which we employ to describe conscious mental acts, such as ideas, purposes, resolutions and so forth, can be applied to them. Indeed, of many of these latent states we have to assert that the only point in which they differ from states which are conscious is just in the lack of consciousness of them. So we shall not hesitate to treat them as objects of physiological research, and that in the most intimate connection with conscious mental acts.

The stubborn denial of a mental quality to latent mental processes may be accounted for by the circumstances that most of the phenomena in question have not been objects of study outside psychoanalysis. Anyone who is ignorant of the facts of pathology, who regards the blunders of normal persons as accidental, and who is content with the old saw that dreams are froth, need only ignore a few more problems of the psychology of consciousness in order to dispense with the assumption of an unconscious mental activity. As it happens, hypnotic experiments and especially post-hypnotic suggestion had demonstrated tangibly even before the time of psychoanalysis the existence and mode of operation of the unconscious in the mind.

The assumption of an unconscious is, moreover, in a further respect, a perfectly legitimate one, inasmuch as in postulating it we do not depart a single step from our customary and accepted mode of thinking. By the medium of consciousness each one of us becomes aware of his own states of mind; that another man possesses consciousness is a conclusion drawn by analogy from the utterances and actions we perceive him to make, and it is drawn in order that this behaviour of his may become intelligible to us.

We limited the foregoing discussion to ideas and may now raise a new question, the answer to which must contribute to the elucidation of our theoretical position. We said that there were conscious and unconscious ideas; but are there also unconscious instinctual impulses, emotions and feelings, or are such constructions in this instance devoid of any meaning?

I am indeed of opinion that the antithesis of conscious and unconscious does not hold for instincts. An instinct can never be an object of consciousness — only the idea that represents the instinct. Even in the unconscious, moreover, it can only be represented by the idea. If the instinct did not attach itself an idea or manifest itself as an affective state, we could know nothing of it. Though we do speak of an unconscious repressed instinctual impulse, this is a looseness of phraseology which is quite harmless. We can only mean an instinctual impulse the ideational presentation of which is unconscious, for nothing else comes into consideration.

We should expect the answer to the question about unconscious feelings, emotions and affects to be just as easily given. It is surely of the essence of an emotion that we should feel it, i.e. that it should enter consciousness. So for emotions, feelings and affects to be unconscious would be quite out of the question. But in psychoanalytic practice we are accustomed to speak of unconscious love, hate, anger, etc., and find it impossible to avoid even the strange conjunction, "unconscious consciousness of guilt," or a paradoxical "unconscious anxiety." Is there more meaning in the use of these terms than there is in speaking of "unconscious instincts"?

The two cases are not really on all fours. To begin with it may happen that an affect or an emotion is perceived, but misconstrued. By the repression of its proper presentation it is forced to become connected with another idea, and is now interpreted by consciousness as the expression of this other idea.

It is of especial interest to us to have established the fact that repression can succeed in inhibiting the transformation of an instinctual impulse into affective expression. This shows us that the conscious system normally controls affectivity as well as access to motility; and this enhances the importance of repression, since it shows us that the latter is responsible, not merely when something is withheld from consciousness, but also when affective development and the inauguration of muscular activity is prevented. Conversely, too, we may say that as long as the conscious system controls activity and motility the mental condition of the person in question may be called normal. Nevertheless, there is an unmistakable difference in the relation of the controlling system to the two allied processes of discharge.

Whereas the control of the conscious system over voluntary motility is firmly rooted, regularly withstands the onslaught of neurosis and only breaks down in psychosis, the control of the conscious over affective development is less firmly established. Even in normal life we can recognize that a constant struggle for primacy over affectivity goes on between the two systems, consciousness and preconsciousness, that certain spheres of influence are marked off one from another and that the forces at work tend to mingle.

## RENÉ DESCARTES

Animal spirit (nervous force) is like an extremely fine fluid, or we might perhaps better say, an active flame. It is continuously generated by the heart and ascends to the brain, which acts as a sort of reservoir, passing thence to the nerves which distribute it to the muscles where its effect is shown in contraction or relaxation. On entering the brain cavities, this force permeates the pores of cerebral matter and thus finds its way to the nerves; the amount of energy flowing through each nerve governs the magnitude of contraction in the muscles to which such nerve leads, and gives rise to movement in the corresponding organs.

This arrangement is similar to the underground piping and fountains in our gardens, wherein, according to the head of the water falling from the tanks, different appliances are set more or less forcibly in motion, and various instruments, whistles, and so forth come into play, in keeping with the way the piping, taking the water to any particular apparatus, is fixed up.

The nerves, in fact, may be compared to such pipes, whilst the muscles and tendons correspond to the machines and fountains at the end of the line, and the nervous energy itself to the water driving each movement. This nervous energy springs from the heart just like water wells up from the earth, whilst the brain acts as the central reservoir for it in the body.

In addition, breathing and other similar natural and customary actions in the body mechanism, depending on the flow of energy, are like the movements of a clock or a mill, kept in action by the stream of water.

External objects, by their mere presence alone, act upon the sense organs and determine which way the machine is to go. If we take into consideration that the parts of the brain are arranged in a manner similar to the hydraulic plant to which we have referred above, we have an analogy to go upon. Let us assume that a stranger enters the cavern without knowing, for instance, that hidden levers in the floor connected with taps cause the water to run through particular conduits.

Our visitor may, perhaps, start making his way towards Diana's Bath, when the soft music of the Pan pipes will strike his ear, and on attempting to proceed towards the spot whence such pleasant sounds seem to come, he approaches the figure of Neptune who starts waving his trident menacingly at him; or, if, on the contrary, he flies in the opposite direction, some sea monster may suddenly start spouting water in his face, and similarly for any sort of arrangement set up according to the fancy of the engineer planning the job.

When a rational mind, say the hydraulic engineer in the illustration given, is in command, it can at will and by reason of its knowledge of the piping with its respective faucets, fittings and accessories bring about different movements by channelling the water gently or vigorously through to the various appliances. In like manner, the rational mind of man — the power of reason—resides principally in the brain, where it plays a part similar to that of our waterworks engineer, by sending energy to the nerves and thereby causing the voluntary movements which take place in the muscles.

We now arrive at a final conclusion. All operations of the bodily mechanism, such as digestion of food, beating of the heart, pulsing in the arteries, feeding and growth of the members, breathing, waking and sleeping, vision, hearing, smell, taste, getting warm, and similar qualities revealed by the organs through the external senses, impressions of ideas through common sense or imagination, retention of such ideas in the memory, internal impulses caused by desire and passion, and lastly, the movements of the limbs, which become so accurately adapted to the purposes put to them through the senses, come from the arrangement of the organs and from no other source whatsoever; just as the working of a watch or the action of an automaton depends on the gears and weights which cause it to go.

It is not necessary, therefore, to assume the existence of some archeus, whether vegetative or sensory, or any other principle whatsoever, apart from the energy contained in the blood-stream, excited by the heat of that fire which flames constantly within the heart and is identical to that which consumes inanimate and inorganic bodies, to enter into the maintenance of activity in the organs.

"Descartes' idea of 'animal spirits' is strange to us — an agent simultaneously spirit and substance. Nevertheless, this concept of animal spirits epitomizes the dilemma we are still in when we try to explain excitation. Is it a material, a chemical influence, or an intangible dynamic discharge? To this very day, we cannot say more than that excitation is a manifestation of life, and that it is a distinguishing and specific property of living matter."

## GEORGE BERKELEY

It is evident to anyone who takes a survey of the objects of human knowledge, that they are either ideas:

- Actually imprinted on the senses, or else such as are
- Perceived by attending to the passions and operations of the mind, or lastly
- Ideas formed by help of memory and imagination, either compounding, dividing, or barely representing those originally perceived in the aforesaid ways.

By sight I have the ideas of lights and colours, with their several degrees and variations. By touch I perceive hard and soft, heat and cold, motion and resistance, and of all these more and less either as to quantity or degree. Smelling furnishes me with odours, the palate with tastes, and hearing conveys sounds to the mind in all their variety of tone and composition. And as several of these are observed to accompany each other, they come to be marked by one name, and so to be reputed as one thing. Thus, for example, a certain colour, taste, smell, figure and consistence, having been observed to go together, are accounted one distinct thing, signified by the name "apple." Other collections of ideas constitute a stone, a tree, a book, and the like sensible things; which, as they are pleasing or disagreeable, excite the passions of love, hatred, joy, grief, and so forth.

But besides all that endless variety of ideas or objects of knowledge, there is likewise something which knows or perceives them, and exercises divers operations, as willing, imagining, remembering about them. This perceiving, active being is what I call mind,spirit,soul, or myself. By which words, I do not denote any one of my ideas, but a thing entirely distinct from them wherein they exist, or, which is the same thing, whereby they are perceived; for the existence of an idea consists in being perceived.

All our ideas, sensations, notions, or the things which we perceive, by whatsoever names they may be distinguished, are visibly inactive: there is nothing of power or agency included in them. So that one idea or object of thought cannot produce or make any alteration in another. To be satisfied of the truth of this, there is nothing else requisite but a bare observation of our ideas. For, since they and every part of them exist only in the mind, it follows that there is nothing in them but what is perceived: but whoever shall attend to his ideas, whether of sense or reflection, will not perceive in them any power or activity; there is, therefore, no such thing contained in them.

A little attention will discover to us that the very being of an idea implies passiveness and inertness in it, insomuch that it is impossible for an idea to do anything, or strictly speaking, to be the cause of anything: neither can it be the resemblance or pattern of any active being. Whence it plainly follows that extension, figure, and motion cannot be the cause of our sensations. To say, therefore, that these are the effects of powers resulting from the configuration, number, motion, and size of corpuscles, must certainly be false.

We perceive a continual succession of ideas, some are anew excited, others are changed or totally disappear. There is therefore some cause of these ideas, whereon they depend, and which produces and changes them. That this cause cannot be any quality or idea or combination of ideas, is clear from the preceding section. It must therefore be a substance; but it has been shown that there is no corporeal or material substance: it remains therefore that the cause of ideas is an incorporeal active substance or Spirit.

A spirit is one simple, undivided, active being: as it perceives ideas it is called the understanding, and as it produces or otherwise operates about them it is called the will. Hence there can be no idea formed of a soul or spirit; for all ideas whatever, being passive and inert, they cannot represent unto us, by way of image or likeness, that which acts. A little attention will make it plain to anyone, that to have an idea which shall be like that active principle of motion and change of ideas is absolutely impossible. Such is the nature of spirit, or that which acts, that it cannot be of itself perceived, but only by the effects which it produceth.

If any man shall doubt of the truth of what is here delivered, let him but reflect and try if he can frame the idea of any power or active being, and whether he hath ideas of two principal powers, marked by the names will and understanding, distinct from each other as well as from a third idea of substance or being in general, with a relative notion of its supporting or being the subject of the aforesaid powers — which is signified by the same soul or spirit.The ideas imprinted on the senses by the

Author of Nature are called real things; and those excited in the imagination, being less regular, vivid, and constant, are more properly termed ideas, or images of things, which they copy and represent. But then our sensations, be they never so vivid and distinct, are nevertheless ideas, that is, they exist in the mind, or are perceived by it, as truly as the ideas of its own framing.

The ideas of sense are allowed to have more reality in them, that is, to be more strong, orderly, and coherent than the creatures of the mind; but this is no argument that they exist without the mind. They are also less dependent on the spirit, or thinking substance which perceives them, in that they are excited by the will of another and more powerful spirit; yet still they are ideas, and certainly no idea, whether faint or strong, can exist otherwise than in a mind perceiving it.

Hence, it is evident that those things which, under the notion of a cause co-operating or concurring to the production of effects, are altogether inexplicable, and run us into great absurdities, may be very naturally explained, and have a proper and obvious use assigned to them, when they are considered only as marks or signs for our information. And it is the searching after and endeavouring to understand those signs instituted by the Author of Nature, that ought to be the employment of the natural philosopher; and not the pretending to explain things by corporeal causes, which doctrine seems to have too much estranged the minds of men from that active principle, that supreme and wise Spirit "in whom we five, move, and have our being."

## RAMÓN TURRÓY DARDER

Those who fancy that the exteroceptor nerves possess the property of inducing changes of state in the consciousness and, further, in the perception of the reason for such changes — a thing which does not take place in the action of the nerves of general or organic sensibility -are wont to attribute nonsensical physiological properties to such nerves. Voltaire looked on this anomalous, hypothetical property as a miracle distinguishing sensations from without upon their appearance within the consciousness.

And so, indeed, it would be — were it so! Inductive reasoning has, however, come into existence to the undoing of miracles, and teaches us that there is no certainty that impressions received in the centres of exteroceptor perception may be deductively classed as peculiar in cause and effect thanks to some mysterious force created by objective considerations. The driving force comes ever from within, from the trophic feeling which starts off by considering the sensations merely as signs of the source sustaining them. Moreover, since such signs do not make their appearance save in the presence of the sustaining source, they are attributed to the latter as their true causal condition. Between the trophic sensation, the external sensation, and the gastric sensation, a profound connection is established through experience: if the first of these notes the absence of something, the second declares the fact openly by means of signs, whilst the third confirms that the signs are a true correspondence with fact.

This is where consciousness begins. A person eating knows with what he may appease his hunger and what are the things from the outside world which have precisely the capacity of satiating his wants. Such a great discovery would never have been made, had there not existed within his sensory centres prior impressions of whose origin he is unaware.

Still, as he observes, through a thousand and one instances, that physiological satisfaction is lacking to the degree that what will allay his hunger does not excite pleasant, tactile or warm sensation in his mouth, or an olfactory impression in his nostrils, an impression of colour in his eyes, and perhaps of sound in his cars, he finishes up by imagining through some inductive process, the terms of which are pre-established in his mind, that these impressions are not born spontaneously in the senses, but that the reappearance thereof is indissolubly linked to the factor advised by the stomach as present when hunger is extinguished.

It is therefore necessary to attribute to the sensory faculties unity, when one observes how the psychotrophic functions are integrated to those of the external sensibility. We are guilty of undue distortion if we assume that the sensorial centres work independently from the start, for we then think that all images and impressions are referable to causes which of themselves have no objective value. If we notice, moreover, that the ingestion of foods is in all cases adapted to the needs of the organism, and thus, if it is surprising and marvellous, for instance, that a hen seeks out the necessary lime-containing substances at the instant it is about to start producing eggs, none the less astonishing is it that all creatures — including man himself — find means of completing their food intake in accordance with nutritional requirements at each particular instant.

If what has been previously mentioned is not taken into consideration, still the fact that so much wisdom and reflection is naturally present will cause us to come out of the difficult situation by attributing all these marvels to a combination of miracles which, in order to give it some name, we shall call instinct. Starting, however, with the idea of psycho-physiological unity of the sensory system of internal trophic processes of acceptance with external receptions, through the operation of trophic experience, the initial and most important operation of intellectual existence, we fill in the blanks of our knowledge and the mystery disappears. Nowadays we have at our disposal standards of comparison hitherto never brought into account, which permit us to approach the most obscure questions from an entirely fresh point of view.

## MIGUEL DE UNAMUNO

Of all the spokesmen for Objectivism, not a single one pays — or rather had we better say, does not wish to pay — any attention to the fact that when a man asserts his ego, that is to say his personal consciousness, he postulates Man himself the actual and concrete entity in the flesh, and not some mere figment of his own imagination, and in so propounding Man as such he also propounds Consciousness, for the only consciousness of which we have real knowledge is that of Man himself.

The world is designed for consciousness, or rather such design, such notion of purpose or rather than notion let us say perception, such teleological perception does not come into being unless there is a consciousness. Consciousness and purpose are basically the same thing. If the Sun were possessed of consciousness, it would doubtless think that it existed for the purpose of giving fight to the universe, but even more than this would it reflect that the universe existed for the purpose of being illuminated by it, so that it should enjoy enlightening the universe and thereby exist. It would be quite right in thinking so!

All this fateful struggle by man to save himself, this eternal craving for redemption, which made Kant give that counterleap from pure to practical reason already mentioned, is nothing but a struggle by the consciousness. If indeed, as some soulless philosopher has asserted, consciousness is no more than a flash in the pan between two eternities wreathed in obscurity, then nothing is more baleful than the fact of being alive.

Some people may see a suspicion of inconsistency in all I am now about to say, sometimes yearning for eternal life and at other times saying that such life possesses nothing of value. Inconsistency? Indeed I think it is! It is the opposition between my heart saying "Yes" and my head saying "No!" Of course there is inconsistency. Who is there that does not remember the Gospel phrase: "Lord, I believe; help thou mine unbelief!" Of course, there is inconsistency, for it is only upon inconsistencies that we live and thrive. Life is doom, and that doom is perpetual struggle with neither victory nor the expectation thereof! All, all is inconsistency!

As you can see, it is all a matter of emotional values, against which reason itself has no chance, because reasons are nothing but reasons and that is the same thing as saying that they are not even facts. There are, indeed, some people who seem to think only with their brain or whatever the organ may be which is specifically detailed for that purpose, whilst on the other hand there do exist some who do their thinking with their whole heart and soul, with their body, their life's blood and the very marrow of their bones, with lungs, belly and everything else about them even to their whole being and life itself. Those who do their thinking merely with their brain turn out to be mere expounders or "professional" thinkers, and we all know what a "professional" is — just a product of the division of labour!

If a philosopher is not a really sincere man, then still less is he a philosopher, for before everything he is then just a pedant, which is to say a mockery of man. The cultivation of any applied science, be it chemistry, physics, mathematics, philology — and even in these cases only very restrictedly and within extremely narrow limits -may perhaps be a matter for detailed specialization; but Philosophy, like Poetry, must be a work of integration and beauteous harmony, otherwise it becomes nothing else but mock erudition, affectation and pretence.

There is one purpose in all knowledge. Gaining knowledge for the mere sake of knowing is not, say what we will, anything but cheerless sophistry. A thing is learned for some immediate practical purpose or in order to complement what we already know. Even a theory which may seem to us most academic, i.e. of the least practical application to our non-intellectual necessities of existence, answers some requirement which, intellectual though it be, is also in very fact a need, whether it is a ground for economy in thought, or a principle of unity and continuity in the consciousness.

But just as scientific knowledge has its aim in further knowledge, so also the philosophy that we may have to embrace has a further extrinsic purpose, all referring to our fate and to our attitude towards life and the universe. To all this, the most fateful problem of philosophy is that of reconciling intellectual requirements with those of the emotions and of the Will. Here lies the breaking point of every philosophy which endeavours to dispense with the eternal and tragic inconsistency that is the basis of our existence. But, do we all face up to this inconsistency?

There is something which, for want of a better name, we shall call the "Fatefulness of Existence," which cloaks a complete conception of life itself and of the universe, an entire philosophy more or less formalized and conscious, that is a perception which, rather than arising from ideas, itself determines them although clearly such ideas may react upon it and corroborate it. This feeling may be possessed and is possessed not only by individuals, but also by entire nations. Nobody has established that man must naturally be cheerful. Consciousness is a heavy burden.

## HENRI BERGSON

This book affirms the reality of mind and of matter, and seeks to determine the relationship between them by studying one definite aspect, that of memory. It is, therefore, frankly dualistic, though on the other hand, it deals with body and mind in such a way as, we hope, to diminish sufficiently, if not entirely, those theoretical difficulties which have always been attendant on dualism, and which cause it to be held in little honour among philosophers despite its being suggested by direct consciousness and accepted by common sense.

These difficulties arise to a large extent out of the idea, sometimes realist sometime idealist, which mankind takes of matter. Our aim is to show that realism and idealism both overstep the mark and that it is a mistake to confine matter to our idea of it, whilst it is equally wrong to make of it something capable of producing perceptions within us though in itself of a different nature entirely. Matter, in our view, is an aggregate of images or pictures, by which term we mean a certain existence something more than what the idealist calls a representation, mental picture or perception, and something less than what the realist calls a thing or object. Our image is found halfway between object and perception, or if you prefer it, between thing and representation. This idea of matter is merely what common sense gives us.

A person unaware of philosophic speculations would be greatly astonished if he were told that an object, which he could see and feel in front of him, existed only in his mind and through his mind and was therefore but a figment of his imagination, or even with greater generality, that it existed solely for a mind, as Berkeley maintained. Such a person would always assert that the object did exist independently of the consciousness perceiving it.

On the other hand, we should astonish him quite as much if we told him that the object were quite different from what he saw in it, having neither the colour credited to it by his eye nor the solidity apparently felt by his hand. Colour and solidity are for the layman qualities within the object and not states of mind; they are part and parcel of an existence independent of our own. Thus it is that as far as common sense is concerned, the object exists in its very self and has its own particular qualities. In this way a picture is conjured up, existing through its own existence, through the presence of the object in question with all its intrinsic features.

Images inform the whole of our mental life, but there are different shades of such mental life. Our psychological existence can be developed on different planes, more or less close to the action, according to the degree of our faculty of attending to life. What is normally held to be a major complication of the psychological condition appears to us as a major extension of our whole personality which, limited generally by action, becomes widened in so far as such limitation is diluted and then, always indivisibly, stretches over a still wider surface. The mental existence is linked inexorably with its concomitant motor in an indestructible unity. In the majority of cases the psychological state overruns the cerebral to an immense extent. The activity of the brain determines only a very small portion of such psychological condition, being translated into movement or locomotion. Every thought is accompanied by images, at least in its origin, which are not portrayed in the consciousness without outlining in the form of a contour or tendency the movements which would bring them into form in space.

Such complex thought depends in its development on the state of the brain at any particular instant, though that is not the end of it. The relation between the mental and the cerebral is not constant, nor is it simple. The cerebral state contains a greater or less part of the mental according to how we tend to project our psychological life, outward through action, or to draw it inward in the form of pure knowledge.

The generality of the idea consists in a certain activity of the mind, in a movement between action and representation, whence it becomes easy for philosophers to locate the general idea in one of the two extremes, boiling it down into words or vaporizing it in recollections; whilst the idea in reality consists in the passage of the mind between one extreme and the other within an indestructible universe.

## Chapter 15

# Functions of the Organs

Observation of the diseased in particular, and of living organisms in general, brings the conviction that there is a correlation of reactions composing life and that such unison is more or less normal or more or less unsettled according to circumstances. This was the view held by the Ancients, notably Hippocrates and Aristotle. The body is a complete entity responding to the variable conditions of existence in an ordered and harmonious manner as a synergetic whole whose activities are directed to the preservation of life.

To explain processes developing in the best possible way in the living body, throughout the ages, recourse is taken nevertheless to the anthropomorphic theory of directive principles endowed with the human properties appertaining to the subject himself — that is man, the curious and contemplative.

### ANALYSIS OF REALITY

What we have to deal with then, is a metaphysical activity which has nothing whatsoever to do with positive science. Still, the whole stage of rational judgment changes when we try to discover what the Universe really is, by considering it as a system based on cause and effect. It then becomes necessary to split things up, to separate out our problems and to study facts by subjecting them to detailed analysis.

Descartes laid down the following principle: "It is wise to split each difficulty under examination into as many separate parts as possible and convenient, in order the better to solve it." Analysis is the great discovery due to Descartes.

"The purpose of analysis is to find out by means of one single truth or a particular fact the principles from which it derives." "Analysis is the means of establishing the truth of the first principles of all knowledge." Roger Bacon and Francis Bacon were the famous forerunners of Descartes.

## FUNCTIONS OF THE ORGANS

Thus, in order to delve into the mysteries of life, it becomes necessary to study the working of every single one of the various parts of the organism, and of each one of its organs. Observing how processes evolve in the body, and bringing about by experiment the reappearance of such occurrences, we learn to associate certain causes with certain effects, and thus come to establish the experimental method as a systematic and scientific instrument which is the auspicious fruit of positive observation.

Harvey reduced the problem of the circulation of the blood to a problem in mechanics. And similarly, many others. Everything in life becomes a mechanism. Francis Bacon in 1620 asserts that experiment is of fundamental importance if we want to acquire scientific knowledge, since experiment enables us to establish the causes determining an occurrence, and to bring about such occurrence when we wish.

Harvey frequently had recourse to Bacon and applied in practice a rigorous method based on strictly experimental data, allowing him to discover with remarkable accuracy many noteworthy physiological facts which have later been confirmed at the end of subsequent great advances in more recent times.

Experimental physiology and everything of practical value in biology dates from that time. Four centuries of labour, classifying the body into parts, studying the morphological peculiarities of each item, and their separate ways of working, is our present heritage. 'Twas a harsh battle between the experimentalists and the vitalists, a battle which became even more harsh with the rise of positivism in the nineteenth century. Last century, as Haldane pointed out in 1922, young physiologists broke away from vitalist traditions and undertook the investigation of the functions of living bodies, piece by piece, with the hope of arriving thus at a knowledge of the whole working of so complex a mechanism. These mechanicists, however, not only demolished vitalism but at the same time they did away with the idea of life itself.

Progress attained through application of analytical methods to the study of life has been stupendous. Another light on biological processes was lit by this method, as rich and valuable in theoretical learning as in practical application.

Modern advances in medicine, physiology and biology are due indeed to the experimental method. It is worthwhile to observe the functions of the various organs and to cause modifications in their development which may be the consequence of changes imposed in the conditions in which such events take place. The supremacy of causal determinism on vital phenomena becomes evident. Claude Bernard spread this view in the course of many highly important writings but particularly in his Introduction to the Study of Experimental Medicine.

Organ by organ, tissue by tissue, cell by cell, the facts of life were explored. The field is bounded by space — in organ, tissue and cell -and by time between the moment a cause commences to act and the instant when its effect is produced, and phenomena are conditioned therefore according to determinism.

Still more is achieved; it becomes possible for cells, and tissues, and eventually the organs also, to continue in existence long after they have been removed from a living creature. From this we come to the conclusion that the parts act and work on their own — like the individual parts of a machine. The whole, whether it be an individual plant or animal, is nothing more in the end than the sum of its parts living in community. The problems of life are solved by analysis, that is, by observation and experiment on the properties of the separate organs.

## FUNCTIONAL REGULATION AND CO-ORDINATION

There are of course certain specialized, particular and local functions worthy of individual study. The truth is, nevertheless, that these do not compose the whole of life, which latter is something fuller and more comprehensive than the use of each one by itself and than the summation of individual parts. In among the common herd of experimenters, who brew up the hotchpot of science, we find a few thinkers of greater vision who warn us, in the face of general opinion, that life also contains other phenomena of wider range than the particularized functions of the organs.

Our attention is first called to the existence of complex mechanisms embracing the functional effect of various organs and then we start the study of inter-organic physiology, the physiology of correlation.

Claude Bernard himself, when referring in 1867 to the internal medium in animals, the significance and importance whereof he placed on record, shows that such internal medium must inevitably keep its properties and composition uniform, effecting this by means of complex physiological adjustments which entail many diverse organs; processes of "continuous and delicate balancing like the equilibrium obtaining in the most sensitive of chemical balances".

It was Bernard also who when reporting on the Annual Award of the Paris Academy, claimed by Cyon on the ground of his discovery of depressor reflexes by cardioaortic receptions, pointed out the importance of the visceral regulative reflexes, by whose intervention functional links became set up among the various evidences of activity in the organs.

Of late years there has been an abundant flow of relevant material, and a number of important books have been published.

Physiological treatises now devote many a page to discussion of the interfunctional processes, which previously had been dismissed with scarcely a passing comment. Cybernetics, the irrepressible study of control mechanisms, arises from the idea of communication, self-regulation, co-ordination, and physiological adjustment between organs, the collective and the individual psyche. The whole face of physiology has changed in modern times.

## PARTIAL AND TOTAL INTEGRATION

Now if this is true in physiology, all the more so do biologists need to make thorough investigations of vital phenomena. The naturalist observes occurrences developing naturally to their ultimate extent; the physiologist carries out experiments, setting down certain conditions and finding the consequences thereof.

His judgment has to be determinist, since its source lies in certain causes leading to certain effects. He contemplates fact by sections and brings his ascertained data into their relevant frameworks. Events present themselves mechanistically and therefore the physiologist is usually a mechanicist. In spite of this, during these latter years there has come about, even in the field of physiology, a reversion to unitary ideas of wholeness, totality and fittening of the functions.

The biologist considers extended series of events which develop without his direct participation. The chain of causation does not thrust itself forward, whilst the synthesis, which stands for life and the opportunity for such events to take place, is exhibited without experimental distortion, but in a precisely natural and factual manner.

Since the field under survey is so vast, problems of biology seem to be more difficult than those posed by physiology, and it is logical that the naturalist should have a greater leaning towards vitalism. This standpoint and that of the physiologist are two distinct criteriological positions explaining the diametrical opposition of their different opinions.

In all ages, biologists studied the reactions of live bodies as if these were complete, that is to say something more than mere isolated mechanisms within the organs. Roux, Whitman, Driesch, and many others, always supported the unitary principle, and it is safe to say that in every vital manifestation, particularly in ontogenetic growth for instance, as previously mentioned in this book, certain cellular and local properties coincide with certain other general properties belonging to the organism as a whole. During progressive multiplication of the zygote by the increasing number of blastomeres, certain automatic differentiations come into operation due to which each cell element of the embryo evolves on its own according to its nature and intrinsic properties and at the same time dependent differentiations also take place through correlation between parts and regulatory processes, as previously mentioned.

Embryogeny continues its development by balancing particular tendencies against general ones. Special inductor, organizatory, evocative, and differentiatory stimulants are necessary to morphogeny, but they must work at any and every instant in agreement with the essential "vital" properties, inherent in each cell and in the whole which is the overall combination of cells.

Form, structure and function make an organization, as Boerhaave wrote, and an organization is something distinct and something more than the mere sum of its separate parts. It possesses properties intrinsic to it as a whole which are not the mere result of a simple addition of the properties of separate elements. The whole exists as a whole and is a real entity quite as evident as any of its factors.

## HOLISM

Starting with Smuts, and Mayer it has become the custom to call general responses holistic, in contrast to meristic local responses. Holism, as it is now called, has recently acquired a certain notoriety, being considered by some as an undefined principle — a brand-new archeus — arising out of the general reactions of organisms, say one of those many theoretical "principles" which time and again have been proposed under the illusion that mere wordy invention of a figment is sufficient to explain a fact. Similar in nature are Monakoff's "hormé," the "libido," and the old-time "neurine," and the rest of the bunch. Unremittingly, time and again whenever events become complex and experimental, determinist analysis becomes difficult there breaks into the field of science some theory of markedly vitalistic colour and distinctive nature.

The "holistic principle" postulates regular working of the organs in the individual and extends even far beyond their direct action. From their presence and activity we should likewise derive the products of human labour in general: civilization, the state, society, culture, arts and sciences. And, according to Mayer, even the very laws of the physical world are subject to the holistic principle, which thus becomes a type of divinity of universal range governing everything and from which nothing can ever escape.

Were this so, then physical laws would have to be deduced from biological concepts and not the reverse as has always been supposed. That would mean coming from complex concepts downwards to the simpler and thence to the simplest, whence the universe would become a holistic aggregate of superimposed totalities each set up upon those below to form a general synthesis — a wholeness of wholenesses. Thence we should get a series of relative entireties ever increasing in their turn, each one complete as far as relates to those of lower rank but being at the same time merely one single part of the next higher. Relations between totalities add up to a combination of all types of force in the universe, which together with the material elements set in motion by such forces make up the substrate of the universe.

Holism, according to Leeman, pertains to a concept of energy whilst existence is an overlaying and synergy of interdependent dynamic fields arranged along the rungs of a ladder of totalities. The basis of these fanciful speculations is well known; it is coordination of the functions, vital unification in the functions and in morphogeny, and fundamentally metabolism. This is the truth as verified by facts, though the conclusions at which we arrive from this starting-point may perhaps be absurd — that depends on any particular writer, and on his followers.

Strict observation of reality teaches us nothing if not that life in the organs develops with its own specific peculiarities, dependent nevertheless upon conditions imposed on and by the organism as a whole with its own particular characteristics in its turn. Thus we learn that the functions are related, added together or crossed in their paths, making up something more than a mere simple addition sum of their parts and forming therefore an organization with everything that the word organization connotes and bears in its train; we learn that the functions develop as for the set purpose of a better fulfilment of themselves and the preservation of life by means of ever increasingly complex mechanisms, perfected and efficient in their action.

Such observation tells us also that the organism is a unit — the individual -wherein local and general responses take place according to stimuli, with more or less extensive help from different parts, whence, that is from general and local reactions one by one, we get the characteristics of the individuality itself.

## ORGANISMAL BIOLOGY

The biology which takes account of general events and their teleological manifestations has come to be called organismal biology. An organismal biology must simultaneously cover causal determinant relationships — which admits no discussion — and more general events which seem at times to be sheer probabilities or contingencies likely to elude the actual possibilities of determinist explanation. It is claimed that such a biology might bring mechanicists and vitalists on to common ground, though, up to the present, organismic synthesis has not been able to dispense with hypothetical conjectures of agency of more or less mysterious principles in the course of biological events, leading thus to still another aspect of vitalism seeking now to become refurbished and brought up to date.

Mechanicism and vitalism are two standpoints of contemplation of the universe, depending on the field of vision, from whence problems are considered, according to the temperament of the student himself so that it becomes difficult to reconcile opinions which are antagonistic from their very roots up. The man of science does not have to demand from events more than they can offer as means of knowledge. Theories, the tools of the sciences, are needful and aid investigation, but they never attain the validity of factual explanations confirmed by practical experiment and experience. It is absolutely safe to say that physiology took on a new complexion at the start of the present century; that the organism must be looked on as a whole with its parts working together harmoniously as a compact entity and in a fitting manner, this being the distinguishing feature between the events of life and those of the purely physical world. This, however, being shown and granted, it still remains somewhat extravagant to state outright that fresh facts give the lie to the reality and governance of a biological causality determining the operations of the living organism.

## BIOLOGY OF THE PARTS AND BIOLOGY OF THE WHOLE

Biology daily discovers fresh prospects and marvellous new domains, wherein a tremendous amount still remains unexplored. Up to the present time, biological investigation has been analytical, studying vital events piecemeal and in isolation. We now know that this dry "meristic" standard means little, and that we must also consider life as a whole, using empirical methods to analyse events as presented naturally by observation, thus bringing reality home to us in its natural state, free from our own interference and a consequent liability to unwitting modification of true conditions.

This does not by any means signify that the experimental method is bankrupt, but only that nowadays we have brought before us vaster, more complex and more difficult problems. The extent and deepness of the task need not dismay us, for the fact that these aspects do vary shows us life by one means or another in its manifold and protean forms. In all these cases we shall have to employ the only means of investigation of the universe available to man, our senses and reasoning, our powers of natural or stimulated observation and the logical arrangement of what we have observed.

Whenever a problem has bristled with difficulties, human intelligence has attempted to solve it at the first shot by attributing the observed facts to the agency of some active principle or principles, projected by the spiritual faculties of mankind — knowledge and will — which are thus satisfied with a metaphysical explanation and not a scientific answer. In the second stage man critically observes the facts and discovers a few causal relationships between events; then the mystery is cleared up and the principle is conjured away. Thus step by step, with all the attendant difficulties, we continue to build up our scientific knowledge, though this does not imply that any science can dispense with the metaphysical atmosphere which stands at the peak of all. Hour by hour, man pierces further into the Universe by dint of blood and tears. With every progress, the immensity of the unknown, whose compass we shall never fully know, becomes more and more evident.

At some not too distant date the flamboyant holistic principle will go the same way as its predecessors. This will happen when the events on which the theory is founded can be interpreted scientifically and not metaphysically by establishing what the conditions are which determine such happenings. It is true that these questions are of immense amplitude and their analysis is correspondingly of extreme difficulty, though this will show in the end that as the parts function so does the whole.

Life is unity, even though its events are of such widespread variety; nevertheless this life in any of its multifarious appearances is ruled by simple and regular laws. There is nothing extraordinary in the fact that the particular activities of the organs operate one by one, being organized anatomically and physiologically, and that there may be unity in vital responses to external or internal stimulations, and further, that the functions answer a set purpose, being directed to the maintenance of the life of the individual and of the species.

## JOSÉDE LETAMENDI DE MANJARES

First of all you must bear in mind that the times are not favourable for organic, monistic or individualistic theories; that those of you who rally to my banner will have tremendous battles with physiologists, pathologists, therapeutists, analysts, particutarists, anarchists, and therefore in order that you shall not be caught unawares, I will from the very start do away with the capital objection, or in better words, the sole one which with many variants will be shot at you regarding the unity of man. "Individual unity," you will be told, "does not exist; man is the result of a combination of parts and a knitting together of forces." Let us consider what this objection is really worth. Right away, regarding its form, this is not an objection, it is not an argument, but a mere unproven statement.

As to its essential nature, if we cannot talk of one man, even less can we speak of one lung. Why? Because the relations of man with the universe are those of contiguity, whilst those of the lung are of continuity. Still more obviously: in order to separate the individual from the environment he lives in, no cutting instrument is required, whilst in order to remove one of his organs from an individual it is essential to cut it out. Well then, where stands unity? Perhaps in the tissue part of the organ? No! Perhaps in the immediate principle? Still no, for the self-same reason of continuity still applies! Is it then resident in the physico-chemical atom? There is some difficulty in asserting this, because the atom, a metaphysical creation of the human mind, is not an object of which we have any positive knowledge.

Thus it is then, that in the analytical or descending scale from the individual downwards, we have no knowledge of any natural unit other than that of the individual. Whence then does harm spring? From the corrupt direction of reasoning in the application of analysis to the science of the individual, whereby it has been held that analysis is itself constituted a scientific advance, it so being that in anatomy and physiology analysis alone establishes the means of attaining progress, which latter consists in the enlightenment of the synthetic idea of man, in such manner as nature shows him to us.

On this point we are getting worse every day, every day we withdraw more and more from such synthetic idea; we have come to a complete loss of any individual concept. From the death of Hippocrates until the advent of Cartesian philosophy in the seventeenth century, that is to say for more than two thousand years, this concept was sustained, for better or for worse, more or less clearly in its expression.

However, since that time, soul and body being divorced from each other by Descartes, the former remaining the plaything of the erudite and psychologists, whilst the latter became the business of doctors and surgeons, the doctors of the Renaissance who fell into the graceless and dangerous extreme of having to base the whole science of anthropology, or shall we say their whole knowledge of man, solely on the investigation of the body, whilst by dint of dissecting and still further dissection, inquiring and still further inquiry, experimenting and still further experimentation, we today have now arrived at so hopeless an elementalism that the most enthusiastic exclusivists, upholders of analysis without end, are already complaining that we are getting lost in a sea of detail, whilst neither in recent physiological treatises nor in those of general pathology.

Does there remain even a shadow of any of those things, such as character, synergy, temperament, diatheses and many others, which antiquity conceived as a theoretical and practical expression of individual unity and which in the event that antiquity perceived them well and interpreted them badly it was the duty of us moderns to have them put right, and not to have suppressed so much as a single line of fact, whilst they remained throbbing with reality in our hospitals and clinical wards.

Consider that by preserving the individualistic method of Hippocrates in making a diagnosis, and from such prognosis taking the patient as a single whole, whose morbose manifestations, consilient or symptomatic, however incoherent they might appear, form the empirical basis of the qualification of the illness and an indication of its probable course, medicine might throughout its vast sphere have been able to advance for all time, not in spite of physiological and anatomical progress, which has been so disturbing of ideas in other directions, but by virtue of such progress.

And if to this we add that advances in the physical sciences, equally upsetting from the clinical viewpoint so far as they claim to be the fountain-head of medical theory, have produced industrial marvels applicable to investigatory and exploratory techniques.

We shall get a full idea of the inestimable benefits medicine owes to modern research.In short, symptoms which in the time of Hippocrates were taken, so to speak, skin-deep from what very naturally was learnt from the patient, are pursued nowadays by the doctor who sounds for them as if they were pearls and coral buried in the deepest depths of the body.

Now you see the conclusions to be drawn from a dispassionate examination:

- That medicine, in all those artistic processes to which the organs lend themselves, has prospered by what they contain of the particular;
- That medicine has flourished in everything relating to diagnosis, prognosis and treatment founded on observation and experience over the whole individual in his clinical unity;
- That medicine in everything relating to the building-up of theory, far from thriving is now in a worse state clinically than what it was in the time of Hippocrates, in so far as the ingenuous and prudent ignorance of this great master, protected by an exquisite sense of unity, has been replaced by a vain and imprudent analytical rashness, not only basing the science of all mankind on the consideration of a single one of its anatomico-physiological elements but also varying even such ground-work in accordance with the deceptive tunes played by every fresh experiment. Ponder deeply on the importance of such aberration; reflect that in medicine, as in any human business, principles are the core of the matter, and there is not a single clinical problem but what, at the least expected moment, a serious question of principle might arise, for though our organs contain a modicum of independent parts, they hold also much of direct dependence on the whole.

## JOHN SCOTT HALDANE

About the middle of last century the younger physiologists broke away from the vitalistic traditions which had been handed down to them, and set about to investigate living organisms piece by piece, precisely as they would investigate the working of a complex mechanism. This method seemed to them to promise success, and was popularized by such masters of clear and forceful expression as Huxley. It is still the orthodox method of physiology, but the old confidence in it has steadily diminished in proportion as exact experimental investigation has shown that the various activities of a living organism cannot be interpreted in isolation from one another, since organic regulation dominates them. The keynote of this book is the organic regulation of breathing and its associated phenomena.

The mechanistic theory of life is now outworn and must soon take its place in history as a passing phase in the development of biology. But physiology will not go back to the vitalism which was threatening to strangle it, and from which it escaped last century. The real lesson of the movement from that time will never be lost.

This book belongs to a transition period, but the transition is forward and not backward. My treatment of the subject may possibly be looked on askance in some quarters as reactionary: for I have been largely influenced by the ideas and work of older physiologists. If, however, I have gone backward, it is only to pick up clues which had been temporarily lost; and all of these clues lead forward — forward to a new physiology which embodies what was really implicit in the old.

The leaders of the mechanistic movement of last century got rid of vitalism, but in doing so got rid of life itself. I have tried to paint a picture of the body as alive. Though the picture is imperfect, others will soon paint it more completely. The time has come for a far more clear realization of what life implies. The bondage of biology to the physical sciences has lasted more than half a century. It is now time for biology to take her rightful place as an exact independent science: to speak her own language, and not that of other sciences.

The endeavour to represent the facts of physiology as if they would fit into the general scheme of a mechanistic biology has led, it seems to me, to the present estrangement between physiology and medicine. Since the time of Hippocrates the growth of scientific medicine has in reality been based on the study of the manner in which what he called the "nature" of the living body expresses itself in response to changes in environment, and reasserts itself in face of disturbance and injury.

The underlying assumption is that organic regulation and maintenance represent something very real, and that only through the study of it can we recognize and interpret disturbance of health, and effectively aid maintenance or restoration of health. I have endeavoured to return to what seems to me the truly scientific Greek tradition, and to give it a form which is not only consistent with modern science and philosophy, but brings physiology and medicine into that close and special relation indicated by the common etymology of the words "physician" and "physiology."

Most of the investigations specially referred to in the book have been carried out on man. It was only by human experiments that the almost incredible delicacy of the regulation of breathing was discovered; and human experiments have revealed to us in other ways how rough many of the experiments on animals, or on "preparations" from the bodies of animals, have been.

Organic regulation, with its all-important relations to practical medicine and surgery, was often entirely overlooked. I hope that the book may contribute towards establishing human physiology in its rightful place, which has been usurped too long by experiments on fragments of frogs and other animals, or on the mere superficial physical and chemical aspects of bodily activity.

## THE INTERNAL MEDIUM

The stability of the internal medium is a primary condition for the freedom and independence of certain living bodies in relation to the environment surrounding them. Physiological mechanisms have to function therein assuring the maintenance of conditions necessary for the existence of the cell elements composing them. For we know that there is neither liberty nor independence in the case of the simplest organisms in direct contact with immediate universal circumstances. The possibility of arranging their own internal medium is an exclusive faculty of organisms which have reached a higher stage of complexity and organic differentiation.

Such stability in the internal medium implies an extremely perfect organism, able continuously to balance outside variations. The greater the freedom of the creature with regard to its external environment, the closer will be, on the other hand, the connection of its cells with such internal medium, which will necessarily have to maintain perfect regularity in its qualities, possible only if it has regulatory processes in operation as precise as the most sensitive chemical balances.

The conditions necessary for existence which such internal medium must find will be, first of all, water, oxygen, warmth and food reserves. The same conditions are essential for the continued existence of even the simplest organism. In a polyplastid creature, maintenance of regularity in the qualities of the internal medium are achieved mainly by the intervention of the nervous system, the agency of the functions producing harmony between the activities of the whole of the body. This is attained thanks to properly co-ordinated physiological mechanisms which consolidate the activities of even the most remote organs.

To explain these remarkable facts, there is no need to resort to the agency of a vital principle opposed to the influence of physical conditions and possibly peculiar to and characteristic of living organisms. It can be shown that regulatory processes consist exclusively of interfunctional physiological mechanisms.

## LUDWIG VON BERTALANFFY

We have made an attempt to think out the implications of the organismic view in the light of modern physics. It need scarcely be said that, in view of the extraordinary difficulty and newness of this problem, we do not wish to defend the foregoing discussion of it in any dogmatic spirit. Our chief aim has been to draw attention to the problem and to urge others to investigate it. We would, in any case, point out that as a method of investigation organismic biology is quite independent of those ultimate decisions which we discussed in the last section. We cannot indicate this pragmatic attitude better than has been done by Woodger:

"If the organism is a hierarchical system with an organization above the chemical level, then it is clear that it requires investigation at all levels, and the investigation of one level (e.g. the chemical) cannot replace that of higher levels. This remains true quite apart from the remote future possibility of expressing the properties of all higher levels in terms of the relations between the parts of the lowest level."

Woodger gives an excellent summary of the reasons why an exclusive attention to physico-chemical explanation is not desirable in biology. To this the reader may be referred for a supplement of what has been said above. The question whether physical concepts at present suffice for scientific biology must be answered in the negative, because neurology, experimental embryology, and genetics — to mention only the more important branches — employ purely biological concepts. To the question whether these concepts will be replaced by physical ones in the future, we must answer: wait and see.

It has often been objected against organicism that the organismic point of view — although of philosophical value - is of no importance for the work of natural science. The organism is something with which the method of natural science cannot work — a hard, smooth, round nut which experimental analysis can neither crack nor lever open at any point without it exploding and vanishing like a Prince Rupert drop.

Now, it is quite true that the non-additive character presents great difficulties to scientific treatment. But precisely the same difficulties are confronting certain branches of physics — and are here successfully overcome. Köhler, for example, points out the difficulties of a mathematical treatment of the structure of electric charges, since it is impossible to determine the charge first in this place and then in that, because the charge at any given place depends upon that at all the others.

Consequently, with the usual additive methods of physics we cannot deal with the problem. The problem must be solved at one stroke as a whole, and this physics has done in an admirable way by means of the theory of integral equations. "No one who has closely studied this part of mathematical physics will ever assert that all physical structures have a purely additive character."

It cannot therefore be said that the "concept of organism" is opposed to scientific treatment. On the contrary we might say that physics has already been dealing with "organisms," with Gestalten, although of a low degree of complication. All that remains of this criticism is the assertion that biology has not yet regarded the organism as a system (as contrasted with an aggregate), and that this is forbidden also for the future. But this is a dogmatism on the side of mechanism which is no better than that of vitalism, when the latter declares that science will "never" be able to explain this or that property of living things. We can therefore summarize the demands of organismic biology as follows:

Since the fundamental character of the living thing is its organization, the customary investigation of the single parts and processes, even the most thorough physico-chemical analysis, cannot provide a complete explanation of the vital phenomena.

This investigation gives us no information about the co-ordination of the parts and processes in the complicated system of the living whole which constitutes the essential "nature" of the organism, and by which the reactions in the organism are distinguished from those in the test-tube. But no reason has been brought forward for supposing that the organization of the parts and the mutual adjustments of the vital processes cannot be treated as scientific problems. Thus, the chief task of biology must be to discover the laws of biological systems to which the ingredient parts and processes are subordinate.

We regard this as the fundamental problem for modern biology.

Since these laws cannot yet be formulated in physical and chemical terms, we are entitled to a biological formulation of them. In our view, the question of a final reducibility of such biological laws is of subordinate importance in view of the foregoing demand.

Even without this final decision, the antithesis between mechanism and vitalism ceases to be a troublesome problem. The mechanist who believes in the possibility of such a reduction, and the vitalist who denies it, can join forces in an attempt to solve this great problem: the establishment of the laws of biological systems.

The investigation of these laws must proceed in two directions. On the one hand, the empirical rules of organic systems must be obtained from the concrete, especially experimental, data. And on the other hand, it must be the final aim of biology to derive the laws of organisms deductively from general assumptions.

## THE CONCEPT OF ENTIRETY AND INDIVIDUALITY

Whatever may be the theory of personality, one thing is certain in practice, namely that there are distinct psycho-physical entities. What is the true essence of such entities? Or, to put it another way, what is the real living nucleus of a psycho-physical entity? Let us examine this question in greater detail.

A dog "possesses" paws, eyes, coat, hair, and so on; an oak-tree "possesses" leaves. What exactly is the "possessor" in these cases? The usual answer to that is of course "the dog," or "the oak-tree" as the case may be. What, however, is the dog or the oak-tree?

Could it be the dog's body, or the oak-tree's trunk shorn respectively of paws, or leaves? Is it not rather in each case a matter of entirety which becomes incomplete the moment its paws or leaves are taken away? Yet, what do we mean, in this context, by these words "entirety" and "incomplete"? Now other complications come into the picture. Suppose instead of a dog, we take a common or garden worm. If we cut this worm's head off, it will grow again. Who "possesses" this power of regeneration, or better still, who "possesses" the power of reproduction of life? Surely not the mere material in the worm's body as such!

We have previously spoken of a deeper analysis of the concept of entirety, but must admit right off that such analysis, namely, a precise definition of the concept, is absolutely impossible, since the meaning of "entirety" or of "individuality" is within everybody's reach but only as an outcome of direct intuition.

It is the same thing with many other ideas. If we try to "define" the meaning of "not," "connection," "so many," for instance, we find that definition is possible only if we make use of the meaning of the very thing we are seeking to define; in other words, we can only get a "definition" by going round in circles.

This is what is happening right now with the concept we are in the course of discussing. It occurs time and time again, except in the field of pure logic, where concepts or meanings are dealt with solely in their quality as such in themselves. The reader acquainted with Kant's philosophy will realize that the so-called "deduction" from the category of entirety or individuality can be deduced from the "table of estimates" if the conjunctive estimate S = P 1 + P 2 + P 3 + . . . + P n be combined with relative criteria. Just the same, I consider Kant's inference to be a type of self-deception, for what is there, pray, in the "table of estimates" which is anything but mere direct intuition?

This digression, however, seems interesting for if, despite everything, we still accept the Kantian thesis, we must reckon also with a possible fresh type of deduction. We may now say that a concept is "entire" when its essence can be destroyed in proportion to the subtraction of any element from the whole. The concept A in such case no longer remains concept A, but becomes replaced by a wide variety of not-A concepts. Suppose A is a "red right-angled triangle"; if we remove the red colour, we are left with only the "right-angled triangle" idea which is no longer the original A concept, but a new one, B. This holds as far as relates to logical individuality or entirety.

Let us now look at the same question of individuality or entirety from an objective viewpoint, in the domain of things or events, which is empirical reality. Things and events are not ideas, but as designated through concepts, even when we evade any metaphysical interference. We know that a table, a steam-engine, a dog, a tree, and the rest form an "entirety." Can this be taken as a definition? Again, we might say, in a purely logical sense, that an "entirety" is something whereof the essence distinguishing it would be destroyed were a part of it removed.

What portion or how many parts have to be taken away from a dog for it to stop being a whole? Say a paw! All right, then; but now consider the dog's coat. If you shave it completely, you can perhaps say that the doggy essence has been destroyed because its essential feature is the very fact of having a coat. Now suppose you cut off three hairs only! Or, if you prefer it, shave off everything but three hairs! How many hairs must you remove before the dog ceases to be a "whole"?

In the earthworm, it will be impossible for us to say that beheading it destroys its wholeness, because the head still exists latent in the de-capitated portion of the body and can be restored through regeneration. We must therefore be satisfied with saying that it is through some kind of instinct that we know what objective entirety may be. This is, however, not enough because some objective standard of entirety still seems to be needed.

We shall get closer to the solution if we say a thing or an event is complete when countless cases or copies of it exist, each one with the same typical shape, which can be used as a standard of comparison against any particular example we possess. Such an answer is still not perfect, since works of art, for instance, may be complete though there is usually only one specimen of each.

We might also say: "an entirety" or "complete whole" is what gets restored when its material structure has been disturbed in some way or other. There are indeed many organisms able to regenerate, as well as artificial products of human activity, industry and art, which can be put together again. Nevertheless cases frequently occur also of living creatures, mammals and others, which though obviously forming an entirety are incapable of repairing any damage they have suffered.

In my opinion, the only valid standard of entirety or individuality would be if we said that "entirety" is something owing its origin to non-mechanical natural agents and is at the same time something rather more than a mere "sum" of its parts. This standard implies that non-mechanical natural agents tend to entirety through their own conditions. "Entirety" however is not the only idea we need for logical understanding of organism.

We cannot do without combination or union of the ideas of entirety and causality, the discussion whereof will lead us to examine the important question which may be considered as the keystone for a logical justification of our vitalism, namely whether different elemental forms of cause and effect are logically possible, or whether causality must always be mechanical, even though we use the word "mechanical" in its widest sense.

For, if causality were wider than summative or mechanist causality, our theory of free will in life would have received some definite logical justification.This is what we are after. Biological concepts have to be subjected to logic, taking this term in its most general sense of a "theory of order," allowing us to conceive empirical reality as "one great ordered whole," wherein every singularity of being or occurrence occupies its own precisely determined right place. In this way, we should realize what I have termed a "monism of arrangement," which will nevertheless continue to be but an ideal, since the structure of reality has two aspects, a blend of causal entirety and fortuitous non-wholeness.

In any event, it is reasonable to ask oneself whether inanimate, non-living nature could also partake of entirety. Henderson, in his famous book The Fitness of Environment, revealed clear pointers on this when he observed the "fitness of the inorganic world for the development and existence of life"; in other words, the pre-established harmony between existence and environment. Let us say, in conclusion, that our present knowledge of entirety and individuality is still a bit shaky, since we lack a stable and suitable standard of judgment. Only personal organism, fortunately, forms an exception. The biology of the person is the prototype of the doctrine of objective entirety and individuality, accurately individualizing causality and evolution.

Human life is a strange reality concerning which the first thing to be said is that it is the basic reality, in the sense that to it we must refer all others, since all others, effective or presumptive, must in one way or another appear within it. The most trivial and at the same time the most important note in human life is that man has no choice but has to be always doing something to keep himself in existence. Life is given to us; we do not give it to ourselves, rather we find ourselves in it, suddenly and without knowing how. But the life which is given us is not given us readymade; we must make it for ourselves, each one his own.

Life is a task. And the weightiest aspect of these tasks in which life consists is not the necessity of performing them but, in a sense, the opposite: I mean that we find ourselves always under compulsion to do something but never, strictly speaking, under compulsion to do something in particular, that there is not imposed on us this or that task as there is imposed on the star its course or on the stone its gravitation. Each individual before doing anything must decide for himself and at his own risk what he is going to do. But this decision is impossible unless one possesses certain convictions concerning the nature of things around one, the nature of other men, of oneself. Only in the light of such convictions can one prefer one act to another, can one, in short, live.

It follows that man must ever be grounded on some belief, and that the structure of his life will depend primordially on the beliefs on which he is grounded; and further that the most decisive changes in humanity are changes of belief, the intensifying or weakening of beliefs. The diagnosis of any human existence, whether of an individual, a people, or an age, must begin by establishing the repertory of its convictions. For always in living one sets out from certain convictions.

They are the ground beneath our feet, and it is for this reason we say that man is grounded on them. It is man's beliefs that truly constitute his state. I have spoken of them as a repertory to indicate that the plurality of beliefs on which an individual, a people, or an age is grounded never possesses a completely logical articulation, that is to say, does not form a system of ideas such as, for example, a philosophy constitutes or aims at constituting.

The beliefs that coexist in any human life, sustaining, impelling, and directing it, are on occasion incongruous, contradictory, at the least confused. Be it noted that all these qualifications attach to beliefs in so far as they partake of ideas. But it is erroneous to define belief as an idea. Once an idea has been thought it has exhausted its role and its consistency. The individual, moreover, may think whatever the whim suggests to him, and even many things against his whim.

Thoughts arise in the mind spontaneously, without will or deliberation on our part and without producing any effect whatever on our behaviour. A belief is not merely an idea that is thought, it is an idea in which one also believes. And believing is not an operation of the intellectual mechanism, but a function of the living being as such, the function of guiding his conduct, his performance of his task.

This observation once made, I can now withdraw my previous expression and say that beliefs, a mere incoherent repertory in so far as they are merely ideas, always constitute a system in so far as they are effective beliefs; in other words, that while lacking articulation from the logical or strictly intellectual point of view, they do nonetheless possess a vital articulation, they function as beliefs resting one on another, combining with one another to form a whole: in short, that they always present themselves as members of an organism, of a structure. This causes them among other things always to possess their own architecture and to function as a hierarchy. In every human life there are beliefs that are basic, fundamental, radical, and there are others derived from these, upheld by them, and secondary to them. If this observation is supremely trivial, the fault is not mine that with all its triviality it remains of the greatest importance. For should the beliefs by which one lives lack structure, since their number in each individual life is legion there must result a mere pullulation hostile to all idea of order and incomprehensible in consequence.

The fact that we should see them, on the contrary, as endowed with a structure and a hierarchy allows us to penetrate their hidden order and consequently to understand our own life and the life of others, that of today and that of other days. Thus we may now say that the diagnosing of any human existence, whether of an individual, a people, or an age, must begin by an ordered inventory of its system of convictions, and to this end it must establish before all else which belief is fundamental, decisive, sustaining and breathing life into all the others.

## INTERNAL FINALITY

Nevertheless, in the group of things we call organisms, such mechanical explanation cannot be claimed until a most precise and detailed knowledge has previously been obtained first of the form and next of the functions of such organisms. Nor can such knowledge be built up without abundant use of the principle of finality.

Indeed, organisms are chunks of reality formed in such a manner that there is to be discovered within them what Kant calls an internal finality or purpose. A thing possesses internal finality when it is itself both the cause and the effect of itself. A tree produces another tree like itself, which is to say, from the viewpoint of species, a holm-oak, for instance, begets itself, is its own origin and cause and its own effect or result.

However so it is also from the individual viewpoint, for the oak evolves and develops its whole concept by internal power; it does not harbour, as a mechanical addition, its material elements but is also nourished on them, converting them totally and embodying them into its own form. In the living organism, the preservation of the whole depends on the preservation and working of its parts, and these in their turn depend on the survival of the whole. If the addition of fresh elements in live organisms is not something extra but an assimilation, similarly a taking of them away is equivalent to death or at least demands that a neighbouring organ shall fulfil by proxy the functions of the organ which has been done away with.

The parts of a live organism are, therefore, organs, that is to say, that far from having no concern with the whole, their form and working must be determined according to the idea of the whole body. The totality of the organic being is an end, whose maintenance is purposed by the parts or organs. An end, comprised in a concept or idea, must decide a priori everything which has to be contained in it.

That, however, is not enough. Any object of human ingenuity, a watch for instance, is at the same time a combination where the idea of the whole determines from beforehand its composition and the shape of its parts. We do not, because of that, say that the watch possesses internal finality or purpose.

In order that a thing may possess internal purpose, that is, for it to be an organism, it is necessary in addition that the complete whole arises in its turn from the shape and working of its parts. These two requirements appear to contradict one another. On one hand, the organism in its whole conception determines the special organs and their working, whilst on the other it is the organs and their operation which beget and preserve the organism.

This contradiction is exactly the principle problem in biology as a science of existence. Existence or continued life is precisely an internal end, a system of forms wherein each part is determinate and at the same time determinant, wherein each part begets the whole and is in turn begotten in accordance with the design of the whole. An organism is an organized being which organizes itself. The watch of our illustration is on the other hand a thing organized by something outside itself.

## FINALISM AND MECHANICALISM

The foregoing statement expounds adequately the part played by the principle of finality in the science of live creatures. Internal finality is not an explanation of life, but a specific character of the latter, a character which has to be explained mechanically.

Biology's problem is to give a mechanical explanation of life. Life, however, shows itself to our observing faculties with an internal purpose. Thus, the knowledge we need to acquire from forms (morphology) and from functions (physiology) will be necessarily founded upon the principle of finality; we shall always have to investigate the why and wherefore of such-and-such form and of such-and-such function.

Live beings, then, appear before our observation as internal systems of ends and means. If, however, biology be limited to a study of forms and functions — morphology and physiology — it would be nothing more than a mere description of what is and what happens, interpreted in a fashion convenient for mankind, but it would not be scientific knowledge nor would it be an explanation of existence.

Biology purposes to find such explanation; a host of modern theories seek to formulate it with more or less success. We do not here have to go into its study. It has been sufficient for us to find the indicative, regulative, auxiliary function which the principle of finality possesses in the study of organisms.

*Encyclopaedia*
*of*
# BIOLOGICAL SCIENCE, TECHNOLOGY AND ENGINEERING

# *Encyclopaedia of* Biological Science, Technology and Engineering

**Volume 2**

**Dr. Albert Shawn**

**ANMOL PUBLICATIONS PVT. LTD.**
**NEW DELHI-110 002 (INDIA)**

**ANMOL PUBLICATIONS PVT. LTD.**
**Regd. Office:** 4360/4, Ansari Road, Daryaganj,
New Delhi-110 002 (India)
Ph.: 23278000, 23261597
**Branch Office:** No. 1015, Ist Main Road, BSK IIIrd Stage
IIIrd Phase, IIIrd Block,
Bangalore-560 085 (India)
Tel.: 080-41723429
*Visit us at*: www.anmolpublications.com

*Encyclopaedia of Biological Science, Technology and Engineering*



First Edition, 2009
ISBN 978-81-261-4121-0 (Set)

PRINTED IN INDIA

Printed at Paras Offset Pvt. Ltd., New Delhi

# Contents

## Vol. 3

# Preface

Biology is the study of life and a branch of the Natural Sciences which studies living organisms and how they interact with each other and their Environment. It examines the *Structure, Function, Growth, Origin, Evolution, and Distribution* of living things. Also, it classifies and describes Organisms, their Functions, and how species come into existence. Four unifying principles form the foundation of *Modern Biology*: *Cell Theory, Evolution, Genetics and Homeostasis.*

Biology as a separate science was developed in the Nineteenth Century as scientists discovered that organisms shared fundamental characteristics. Biology is now a standard subject of instruction at schools and universities around the world, and over a million papers are published annually in a wide array of biology and medicine journals.

Most Biological Sciences are specialized disciplines. Traditionally, they are grouped by the type of Organism being studied: *Botany, Study of Plants; Zoology, Study of Animals; and Microbiology, Study of Microorganisms*. The fields within biology are further divided based on the scale at which organisms are studied and the methods used to study them: *Biochemistry* examines the fundamental chemistry of life; *Molecular Biology* studies the complex interactions of systems of *Biological Molecules; Cellular Biology* examines the basic building block of all life, *Cell;* Physiology examines the *Physical and Chemical Functions of the Tissues and Organ Systems of an Organism;* and *Ecology* examines how various organisms and their environment interrelate.

Author

# Chapter 16

# Structure and Functions of Animals

The study of development has for some years suffered from an inferiority complex. As an area of interest it has held its rightful place alongside genetics and evolution and, in fact, this is the very cause of the complex. Ever since Darwin there has been a general satisfied feeling among biologists that the basic outlines of the mechanism of evolution are clear and understood. Beginning with Mendel, and certainly ever since the work of Morgan and his group in the second decade of this century, there also has been a feeling that the basic principles of genetics are known.

But development is considered the helpless third child that never is capable of growing and maturing properly; in fact it seems to remain in a perpetual state of disorderly puberty. How many times have I heard the rather wistful refrain that nothing really important in the study of embryology has turned up since the time of Spemann, or Roux, or Aristotle. The despair has placed development in the position of at least one of the great unsolved problems of biology. Matters are not improved by the close identification of development with cancer.

The nature of cancer is not known nor can it be cured in any very rational or specific way, and the hope expressed in many dollars is that, like anthrax, it will suddenly let all its secrets out in a dramatic way and will be cured in the fashion antibiotics cut down bacterial diseases. It is nothing but human nature to wait for the great flash of illuminating lightning that will not only solve the cancer problem, but the problem of development along with it.

Yet I think those who are actively working on the problem of cancer or those who are studying normal development have little hope for such a flash. Rather, they see the problem as exceedingly complex, with many facets and therefore possibly many answers.

One could argue that this pessimistic feeling comes only because they are so close to the details of their experiments that they cannot see the possibility of anything larger, any theory of development of the satisfying and general character of the theories of evolution and genetics. However, if one stands back a few paces, as we shall attempt to do here, the conclusion remains the same: there does not seem to be any place for a new, grand, and unifying scheme.

This does not mean that there is no interest or profit to come from studying development. But, it does mean that if development is put in its proper perspective, especially with relation to genetics and evolution, then we might cease to suffer any complex of inferiority, cease to expect any single, monumental solution, and have a clear picture of where to find healthy progress.

The first problem is that the trilogy of which we have spoken, evolution, genetics, and development, may in itself be misleading because, when these three subjects are placed together, the natural implication is that they can be equated, like an apple, a pear, and an orange because they are all fruits. A moment's reflection will show that this is far from the case, and, again by analogy, it would be more like describing the essentials of the functioning of an automobile by listing fuel, mechanical parts, and combustion.

These three are not in any one class; the only thing that brings them together is that together they give at least a partial description of an automobile in operation. By the same token, evolution, genetics, and development combined, give a partial description of the living world and now let us examine how these fit together. This will necessitate the saying of some very simple and obvious things but perhaps this can be justified by our object of putting the area of development in its proper setting.

## EVOLUTION

By evolution we mean here its mechanism which is, of course, natural selection. If selection is to operate, two things are required, variation and reproduction; variation to give some basis of choice and reproduction to provide new individuals to harbor the variation. So we immediately see, in our first glance, a relation between the three major areas, for variation is the subject of genetics and reproduction in the sense used here is the subject of development.

It is useful to think of the variation in terms of adaption for this implies variation which has been satisfactory in a particular environment. In examining any one organism, be it a plant or an animal, the point that needs emphasis is that all phases of its life history are adap tive; there is no activity or structure that can escape the fine-toothed comb of natural selection.

This statement does not seem very remarkable until one considers the details of what it means. It means in the first place that the adult, in all its structure and in all its activities, is adapted to its surroundings. Perhaps so much emphasis has been placed on this obvious fact that other equally important ones have been unintentionally neglected. For instance the life span of an organism is adaptive as Weismann was the first to emphasize. Senescence, therefore, is a controlled variation. On the other end of the life history, the whole of development is adaptive.

Development is so often thought of as simply a means to an end, the adult, that its own adaptive significance is neglected. But each stage, each step is adapted to its immediate environment as well as being full of the proper potentialities for future adult adaptations.

After all, there is no more reason why the preparation for maturity cannot be as subject to variation and selection as the period of decay and senescence. This is directly implied in the old idea of Minot who considered senescence to start at the beginning of development and that the whole life span is one of decreasing rate of senescence.

The emphasis is the correct one: an individual organism is not merely an adult but it is the whole life history. This also means that gametogenesis and all aspects of the sexual system are adaptive as well. Again, this is an old idea but important here because it emphasizes the closeness of the relation between evolution and genetics. In Darlington's phrase there has been an evolution of genetic systems or, in other words, the very methods of producing variation are themselves capable of variation and adaptation. It is like a toolmaker who improves the particular tools he uses in his trade.

## GENETICS

Sexuality, which is the most common and successful method of handling variation, is peculiar, however, in that the amount of change that has occurred during the course of evolution is negligible when compared to the changes in other aspects of the life history. For instance, chromosome structure and the basic mechanism of meiosis and fertilization are fairly uniform from the most primitive to the most complex animals and plants. The reason no doubt is that it is such an extremely successful system that any basic change would be inadaptive. It has been preserved and changed only in minor ways because of its great selective advantage.

There are also some well-established alternatives to sexuality as a means of handling variation. Among bacteria we have transduction and there is also the fact that, by direct mutation alone, microorganisms, which have short rapid generations, can have a reasonably effective variation system.

Another alternative that should be mentioned is heterocaryosis, found especially among fungi, for here there is no recombination on the chromosome level but there can be recombination of nuclei in that the spores can have different combinations of dissimilar nuclei to start new variants. In fungi, also, there is a special kind of mitotic chromosome recombination which Pontecorvo has described and termed parasexuality.

In some peculiar organisms such as the cellular slime molds where there is an aggregate of cells which may be of a diverse genetic constitution we may have a parallel to heterocaryosis and, for this reason, it has been termed heterocytosis. Finally, there are many known instances where the inheritance is handled in the cytoplasm rather than in the nucleus and here, as with the single gene mutation mentioned above, we must expect no recombination, but single direct passage of possibly mutable entities from mother to daughter cell.

All these methods of handling variation, from the sexual ones to the various simpler types, have one thing in common. They require that the key events occur in single cells. This is especially striking in the case of true sexuality for meiosis and fertilization involve the formation of two haploid nuclei fusing to form one diploid zygote nucleus. But, it is equally true of other systems to varying degrees.

This point takes on significance when one recognizes the fact that many organisms, indeed all those which are considered to have a true development, are multicellular (or multi-energid).

The reason for this is no doubt that there simply has been, for these forms, an adaptive advantage to an increase in size. Such being the case, we are faced with a paradox: variation must be transmitted and handled by single cells while the adult is multicellular. The inevitable result is clearly a development and therefore one can say that development is the result of reproduction and size.

## DEVELOPMENT

If we were to summarize the argument thus far concerning the relation of evolution, genetics, and development one could say that evolution through natural selection is the over-all controlling factor in all change. This change is dependent upon self-perpetuation with variation and, therefore, one of the first steps in evolution has been to set up an effective genetic system, an effective system of handling and transmitting variation. It so happens that all the genetic systems, perhaps partly because they evolved at an early stage in evolution, involve single cells and single nuclei.

Therefore, the variation transmission stage is one having small units. Because there has also been a selection pressure for large organisms, the inevitable result is that the gap between the small reproductive cells and the large adult must be filled in by development. But, each step in this development is in itself adaptive to its particular environment. So the evolutionary mechanism lays down the ground rules where each alteration is closely scrutinized by selection; the genetic mechanism, which itself is molded by selection, produces the currency, the method of producing variation so vital to selective processes; and, finally, development is merely a stage in the life history as the variations of evolution become more complex.

To put it another way, one cannot produce a large complex organism from a single zygote cell in one magic explosion. The very fact that it is so large and intricate means that it must be built slowly with care and system. It is no different from the problem of building a chicken coop compared with that of building a skyscraper. Evolution, genetics, and development do closely tie in one with another, but they each are of a fundamentally different nature; they are complementary rather than supplementary.

## EPIGENESIS

To come now specifically to the problem of the elaboration of the egg to a large adult, we must first keep in mind one of the oldest lessons of embryology. It is simply the idea that one step, one set of conditions leads to another, and that developmental progress is a causal sequence of events. This is the epigenesis of Aristotle and William Harvey; this is the causal embryology of Wilhelm Roux in his Entwicklungs-mechanik.

In modern times, the only significant addition that has been made to this basic scheme of sequential events is that they are controlled and guided by the activities of the genes. Development is now thought of as a period of particularly active gene action. The conclusion is indeed obvious if one remembers that the genes are all in the zygote, along with any cytoplasmic factors that may be of hereditary importance and they are the determinants of the ultimate mature structure. They guide all the steps from fertilization to decrepitude.

Perhaps the reason we think of development as unsolved is that it was in fact solved in basic outline so long ago and the solution is so simple that it has escaped us. When one thinks of its relation to genetics and evolution, and when one thinks of its unfolding from egg to adult as we have just done, then can we not say that the fundamental theoretical framework of development is known and understood; it is an epigenetic system.

There is no doubt that this is a correct statement but there may be some question in the minds of biologists as to whether this really represents a solution. Of course, the word "solution" in itself is bound to be unprecise and contusing because we are not dealing with an algebra problem or a theorem of Euclid.

What we mean by "solution" here is some vague notion of mental satisfaction to the enquiring scientist. As far as development is concerned there are still things he wants very much to know. It is the hope of any experimental biologist that his discoveries will lead to far-reaching conclusions. The contention here is that we might predict that the conclusions will never be of the magnitude of Mendel's laws or Darwin's theory of natural selection; the generalization of that importance lies in the epigenetic theory of development. But there is ample room for lesser generalizations within this framework. It will now be our task to attempt some understanding of what type of generalizations these might be.

## STIMULUS-RESPONSE SYSTEMS

With this in mind, it will be helpful to think of organisms as reaction systems, that is, systems capable of responding to stimuli. To see this clearly one can think of many non-living systems which are able to respond. Besides complicated machines which can be activated, for example, by photocells, one need only turn to the simplest chemical solution containing substances that can combine. In this case, the stimulus, or the activation energy of the physical chemist, might be light or heat or even physical movement, and the chemical components will react or respond. In some responses the stimulus may be highly specific while, in others, a slight energy change of any sort may produce the reaction.

It is not usual in chemical reactions or many types of biological reactions to use the terms stimulus and response: more often the biologist guards these terms specifically for nerve or muscle physiology. But they are so generally useful that here they will mean any system that is capable of reaction to either a specific or a general source of activation energy.

It is helpful to gain perspective of biological processes by attempting to imagine how they might have arisen in the origin of life. In the case of stimulus-response systems, one of the first major steps must have been to develop a system that keeps going, a system that is an energy machine, a machine that keeps converting energy in a steady state. This is a very different proposition from a chemical solution that reacts once and then no more. A living system has a continual set of stimuli and a continual set of responses so that, in essence, the motor keeps running.

To achieve this kind of stability it is necessary, even in the case of the simple maintenance energetics of an organism, to have a complex set of interlocking stimuli and responses. It is really only through complexity that any kind of a steady state can be achieved.

Along with the complexity, another factor enters into the making of a primitive organism in the origin of life. It is that the complex stimulusresponse system must be packaged and be physically defined—it cannot spread out indefinitely into the environment any more than can a clock or an automobile. The cell is of course considered to be this metabolic or energy unit.

The next step is for a system of variation and reproduction. Both of these require further and more elaborate stimulus-response systems. In reproduction we must have the machinery to synthesize and duplicate parts, and in variation we must have machinery to allow orderly and controlled change.

Before proceeding to a discussion of development in these terms let us pause to examine the implications of what has been said. In the first place, we have pointed out that metabolism, reproduction, and variation can be thought of in terms of a sequential series of chemical reactions. The important thing is that they are controlled and this is achieved by a number of factors. First of all there is the packaging, the unifying and isolating of the components of the machine. Second, by making many of the stimuli and responses specific, they can only proceed in one way.

This specificity can be achieved by specific chemicals (i.e., enzymes) and by localization of the materials in specific sites within the cell package. Third, the most important feature of control or stability is having many parts, many steps, many sequential stimuli and responses. This is an insurance against catastrophe and error.

It is particularly easy to find examples of this principle in the world of mechanical and electronic machines; the more control devices and feedback systems, the greater the chances of smooth and uneventful operation. Therefore, with the notion of sequential stimulus-response systems, we are proposing to look at life as a communication system.

## DEVELOPMENT AS A SEQUENTIAL STIMULUS-RESPONSE SYSTEM

Development, which comes from having small gametes and zygotes and a large adult, is necessarily a period where there must be violent changes, changes which can all be conveniently thought of in terms of our stepwise stimulus-response systems. In this we are saying nothing new, but merely using different words to say that development is an epigenetic process where one condition leads to another, and does so in a channelled and controlled fashion.

By a different route we have again arrived at our ancient but large generalization about development. Does this mean that to understand development completely we must methodically unearth the details of each step so that they finally can be put together like a string of beads upon a necklace?

We have said that development is a series of interlocking stimulus-response events; what else is there to do but to reconstruct each event? Is this not the reason why those working in the field have the strong notion that there is not one but many answers? Before answering these questions let us glance at this approach and see what progress has already been made along these lines.

## CHEMICAL EMBRYOLOGY

The first bravely to recognize and use this approach are the "chemical embryologists." After all, the steps are primarily chemical events and their elucidation must therefore be by chemical means. The progress in this field, along with biochemistry in general, has been one of the great advances in the last few years, but the unfortunate thing is that, with all our new knowledge, our understanding of development has remained distressingly static. Part of the difficulty is simply that it is a new field and the techniques alone make such rapid strides forward that, in some cases, the work of a few years back is wholly out of date and must be repeated.

Another problem is that often it is possible to discover beautiful changes in specific chemical constituents from one stage of development to another, but it is not possible to correlate these changes with morphological events in any causal fashion. A classic example of this is the case of fertilizin and antifertilizin discovered by F. R. Lillie in the early part of this century. Despite the large amount of work done on these substances by many workers since then, we still do not know with any certainty what role they play, what step they mediate in normal fertilization.

A further difficulty is that often the biochemistry itself is not wholly understood and, therefore, its application to development must necessarily wait. Perhaps the most alarming of all thoughts on this matter is when one considers how many chemical steps there must be between the beginning and the end of development. Of course it might be possible to generalize to some extent and it would indeed be helpful to discover that certain types of reactions are associated with certain types of developmental events. In any case, it is hoped that the chemical embryologist will not be discouraged, for, ultimately, it is imperative that we have the kind of information he seeks to supply.

## BIOCHEMICAL GENETICS

Another related approach of increasing importance is that of the biochemical geneticist. His interest, which is really identical, is of geneaction during development and he has the advantage of attacking the problem from two sides simultaneously: he has the genetic information as well as the chemical information of the developing organism. The fact that these studies have largely been carried out in molds and bacteria rather than vertebrates is of no disadvantage as far as the general problem of development is concerned. In simplest terms, the biochemical geneticist seeks to determine the steps involved in the synthesis of a particular substance.

He discovers various mutants that will block the synthesis and then by adding different precursors to the medium he can ultimately find a whole chain of events leading to the formation of the substance. In some cases, the genes involved are found on various chromosomes, but there are some striking cases where the genes are in a sequence on a chromosome—the same sequence as the synthetic reactions.

It is perhaps somewhat confusing to equate the synthesis of one substance with development, but basically this is reasonable for, after all, development must be a series of syntheses one following the next. Also, it should be said that recently the attention of biochemical geneticists has turned to the correlation of gene-controlled chemical changes with morphological changes.

Therefore, the interest is the same as that of the chemical embryologist but the scope of the studies is extended to examine not only the chemical changes during development, but their genetic control as well.

## NUCLEO-CYTOPLASMIC RELATIONS

Still another approach to the problem, and this one is on a less detailed level than the chemical level, is to attempt to determine the respective roles of cytoplasm and nucleus in controlling the course of development. Nowadays, it is not so much whether all the factors lie within the chromosomes or whether the egg cytoplasm also makes an appreciable contribution, although this question still needs further resolution, but rather it is how the nucleus behaves during the course of differentiation.

The old concept was that the nucleus remains the same in all tissues of the differentiated adult, largely because it seems to undergo little or no morphological change; the latter is confined entirely to the cytoplasm. But, in recent years, the emphasis has shifted partly because of the work of Beerman and Pavan who showed that the chromosome in developing larvae of certain insects show puffs in different regions correlated with different areas of the embryo, and the implication is that the chromosomes are behaving differently in a regional fashion.

The main reason, however, for the shift in emphasis comes from the important work of Briggs and King who have shown by ingenious transplantation experiments that the nuclei of frog embryos, after a certain stage, become altered and that this alteration is fixed and will pass to the daughter nuclei of one parent nucleus.

It is unfortunately not yet clear what are the nature of these alterations, nor is it known what are the real implications of these facts as far as the mechanism of development is concerned.

But it does seem to be a bit of information of great potential that might lead to significant new insights into development. It could be the beginning of a more complete understanding of how the signals of the stepwise stimulus and responses pass from the nucleus into the differentiating cytoplasm.

Thus far, all the approaches to the problem of development have to varying degrees attacked detailed steps.

As has already been said this is an absolute necessity, but it must nevertheless be admitted that the result to date has been depressingly small as far as elucidating all of development is concerned.

To put it baldly, the biochemical geneticist, after some years of work, may piece together five enzymatic steps in the synthesis of one substance. During the development there may be a thousand or many more of such steps. The advance has been more in how the steps occur, for surely their pattern might be a recurring one, but still the whole problem seems gigantic alongside the small chips of progress.

## MORE GENERALIZED APPROACHES

There are, however, other approaches to the problem of development than reconstructing each step. They involve trying to see patterns, trying to outline the structure of the stimulus-response sequential communication systems. The work on the respective roles of the nucleus and the cytoplasm is already in this direction, for certainly this is on a more general level than the chemical steps themselves. But this is still not at the stage where it can give us any useful generalizations about development; generalizations that give feelings of greater understanding and that connect many isolated facts together.

One approach on the more general level is that of Waddington who has met the problem of the complexity of development head on and attempted a method of dealing directly with it. This is not an experimental approach but rather the seeking of a reasonable perspective, and therefore it has no possibility of providing any sort of solution to specific problems. As Waddington himself suggests, it is helpful because it "provides a rough and ready picture of the developing embryo," and he adds that it cannot be interpreted rigorously.

The picture he has in mind is his "epigenetic landscape," which is a valley running downhill, of varying depth and branching, down which the developmental process proceeds. The valley emphasizes the fact that the process is controlled in its direction, or to use his term, it is "canalized." He also has suggested the word "creode" to emphasize the fact that this valley is a special kind of flowing equilibrium system which restores itself after a disturbance.

With this scheme in mind, he proceeds to speculate as to how the whole epigenetic landscape is under the influence of genes and that evolutionary change is related to development by virtue of the complex relation of the genes to the system of epigenetic canals.

If his picture were translated into the terms that have been used in this essay, the valleys are the stepwise stimulus-response sequences. The fact that they are specific and controlled is equivalent to the stability implied by the term creode. The reason for emphasizing the sequential communication system, as has been done here, will now be evident because it allows us to answer the question posed before: as to whether or not it is possible to make any useful generalizations about development below the level of the epigenetic generalization, but above the specific details of each chemical step.

## TRANSCENDENT STIMULUS-RESPONSE COMMUNICATION SYSTEMS

Up to this point, we have made the tacit assumption that the steps of development are all minute increments. The implication has been that these gene-controlled changes occur within cells and that they have been operating on a minuscule level. It is easy to show that this is not necessarily the case and, in fact, the broader the sweep of the communication system, the greater is its importance as far as our understanding of development is concerned in the search for a satisfying solution.

Clearly, the smallest communication steps must occur within cells, perhaps even within different parts of the nucleus or between different parts of the cytoplasm, and certainly, as we have already discussed, between the nucleus and the cytoplasm. But, over and above these intracellular stimulus-response systems, there are ones that transcend the cells and constitute an intercellular communication system. Since these are particularly important during development they will presently be discussed in greater detail, but now I should like to point out the fact that one can extend this argument and show that there are also examples of communication systems between parts of organisms, such as between organs. In fact, one may continue even further with this progressive extemalization of stimulus-response systems and find examples between organisms in a kind of extension of development.

Examples within a species are easily found in the relations between males and females during courtship or between young and parents during the period of parental care. Some of the most striking examples come from social animals such as the various types of social insects.

There are also excellent examples of close communication between different species such as the factors which lead to commensalism or any kind of mutualism. It is not within the scope of this essay to consider these matters in any more detail but the point that needs emphasis is that specific, stepwise, gene-controlled stimulus-response systems can operate on many different biological levels.

Now, if one asks what have been the important advances in the study of development in the present century, it is possible to show that they each involve the elucidation of some larger, transcendent communication system. The fact that these communication systems sweep across cell boundaries and integrate the developing embryo automatically provides us with information of a more encompassing sort; it provides us with satisfying factual generalizations about development. In other words the contention here is that if one considers the most striking discoveries in the field of development, they all consist of extracellular communication systems operating between parts of the embryo.

## INDUCTION

The most famous example is the discovery of induction by Spemann. It makes no difference that, in the beginning, he first ascribed too many powers to the organizer in the dorsal lip of the blastopore. He showed that one part of the embryo stimulated another part and the importance of this contribution was immediately felt and recognized. The refinements that followed, those of Holtfreter and others, which led to the discovery that the stimulus was chemical, were of great if not equal importance because they also bore upon the elucidation of a major stimulus-response system in the embryo. The only agonizing part of the tale is the search for the identity of the chemical inductor substance, the evocator, but this is agonizing because its identification seems to be something so desirable, so worth knowing and yet so elusive.

Even in very recent years, contributions on the nature of induction are in turn ranked as important studies and we could list, among others, the contributions of Holtfreter who has emphasized the role of cytolysis, Barth, Niu, and Twitty who have studied induction in explants, Yamada, Toivonen, and Nieuwkoop who have independently examined the nature of regional induction.

There has also been the extremely interesting work of Grobstein on the matter of contact between the induction source and the responding system. This is a shamefully abbreviated list of the many current contributions to the subject of induction but its purpose is to emphasize that this phenomenon has been one of central concern in the study of development, and my contention is that the reason for this is that induction is a major bit of communication between parts of the embryo.

As a corollary in the study of development in plants there is the discovery of the plant growth hormone by Went and others which acts as a stimulus, an inductor. It also is produced in one part of the plant, capable of movement through the plant, and capable of inducing responses in different parts of the plant. This discovery can certainly be favorably compared with induction in animal embryos. Each has produced comparable interest among biologists, and each has been continuously pursued since its inception, and each is an extracellular communication system between parts of the embryo.

## GRADIENTS

Gradients are another transcendent communication system in developing organisms. We owe our awareness of gradients to Child who not only demonstrated their existence in many ways but showed how universal they were in developing animals and plants.

Because Child attached such special importance to the metabolic aspect of gradients, he perhaps never received his due credit, but, as time progresses, we seem to be gaining a better perspective of his contribution.

Rather than emphasize metabolic rates, we would now prefer to emphasize the gradation of particular chemical constituents. Inevitably, such gradations lead to differences in parts and these in turn provide a step towards ultimate morphogenesis and differentiation. The gradients transcend the cells and, by the production of local differences, there are differences in stimuli and therefore responses. Again it is a major communication system that pervades the whole embryo. Its importance is unquestioned and although it has not produced as many studies as the subject of induction, the double gradient system of developing sea urchins has, in the work of Hörstadius and others, provided valuable material for experimental analysis.

## COORDINATED CELL MOVEMENTS

As a final example of important modern work concerned with extracellular communication, the studies on the coordination of cell movement during the so-called morphogenetic movement of development are of special interest.

First, there were the pioneer discoveries of Harrison and, particularly, Weiss on the sensitivity of the moving cells to the substratum. Later came the notable observations of Holtfreter of the coordination of this movement by a surface coat in the amphibian embryo; all the surface cells are attached to it and it binds, leads, and guides the cells so that they produce the beautifully-controlled movement of gastrulation.

In this case, the stimulus is a physical one of a binding to the surface, and the response is by the contact guidance recognized by Harrison and Weiss. The most recent development along these lines is the work of Weiss and Andres, Townes and Holtfreter, and particularly the work of Moscona on the dissociation of cells in the embryo similar to the sponge dissociation experiments of Wilson.

The point is that the mixed-up cells wander about in masses of tissue of varying sizes so that ultimately they find their proper place according to their future fate. Here the communication is distinctly between cells and, in some way, they can stimulate and respond to one another as they touch each other so that like cells adhere and unlike cells pass on. The implication of this work as far as normal development is concerned is still uncertain, but the resolution of this problem in cell communication is clearly important and will be eagerly pursued.

These examples of major communication systems in developing organisms are by no means complete. For instance, there is considerable interest in cases of chemotaxis during the course of morphogenetic movements. Also, one might make the prediction that totally new and unsuspected stimulus-response systems that transcend cells will be discovered and carefully pursued.

This is, it would seem to me, the level where the most fruitful generalizations about development are likely to continue to appear. I find it hard to imagine that the study of development will ever spawn a Darwin or a Mendel, but, within the framework of epigenesis, there are many fascinating and intriguing puzzles to be elucidated. These necessarily will involve the untangling of the stepwise stimulus-response communication system. This should not mean that the subject is now cut and dried even though to some extent the course is set. There may be many delightful surprises when we discover the steps themselves. The purpose of this essay is not to discourage, but solely to put development in its proper setting;

to see what it consists of, and to see its relation to genetics and evolution.

## LUMINESCENCE

There are certain caves in New Zealand where the ceilings are covered with thousands of glow-worms, each dangling a long luminescent thread. Any sudden noise causes the larvae to turn off their lights, almost as one. After a brief wait, a few of the bolder insects will switch on, followed by a few more, and suddenly the cavern will again be aglow with an eerie light.

The New Zealand glow-worm represents just one of a surprisingly diverse group of plant and animal species which exhibit bioluminescence —the biological emission of light from living organisms. Many animals manufacture their own light—usually in shades of blue but in some cases in red, orange, yellow, green or white—but none are higher in the animal kingdom than the fishes.

Although bioluminescence (in fungi) was known to Aristotle and studied by such notables as Francis Bacon and Robert Boyle, it was not until 1887 that the mechanism behind this phenomenon was discovered. In that year, physiologist Raphael Dubois extracted juice from clams that were graced with light. He discovered that the juice contained two substances which, when mixed, were responsible for the glow. One he named luciferin after Lucifer, the light bearer, and the other he called luciferase to indicate that it had the properties of an enzyme.

Recently, and under the aegis of the National Science Foundation, the pace of research in bioluminescence has quickened. This renewed interest stems from the discovery that the processes a firefly employs to convert chemical energy into light can provide a sensitive instrument for biologists probing the secrets of cell metabolism.

One of the men most deeply involved in current research on bioluminescence is Dr. William McElroy of Johns Hopkins University. With the assistance of Baltimore area school children, who were paid 25 cents for every 100 fireflies they collected, he was recently able to isolate luciferin, establish its structure, and synthesize it.

Dr. McElroy found that the firefly emits light extremely efficiently, producing approximately one basic unit or quantum of "cold" light for each luciferin molecule oxidized. This means that the firefly is able to convert almost all of the chemical energy of the bioluminescence reaction into light, wasting very little as heat. By comparison, an incandescent bulb dissipates most of its electrical energy as heat.

In the firefly bioluminescence results when luciferin combines with oxygen in the presence of luciferase and ATP—an energy-yielding compound. An unusually large amount of energy is liberated in this reaction, enough to convert the luciferin-luciferase complex from its lower energy ground state to an electronic configuration of higher energy, the excited state.

The molecule in the excited state may return to the ground state by losing its energy as a photon of visible light. Researchers are finding that, by borrowing the firefly's system of creating light, they are able to measure minute quantities of ATP in the laboratory. With this extremely sensitive technique they are studying the role of ATP in the mechanism of muscle contraction, nerve and brain function, photosynthesis, and many other biological systems. The enzyme luciferase in addition to serving as a catalyst for the bioluminescent reaction, also appears to control the color of the light.

This may be a clue to what goes on when the larva of the South American beetle, *Phrixothrix*, is disturbed and begins to crawl. When the larva is at rest at night, it glows red at one end; but when it starts to move, 22 small green spots arranged along the body suddenly light up.

With every light source ablaze, the larva's appearance strongly suggests its common name—the railroad worm. Recent studies have revealed that the color of bioluminescent light in many animals depends on slight alterations in the structure of luciferase (a protein). It may be, therefore, that the railroad worm uses one form of the enzyme when only glowing red, and brings a second form of this enzyme into play when it starts making tracks away from a disturbance.

Among the roster of organisms with night lights may be found bacteria (only salt water, aerobic types have been found glowing), toadstools, ocean worms, sea pansies, jellyfish, squid, sponges, crustaceans, snails, millipedes and fish. These bioluminescent animals employ various systems to convert chemical energy into light.

Consider, for example, the single-celled, flagellated organism *Gonyaulax*—one of many organisms responsible for the "red tides" of the ocean, and the phosphorescent glow stirred up by the wake of a ship. When the water around them is disturbed, these creatures emit flashes of light lasting some one-tenth of a second.

Recently Dr. J. W. Hastings of the University of Illinois, working under an NSF grant, discovered luminescent crystalline particles within the bioluminescent organism, *Gonyaulax polyedra*. Isolated particles, referred to as "scintillons," can be made to emit a luminescent flash that resembles the flash of the living *Gonyaulax*.

These puzzling scintillons react with neither the enzyme (luciferase) nor the substrate (luciferin) extracted from the organisms. Luminescence is triggered simply by increasing the acidity of the particle solution in the presence of oxygen. The flashing of these isolated crystalline subcellular particles constitutes a previously unknown bioluminescent system. Although he has isolated a typical luciferin-luciferase system within this same organism, Dr. Hastings believes that the scintillons are responsible for the characteristic flash of the living *Gonyaulax*.

Dr. Hastings discovered the scintillons while investigating the biochemical mechanisms involved in various clock-like rhythms exhibited by *Gonyaulax*. When exposed to alternating 12-hour periods of light and dark, cultures of *Gonyaulax* emit about 60 times more light when stimulated at the middle of the dark period than during the light period.

Also, the amount of luciferin and luciferase in each cell reaches a maximum at the middle of the dark period. Dr.

Hastings found that this biological rhythm is inhibited by a certain chemical known to block RNA synthesis. Thus, the cell's ability to synthesize RNA may be a crucial variable affecting the biological clock mechanism of bioluminescence.

Bioluminescence apparently plays a significant role in the lives of many marine organisms. Specifically, it may affect migratory activities, or speed and direction of swimming under different conditions.

With the aid of the bathy-photometer, a very sensitive instrument for measuring light, Dr. George L. Clarke (Woods Hole Oceanographic Institute) has obtained records of bioluminescent flashes at depths as great as two and one-third miles. A luminescence camera was used to photograph a rare jellyfish at a depth of 3,000 feet; the jellyfish provided all the illumination needed.

## A VESTIGIAL MECHANISM

Recent investigations continue to turn up still other chemical variants of the bioluminescent reaction. Dr. Frank Johnson (Princeton University) has discovered a new type of luminescent system in the jellyfish, *Aequorea,* which requires no oxygen or luciferin, as such, for the emission of light. In this system, a single protein component emits light in the presence of calcium. Using the light from this unusual bioluminescent system as a quantitative measure of the presence of calcium is the most sensitive method known for detecting trace amounts of calcium in various biological and other solutions.

However bioluminescence is produced by an organism, it may be that they are all carry-overs from a time many millennia ago when free oxygen first appeared in the atmosphere. A theory by Dr. McElroy and his colleague Dr. Howard Seliger (also at Johns Hopkins University) jumps off from the generally accepted assumption that life developed in the oceans before there was any oxygen in the air. The first organisms were therefore anaerobic, and the oxygen which was later freed from the water and soil would have been toxic to them.

These primitive animals must have possessed an energy transport system, and this system was best suited for eliminating the oxygen by converting it to water. The conversion would liberate enough energy to excite organic molecules to emit light. Even after the evolution of aerobic animals, the oxygen-removal system was retained by some organisms as part of the over-all electron-transport system still needed by the body. This theory gains credence from a report that, in most cases studied, bioluminescence appears to be produced by a nonessential enzyme system.

But even if bioluminescence relies on vestigial mechanisms, the reaction promises to be of great aid to biologists. Because each flash is easily observed proof of the transfer and liberation of energy, researchers are finding the phenomenon an ideal tool for monitoring the vital process by which the body converts food into energy. Also, the development of highly sensitive devices for measuring the emission of light, may permit this enzyme-catalyzed reaction to be used as a very discriminating indicator of the effects of various drugs or chemical inhibitors of cell activities.

## LIFE'S MYSTERIOUS CLOCKS

One of the greatest riddles of the universe is the uncanny ability of living things to carry out their normal activities with clocklike precision at a particular time of the day, month and year. Why do oysters plucked from a Connecticut bay and shipped to a Midwest laboratory continue to time their lives to ocean tides 800 miles away? How do potatoes in hermetically sealed containers predict atmospheric pressure trends two days in advance?

What effects do the lunar and solar rhythms have on the life habits of man? Living things clearly possess powerful adaptive capacities—but the explanation of whatever strange and permeative forces are concerned continues to challenge science. Let us consider the phenomena more closely. Over the course of millions of years living organisms have evolved under complex environmental conditions, some obvious and some so subtle that we are only now beginning to understand their influence. One important factor of the environment is its rhythmicality. Contributing to this rhythmicality are movements of the earth relative to the sun and moon.

The earth's rotation relative to the sun gives us our 24-hour day; relative to the moon this rotation, together with the moon's revolution about the earth, gives us our lunar day of 24 hours and 50 minutes. The lunar day is the time from moonrise to moonrise.

The moon's arrival every 29.5 days at the same relative position between the earth and the sun marks what is called the synodical month. The earth with its tilted axis revolves about the sun every 365 days, 5 hours and 48 minutes, yielding the year and its seasons.

The daily and annual rhythms related to the sun are associated with the changes in light and temperature. The 24.8-hour lunar day and the 29.5-day synodical month are associated most obviously with the moondominated ocean tides and with changes in nighttime illumination. But all four types of rhythms include changes in forces such as gravity, barometric pressure, high energy radiation, and magnetic and electrical fields.

Considering the rhythmic daily changes in light and temperature, it is not surprising that living creatures display daily patterns in their activities. Cockroaches, earthworms and owls are nocturnal; songbirds and butterflies are diurnal; and still other creatures are crepuscular, like the crowing cock at daybreak and the serenading frogs on a springtime evening. Many plants show daily sleep movements of their leaves and flowers.

Man himself exhibits daily rhythms in degrees of wakefulness, body temperature and blood-sugar level. We take for granted the annual rhythms of growth and reproduction of animals and plants, and we now know that the migration periods of birds and the flowering periods of plants are determined by the seasonal changes in the lengths of day and night.

In a similar fashion creatures living on the seashore exhibit a rhythmic behavior corresponding to the lunar day. Oysters and clams open their shells for feeding only after the rising tide has covered them. Fiddler crabs and shore birds scour the beach for food exposed at ebb tide and retreat to rest at high tide. The reef heron, though living many miles inland, appears to know when low tide will occur and leaves home each day just at the proper time to take advantage of it.

Synodical monthly breeding rhythms, geared to particular portions of the year, phases of the moon, and times of solar day, are common among animals inhabiting the sea. These rhythms assure that the eggs and sperm of a given species will be available at the same place at the same time. Each species has its own characteristic breeding time, which is often so precisely scheduled that we can accurately predict its occurrence.

If, for example, we should go to the water's edge in Bermuda about an hour after sunset during a three- or four-day period of a summertime full moon, we would witness within a few minutes the mating display of the Atlantic "fireworm." At that time the females swarm to the surface from their burrows in the coral rock and luminesce brilliantly until joined by the males.

At the same location, just before midnight either three to four days before or two days after a new moon, the swarming of the males and females of the glassy transparent shrimp, *Anchistioides,* takes place. Palolo worms of the Southwest Pacific swarm in huge numbers on the nights of the third quarters of the October and November moons, liberating their reproductive elements into the sea water just as the dawn breaks.

The breeding behavior of a small California fish, the grunion, is also exquisitely timed. Just after the moment of high tide, on nights from April through June, when the tides are at their monthly highest, these fish arrive at the beach in large numbers and ride the waves onto the sand. The fish quickly dig pits into which they discharge their eggs and sperm. Thus the new generation is able to develop over the period of a month without being prematurely washed out by the surf of the ordinary high tides.

Reproductive rhythms also occur in certain seaweeds. *Dictyota*, a brown alga, produces eggs and sperm on a monthly schedule. All the plants in one area may be synchronized to one phase of the moon, while those in another area will be regulated to another phase. The particular phase appears to be determined in some manner by the local tides.

Lunar reproductive rhythms are not restricted to sea dwellers. The reproductive cycle of the human female, averaging exactly the synodical month, indicates that here, too, exist both solar and lunar rhythms.

Though it might appear that such rhythms are merely the responses of organisms to rhythmic changes in light, temperature or the ocean tides, this is far from being the whole answer. For when living things, ranging from the single-celled Paramecium to flowering plants and mammals, are removed from their natural habitat and placed under conditions where no variations occur in any of the forces to which they are generally conceded to be sensitive, they commonly continue to display the same rhythms they displayed in their natural environment.

The fiddler crab, for example, normally darkens by day and pales by night, runs actively at low tide and rests at high tide. When removed from the beach to laboratory isolation from light, temperature and tidal changes, the crab continues to behave in synchrony with his fellows still free on the beach. The crab somehow possesses the capacity to measure accurately and simultaneously both solar-day and lunar-day intervals without the stimulation of light, temperature and tidal changes.

Seeds persist in their annual sprouting cycle under similar laboratory conditions. This persistent adherence to rhythms with sun- and moon-related periods under conditions in which the organism is isolated from any obvious manifestation of these time cycles strongly suggests that an inherent clock system is probably a universal attribute of life.

Although living clocks appear to function simply and precisely in their regulation of organisms in nature, they present baffling difficulties when they are studied in the laboratory. The inquiring biologist faces a problem comparable to that of an observer in space who tried to figure out the nature of man's artificial clocks by observing the activities of a coastal industrial town. In the daily activity rhythm, some persons would be early risers, others late. Some, the clam diggers, would appear to have a lunar-day timer. The employees of a factory with three equal shifts working around the clock, would appear to have no daily clocks. The general population would appear to have a seven-day clock reflecting no natural terrestrial period.

In studying the living clocks in nature we are confronted with a similar observational limitation; we can analyze the clocks only by observing the rhythmic phenomena they time. We must, in other words, work from what the clock does back to the clocks themselves. This sort of inference is simply illustrated in a study of the common fruit fly.

In its natural habitat, the adult fruit fly normally emerges from its pupal case about dawn. If this process is subjected to laboratory control, with the eggs being laid and allowed to develop in continuous darkness, the young flies will emerge at any time of the day. If the maggot-larvae hatched from the eggs are subjected to even a single light flash during this controlled period of darkness, however, the flies, when they emerge, will come out of their pupal cases at the same time of the day that they were exposed to the light flash.

This suggests that fruit flies have operating twenty-four-hour clocks; under natural conditions the clocks of the developing flies are set to local sun time while they are still maggots and alert the flies for emergence at dawn. In much the same way, the lunar-day activity rhythms of such creatures as oysters and fiddler crabs correspond to the tidal times of their local shores.

The adaptiveness of the basic timing system may be illustrated with organisms as unlike as beans and bees. The leaves of beans, which rise and fall in a regular daily sleep rhythm when the plants are kept in a constant dim illumination, may be induced to rise momentarily from their lowered position by a brief light stimulus. This induced behavior will continue day after day at the same time, without further stimulus, if the plants are kept in the same environment. Similarly, honeybees trained to come to a sugar-water feeding station at one or two arbitrarily selected times of day, will persist in this same twenty-four-hour food-seeking schedule for a few days even if the food is no longer provided.

This adaptiveness of daily rhythms is most useful when living things are moved rapidly to a new time belt. When, for example, a person flies from California to England, he arrives in England with his physiological rhythms of waking, body temperature, blood-cell count and hormone secretion still adjusted to the local time of California.

It takes at least eight or ten days for these physiological processes to adjust to the new local time. The various processes change at different rates; time of waking shifts fast, requiring only two or three days, while other rhythms shift more slowly. Hence there appears to be a period during which various bodily processes are shifting gears relative to one another. Recent studies by Dr. Mary Lobban and by Dr. Janet Harker, both at Cambridge University, suggest some possible consequences of having our rhythms out of their ordinary relationships with one another.

Doctor Lobban found evidence of stress in some human subjects who, during the protracted daylight of the arctic summer, lived by watches adjusted to indicate recurrent "days" of unnatural lengths.

Doctor Harker, working with cockroaches, discovered she could produce tumors by transplanting a hormone-producing neural element, the subesophageal ganglion, from animals with their running-activity rhythms set to the usual time of day into animals whose rhythmic changes had been artificially reset (by reversing the natural light-dark cycle) to be twelve hours slow.

It is well established that some animals navigate by the sun, the moon or even the constellations, orienting themselves relative to these objects at an angle which changes systematically with the rotation of the earth. Birds or bees also alter their usual direction of orientation when their natural clocks are reset by cycles of light and darkness to indicate a different time zone.

One can even predict the new direction of orientation from the interval by which the timing rhythm has been reset. This was dramatically illustrated in a recent study by Dr. Max Renner of the University of Munich. Honeybees were trained on the East Coast of the United States to fly northwest to a feeding station at one P.M., Eastern Standard Time. The trained bees were then taken by airplane to the West Coast. The next day, in California, the bees went seeking food about ten A.M., Pacific Coast Time. They were still on Eastern Standard Time. But they adhered to their previously learned sun angle, and now with the morning sun flew off in a different direction-southwesterly. In both cases the sun was, so to speak, in their left eye.

Daily changes of both light and temperature are primarily responsible for resetting natural rhythms to local time. The twenty-four-hour rhythms in an organism's sensitivity to light and temperature facilitate this resetting process. If a plant, for example, is exposed to light during the hours of darkness (when it is sensitive to light), the plant will immediately reset its twenty-four-hour rhythm of light sensitivity.

If the plant is then returned to its natural environment of uninterrupted periods of light and darkness, this adjusted rhythm will continue to be reset by a small amount each day until the plant's light-sensitive and light-insensitive periods match the appropriate periods of night and day. In a similar fashion a mouse's rhythm of light sensitivity is set so that its activity and its running occur at night.

Each species seems to utilize such characteristic sensitivity rhythms. It also seems probable that these same twenty-four-hour cycles of light and temperature responsiveness in some way assist living creatures to measure the changing lengths of night and day and thus adapt themselves to natural annual cycles.

When some animals and plants are kept under controlled conditions of unchanging light and temperature, their daily rhythmic activities are observed to occur, in some cases, slightly earlier, in others, a little later each day. This produces activity rhythms with periods which vary from the twenty-four-hour cycle. These latter rhythms also vary somewhat in their periods with changes in temperature and illumination and from one individual to another. Such variations from the twenty-four-hour cycle introduce a fundamental question. Are these variations the product of poor-quality living clocks which run fast or slow? Or are the clocks themselves precise—with the apparent inaccuracies ascribable to some other cause?

We think we have the answer to why the rhythms often appear to run fast or slow under these unchanging conditions. To the organism placed in the laboratory at a constant level of light and temperature these two factors will continue to have the most impact during the sensitive phase of its twenty-four-hour cycle of responsiveness to the environment. These two factors in effect will appear, therefore, to the organism to show a daily variation, but now the light and temperature cycles will seem to have become inverted.

Thus, during the sensitive period of its cycle—which normally falls at night—it will interpret the increased effectiveness of the light and temperature as indicating daytime. Employing the same splendidly adaptive machinery it uses in nature to reset its rhythm until the sensitive portion comes to fall in the darker, cooler nighttime, the organism keeps resetting its sensitivity rhythm a little forward, or backward, regularly each day in a futile attempt to adjust to the illusory "day-night cycles."

This simple, reasonable hypothesis, compatible with all our current knowledge, makes it probable that the living clocks are always precise in their time keeping. The alternative interpretation—that the inaccurate rhythms persisting under controlled conditions reflect inaccuracies in living clocks—is most improbable in view of the precision of the clocks under natural conditions where such resettings would not occur.

## BASIC CLOCK QUESTION

Scientists are reluctant to credit a phenomenon for which they see no plausible explanation. As a consequence, evidence for living clocks was for a long time essentially ignored by most biologists.

Recently, as our knowledge has become more refined, the phenomenon of timed rhythms was encountered so frequently by investigators in so many areas of biology that the possession of clock systems by living things became tacitly accepted. The problem of the nature of the clock systems had then to be faced.

Man-made clocks are of two general types—those with intrinsic timing and those with extrinsic. Examples of the intrinsic type are the hourglass, the pendulum and hairspring-balance clocks. Intrinsic clocks possess independent timing capacity and are useful any time and any place. Extrinsic clocks, like the sun dial and the electric clock, have no independent timing capacity but depend upon an inflow of timing information.

The sun dial depends upon the sun's shadow, the electric clock upon the sixtycycle alternating current. Extrinsic clocks are sometimes referred to as repeaters. The basic question concerning the clocks of living things is whether they are intrinsic or extrinsic or whether, perhaps, they are both. This question has long been debated by students of biological rhythms. The intrinsic—clock hypothesis has been generally favored over the years.

This seemed the simplest hypothesis to account for the persistence of behavior of rhythms in unvarying light and temperature—especially the apparent inaccuracies and individually differing periods of the rhythms, their adaptiveness and their persistence in organisms transported long distances eastward or westward.

The extrinsic hypothesis, on the other hand, was questioned because it demanded that living things be sensitive to still unidentified, subtle pervasive factors. Such an hypothesis was radically opposed to the universally accepted concept of "constant conditions" in the laboratory-which postulated that under controlled laboratory conditions we could truly isolate an organism from fluctuations of every factor of the environment to which the organism is sensitive.

The organism was deemed insensitive to such factors as magnetic and gravitational changes.

Since the rhythms persist in meticulously controlled "constant conditions," the working hypothesis of most investigators has been that the living clocks are intrinsic. The lengths of their natural periods, it was thought, are inherited and a consequence of fully independent cyclic biochemical transformations which would someday be explained solely in terms of the principles of physics and chemistry. But every investigation based upon this assumption found the living clocks unorthodox in terms of any ordinary biochemical scheme.

Though all ordinary metabolic processes are greatly speeded up or slowed down by raising or lowering temperature, respectively, the periods of the clocks seemed generally independent of temperature.

Crabs, as indicated by their color change and running habits, continue to measure accurately the periods of the day and the tides whether they are kept at 54° F. or 77° F.; dried seeds, as shown by testing samples periodically, display the same annual rhythm in their capacity to germinate whether they are stored in a freezer at -8° F. or in an incubator at 113° F. Similarly, none of the drugs known to alter the rate or character of metabolic changes seem to interfere with the accuracy of the basic timer. Clearly, if the timer is intrinsic, it is a most extraordinary mechanism. But this is exactly what one would expect of any good clock. And if the experimental conditions in which these rhythms persisted were really constant for the organism, the timer had to be intrinsic.

On the other hand, if it were proved that laboratory conditions hitherto presumed constant for the organisms were really not constant and that the organisms continued to receive outside rhythmic stimuli despite laboratory controls, a potential timing signal would be available for an extrinsic clock. And such an extrinsic clock could account readily for all the observed properties of the rhythms, including temperature and drug immunities, and the long rhythmic periods which are the same lengths as the geophysical ones.

## EXTRINSIC RHYTHMICALITY

It has recently become evident that organisms, even when hermetically isolated under so-called constant conditions, still derive information as to the geophysical rhythms from their environment. Such information must be transmitted by highly pervasive forces hitherto ignored by biologists. Let us briefly consider the evidence for this.

Oysters, collected in New Haven Harbor, Connecticut, were shipped to Evanston, Illinois. Kept in pans of sea water in a dark room, they continued for a time to open their shells widest when it was high tide in New Haven waters. By the end of two weeks the New England oysters had reset their rhythms to open their shells widest at the moon's zenith and nadir positions with reference to Evanston.

The oysters maintained thereafter this new schedule throughout the month they were observed. The zenith and the nadir positions of the moon—the two lunar positions on opposite sides of the earth which give rise to twice-daily high tides-are the periods of the moon's maximum gravitational effects upon the atmospheric tides over Evanston; this same gravitational effect would produce high ocean tides if Evanston were a coastal city.

Recently, from simultaneous studies of fiddler crabs collected from beaches with differing local tidal times, we have learned that these creatures are also able to reset their tidal rhythms of running activity to accord with lunar zenith and nadir, despite absence of any obvious cues as to these times. Obviously some subtle atmospheric fluctuation related to local lunar time is able to substitute for local ocean tides and reset the periods of maximum activity. The same factor must continuously signal the lunar periods.

Living things as different as fiddler crabs and potatoes continue indefinitely to display both solar-day and lunar-day rhythms in their metabolism while they are subject to constant conditions in a laboratory.

These two rhythms co-operate to provide both species with a synodical monthly rhythm. Plainly, the artificially isolated organisms continue to derive a wealth of information about outdoor atmospheric rhythms. Some of the evidence available is quite fantastic.

The solar-day tides of the atmosphere are expressed in rhythms of barometric pressure. The atmospheric tide rises in the morning, reaching its highest point about ten o'clock, and then falls to its lowest in the afternoon. The daily rhythms of organisms, even when they are sealed off from such pressure changes, are somehow associated with these daily tides of the atmosphere and their regular modification by lunar influences.

Nature provided us with a ready means for discovering this relationship. The daily tides of the atmosphere in temperate latitudes are continuously distorted by large irregular, weather-associated pressure changes. We know that both potato plants and crabs follow some outside factor reflecting the daily atmospheric tides because the daily cycles in their rate of oxygen consumption continuously reflect significantly the unpredictable distortions in the daily pressure cycles.

Since it is inconceivable that living things are provided with a detailed program of all the erratic weather changes which are to occur while they are sealed under conditions including unvarying light, temperature, humidity and pressure, we must conclude that information reaches them continuously.

All of numerous species of animals and plants which have been studied reflect, in their metabolic processes between five and seven A.M., the amount the barometric pressure changed between two and six o'clock that same morning. Their metabolism from five to seven P.M. reflects the amount the pressure changed between two and six o'clock that same afternoon. These times are highly specific; no comparable relationships are found at any other time of day.

In both winter and summer the atmospheric tide tends to rise in the morning; at the latitude of Chicago, the time of lowest tide occurs about two P.M. during the coldest months and about seven P.M. during the warmest months, and so the afternoon pressure tends to fall during summer and rise in winter. The metabolism of potato plants, even when they are sealed off from pressure changes, conforms to this annual pressure change.

A by-product of this study has been the astonishing discovery that the late-afternoon biological activity of nearly all living things, while only remotely related to mean daily barometric pressure of the same day, is intimately related to that of the second day thereafter. These organisms seem to be "predicting" the atmospheric-pressure trends. The biological explanation of the riddle is this: the afternoon pressure change which the organism's late-afternoon metabolism reflects is itself quite inexplicably tending to predict the barometric pressure trends two days in advance. Imagine the predicament of a meteorologist, sealed away like potatoes, and faced with the problem of weather prediction.

Fluctuations in the numerous factors in the earth's atmosphere are variously interdependent. The discovery that the metabolism of living things, even under hermetic seal, reflects erratic changes in such rhythmic environmental factors as outdoor air temperature, cosmic radiation and general background radiation was, therefore, not surprising. Collectively, these relationships, continuing year after year, compel us to conclude that the organisms, even in so-called constant conditions, are not fully shielded from their rhythmic geophysical environment.

Within the past year my associates and I have implicated magnetism as one subtle factor related to the rhythms. Terrestrial magnetism is known to fluctuate rhythmically with the solar and lunar periods. Additionally, the earth's magnetic axis, at an angle with its polar axis, wobbles as the earth rotates and produces a movement of the magnetic field in relation to the polar axes.

Snails, when oriented geographically, were found to display solar- and lunar-day rhythms in their tendency to veer from a true southward path, even while in presumably unchanging conditions. Experiments with magnets further proved the snails could perceive very weak magnetic fields. In addition, the snails' response to a bar magnet showed both solar and lunar rhythms.

The snails were able also to distinguish directions of magnetic fields and, therefore, they possess a magnetic compass. The snails' perceptive system for a magnetic field behaves like two rotating "directional antennae," the rotation of one is related to the sun, the other to the moon. For fields of equal magnetic strength, the field parallel with the snail's body was most effective when the sun or moon was above the horizon, while a field at right angles was most effective under other conditions. Hence, the magnetic receiver behaved like a living compass geared in with the living-clock system. This compass may serve as a navigational instrument.

These magnetic-field studies have introduced a subtle and pervasive geophysical factor with which biologists must now reckon. It seems highly probable that other animals will be found similarly armed with both a clock and a compass. We have, furthermore, experimental evidence suggesting that both flatworms and snails perceive changes in electrostatic fields, the fields surrounding electrically charged bodies. And, as with magnetic fields, the response seems to be regulated by the living-clock system.

The weight of all the evidence suggests that living clocks depend upon some universal-time, geophysical rhythm—one with simultaneous world-wide changes. Magnetic and electrostatic sensitivities support this theory, since these forces are known to fluctuate on a universal scale. If living things truly possess such a universal timer, they have always available what man terms Greenwich time and uses for purposes such as navigation and astronomy—a means for pinpointing their location on the earth's surface. Interestingly, studies of bee, bird and fish navigation have proved that these creatures are acutely aware of the heavens.

Experimental evidence suggests, therefore, that life, time and space, in the range of their terrestrial dimensions, are very intimately interrelated. It appears that the forces regulating the life processes are dependently related to their counterparts of the outside physical environment. The living organism is a diminutive oscillating system with periods paralleling those of the physical environment.

What kind of timing system did nature fit into the microscopic dimensions of single cells—a timer which could reproduce so unerringly the long natural periods?

Evidence suggests the primary timing system to be the movements of the sun, moon and earth. Nature provided means by which this timer could simultaneously serve the rhythmicalities of both living organisms and their environment.

Ingeniously the timing bridge to living systems was fashioned, not in terms of such variable and biologically potent forces as light and temperature (to which organisms must respond in specifically adaptive fashions), but in terms of more stable forces demanding little or no specific adaptive response and simultaneously so pervasive that no living thing would ever normally be deprived of their influence. Only with such provisions could living clocks become the loyal servant rather than the domineering master of life.

## AVIAN ADAPTATIONS

Birds are doubtlessly the most highly specialized of the vertebrates. Wings and feathers, two remarkable adaptations by themselves, have combined with other, more subtle changes to produce a highly efficient flying machine. In almost every conceivable way, the avian body is "designed" for flight, incorporating lightness, strength, and efficiency into a rigid, compact body with a low center of gravity.

The most distinctive characteristic of birds is their feathers, which among other functions, insulate the bird from the elements, streamline its body, and provide a lifting surface over the wing. Feathers are as varied in structure as the jobs they perform, but the long tail or wing feather is usually considered typical. It consists of a central stem, or rachis, and a broad portion or vane.

The vane is composed of small branches or barbs, with numerous overlapping small barbules. Tiny hooks lock the individual elements of the feather into a sturdy, flexible unit. If a feather should become disarranged, preening by the bird locks the elements back into place.

## THE FRAMEWORK

A comparison of the bird's skeletal system with other, more generalized vertebrate skeletons, shows how specialized the bird's has become. As the bipedal form of locomotion freed the forelimbs for modification into organs of flight, the loss of some bones followed, with the fusion of others. As illustrated in *Flight* the large bones became hollow, and strengthened with internal struts. Air sacs, extensions of the bird's extensive respiratory system, invaded some of the bones, adding to their buoyancy.

In flight, the body of the bird hangs from the wings, an arrangement which is reflected in the strong shoulder girdle and rib cage.

Unlike those of most other vertebrates, the ribs of birds have no cartilage, but are fully ossified and connect the backbone with the sternum, or breastbone, forming a rigid cage. The backbone itself is fused for most of its length, with a small freely movable tailpiece, or pygostyle, which controls movement of the tail feathers.

## THE FLIGHT MECHANISM

Changes in the skeletal system have produced forelimbs that serve efficiently as wings. As described by Wesley E. Lanyon in *Biology of Birds* most of the bones of the hand have become fused, with the loss of the first and fifth digits. The second digit may be moved independently, and, along with the feathers attached to it, it is an important structure in flight.

The breastbone of all flying birds is provided with a large keel, which serves for attachment of the massive wing muscles. In general, the size of the keel is an indication of wing power, and flightless birds, such as the ostrich, lack this adaptation.

Undoubtedly because of the need for a low center of gravity, both the muscles for raising, as well as those for lowering the wing are located on the keel, eliminating all large, weighty muscles from the back. The overlying breast muscle, *pectoralis major*, is attached to the humerus, or upper wing bone, and serves to pull the wing downward. Underlying the *pectoralis major* is the *pectoralis minor*. A tendon extends from this muscle, passing over the shoulder to its insertion on the upper side of the humerus. Acting by a pulley-type mechanism, a contraction of the *pectoralis minor* elevates the wing.

## ENERGY FOR FLIGHT

Skeletal and muscular systems alone, of course, will not provide power for flight. Birds have evolved extremely efficient mechanisms for providing themselves with fuel and oxygen. Their lungs, for example, are unlike those of other animals; they are relatively inelastic and have no diaphragm to help bring in the air.

A number of large thin-walled air sacs are situated throughout the body, among the internal organs and muscles, and within the hollow bones. Besides invading the wing bones, sacs usually extend into the femur (the large bone in the leg), the ribs, and also the vertebrae.

In screamers and hornbills the air sacs are so extensive that they penetrate even into the pygostyle, the tips of the wing bones, and the toes. The air sacs are connected to the lungs by rather large diameter tubes, where they meet with a number of smaller, interconnecting passageways. Lungs of birds, therefore, do not have dead end spaces, such as do the lungs of mammals, for instance.

## EFFICIENT RESPIRATION

Breathing is accomplished by muscular movement of the thoracic region, and is aided by wing movements during flight. These movements force air in and out of the air sacs, thereby continuously ventilating the lungs. Unlike lungs with dead spaces, the air in the lungs of birds is completely changed during each respiratory cycle.

The transfer of oxygen and carbon dioxide occurs in the lungs them selves, but the air sacs are not entirely without physiological function. When large amounts of heat are generated, as in flight, the air sacs aid in cooling the bird. Since they have no sweat glands, birds must rely on evaporation from these moist internal sacs for cooling.

In response to the high respiratory requirements imposed by flight, the avian heart is relatively larger in size, beats faster, and pumps more blood per unit time, than the mammalian heart. Complete separation of oxygenated and unoxygenated blood is maintained by the double circulatory system characteristic of all homoiotherms ("warm-blooded" animals). According to Josselyn Van Tyne and Andrew J. Berger in *Fundamentals of Ornithology* the basal heart rate of birds varies from one species to another, but is generally high; 135 beats per minute in the mourning dove, to 615 in the ruby-throated hummingbird.

## REDUCTION IN WEIGHT

Economy of weight must be maintained in spite of the high fuel consumption needed for flight. This apparent dilemma has been resolved in the highly efficient avian digestive system, which digests food rapidly, and immediately eliminates wastes. Experiments performed on song birds have shown that artificially stained food (grain, fruit or insects) passes through the digestive tract in about one and one-half hours. Various fruits have been shown to pass through the tract of a cedar waxwing in from 16 to 40 minutes.

The horny bill and muscular stomach, or gizzard, have taken the place of teeth in seed and vegetable-eating birds, serving to mechanically break down the food. In carnivorous and carrion feeders, the gizzard is poorly developed. Bones, fur, and feathers are regurgitated from the mouth in pellet form, without burdening the rest of the system. When the barn-owl, for instance, preys on small mammals, it swallows its prey whole. A neat pellet, consisting of bones and fur, is later ejected from the mouth. A short large intestine rapidly eliminates digestive wastes.

Adaptations resulting in weight reduction have involved every part of the avian body, including the urinary and reproductive systems. Most adult female birds have only a single (left) functional ovary. The rudimentary right gonad remains as a potential ovary, testis, or ambisexual organ. Birds also have no urinary bladder, thus eliminating the weight of stored urine. Rather, the concentrated urine, high in uric acid content, is mixed with the feces in the common urogenital chamber, or cloaca, and voided in the semi-solid state.

## NERVOUS CONTROL

The nervous system has also changed, in keeping with the specialized needs of an aerial way of life. Associated with the importance of equilibrium and coordination in flight, the cerebellum is highly developed. Also well developed are the semicircular canals, which help to provide the equilibrium needed in flight. Greatly enlarged optic lobes are located at the sides of the brain, rather than roofing the midbrain as in reptiles and mammals. Eyes of birds are disproportionately large, and are not spherically shaped, as in man. Their flattened shape saves considerable space and weight, with no loss in optical efficiency.

Of course, there have been many variations upon the basic theme of bird structure. Numerous variations in wing and body structures coincide with particular flying abilities; soaring and gliding as well as powered flight. Their flying ability made the birds the masters of the sky for centuries before man's machines invaded their realm. Even in this time of supersonic speeds, though, the near perfection of birds is a source of wonder.

## Chapter 17

# Main Parts of Blood

Blood is a tissue made up of cells in suspension and of plasma containing substances in solution. Forty-five percent of blood volume consists of the formed elements (red cells, white cells, and platelets). Fifty-five percent of blood volume is composed of plasma, which consists mainly of water, some proteins, and the various mineral salts necessary to life.

## HOW IS BLOOD MANUFACTURED

Blood is assembled by the body from many different sources. With the exception of injections that may be necessary for treatment, all constituents of blood ultimately come from substances taken by mouth.

The plasma is formed from water, proteins, and salts, all of which are absorbed from the digestive tract. The proteins are modified by the metabolism of the body into the type needed for the human species.The cells of the blood come from diverse sources. The various organs and tissues that collaborate in the manufacture of blood cells are known as the hemopoietic (blood-forming) system.

This includes the bone marrow, spleen, lymph glands, stomach, liver, and the reticuloendothelial system (a network of special tissue cells plus the lining of the blood vessels). The reticulo-endothelial cells are widely distributed throughout the body, but the most important sites are the lining of the bone marrow and spleen.

They are phagocytic; that is, they can ingest micro-organisms and foreign particles of all sorts, thus removing them from the circulating blood.

## FUNCTIONS OF THE BLOOD IN THE BODY

Blood, a complex fluid serving many purposes in the human body, preserves the internal environment required for normal activity of the living cells. Following are some of its main functions:

Respiratory. Blood transports oxygen from the lungs to all the body tissues, and simultaneously carries carbon dioxide from these tissues to the lungs, where it is expired. (External respiration.)

Nutritive. Blood conveys food materials from the alimentary canal and food depots to all the tissues of the body.

Excretory. Blood assists in the removal of waste products of metabolism by conveying them to the kidneys, where they are excreted in the urine.

The maintenance of the water content of the tissues. Fluid lymph, containing oxygen and food materials, leaves the vascular channels and comes in contact with the tissues. Here it picks up carbon dioxide and other waste products and then flows back into the vascular channels. (Internal respiration.) The complex chemical processes of the body are enabled to take place because of the high dissolving and ionizing property of water in the blood and lymph.

Regulation of body temperature. The body owes its ability to regulate temperature largely to the water content of blood and of the tissue fluid. Heat from the deeper regions of the body is brought to the skin and lungs by the blood stream and dispelled.

Protective and regulatory. Blood and lymph contain complex chemical substances such as antibodies, antitoxins, and lysins—all of which protect the body against injurious substances. The circulating blood also brings hormones from the ductless glands into direct contact with the tissues that respond to such stimuli.

## HOW MUCH BLOOD IS IN THE BODY

The average healthy individual has a blood volume equal to about 8 percent of his body weight. Thus the mean blood volume for a person weighing 110 pounds would be 4,400 ml.;

for a person weighing 150 pounds, it would be 6,000 ml.; for a person weighing 200 pounds, it would be 7,600 ml. As 1,000 ml. is equal to 2.1 pints, it will be easy for one to estimate the number of pints of blood in his body. The volume for men will be slightly higher than that for women of the same weight because men have an agerage of 7½ percent more red cells than do women.

The amount of blood in the human body has been the subject of careful study. The methods used to calculate blood volume are based on this fact: When a known amount of concentrated substance is introduced into the blood stream, the extent to which the concentrate is diluted is a measure of blood volume. Substances used must be harmless in small quantities and easily identifiable in blood samples. Depending upon the type used, they attach themselves either to the red cell or to a plasma protein. Red cell volume has been measured by tagging the red cells with radioactive molecules; plasma volume has been measured by using a harmless blue dye. Information gained in this way can be used to calculate the whole blood volume.

The blood volume may vary within certain normal limits, depending upon the needs of the body. At rest, the circulating volume is at a minimum. But the blood volume may be increased by several conditions, including hot weather, high altitudes, muscular exercise, emotional excitement, and pregnancy. On the other hand, an abnormal decrease in blood volume may occur during anemia, hemorrhage, loss of plasma due to extensive burns, or by various forms of dehydration.

## RED CELLS

Red cells, or erythrocytes, are small, solid, disc-shaped particles whose main function is to transport oxygen from the lungs to the tissues and to transport carbon dioxide from the tissues to the lungs. The red cells contain hemoglobin, a chemical substance that has an amazing affinity for oxygen and carbon dioxide.

Compared to other cells in the body, red cells are incomplete and cannot reproduce themselves because they lose their nuclei shortly before being released from the bone marrow into the blood stream.

However, they are considered to be "alive" in the sense that they are able to perform their functions in the body for as long as 120 days. After that time they wear out and are then removed from circulation and destroyed by the spleen. The various components of protein and iron are salvaged for reuse by the body.

Red cells are created in vast numbers, mainly in the marrow of the short, flat bones of the body. It is estimated that they are worn out and removed from the circulation at the rate of 10 billion red cells an hour. The means by which the bone marrow must produce this same amount in the same period of time is part of the marvelous balance of supply and demand carried on by the normal physiology of the body.

Red cells appear fragile in structure under the microscope but are actually very durable. They must be able to withstand a good deal of hammering about while being pumped through the blood vessels, and are able to squeeze through a capillary, which has a smaller diameter than the cell itself. In order to do this, the red cells must fold and bend in a flexible manner and be able to repeat the process many times.

While in the lungs the red cells are exposed to oxygen for only a brief time, but, owing to a chemical balance in the surrounding plasma, they become saturated with oxygen in 1 second. These oxygenated cells are then carried from the lungs to the heart and pumped out into the body. In the tissues, a delicate chemical change takes place in the plasma, and the oxygen is released for use. The carbon dioxide, created by the metabolism of the body cells, is picked up immediately by the red cell in exchange for oxygen and carried to the lungs. Here, another chemical change takes place in the plasma, a change enabling the red cell to release the carbon dioxide and almost instantly pick up more oxygen. This cycle is constantly repeated.

Red cells are very small, measuring about 7.5 microns in diameter and 1.95 microns in thickness. (A micron is about 1/25,000 of an inch.)

It would require approximately 3,000 red cells lined up in their widest diameter to cover the space of an inch. Yet, because there are so many red cells in the human body, they would encircle the earth four times if lined up in a beadlike chain. In spite of the smallness of the individual red cell, the surface area of the combined cell mass is very large. In a man of average size, the surface area of the red cells represents a space as large as the surface of a football field.

Because one-fourth of the circulating blood is in the lungs at any one time, a combined cell surface area of 1,200 square yards is available to air. In 1 second, about 370 square yards of red cell tissue surface is exposed to the air, binding the oxygen to the hemoglobin. Only through these unique properties of the red cell, can manwith his relatively large size and weight-obtain enough oxygen to satisfy the needs of his complex system. From the heavy work of the muscles to the delicate functions of the brain, all cells and tissues depend on oxygen for life.

## WHITE CELLS

The white cells, or leukocytes, are the protective cells of the blood stream. They are two or three times as large as the red cell and possess a nucleus. In a normal person, there are about 7,000 white cells per cubic millimeter* of blood (or about one white cell to 600 or 700 red cells).

There are several varieties of white cells, and each variety has a different function. When a drop of blood is placed on a glass slide, spread into a thin film, and allowed to dry, it may be stained with a dye so that each type of white cell has a distinctive appearance and color. White cells may thus be classified into two major groups: those that have easily visible granules in the cytoplasm (the body of the cell exclusive of the nucleus) and those that do not.

The most common type of white cell in the circulating blood has granules and is thus called a granulocyte. This variety comprises 60 to 70 percent of the white cells. The granulocvtes have the ability to migrate, like an amoeba, along the wall of a blood vessel and out of the capillaries into the body tissues, to fight bacteria. Also like an amoeba, they are able to engulf foreign substances. (This property is common to the granulocytes, monocytes, and reticulo-endothelial cells—all sometimes referred to as phagocytes.) When the body is threatened by infection, the granulocytes quickly gather at the point of invasion and set up a first line of defense. As an example, when you prick your finger and the bacteria enter and start to multiply, the body will mobilize many white cells at that point. They call sometimes be seen as a yellow accumulation called pus.

Lymphocytes, a second type of white cell, comprise 25 to 35 percent of the white cell population. These cells—formed by the spleen, lymph nodes, and other lymphoid tissue of the body—do not appear to be so aggressive in combating acute infection, but cluster in the tissues where chronic infection exists. There they form a barrier against the spread of infection to other parts of the body. Lymphocytes may undergo certain chemical changes that are protective in nature, and they may have some function in the development of immunity.

Monocytes, a third type of white cell found in the blood, are larger than the other white cells and relatively few in number. Like the granulocyte, they have the ability to migrate to the site of infection and ingest bacteria. Like a lymphocyte, they tend to wall off the infected area. During the healing process, they are capable of growing into another type of tissue repair cell, strengthening the weakened area.

The disease of leukemia is an abnormal, unrestrained overgrowth of white-cell-forming tissue. In other diseases, there may be an increase or decrease of white cells in the circulating blood. By obtaining a laboratory count of the number of white cells in a cubic millimeter of blood, a doctor has a valuable clue as to the state of health or disease of the body.

## PLATELETS

Platelets are small, colorless, cell-like bodies that have an important function in the clotting of blood. They are a source of thromboplastin, a substance that is one of the links in the chain of clot formation. Platelets are liberated into the blood stream as fragments of a large parent cell in the bone marrow. They are smaller than the red cell and fewer in number (one platelet to 10 to 20 red cells). The normal range in platelet counts is quite wide, varying from 200,000 to 400,000 per cubic millimeter of blood.

## PLASMA

Plasma is the fluid part of the blood in which the red cells, white cells, and platelets are suspended. It is composed of 92 percent water and 7 percent proteins; the remaining 1 percent consists of fat, carbohydrates, and many mineral salts necessary for life. It also contains hormones, vitamins, and enzymes. Many of the plasma proteins have specific uses in the treatment of disease.

The most well known of these plasma derivatives are gamma globulin, serum albumin, and fibrinogen. Fresh frozen plasma is also prepared for the treatment of certain bleeding problems, especially hemophilia.

Plasma is separated from whole blood by centrifugation; the red cells, being heavier, settle to the bottom of the blood-collecting container. The plasma can then be drawn off into separate containers and processed as fresh frozen plasma or sent to commercial firms for fractionation into specific derivatives.

## BLOOD DERIVATIVES

Blood derivatives are those substances that can be separated from whole blood and used for specific purposes. Plasma itself is a component of whole blood. From the plasma there are a number of proteins that can be separated, and these fractions are usually called derivatives. Serum albumin, which in the body maintains plasma volume, is used for shock, low blood protein, and edema (the loss of water from the blood stream into the tissues). Antibodies against disease are concentrated in the gamma globulin.

Fibrinogen is one of the proteins essential for the formation of a blood clot. In addition to serum albumin, gamma globulin, and fibrinogen, a number of other plasma proteins are under study. These may prove valuable additions to the derivatives mentioned above. The red cells left in the bottle after the withdrawal of plasma can also be used to great advantage in certain diseases.

## BLOOD CLOT

The clotting of blood is the result of a delicate and complex series of biochemical events, some of which are not fully understood. This mechanism must satisfy two rigorous requirements if life is to be sustained:

- Blood must never clot while it is performing its functions within the intact circulatory system;
- Blood must always clot when any damage occurs that causes a break in the blood vessels. The body must be able to vary this complex response to bleeding ranging from a slight cut to a massive hemorrhage.

When tissues are damaged, thromboplastin is released from the nearby cells and platelets. This initiates a chain of reactions in which calcium plays an important part. The end result is that the soluble protein (fibrinogen) is converted to a gelatinous solid (fibrin).

Fibrin forms in threadlike strands, making a webbed mesh in which the sticky platelets, red cells, and white cells are trapped, forming a jellylike clot. As the platelets distintegrate, they cause the fibrin web to shrink and have a constricting action on the walls of the torn blood vessel, causing it to clamp down and narrow the tear through which blood is escaping.

This knowledge of clot formation is applied in the collection of blood for transfusions. The calcium, which plays a key role, is inactivated when it is mixed with a citrate solution in the blood container during collection.

## ANTIBODY

An antibo·ly is a substance in the plasma that reacts in a highly specific way with a foreign substance (called an antigen) when it is introduced into the body. Although this reaction is usually protective, it may sometimes result in injury to the body.

As an example of its protective function, if the body is invaded by a virulent strain of bacteria, two main lines of defense are set up. The white cells attempt to wall off and engulf the invaders; the bacteria, being a foreign substance, are antigens, and stimulate the reticulo-endothelial cells of the body to produce an antibody that will assist in repelling them. This antibody, contained in the globulin fraction of the plasma, may act on the bacteria in one of several ways. It may dissolve them; it may cause them to clump so that they cannot circulate freely through the blood stream; it may soften their protective membrane and inactivate them so that they will fall easy prey to the phagocytic white cells.

So it is that this type of antibody forms a protective agent against disease. Furthermore, as most of us are exposed to a large number of disease-causing microorganisms during our lifetime, we develop a concentration of various antibodies that may remain in the blood stream for a long time, even after we are well.

If we donate blood, the globulin fraction containing these protective substances can be separated from the rest of the plasma components and concentrated to about 25 times its normal strength. This concentrated substance, called gamma globulin or immune globulin, can then be injected into others to help protect them against certain diseases.

The antigen-antibody reaction also has an important application in giving blood transfusions. Red blood cells, if they are not the proper group and type, may act as antigens in the blood stream of the recipient. In this case, oddly enough, the antibodies in the A, B, AB, and O groups do not have to be developed by the introduction of red cells but exist in the blood stream in the natural state. In contrast to this, the antibodies to the Rh factor do not exist in the natural state and must be developed by a transfusion of incompatible Rh-type cells.

In the case of the ABO groups or the Rh types, the incompatible antigenic red cells are attacked by the circulating antibodies and caused to clump, or agglutinate, in the blood stream. If this reaction is strong enough, the red cells may be broken up, or hemolysed, by the antagonistic antibody. If enough of the transfused, incompatible red cells are clumped or broken up in the recipient's body, they have a damaging effect on the tubules of the kidney, and serious illness or even death may result from the inability of the kidney to excrete waste products. The steps by which these hazards are avoided will be discussed in a later section on laboratory work.

## BLOOD GROUPS

Blood groups are inherited patterns into which human blood may be divided for scientific purposes. There are four major blood groups: A, B, AB, and O. * Groups A, B, and O were discovered by Dr. Karl Landsteiner in 1900, and a fourth, AB, was described by another scientist in 1902. Although the percentage of people belonging to each group tends to vary in different countries (as do other inherited characteristics, such as color of eyes and hair), the following distribution represents the general spread in this country: 45 percent—group O; 40 percent—group A; 10 to 12 percent —group B; 3 to 5 percent—group AB.

About the turn of the century, there was great interest in learning how to use blood transfusions as part of the practice of medicine; but doctors were held back by a forbidding obstacle: while about half the transfusions given to treat illness had a remarkable therapeutic effect, the other half seemed to make the patient much worse and sometimes resulted in death.

In seeking the clue to this riddle, Landsteiner performed a series of tests in the course of which he placed the red cells of one person in suspension in the serum of another person. From this crucial experiment, he noted that in a number of the combinations of red cells and sera, the red cells remained undamaged and freely suspended; in other combinations, the red cells were clumped together in sticky clusters—with resultant damage to the cells.

From this evidence, he deduced that certain combinations of cells and sera were incompatible and that it was these unfortunate combinations that were causing illness and death in some patients following transfusion (transfusion reaction).Further investigation of the factors causing clumping, or agglutination, of the red cells when exposed to incompatible sera revealed that there are two main antigenic substances (agglutinogens) in the red cells and two main antibodies (agglutinins) in the plasma. The antigens in the red cells were named A and B, and the antibodies in the plasma were given names to correspond with the group of red cells that they would agglutinate.

Therefore, the plasma antibody causing group A cells to clump was called anti-A; the plasma antibody causing group B cells to clump was called anti-B.

As noted in the discussion on antibodies, the anti-A and anti-B substances are contained in the globulin protein of the plasma. These basic facts enable us to understand the four simple combinations that are the bases for the major blood groups:

- *Group A blood:* Has A substance in the red cells, and anti-B substance in the plasma.
- *Group B blood:* Has B substance in the red cells, and anti-A substance in the plasma.
- *Group AB blood:* Has both A and B substance in the red cells, and neither anti-A nor anti-B substances in the plasma.
- *Group O blood:* Has neither A nor B substance in the red cells, and has both anti-A and anti-B substances in the plasma.

It will be noted that each blood group is named for the type of antigenic substance that is an inherited characteristic of the red cell.

## DETERMINATION OF BLOOD GROUPS IN THE LABORATORY

Blood groups are determined in the laboratory by a simple technique utilizing the known facts about the A and B substances in the red cells, and the anti-A and anti-B substances in the plasma. On the left side of a clean glass slide, the technician places a drop of serum known to contain anti-A substance; on the right, a drop of serum known to contain anti-B substance. He then places a drop of blood (obtained from the person whose group he wishes to determine) into each of the specimens of anti-A and anti-B sera. After a minute or two, the technician can observe one of four possible events take place by which he can determine the blood group:

- If the anti-A serum alone causes the red cells to clump, the blood sample is labeled *group* A.
- If the anti-B serum alone causes the red cells to clump, the blood sample is labeled *group* B.
- If *both* anti-A and anti-B sera cause the red cells to clump, the blood sample is labeled *group* AB.
- If *neither* anti-A nor anti-B serum causes the red cells to clump, the blood sample is labeled *group* O.

The clumping, or absence of it, is clearly visible to the naked eye in a few moments. Thus, in a well-organized laboratory the technician can determine the groups of a large number of samples in a few hours' time.

For safety, the results of blood group determination are checked in the laboratory in another procedure called proof of blood group.

This utilizes samples of known group A and group B red cells, against which samples of the unknown serum are tested. Once again, four observable results may be noted: the group A cells alone may be clumped, the group B cells alone may be clumped, both group A and B cells may be clumped, or no clumping at all will occur.

One can easily work out the blood group that would occur with each of these events. Still another use is made of these facts when the bottle of blood, properly identified as to group, reaches the hospital and is ready to be transfused into the injured or ill patient. Even though the blood is the same group as the patient's, another precaution must be taken to ensure that bloods of donor and recipient are compatible.

This is a laboratory test, very similar to those described above, known as crossmatching. A drop of the donor's red cells is mixed with a drop of the recipient's serum; this is known as the major side of the crossmatching test.

The mixture of recipient's cells with donor's serum is known as the minor side of the test. If agglutination occurs on the major side, the blood is incompatible and must never be given because the donor's cells would be agglutinated and hemolyzed in the recipient's blood stream, causing dangerous transfusion reaction.

If agglutination occurs on the minor side, as it will when group O blood is given to A, B, or AB recipients, it does not mean that the blood cannot be given. At the present time, however, it is the practice to give group-specific blood (group A to group A recipients, B to B recipients, and so on) wherever possible and to use group O donors as universal only in emergencies.

## HOW ARE BLOOD GROUPS INHERITED

Blood groups are inherited as dominant characteristics according to the Mendelian law: for every inherited characteristic, including blood group, there is a pair of genes, one contributed by each parent. The blood group that a child may inherit is therefore dependent on the combinations resulting from genes contributed by both parents. Owing to intermarriage of people with different blood groups, a variety of combinations can result; but if the parents' blood groups are known, those possible for the children can be predicted.

## MEDICOLEGAL IMPLICATIONS OF BLOOD GROUPS

Blood groups are inherited and never change. For this reason, they may have legal importance in identifying bloodstains or determining parentage. But the evidence so provided is useful only in a negative sense.

For example, if a murder suspect had his clothing stained with group A blood, and the victim was of group O, this is evidence that the suspect was not stained by the victim's blood. However, even if the stains proved to be group O, it still would not be conclusive evidence, for the suspect may have had his clothing stained with the blood of any person who was group O.

This logic also applies to paternity cases. The determination of paternity in disputed cases is very complicated. Such cases must be studied by experts in forensic medicine.

## THE Rh FACTOR

The Rh factor is an inherited antigenic substance in the red cells (similar in nature to the group A and B substances) that occurs in 85 percent of the white population of this country.

Those who possess this factor are said to be "type Rh positive." Those who do not possess it are "type Rh negative." It occurs in about the same percentage in all blood groups; that is, of 100 group A bloods, approximately 85 will be Rh positive and 15 Rh negative; of 100 group B bloods, approximately 85 will be Rh positive and 15 Rh negative. These same percentages hold for all four blood groups.

The Rh factor was named in 1940 by Landsteiner and Wiener, who were doing work with the red cells of the rhesus monkey. They discovered that the antiserum that agglutinated the cells of the rhesus species also agglutinated 85 percent of the donors they examined; it failed to agglutinate the other 15 percent.

The word rhesus was shortened to Rh in discussions of this new factor and thus entered the literature. In 1939 Levine and Stetson reported a severe transfusion reaction in a woman delivered of a dead baby, and discovered that her blood would agglutinate 80 percent of donors who were of her own blood group.

They assumed that the infant had inherited from the father a blood factor that was not present in the mother's blood. Small quantities of the fetal red blood cells had seeped through the placental membranes and caused antibody production in the mother.

The brilliant scientific work by which these groups of research workers correlated their separate findings and laid the foundations for the important work on the Rh factor is worthy of additional reading.

## SIGNIFICANCE OF THE Rh FACTOR

The Rh factor is of significance in blood transfusion and may be significant in childbirth. The Rh substance can cause a reaction if injected into the blood stream of a person who is Rh negative; that is, his body will have a tendency to break down the Rh-positive cells. This may clog the kidney with the residue of broken-down red cells and, if severe, can result in death.

However, the first tranfusion may not cause this reaction because the recipient may become only sensitized; in other words, he develops anti-bodies—but not enough to produce severe symptoms. However, having been sensitized, he might suffer a severe reaction from subsequent transfusions.

In pregnancy the Rh-positive child of an Rh-positive father may be carried by an Rh-negative mother. In a small percentage of these cases, Rh-positive cells from the infant's blood stream cross the placental barrier membrane and seep into the mothers' circulation.

Because these cells of the infant carry an antigenic substance, Rh, they may evoke in the mother's body an antibody response against the cells. This has little effect on the mother because the amount of the infants' cells in her blood stream is very small compared with the volume of her circulating blood.

However, if the antibodies to the Rh-positive cells generated by the mother are carried back into the blood stream of the infant, they may do a great deal of damage. The concentration of the antibodies in the small body of the infant is so great that it may damage a large percentage of the red cells.

Depending on the degree of severity with which this happens, the infant at birth may be merely jaundiced, a condition that may clear up in a few days; he may be quite ill, and to save his life may require an exchange transfusion (removal of baby's blood and simultaneous replacement with Rh-negative blood). In a very few cases, the baby is born dead.

Because the Rh factor has received a great deal of publicity during the past few years, many women fear the possibility of giving birth to an infant affected by hemolytic disease of the newborn (erythroblastosis). Actually, much of this apprehension is unfounded. It should be understood that Rh is a normal inherited characteristic, just as color of eyes and hair.

Again, marriages with an Rh-negative woman and an Rh-positive man may occur in approximately 13 percent of marriages (15% x 85% = 13%). Most women who are Rh negative do not produce anti-Rh antibodies, and those who do usually have at least two babies that are unaffected. Even in marriages with an Rh-negative woman and an Rh-positive man, only 1 in about 26 has an infant affected, and many of these children only mildly so. With the present knowledge about blood, there is a much better outlook for detecting any Rh incompatability and treating the affected infant.

## BLOOD OF DIFFERENT RACES CAN BE USED INTERCHANGEABLY IN TRANSFUSIONS

Although there are differences in the percentages of blood groups and types among the various ethnic groups, the structure and function of blood cells and plasma are essentially the same among all peoples and, with due regard for group and type compatibility, may be used interchangeably throughout the human species.

## ANIMAL BLOOD CAN NOT BE USED FOR HUMAN TRANSFUSIONS

Blood is highly species-specific. Red cells injected from one species to another act as a foreign protein and therefore undergo rapid agglutination and hemolysis, with the consequent transfusion reaction causing damage or death to the recipient.

## BLOOD CANNOT BE MADE SYNTHETICALLY

Blood can be produced only in the body and cannot be made synthetically. The oxygen-carrying red cells needed to sustain life must be produced in the body or given to it via transfusion.

Fortunately, however, some substances can be used in limited amounts per patient in place of plasma. They are called plasma volume expanders or plasma extenders. Among these is Dextran, a carbohydrate material adjusted to a molecular size similar to that of a molecule of albumin. It should be emphasized that blood is a fluid for which there is no substitute.

The synthetic plasma volume expanders have a valuable role in preventing death from shock for the first few hours after injury until whole blood becomes available.

# Chapter 18

# Structure and Functions of Higher Plants

## PHOTOSYNTHESIS

The most important function of a leaf is *photosynthesis*. This term comes from the Greek word *photos*, meaning "light," and *synthesis*, which means "putting together." In photosynthesis, carbon dioxide and water are put together to make sugar and starch, using energy from light. A few simple experiments can convince us that this is true. You can cut your name in a piece of aluminum foil. Use the foil to cover one of the two primary leaves of a young bean plant growing in a pot.

Let the plant photosynthesize for a few days by keeping it under the same good conditions under which it grew before you started the test. Now, cut off both leaves and plunge them into boiling alcohol. (Caution: Do not boil alcohol over an open flame).

You will notice that the green pigment dissolves in the alcohol. The decolored leaves should then be removed from the boiling alcohol and placed in a shallow dish containing a dilute solution of iodine in potassium iodide. After a while, you will notice that the leaf that remained uncovered turns blue. This indicates the presence of starch. In the other leaf, only your name appears in blue. Starch was formed only in the portion of the leaf that received light.

Another test can be made under a tightly sealed Bell jar. A large pickle jar will do, too. In the jar is placed a dish with sodium hydroxide solution, which will capture the carbon dioxide out of the air. Let a bean plant remain in the sealed jar in the light for a few days. When the plant is removed from the jar and the leaves tested for starch with the iodine test, no starch will be found. This time carbon dioxide was lacking. To provide a control for the experiment, put a bean plant in another jar without sodium hydroxide. The leaves of the control plant should have starch in them at the end of the test.

## CHEMISTRY OF SUGAR

We have seen that during photosynthesis a plant must have light and carbon dioxide in order to form starch. For a deeper understanding of photosynthesis, we must go to the level of atoms and molecules again. Starch is simply many molecules of sugar put together end to end in a chain-like arrangement. The sugar molecule is built of atoms of carbon, oxygen, and hydrogen. The hydrogen (H) and the oxygen (O) occur in the same proportions as in water ($H_2O$).

On the surface it appears that sugars ($CH_2O$) are composed of carbon (C) and water. This is why the class of chemicals to which sugars belong is known as carbohydrates, from the Latin *carbo* (coal), and the Greek *hydor* (water): coal combined with water.

Starch is a carbohydrate and so is cellulose. When the sugars are linked together in one way, a coil, starch, results. When the sugars are linked together in a slightly different way, the straight chain of cellulose results. You see starch every day in bread and potatoes; you see cellulose in the cotton of your handkerchief.

Early experimental scientists did not know the chemical composition of sugar. Nor did they know the fundamental significance of sugars in nature. Jan Baptista van Helmont, a physician and chemist who lived from 1577 to 1644 near Brussels, was one of the first to experiment with plants. He placed in a pot a certain quantity of earth which, when highly dried, weighed 200 pounds.

A willow branch weighing five pounds was then set in this pot that was protected from dust by a cover, and watered daily with rainwater. In five years' time the willow had grown large and strong and had increased in weight by 164 pounds. The earth in the pot, when once more dried, showed a loss of about two ounces. Van Helmont concluded from this experiment that the considerable increase in weight in the plant had been gained entirely at the cost of the water, and, consequently, that all the materials in the plant, though distinct from water, nevertheless came from it.

## CHEMISTRY OF PHOTOSYNTHESIS

Van Helmont did not realize that plants take up carbon dioxide from the air. The chemical symbol for carbon dioxide is $CO_2$, signifying that each molecule is composed of one atom of carbon and two atoms of oxygen. There is only 0.03 percent of this gas in the atmosphere. It is put in the air continuously by respiration, as well as by combustion and decomposition of wood, gasoline, and other natural products. It is removed from the air by plants and built again into natural products via photosynthesis.

After physicists had learned to make radioactive atoms, biologists used them to trace the paths of chemical elements through the plants. When photosynthesis was studied in this way, some interesting new information was obtained. This information can be written in chemical shorthand:

$$CO_2 + 2H_2O \xrightarrow[\text{chloroplast}]{\text{light}} (CH_2O) + O_2 + H_2O$$

| carbon dioxide | water | | carbo-hydrate | oxygen | water |
|---|---|---|---|---|---|

You will notice that the equation is balanced. There are equal numbers of the same atoms on each side of the arrow. The equation signifies that carbon dioxide and water yield carbohydrate, oxygen, and water in the leaf. A carbohydrate is represented in the equation by $(CH_2O)$. More accurately, this should be $(CH_2O)n$. The $n$ indicates that carbohydrate molecules are multiples of $CH_2O$. For ordinary sugar with its 12 carbon atoms, $n$ would be 12. For simplicity, we will ignore the n in this discussion.

The carbon atom from carbon dioxide becomes incorporated into carbohydrate. The oxygen in the carbohydrate comes from carbon dioxide. On the other hand, the free oxygen formed during photosynthesis is liberated from water. The hydrogen in the carbohydrate also comes from water. Notice that in photosynthesis water is consumed as well as produced.

The release of oxygen during photosynthesis can be easily demonstrated with a branch of *Elodea*. This branch is put in a tall glass jar* with the cut end of the stem up and the top down. When left undisturbed in strong sunlight, bubbles come from the stem and rise to the surface. These are oxygen bubbles. The gas can be trapped in a test tube and identified.

## DUAL NATURE OF PHOTOSYTHESIS

The convenient equation of photosynthesis that was just shown is known as an over-all equation. It indicates what products go into a reaction and what finally comes out. It tells nothing of what goes on in between. The mechanism of photosynthesis is far, far more complicated than this simple equation indicates.

There are two main steps. The first step needs light and can be called photosynthesis proper. During this phase, oxygen is liberated and light is required to liberate the oxygen from water. However, no carbohydrates are formed as yet. Only intermediate chemicals result; these are loaded with the sun's energy.

These chemicals are then used in the second step, which does not require light, and is sometimes called the "dark reaction." This second major step, during which carbon dioxide is turned into carbohydrates, has recently been worked out in great detail by Professor Melvin Calvin, of the University of California. In 1961, he received a Nobel prize in chemistry for this feat.

## BREAKTHROUGH

The first step involving the reactions that require light is just now beginning to be understood. For many years, scientists had tried to make photosynthesis work outside of the cell, but they succeeded only in making chloroplasts produce oxygen in the light. Carbohydrates were not formed.

The major breakthrough finally came when Professor Daniel I. Arnon and his co-workers at the University of California succeeded for the first time in causing chloroplasts to photosynthesize outside of the cell. They found new ways by which the cell could be broken down more gently.

Dr. Arnon's work proved that the whole process of photosynthesis takes place exclusively in the green bodies of the cell, the chloroplasts. Not only do the light reactions take place there, but also the dark reactions.

When the electron microscope became available, it disclosed in great detail a number of tiny bodies inside cells. When our knowledge of biochemistry was combined with the knowledge of electron microscopy, it was found that the vital functions of life are located in these special bodies, which are called *organelles*. We now recognize that in a cell, organelles perform certain jobs, just as the organs of the human body have special functions.

## ENERGY CONVERTERS

There are two types of organelles whose function it is to convert energy from one form to another. The *chloroplasts* convert light energy to the chemical energy of food. The *mitochondria* change the chemical energy of food to the chemical energy of ATP. The full name of this chemical is *adenosine triphosphate*. It has three phosphate groups. ATP is a magic word in the biological world. Just as money is necessary in the economic world, ATP is the currency used in metabolic activity. Nothing gets done inside the plant or animal unless ATP is available.

ATP is made from food in the mitochondria of your active cells. The oxygen of the air you need for respiration is used by the mitochondria to slowly oxidize the food. And during this process of *oxidation*, the stored energy of food is changed to the quickly available energy of ATP. This is the essence of respiration. You eat because you need ATP. When you jump quickly, it is ATP energy that makes you move. It would take far too long to extract the needed energy from sugar for immediate use.

When ATP is formed, one of its phosphate groups, the last one in the molecule, is pushed into the rest of the molecule like a loaded spring in a BB gun. The ATP molecule is loaded indeed! Loaded with energy that can be made available instantly.

For many years it was thought that the mitochondria were the only organelles that could make ATP, but Dr. Arnon discovered in 1954 that chloroplasts could do it too. It is now generally accepted that plants can generate ATP in two ways: by photosynthesis and by respiration. Photosynthesis takes place in the chloroplasts, and respiration in the mitochondria. Man and the animals, however, can only generate ATP by one process, by respiration in the mitochondria. Strictly speaking, there is another process, called *fermentation*, that also yields ATP. In fact, this is the only way in which fermenting yeast cells can obtain ATP.

You may be surprised to find out that the ATP of humans, animals, and plants is the same, and that these organisms have comparable mitochondria. The chemistry of respiration is the same in all living things. The reason for this appears to be that all living things are built on one and the same pattern.

## PHOTONS AND ELECTRONS

The brilliant work of Dr. Arnon and his associates has helped clarify the way chloroplasts make ATP. To understand what happens when light strikes chlorophyll, let's first consider the nature of light. Light can be pictured as a stream of bullets, tiny energy packages called photons. The sun shoots a stream of these at us all the time. During photosynthesis, the stream of photons hits the chlorophyll inside the chloroplast.

When photons hit chlorophyll, the impact throws electrons ($e^-$) out of their "track." These electrons may fall back into their track or they may be captured by other molecules—either other molecules of chlorophyll or other kinds of molecules just right for the capture of electrons. Still different molecules may then capture these electrons and the electrons may wander farther and farther away from home.

When an electron is knocked out of the chlorophyll molecule by photons, a hole is left in the molecule. Electrons have a negative charge, and so the more electrons that are shot out of the chlorophyll, the more lectrically unbalanced the chlorophyll becomes. It becomes electrically positive ($^+$). Such a positive charge attracts negative electrons to it; soon electrons start flowing back into the chlorophyll molecule.

What we have is an electron pump: electrons leaving chlorophyll and electrons returning to it. In other words, there is an electric current going around inside the chloroplast. What actually happens in a leaf during photosynthesis is that the light energy of the stream of photons is converted to the electrical energy of a stream of electrons. In the chloroplast this electrical energy produces ATP.

In recent years, we have learned many details about how the photosynthetic process goes on inside the chloroplast. Some of our understanding is still based on guesswork; yet the outlines are becoming clear. The theory of electron flow as a result of light rests on strong experimental evidence. For instance, scientists took a dried film of the light-catching organelles of photosynthetic bacteria (something like the grana of higher plants) and tried to pass an electric current through it.

The resistance of the film was so high that little current passed through in the dark. However, upon illumination of the film, current passed through readily. In other words, only when light was shining on it did there appear to be freemoving electric charges in the photosynthetic apparatus.

## ELECTRON CARRIERS

Another part of photosynthesis that is fact and not fiction is the presence of *electron carriers*. These are the compounds that conduct the electron toward chlorophyll to fill the positive holes. There are also electron carriers that conduct the highly activated electron away from chlorophyll. *Cytochrome* is an iron protein that conducts electrons toward chlorophyll. *Ferredoxin* is another iron compound that carries electrons away from chlorophyll. The properties of the electron carriers are determined by their electrical characteristics.

You will note that cytochrome in the chloroplast has a potential of +0.4 volt, while ferredoxin has a potential of -0.4 volt. In other words, during photosynthesis there is a potential difference of 0.8 volt between cytochrome and ferredoxin. Thus an uphill current flow occurs during photosynthesis.

## PUTTING ELECTRONS TO WORK

Energy of an electron flow can be converted to chemical energy of ATP. Electrons can do more. When an electron combines with a molecule, chemists say that the molecule that accepts the electron is *reduced*. The extra negative charge often attracts a proton.

A proton is nothing but a positively charged hydrogen atom ($H^+$). Protons are found in water and are available throughout a plant. The highly activated electrons in photosynthesis can do many things, although usually not all in the same plant.

For example, in photosynthetic bacteria that live in the mud of ocean bays, a highly energized electron can combine with a proton and form ordinary molecular hydrogen ($H_2$). Then in some of these bacteria and in some blue-green algae, the highly charged electrons of photosynthesis can reduce free nitrogen of the air ($N_2$) to ammonia ($NH_3$). This process is known as *nitrogen fixation*.

All plants need nitrogen, yet the vast pool of nitrogen gas in the air is not available to many higher plants. They can use only nitrogen *compounds* such as ammonia. Lower forms of plant life, such as blue-green algae, pave the way for the growth of higher plants by producing ammonia from nitrogen of the air. Blue-green algae enrich rice fields, fertilizing this crop free of charge.

When a volcanic island blows up and all life is destroyed, the first thing to establish itself after the island has cooled off are these pioneers, the blue-green algae. The ammonia they produce from the nitrogen of the air makes possible the subsequent growth of other plants. Even in Antarctica, blue-green algae carry on their work using the high-powered electrons made in photosynthesis.

In higher plants, the high-powered electron is used another way. Here, the reducing power of the electron is preserved in a compound known as triphosphopyridine nucleotide, or TPN, for short. TPN resembles the molecule of ATP. Both have the adenine base, both have ribose sugar, and both have phosphate.

That is just the bulk of the molecule. The "business ends" of the two molecules differ greatly. ATP, as you will recall, has a high-energy phosphate bond (the coiled spring) at its end. TPN has a nicotinamide group with a positive charge at its end. This charge attracts the electron from photosynthesis. Ferredoxin, the electron carrier, hands over its charge to TPN. TPN then becomes negative and at the same time attracts a proton, forming TPNH, the reduced pyridine nucleotide. TPNH, in turn, can reduce other chemicals; for example, carbon dioxide to carbohydrate.

Now let us turn to another aspect of electron flow. The very fact that a potential difference exists and that electrons move implies that work can be done. Early physicists who studied electricity recognized the fact that electricity behaves much like water. Electric potential is very much like the potential of water locked in a lake high in the mountains. When allowed to run downhill, the water can do work: move heavy boulders, turn millwheels and electric generators.

Similarly, when electrons run downhill they can do work. In the living cell, the most spectacular thing they do is operate the mechanism that produces ATP. This mechanism compresses the spring at the end of the ATP molecule when it puts a molecule of inorganic, free phosphate on the end of ADP (adenosine diphosphate). ADP has only two phosphate groups instead of the three of ATP.

| ADP | + | $P_s$ | $\longrightarrow$ | ATP |
|---|---|---|---|---|
| adenosine | | phosphate | | adenosine |
| diphosphate | | (inorganic) | | triphosphate |

Exactly how flowing electrons accomplish this conversion of energy is not yet clear, but the process certainly involves the membranes of the organelles (the grana). Photosynthesis, respiration, and most biochemical reactions are intimately tied to the minute structures of the cell. Studying these reactions without taking into account the function of the fine, delicate structures of the cell does not give a full understanding of life's processes. It is very much like trying to understand the operation of an automobile engine by considering only the reaction of burning gasoline.

## ASSIMILATORY POWER

Remember that a stream of photons (light) sets in motion a flow of electrons inside the grana of the chloroplast. The chloroplast, then, is much like the photocells used in spacecraft and in cameras. In plants, as a result of this electron flow, a substance with strong reducing power, TPNH, is formed and also a substance with a high-energy content, ATP. These two are the only direct products of photosynthesis proper. They contain the energy of trapped light.

Together, TPNH and ATP are called assimilatory power. With the aid of these two substances, the cell can carry out many reactions without requiring any more light. The most important of these reactions is the reduction of carbon dioxide to sugar and starch. This is often referred to as the Calvin cycle after Professor Melvin Calvin who, with his associates, worked out this mechanism in painstaking detail.

The sugar-phosphate combination formed in photosynthesis is not an ordinary sugar of 6 or 12 carbon atoms, but one with only 5. It has become supercharged by combining with the energy compound ATP, and a 5-carbon sugar *di*phosphate results. This compound is so reactive that it can combine with carbon dioxide of the air. It may be that an intermediate 6-carbon molecule is formed that is so unstable it breaks down immediately into two 3-carbon molecules of PGA, or phosphoglyceric acid. At any rate, PGA is the first stable compound found with carbon dioxide built into it.

Now the reducing compound TPNH and more ATP enter the cycle and reduce the acid (PGA) to a 3-carbon sugar. Two 3-carbon sugars are converted to 6-carbon sugar. At this stage, these sugars are still combined with phosphate. These can yield starch and sucrose or undergo further modification to again yield the 5-carbon sugar phosphate with which we started. The cycle goes around and around as long as ATP, TPNH, and $CO_2$ are fed into it. The result is stable energy-storage products, the foods starch and sugar, that figure so prominently in animal and human nutrition.

## PHOTOSYNTHESIS IN HIGHER PLANTS

Until very recently, photosynthesis was not sufficiently understood to enable us to describe the process in any detail. Two or three years ago, it would not have been possible to put together very many pieces of this jigsaw puzzle. The pattern is still not complete, but the outlines are clear. It is now even possible to follow in higher plants the course of an electron during photosynthesis. Chlorophyll acts as a pump, operated by the energy of light.

Chlorophyll pumps an electron "uphill" from cytochrome and delivers it to ferredoxin which, in turn, hands it over to TPN. With the addition of a proton ($H^+$) from water, TPNH is formed. Where did the cytochrome obtain its electron? Higher plants get theirs from water. The electrons of water have a very low energy content, a potential of only +0.8 volt, and must be pumped uphill. Recently it has become clear that another pigment, probably chlorophyll b, delivers the electron to another carrier, *plastoquinone*.

Plastoquinone is not an iron compound like cytochrome and ferredoxin. It is related to Vitamin K. By the way, the nicotinamide of TPN is nothing but a vitamin-B compound. These vitamins are working parts of the plant. Plants can make these compounds, but animals cannot. That is why you need vitamins in your diet.

After plastoquinone receives an electron from water charged by chlorophyll *b* and light, it turns over its electron to cytochrome. On this "downhill" run, the electron flow furnishes the energy to operate the ATP-producing mechanism. The electron flow is coupled to the ATPproducing mechanism, the way the generator that charges the battery in a car is coupled to the engine. This downhill run from plastoquinone to cytochrome may not be the only place where ATP is generated. There may be as many as two additional sites where the generation of ATP is coupled to photosynthetic electron flow. Let us return to the water from which the electron came. Water normally is dissociated to a certain degree into protons ($H^+$) and hydroxyl ions ($OH^-$).

$$H_2O \rightleftharpoons H^+ + OH^-$$

It is this proton that is used in the production of TPNH.

$$2e^- + TPN^+ + H^+ \longrightarrow TPNH$$

The OH- ion is robbed of its electron by the pull of the chlorophyll-b electron pump and the rest goes to oxygen and water.

$$40H^- \longrightarrow 4e^- + 2H_2O + O_2$$

Photosynthesis is the source of most of the oxygen of the air. Oxygen is a mere by-product of the process, but its production changed the course of evolution. It made possible the development of creatures like ourselves, dependent on the oxygen of the air.

## RESPIRATION

### The Release of Energy

Photosynthesis results in the conversion of light energy to chemical energy. This chemical energy is incorporated into the bonds of synthesized compounds, primarily six-carbon sugars. The energy present in such compounds is essentially stored energy, which can be released only by breaking the chemical bonds that hold the molecule together.

The energy present may be released by breaking all the bonds at the same time. This type of energy release occurs when a piece of wood burns. An indication of the amount of energy involved is given by the heat and light evolved during the combustion. Biological systems effect the release of energy in a similar yet far more subtle way. Each bond may be broken individually, and the released energy is then trapped by suitable receptors in ways that permit the future transfer of this energy to other compounds.

The receptors of the released energy are the same as those involved in energy transformations in photosynthesis: ADP and PN. The energyrich compounds that result are also the same as in photosynthesis: ATP and $PNH_2$. *These compounds are extremely important to the cell.

They are the negotiable energy currency of the cell, the currency that must be available for investment in the building and maintenance of the cell. The systematic breakdown of a molecule such as glucose is a complex undertaking, and in the cell it is accomplished by scores of reactions. Each of these reactions is controlled by a specific enzyme, and the entire process involves many enzymes acting in teams.

## ENZYMES

Enzymes have been termed organic catalysts and this is an accurate, if limited, statement. A catalyst can be defined as a compound that speeds up a chemical reaction and yet remains unchanged at the end of the reaction. In the case of the enzyme *amylase,* the reaction that is catalyzed is the breakdown of starch to individual glucose molecules. Some breakdown would occur, if enough time were available, without the presence of amylase.

But in the presence of amylase the reaction is rapid and large amounts of glucose appear. Amylase does not change the course of the reaction. Moreover, although amylase is involved in the reaction, it is unchanged at the end of the reaction.

Enzymes lower the activation energy of reactions in which they are involved. Molecules will break down if their energy is increased to a point where the molecules are no longer stable. The energy needed to achieve this point is called the *energy of activation.* As the temperature is raised and more energy is available, the reaction rate increases. Enzymes, by lowering the activation energy of the reaction, permit reactions to occur at normal environmental temperatures. How enzymes are able to permit such reactions is unknown.

All enzymes are proteins. They are, with few exceptions, large molecules composed of hundreds of amino acids. The amino acids are joined together in long, elaborately coiled chains. Some enzymes consist solely of proteins, while many consist of two parts. The larger of these two parts is inevitably a protein molecule. This does not mean that all proteins are enzymes; but a part of every enzyme is a protein. The second part, called the co-enzyme, may be a metal such as iron, manganese, or copper, or a vitamin such as thiamin, riboflavin, or nicotinic acid.

When an enzyme consists of two parts, both parts must be present for the enzyme to function. The fact that enzymes are proteins and therefore highly complex molecules means that most enzymes are easily damaged. Any treatment that will break down or alter (denature) proteins will destroy the enzymes' activity. Thus, while an increase in temperature will increase the rate of an enzymatic reaction, too high a temperature denatures the protein and the enzymatic reaction ceases. Any number of other changes in the immediate environment of the enzyme molecule may result in the loss of activity.

The enzyme molecule is believed to combine with the substrate molecule. This combination is short-lived and, when it breaks up, the enzyme molecule remains intact while the substrate molecule is changed. Enzymes are highly specific. This specificity is believed to be based on the spatial configuration of the surface of the enzyme molecule. Enzymes can be inhibited by compounds that closely resemble the substrate but that are inactive in the reaction. A well-known and important example of this type of enzyme inhibition involves the enzyme *succinic dehydrogenase,* which acts on the substrate succinic acid.

Another compound, *malonic acid,* closely resembles succinic acid but is different enough so that the enzyme while combining with the malonic acid does not change it. The malonic acid, however, occupies the reactive site on the enzyme molecule and prohibits the attachment of succinic acid. This type of inhibition is termed *competitive inhibition.* Other types of enzyme inhibitors act by combining or changing various parts of the enzyme molecule so that the reactive site is destroyed. A large number of enzyme inhibitors have been discovered and their use has led to the elucidation of many aspects of enzyme action.

Each enzyme mediates only a limited number of reactions—in most cases, only one. Thus, an enzyme involved in a reaction that results in the loss of a hydrogen atom from one compound is different from an enzyme involved in a similar type of reaction with a different compound. Consequently, there are hundreds of enzymes in the cell, each acting on a specific compound or substrate. To understand the functioning and growth of a cell, we must understand the factors controlling enzyme formation.

A formidable body of information indicates that enzyme synthesis is controlled by genes. Many of the details of the nature and method of this control are now understood.

A feature of many enzymes is that they function in teams. In such an arrangement, the end product of one enzymatic reaction becomes the substrate for the next enzyme in the series. This pattern can be repeated dozens of times. The $CO_2$ -fixation cycle discussed in relation to photosynthesis is an example of enzymes working in a team. Similar arrangements of enzymes are found in the breakdown or respiration of carbohydrates.

## RESPIRATION

Respiration is the biologically controlled breakdown of energy-containing substances, such as carbohydrates, fats, and proteins, with the release of energy. As such, it is the reverse of photosynthesis, in which energy is used to synthesize compounds. The major pathway of respiration in the cell can be broken into two parts:

- *Glycolysis* and
- The *Krebs cycle*. In the first part, glycolysis, glucose is eventually split into two pyruvic acid molecules, each having three carbon atoms. At this point, the second part of respiration begins; and after one carbon is removed, the remaining two-carbon compound enters the Krebs cycle. During the Krebs

cycle, the two-carbon compound is converted to $CO_2$. This part of respiration is a cycle because the reactions are organized so that the chemical compound *oxalacetic acid*, which accepts the twocarbon piece entering the cycle, is regenerated at the end of the cycle and can accept another two-carbon piece.

The two parts of respiration differ in a number of important features. In the absence of oxygen, glycolysis can occur but the Krebs cycle cannot function. If no oxygen is present, the pyruvic acid produced at the end of glycolysis is converted, with the loss of one carbon atom, to ethyl alcohol. In this case, the process is *fermentation*.

Another important difference between glycolysis and the Krebs cycle is the amount of energy made available to the plant cell in the form of ATP molecules. The important consideration is not the total number of ATP molecules formed but the net number available at the end of the reactions. This is because some energy in the form of ATP must be invested if the reactions are to occur. During glycolysis, only two ATP molecules are obtained per glucose molecule respired, while in the Krebs cycle the net gain is 38 ATP molecules per glucose molecule.

## ANAEROBIC RESPIRATION

$$C_6H_{12}O_6 \longrightarrow 2C_2H_6O + 2CO_2$$

This equation is the over-all reaction for fermentation. It states merely that one molecule of glucose is converted to two molecules of ethyl alcohol and two molecules of carbon dioxide. This is the merest outline of the reactions involved. These reactions show a minimum of some ten steps between glucose and ethyl alcohol.

The first step in fermentation is the addition of a phosphate to the glucose molecule. Shortly after, a second phosphate is added. These steps are carried out at the expense of two ATP molecules. The end product is a six-carbon sugar with two phosphates attached, called *fructose-1, 6diphosphate*. This is split into two three-carbon compounds and, in the reactions that follow, two ATP molecules as well as two $PNH_2$ molecules are generated—one each per three-carbon compound. As the reactions continue, pyruvic acid is produced, which results in the formation of an additional ATP per three-carbon fragment. Finally, as a necessary consequence of the absence of oxygen, the electrons from the $PNH_2$'s are used to produce ethyl alcohol. This final reaction results in the formation of one $CO_2$ per pyruvic acid molecule.

If a balance sheet is compiled to show the energy of the compounds involved, it will show that, for each glucose molecule used, four ATP molecules and two $PNH_2$ molecules are produced at the expense of two ATP molecules and two $PNH_2$ molecules. The net gain is thus two ATP'S. There is considerable energy remaining in the ethyl alcohol—energy that can be released only by different metabolic pathways.

The basic reactions involved in anaerobic respiration show only the fate of the carbon atoms and the net changes in ATP. The first stages of anaerobic respiration are very similar to part of the $CO_2$-fixation cycle of photosynthesis. The direction of the reactions is reversed in the two processes. Except for this and differences in the details of the steps, the reactions from fructose-1, 6-diphosphate to PGA are essentially the same.

The entire process of fermentation can be carried out in the complete absence of oxygen. Fermentation is the major energy-producing system of many micro-organisms. Commercially, the most important of these organisms are the yeasts. The production of alcohol through fermentation has been discovered by almost every civilization, and the art of making alcoholic beverages has been practiced for centuries.

## AEROBIC RESPIRATION

$$C_6H_{12}O_2 + 6O_2 \longrightarrow 6CO_2 + 6H_2O$$

In the presence of oxygen and the necessary enzymes, the breakdown of glucose is complete and occurs as outlined in the equation above. For every molecule of glucose used, six molecules of $CO_2$ and six molecules of $H_2O$ are produced. During the course of this reaction there is a net production of 38 ATP molecules produced and made available to the plant.

Aerobic respiration, as noted earlier, can be described as having two parts. During the first part, the glucose molecule becomes involved in a set of reactions that yields two pyruvic-acid molecules. This part is identical to the similar reactions that occur in anaerobic respiration. At this point, however, the fate of the pyruvic acid differs depending on the presence or absence of oxygen and the appropriate enzymes.

In the presence of oxygen, a carbon atom is removed from the pyruvic acid and an activated two-carbon compound is formed. This compound unites with a four-carbon compound, *oxalacetic acid,* forming citric acid, which contains six carbon atoms.

Then, through a series of reactions, two molecules are released and oxalacetic acid is formed. For each $CO_2$ molecule produced, six ATP molecules are made, so that, for each twocarbon fragment entering the cycle, 12 ATP molecules are made available to the cell.

The total energy relations in aerobic respiration are more involved. The aerobic cell receives more energy from the first part of respiration than does the yeast cell for the same series of reactions. This results from the fact that in the yeast cell the $PNH_2$ that is produced is used later in the formation of alcohol. In the aerobic cell, the energy of this $PNH_2$ is converted to three ATP molecules.

Thus the net gain of ATP in the yeast is two ATP molecules, while in the aerobic cell it is eight ATT molecules per glucose molecule. An additional six ATP molecules are gained between pyruvic acid and the Krebs cycle. The total number of ATP molecules produced per glucose molecule used is 38 (8 + 6 + 24 = 38). Aerobic respiration is thus a much more efficient process with regard to energy release than is fermentation.

## DIRECT OXIDATION PATHWAY

The major pathway of aerobic respiration appears to be through glycolysis and the Krebs cycle. Another pathway, known as the direct oxidation pathway or the pentose phosphate shunt mechanism, is known to exist in both plants and animals. More involved in some ways than the glycolysis-Krebs-cycle reactions, it has the same over-all equation:

$$C_6H_{12}O_6 + 6O_2 \longrightarrow 6CO_2 + 6H_2O$$

The first steps are the same as in glycolysis, but one $CO_2$ and a fivecarbon sugar are then formed. This sugar is broken down through a rather complex set of reactions involving a number of three-, six-, and sevencarbon compounds. Ultimately, for every six glucose molecules entering the cycle, six five-carbon sugars, six $CO_2$ molecules, and 12 $PNH_2$ molecules are produced.

The energy of the $PNH_2$ can be converted to 36 ATP molecules. The reaction is, therefore, nearly as efficient as the glycolysisKrebs-cycle pathway. The five-carbon sugars produced are important in the synthesis of ATP and various other cellular constituents. Moreover, these same reactions play a very important role in the $CO_2$ -fixation cycle of photosynthesis.

## ELECTRON TRANSFER AND OXYGEN UPTAKE

The presence of oxygen is necessary for aerobic respiration to take place. The uptake of oxygen during respiration is a consequence of the transfer to oxygen of electrons from the various compounds involved in the metabolic breakdown of glucose. This transfer of electrons results in a reduction of oxygen gas to water. The fact that electrons have varying amounts of energy was stressed in the discussion of photosynthesis.

During photosynthesis, electrons are raised to higher energy levels and their energy is ultimately incorporated into the bonds of carbohydrate molecules. During respiration, electrons pass through a series of reactions that take them from high energy levels to lower energy levels. At the lowest level, the electron combines with oxygen and $H^+$ ions to form water. At the intermediate levels, the energy released is transformed into high-energy phosphate bonds by the conversion of ADP to ATP.

At six places in the glycolysis- Krebs-cycle pathway, a pair of electrons is released. These electrons are accepted by PN, which becomes $PNH_2$. Then, through a series of cyclic reactions, the electrons move down a series of compounds. During this passage, the electrons lose energy and ATP molecules are formed. The compounds involved in these reactions are PN, a *flavo-protein,* and a series of *cytochromes.* The latter contain iron. The final step in the electron-transfer system is the reduction of $O_2$ to water mediated by the enzyme *cytochrome oxidase.*

## THE METABOLISM OF FATS AND PROTEINS

This discussion has stressed the metabolism of carbohydrates, and glucose in particular; however, most of the compounds present in the cell can be respired. Most types of carbohydrates can be and are used. Similarly, fats and proteins may also be respired.

Both fats and proteins can undergo a series of reactions that eventually reduce them to two-carbon fragments, which can enter the Krebs cycle. The utilization of fats and oils supplies much of the energy used during germination. Fats are a highly efficient means of storing energy, since on a comparative-weight basis they have more stored energy than the carbohydrates.

Proteins can be utilized as an energy source, but they are not normally used in this manner. In general, only after other energy sources are exhausted are the proteins respired.

## SITE OF RESPIRATION IN THE CELL

The fact that the chloroplasts were the site of photosynthesis in the cell was known long before the process was understood. Conversely, the mechanism of respiration was known in considerable detail before the site of the process was determined in the cell.

The major breakthrough in the search for the site of respiration in the cell came with the development of the methods of cell homogenization and separation of cell parts by differential centrifugation. During differential centrifugation, the solution containing the cell parts is rotated rapidly at varying speeds. The parts separate because the heavier parts settle to the bottom of the tube first and then the lighter ones follow.

From studies like these and from other lines of evidence, it has been established that the enzymes of the Krebs cycle are found associated with the mitochondria. The enzymes of glycolysis, on the other hand, are found primarily (although not exclusively) in the soluble portion of the cytoplasm.

The localization of the enzymes of the Krebs cycle in the mitochondria was one of the first and greatest discoveries using cell fractionation procedures.

## PHOTOPERIOD

We now know that the flowering of many angiosperms is controlled by two major factors of the environment, photoperiod and temperature. The discovery of photoperiodism by Garner and Allard in 1920 was the accidental result of an attempt to propagate a mutant type of large-leaf tobacco, called Maryland Mammoth, which had arisen by chance as a single individual in a field of other tobacco plants. As the season progressed, the original type flowered profusely, but the Maryland Mammoth did not.

Wishing to obtain seeds of this valuable new type and fearful that the plant might not flower before the autumn frost, these investigators removed the plant from the field and transferred it to the greenhouse. Despite every urging, however, the plant steadfastly refused to initiate floral organs until approximately mid-December, many months after the normal plants had completed seed production successfully.

When the seeds of the self-pollinated Maryland Mammoth type were planted in the field the next year, this behavior pattern was repeated, i.e., the plants grew vigorously in the field, failed to flower together with the original type, but did flower when removed to the greenhouse and maintained until about Christmas time.

An analysis of the various factors that could possibly be responsible for this behavior finally led Garner and Allard to the inevitable conclusion that the plant was flowering only during the very short days characteristic of the northern hemisphere at Christmas time. They discovered that flower initiation could be induced at will by transferring Maryland Mammoth plants to special chambers in which the length of day could be artificially shortened. They named this response of plants to length of day *photoperiodism*. Maryland Mammoth tobacco, which flowers only if the day length is reduced below a certain critical value, was called a *short-day plant*.

Other plants of this type are soybeans and chrysanthemums. Another class of plant is the *long-day* type, such as spinach and certain cereals, in which flowering occurs only if the day length exceeds a certain critical value. Finally, there is a class of plants called *day-neutral plants* in which photoperiod does not exert a major effect on the time of flower initiation. An example of this type is the tomato plant, in which floral primordia are initiated at particular nodes when the plant has obtained a particular size. This situation is not amenable to control by photoperiod, although the tomato may be induced to premature flowering by certain synthetic chemicals that are related to auxins.

In the years since 1920, considerable work has been done to determine how photoperiod exerts its effect on the flowering of the plant. It has been unequivocally demonstrated that the leaf is the receptor for the photoperiodic stimulus. For example, if a single leaf of a Maryland Mammoth tobacco plant is enclosed in a black bag and given the appropriate short photoperiod required for floral initiation, the terminal bud some distance away from the leaf will initiate floral primordia. This separation in space of the photoperiodic receptor (leaf) from the region of response (bud) makes it necessary to postulate some connecting link between the two.

Since it can also be demonstrated that a plant or plant part exposed to a short day can transmit its florally initiated state to a receptor plant or plant part maintained vegetatively on a long day, it is clear that some hormonal substance, tentatively named *florigen*, is produced in the photoperiodically stimulated leaf. This stimulus can be transmitted from a donor plant to a receptor, provided there is tissue union between the two grafted plants. The transmission of the stimulus is prevented by steamgirdling of the petiole or by other means of interrupting phloem transport. These facts clearly imply the production in the photo-induced leaf of a yet unknown substance, which is transported via phloem to the terminal bud, where it influences the meristem to favor the production of floral primordia rather than vegetative organs.

Recently it has become clear that many long-day plants may, if supplied with gibberellin, initiate floral primordia under an otherwise unfavorable photoperiod. In this group of plants, there appears to be some connection between gibberellin and florigen. Gibberellin, however, does not promote flowering in short-day plants. Since there is very good evidence from grafting experiments that the florigen of long-day and of short-day plants is at least functionally equivalent, if not chemically identical, the exact nature of the relation between gibberellin and florigen is not clear.

In still other plants, such as the pineapple and the litchi, floral organs can be initiated by the application of certain synthetic auxins. In the series of events leading to the production of reproductive organs, therefore, various substances may well become limiting in different plants. This substance may be gibberellin in some plants, auxin in other plants, and perhaps additional, possibly unknown, substances in still other plants.

One interesting generalization that has developed out of the study of these photoperiodic phenomena is that most plants respond not to the length of the light period but to the length of uninterrupted darkness. Thus, a so-called short-day plant is really a "long-night" plant that requires an uninterrupted dark period of a certain minimal duration for the initiation of its floral primordia. In the same way, a long-day plant is in reality a "short-night" plant, that is, a plant that will flower only if the night period is not longer than a certain critical maximum.

The effective period of darkness can be interrupted by the simple expedient of making it a bit too short (even a few minutes will do), or by interposing a brief flash of light in the middle of the dark period. For example, in the short-day cocklebur plant, flowering will occur in a regime of 15 hours of light and 9 hours of dark. If the 9-hour dark period is shortened appreciably, to about 8½ hours, flowering will not occur.

But one single cycle of exposure to 15 hours of light and 9 hours of dark will suffice to induce the plant to initiate floral primordia, even though photoperiods unfavorable for floral initiation are immediately re-imposed. Such a phenomenon is referred to as *photoperiodic induction*. If the single long dark period of 9 hours is interrupted at its midpoint by a flash of light, the plant will not flower. Certain chemical processes that are very sensitive to minute quantities of radiant energy must thus be proceeding in darkness in the leaf.

If a quantum of absorbed light impinges on the course of these reactions, the entire sequence of events is wrecked and the plant must start over again. With the long-day plants, the situation is just the reverse, i.e., the interruption of an unfavorably long dark period by a brief flash of light will lead to floral initiation. These two types of plants seem to possess the same kind of photoperiodic mechanism, but they somehow work in reverse fashion.

By the use of the action spectrum technique, it is possible to obtain some notion of the wavelengths of light that are effective in interrupting or promoting flowering when administered in short flashes. We should mention here that the same kind of light that inhibits the flowering of short-day plants will promote the flowering of long-day plants. Recent experiments with monochromatic light have revealed that many types of plants respond best to red light in the region near 660 mì.

It has also been surprising to find that the effect of red light may be instantaneously and completely negated by the subsequent application of what is called near-infrared light, or "far-red" light, in the region of 730 mì. These experiments indicate that there is a pigment in plants present in two forms, a red-absorbing form and a far-red-absorbing form. In the shortday plant, the absorption of red light in the middle of the long inductive dark period leads to a negation of flowering, while the absorption of farred after the red leads to the repromotion of flowering.

The control of flowering thus appears to be a resultant of the ratio of the two forms of this pigment in the plant. So far, we do not know exactly what this pigment is, although extracts of it that show reversible spectral changes on irradiation have been obtained from several plant tissues.

For example, the administration of red light has resulted in a decreased absorption of red near 660 mì and an increased absorption of far-red near 730 mì, whereas administration of far-red has resulted in the reverse changes. Our best guess is that this light treatment is causing a reversible chemical change in the effective pigment molecule and that these changes are determining the course of development of the plant. The yet uncharacterized protein pigment has recently been named *phytochrome.*

The discovery of the existence of this reversible photoreaction governing flowering has clarified several other perplexing problems in plant physiology. For example, it is well known that the germination of many seeds is affected greatly by light. Seeds of the Grand Rapids variety of lettuce will not germinate at all when placed in darkness on moist filter paper at room temperature, but the administration of minute quantities of red light will result in prompt germination. If the red-light-treated seeds are promptly irradiated with far-red, the effect of the red is completely canceled out and the seeds remain dormant.

Here again, the growth of the plant is apparently controlled by a two-way switch mediated by the unknown red-far-red pigment. With other seeds, such as the California poppy, germination is inhibited by light. Here, too, the redfar-red system is involved; red light promotes and far-red inhibits germination. The inhibition, rather than promotion by white light in these seeds, is the result of a reverse differential sensitivity to the two regions of the spectrum.

We have already mentioned that the germination of certain seeds can be greatly promoted by the plant growth substances gibberellin and kinetin. Indeed, a portion of the light response of these seeds may be interpretable in terms of the alteration of the levels of such substances within the seeds. This, however, has not yet been experimentally demonstrated.

The same controlling red-far-red morphogenetic photoreaction can be seen to operate in *etiolation* (growth in darkness) of stems and leaves. A seed germinated in total darkness gives rise to a seedling with a very long, slender, unpigmented stem and to scale leaves that never expand greatly. If we analyze the visible spectrum in terms of its ability to transform the etiolated plant into a normal plant, we find that red light is again most effective and that its effect can be prevented by far-red light administered after the red.

The exact nature of the response, however, depends on the tissue. If the red light is given to stem tissue, that stem will then be greatly inhibited in its growth, but a leaf exposed to the same red light will be promoted in its growth. The responses of both stem and leaf will be completely prevented by the application of far-red light after the red. Therefore, we see that the response to visible radiation is pre-determined by the differentiation processes that have given rise to the specific types of cells.

The red-far-red morphogenetic photoreaction described above is certainly of great importance in the life of the plant, affecting such diverse processes as the germination of seeds, the growth of roots, stems, and leaves, and the initiation of floral primordia. Clearly, increased knowledge of the nature of phytochrome and of its mode of action is greatly needed and must be an important aim of research in plant physiology.

## GROWTH HORMONES

In addition to the water, light, carbon dioxide, and various other materials absorbed by a plant from its environment, other chemicals are required for plant growth. These substances, called *hormones,* are generally needed in only infinitesimally small quantities, and in most instances they are produced in adequate amounts by the plant itself. By appropriate experimental techniques, we can deplete their supply in the plant, demonstrate their existence, and deduce a great deal about their nature. A hormone is a substance produced in very small quantities in one part of an organism and transported to another part where it produces some special effect. Two major classes of growth-regulatory hormones have been shown to exist in most, if not all, higher plants. These are the *auxins* and the *gibberellins,* to which we now turn our attention.

## THE AUXINS

The auxins are a group of substances produced by the growing apexes of stems and roots. They migrate from the apex to the zone of elongation, where they are specifically required for the elongation process. If the tip of a rapidly growing stem is removed, the growth in the region below the cut will slow down very quickly, and within several hours or days, depending on the type of plant, it will come to a complete halt. If the removed tip is replaced, growth of the stem continues almost normally, showing that some influence emanating from the tip is conducted across the wound to the growing cells.

If the tip is placed on a block of gelatin or agar for a period of several hours and the block without the tip is then transferred to the cut stump of a decapitated stem, the block will partially substitute for the tip in facilitating growth of the subjacent regions.

From this experiment, we deduce that a substance, called *auxin,* moves from the tip to the block and from the block down to the base. Extensive chemical work of the last several decades has revealed the existence of many substances with auxin activity. Several of these have been isolated in pure form and have been shown to be native plant growth hormones. The most common of these substances is the simple material indole-3-acetic acid, which is probably derived from the amino acid tryptophan.

The amount of auxin in any plant tissue may be determined by extraction with some solvent, followed by the application of the extract to some tissue that will respond quantitatively to the auxin contained in it. Normally, the tissue is placed in diethyl ether at a temperature near 0°C and gently shaken for a period of 2-4 hours. This ether extract is then concentrated and, when reduced to a small volume, is incorporated into an agar block that is then placed asymmetrically on the decapitated stump of the auxin-sensitive organ. Traditionally, the leaf sheath, or *colepotile,* of dark-grown oat plants has been used.

In this plant, the asymmetrically placed auxin enhances the growth of the tissue only directly below it. This unequal growth on the two sides of the coleoptile causes a curvature of the organ that is directly proportional to the amount of auxin incorporated into the block. Thus, to determine quantitatively the amount of auxin in an unknown organ, the extract is made, the curvature is measured, and this curvature is compared with curvatures produced by known quantities of auxin in another experimental series. This technique of using the response of an organism to measure the amount of a chemical in an extract is called *bioassay.*

The curvatures produced by the unilateral application of auxin to plants bring to mind the curvature of various plant organs toward or away from light or gravity. In fact, we now know that such curvatures (called *tropisms*) are due to the asymmetrical distribution of auxin in the organ involved. For example, if an oat coleoptile is subjected to a low intensity of unilateral light, it will curve toward that light (*phototropism*).

This curvature results from the fact that the side near the light has had its growth somewhat depressed by the light while the growth on the side away from the light has been accelerated. If the tip of a unilaterally exposed coleoptile is removed and the amount of auxin of the two halves (light and dark) assessed by the curvature test described above, invariably the side away from the light will have about twice as much auxin in it as the side toward the light. Plant physiologists have therefore concluded that light acts in producing curvature by affecting the distribution of auxin in the organ; this auxin concentration, then, controls growth.

Similarly, a stem laid prostrate can be shown, after some time, to accumulate more auxin on the lower surface than on the upper surface. This results in an accelerated growth below and an ultimate curvature upward (*geotropism*). The growth downward of a prostrate root is a consequence of the different auxin sensitivity of the root. In the prostrate root, as in the prostrate stem, auxin accumulates below, but since in the normal root the auxin concentration is already optimal, or supraoptimal, this greater concentration of auxin on the lower side leads to depressed growth, and thus to a downward curvature of roots.

So far we have been discussing only those aspects of auxin action pertaining to cell elongation. Auxin may also initiate or promote cell division. For example, in tissue cultures of normal cells excised from stems or roots, cell division is entirely dependent on the application of auxin to the medium. Similarly, the initiation in the spring of cambial activity in trees can be shown to be controlled by auxin diffusing downward from developing buds. In addition, the formation of branch roots and adventitious roots from the pericycle region of stems or roots can be initiated by the application of auxins.

In this mitosis-inducing activity, auxin apparently works together with other substances such as the constituents of nucleic acids. In fact, a substance called *kinetin* has been isolated from autoclaved yeast nucleic acid and has been shown to be active in promoting cell division of plant cells in the presence of auxin. Substances resembling kinetin in chemistry and physiological action have been shown to occur in seedlings, dividing cells, fruitlets, and in coconut milk and other liquid endosperms. Such substances are referred to collectively as *kinins*.

In addition to its roles in promoting cell division and cell elongation, auxin has other correlative effects on the growing plant. For example, auxin determines whether the apical bud or a lateral bud on a stem will develop. In the intact stem of many plants, only the apical bud can grow. Removal of the apical bud, however, results promptly in the growth of one or several of the buds lower down. If the tip is removed and the cut surface covered with an auxin paste, the buds lower down will continue to be inhibited. From this type of experiment, it has been deduced that auxins inhibit lateral bud growth.

Auxins are also important in regulating the fall of leaves and fruits from plants. A leaf blade is attached to a stem by means of a petiole that persists during the growing season but falls off at some time later in the year. As long as the leaf blade produces adequate quantities of auxin, the attachment of the petiole to the stem is firm.

If, however, the leaf blade becomes deficient in auxin production, the petiole forms at its base a special layer of cells called the *abscission layer*, which is mechanically so weak that the leaf blade is easily caused to fall fròm the plant by a breeze or by mechanical irritation.

This knowledge has been put to good use in agriculture. For example, where it is desirable to retain leaves or fruits on a tree (as in apples and oranges), it is merely necessary to spray the tree with a dilute solution of 2,4-dichlorophenoxyacetic acid or some related auxin. This simple process has saved millions of dollars for orchardists whose fruits normally fall off the tree when they are not yet ready for harvest. Similarly, by the production of chemical analogs that antagonize the action of auxin, leaves or fruits can be causd to fall prematurely from a plant. This fact greatly aids mechanical cotton picking, for instance.

In the pineapple plant, one auxin, *a*-naphthaleneacetic acid, has the remarkable effect of promptly inducing the onset of flowering. Although the mechanism of this effect is obscure, its value in pineapple agriculture is obvious. The plants can all be grown to a uniform size and naphthaleneacetic acid applied at any desired time. The fruits will then develop uniformly, making mass methods of harvest possible.

In another important economic activity, auxin is applied to the pistil of the flower to produce artificial, or *parthenocarpic* fruits. Normally, most fruits are formed as a result of pollination and fertilization of the ovary of the flower and of the subsequent growth stimulation produced in the ovary by the fertilization process. If, however, no pollen reaches the pistil, the development of the ovary into a fruit can be stimulated by the application of fairly large quantities of auxin-type materials.

One final aspect of auxin action should be noted here. In a lawn containing dandelions, the application of 2,4-D kills the dandelions while leaving the grass intact. Similarly, the application of 2,4-D to a cornfield containing the noxious bindweed will kill the bindweed, leaving the corn intact. As a result, chemical weeding has been substituted for the laborious, injurious, and expensive practice of mechanical cultivation. The savings from such practices are incalculable.

## THE GIBBERELLINS

Another group of important plant growth hormones, also discovered through a series of accidents, is the gibberellins. Gibberellic acid, when applied to plants, produces tremendous hyperelongation effects on stems. At the moment, the most important agricultural use of the gibberellins is in the grape industry, where the application of gibberellins to seedless grape clusters results in the retention and development of a greater number of grapes, and in fruit much larger than normal.

For a relatively small expenditure of time and money on the chemical, the farmer obtains a much larger crop. Gibberellin is also useful in celery growing, where it produces larger, more succulent plants in a shorter time. It can also be used to stimulate seed germination and growth of early seedling grass, including the barley that is used as malt in the brewing industry. Its potential in agriculture is very great, and is only now beginning to be explored.

## THE AGRICULTURAL SCIENCES

The harvest from 311 million U.S. crop acres and many more millions of acres of pasture fed 92 million Americans in 1910. Today, fewer acres provide an even greater supply of food for nearly 200 million.

Our abundance of food, more appetizing, nutritious, and convenient today than ever before, can be attributed in large part to progress in the agricultural sciences. The animal, plant, and soil sciences, food science, agricultural engineering, and other biological and physical disciplines have made it possible to produce more and better food. Agricultural economics and other areas of social science have contributed to maximizing return on investment in agriculture, whether the investment has been in production goods or human resources.

The agricultural sciences are products of the synthesis of evolving disciplines characteristic of the world of science. Through free exchange of ideas, concepts, and research-derived principles, the agricultural sciences have emerged as distinct entities from the biological sciences, mathematics, physics, chemistry, and the social sciences, during the past century. Though now highly developed and rapidly maturing, the agricultural sciences continue to serve as a vehicle for exploring new areas and catalyzing an onrush of changing technology as new problems are encountered.

Increasingly, agricultural research is delving into the nature of life itself. Work in progress on blood antigens offers hope that we may learn more about the manner in which hormones and genes act in the formation of those antigens. Basic studies in plant physiology are leading ever closer to an understanding of the processes in photosynthesis. Agricultural scientists are studying viruses to learn how they develop, mature, and reproduce themselves.

Although the relationship of the agricultural sciences with the biological sciences is particularly close, the importance of the physical sciences should not be minimized.

Further, all life exists in time and space within an environment of land, water, air, climate, and other components, both animate and inanimate. It is man's challenge to make productive use of living things within their environments, as those environments currently exist or as they may be modified. This, then, is the broad realm of the agricultural sciences. The principal architect in this realm is the agricultural scientist.

He may be an animal geneticist, physiologist, or behavioral scientist; or he may be a plant geneticist, pathologist, virologist, horticulturist, or agronomist. He might also be a biochemist, microbiologist, entomologist, agricultural engineer, agricultural economist, or yet another one of the many kinds of biological, physical, or social scientists. Use of the term "agriculturalist" to describe him is no longer appropriate except in a generic sense.

## THE PRINCIPLE OF INTERACTION

Regardless of his specialty, the agricultural scientist is affected by the principle of interaction, probably to a higher degree than any other scientist. There is a natural and necessary interaction among the agricultural sciences, and between agriculture and cognate disciplines.

Agricultural scientists cannot improve crop varieties, for example, without using and perhaps advancing our knowledge of plant genetics. But a superior variety cannot produce to its inherent capacity unless advances in soil chemistry and soil physics make possible advanced soil fertility practices which will, in turn, provide better nutrition for the improved strains of crops. Similarly, greater milk output, to use an example, depends on improved genetic potential of dairy cattle. This potential will not be realized, however, unless nutritional requirements as well as physiological responses of the dairy cow are thoroughly understood.

It is only after scientists have improved yields from crops *and* animals and have studied pertinent economic factors that we can produce larger quantities of commodities such as milk at a lower cost per unit. Even then this opportunity will be denied unless agricultural scientists provide the means to control insects and diseases which attack plants and animals and impair the flavor or quality of milk or of the many food products which include milk as an ingredient.

## INTERACTION IN AIR POLLUTION STUDIES

Air pollution studies provide still another example of interaction involving agricultural scientists. Contaminated air causes more than poor visibility, unpleasant odors, eye and throat irritations, and ash deposits on buildings and cars. It also injures vegetation—to the extent of 10 million dollars annually in just one metropolitan area.

A team of plant pathologists at Rutgers University learned that airborne fluorine damaged the foliage of peach trees and more than 30 other kinds of vegetation in a certain industrial area. Recently, they identified aldehydes as a toxic component of the air pollutants being studied.

Agronomists and horticulturalists at Rutgers are assisting the pathologists in developing cultural and chemical control practices. Chemists are making further identification of atmospheric compounds. Meteorological data have made possible correlations among the concentrations of the pollutants and several atmospheric factors, including wind speed and direction, and barometric pressure.

Preventive anti-oxidant sprays are being tested. Geneticists and plant breeders are at work on the development of plant varieties that are resistant to air-borne phytotoxins, since species and varieties have been shown to vary in degree of susceptibility to pollutants. Plant physiologists are seeking to explain the mode of action of the toxins' activities in plants.

Perhaps soon, as the plant pathologist leading the Rutgers research team proposed before the Senate Subcommittee on Air Pollution, a research center will be formed to serve the Eastern United States by bring ing together "sufficient diversification of skills to allow delving deeply into this 20th century challenge."

## THE AGRICULTURAL DISCIPLINES

A scientific discipline, to deserve the title, should be a distinct field of objective investigation which develops a systematic body of verifiable facts and general laws. While agricultural disciplines such as animal science and plant science rely heavily upon knowledge from other fields of science, they also meet these criteria. It is not argued, however, that agricultural engineering, to use an example, is *agriculture* and not *engineering*, or that agricultural economics is *agriculture* and not *economics*.

The important point is that agricultural engineering, agricultural economics, and other disciplines exist as distinct entities to contribute to agricultural progress. In this paper, we explore the nature, achievements, and challenges of six such disciplines-soil, plant, animal and food sciences, agricultural economics, and agricultural engineering. *This is not claimed to be a complete list of scientific disciplines devoted entirely to agriculture.* Some would subdivide our categories. Plant science, for example, includes (agronomic or field) crop science, as well as the study of horticultural (fruits, vegetables, ornamentals) and range crops.

Some would argue that forestry should be a part of plant science, or at least be included in the agricultural sciences, and cite such justifications as the large acreage of farm woodlots. We are content to acknowledge that forestry is closely related to agriculture-especially because of the concept of multiple land use, and because of common interests in several disciplines including soils, microbiology, and economics-and we are content to state that the significance of this relationship transcends any possible importance of defining forestry as a part of, or separate from, agriculture. We have merely elected not to include forestry in this paper.

Similarly, animal science can be divided into meat animals, dairy, poultry, and horses and-expanding beyond traditional agriculture-can include fish and wildlife. We have chosen not to include these two areas which, from an agricultural viewpoint, have a relationship to animal science somewhat similar to forestry's relationship to plant science. Animal science and veterinary science overlap to the extent that each is concerned with the physiology and pathology of large, domestic animals. But animal science is concerned with many other aspects of large, domestic animals and veterinary science, of course, also includes pets and laboratory animals. Again, we have chosen not to discuss veterinary science.

There are other classifications of the agricultural sciences in addition to the one which we have selected. One example is seen in the field of agronomy, which is a mechanism for combining two closely related disciplines—(agronomic or field) crop science and soil science.

Finally, it is not intended to minimize, by omission, other disciplines which contribute heavily to agricultural research but which are not oriented *exclusively* to agriculture. Examples are biochemistry, entomology, plant pathology, microbiology, and meteorology, each of which occasionally has the adjective "agricultural" preceding it. Economic entomology, for example, is readily identifiable; it concerns itself with beneficial as well as harmful insects, has its own society, and publishes a journal.

## SOIL SCIENCE

Soil scientists are responsible for leadership in the never-ending battle of rebuilding, maintaining, improving, and conserving our nation's soil.

A considerable portion of the knowledge of soil science involves the application of chemistry, physics, microbiology, and geology to the biosystem we call soil. The contribution of the basic sciences is seen in the phenomenon of cation exchange, discovered years ago. It was soil scientists, with strong training in physical chemistry, who explored the nature of the exchange phenomenon and built a background of basic information which has made important contributions to our understanding of the function of synthetic exchange materials. As a consequence of their investigations, we have new understandings of the mechanisms involved in nutrient absorption by both plants and microorganisms.

Specialized fields such as soil genesis and soil classification, though rooted in fundamental science, have developed a large body of knowledge which is unique to soil science. An illustration of the use of knowledge about soil classification is seen in interpretations from soil classification surveys. These can be used to predict productivity, erodability, suitability for irrigation or drainage, fertilizer and lime requirements, and best uses —including crops, wildlife, and forests.

In recent years soil scientists have been refining methods of examining the soil and defining the reactions that go on within it. The importance of the diffusion process in physical-chemical mechanisms in the soil has resulted in research on the movement of ions, liquid, and gases in the soil.

As soil scientists work toward complete understanding of the earth's land surface mantle, they encounter special kinds of soil problems such as irrigation and reclamation of saline and alkaline soils. Others include the problems associated with the soils of forests, watersheds, wildlands and deserts, and the complex variations in soils induced by different climates, vegetative covers, and underlying parent rocks.

## PLANT SCIENCE

Plant scientists, often in cooperation with other scientists, have made it possible to control epidemics of plant diseases, insect attacks, and severe weed competition. They have developed plant varieties adapted to specific environments and have determined plant nutritional needs so that yields could be increased far above those believed possible a few decades ago.

Plant scientists have added to our pleasures. The discovery of the influence of the length of night on flowering plants makes it possible for us to enjoy floral plants any time of the year instead of only during a certain season. Plant scientists now are engaged in isolating and identifying the enzymes and hormones responsible for flowering and other developmental processes.

Years ago, plant scientists produced hybrid corn. Today, studies in biochemical genetics of corn and other crops are leading to a much clearer idea of how genetic material functions in transmitting a cell's likeness to new cells. Scientists are studying subcellular differences which cause certain plant strains to be susceptible to diseases and some weeds to be resistant to herbicides.

Plant science research illustrates the fact that the solution of many of mankind's basic problems often derives from the efforts of scientists in rather diverse fields of interest. Two such fields that are important to plant science are biochemistry and biophysics. Although photosynthesis cannot yet be carried out continuously in entirely artificial systems, we do have increasing knowledge of how the process operates. Biophysicists have studied and largely explained the intricacies of the energy transfer. Biochemists have traced the chemical pathways involved.

Plant physiologists and other plant scientists have contributed through their knowledge of cell structure and function. The pioneers of future discoveries concerning photosynthesis may represent any one of several areas of science or a combination of them.

## ANIMAL SCIENCE

Man's dependence upon animals as sources of food, fiber, and power traces to the earliest chronicles of history. Despite the revolution in power sources in the well-developed countries of the world, animals are perhaps more important to the health and welfare of mankind today during any earlier period.

Foods of animal origin provide high quality protein together with vitamins, minerals, and associated dietary factors. Meat, milk, and eggs not only contribute much to the health of our nation but also to that of many other peoples who are considered to be the best fed populations of the world.

Wool, mohair, leather, and other fibers have long been staples of our American economy. Although the horse no longer provides our power and transportation, its popularity for recreational activities is at an all time high.

Animal science embraces all the disciplines in the biological and physical sciences which influence animal life. The principles underlying these disciplines have been employed to select and breed increasingly more desirable animals, to better understand reproductive physiology and thereby improve fertility and enhance productivity, and to provide well balanced and adequate rations which have decreased death losses of the young and provided the nutrients for fast and efficient growth and production.

One agricultural research project has determined that a riboflavin deficiency in chicken eggs is caused by an autosomal recessive gene, and studies are under way to determine the mechanism by which the gene produces its effect. Another searches for the neural factors which control pituatary functions that are related to ovulation in the female. Still another seeks to learn the amount and seasonal fluctuations of thyroxine production in a cow, and the effects of thyroxine injections.

All are exploring fundamental questions that, when answered, will result in more and better animal products at lower cost.

## FOOD SCIENCE

Food science is important at every step in food processing, beginning with the raw product and terminating with consumer acceptance of the purchased commodity. Food science is increasingly being considered a part of a broader field of nutritional sciences, consisting of nutrition and dietetics as well as food science.

The food scientist uses the principles of chemistry, genetics, microbiology, physics, mathematics, and engineering in evaluating the quality attributes of foods. These include flavor, color, texture, nutritive value, and wholesomeness. He studies food processed by various methods such as heating, freezing, dehydration, fermentation, or irradiation. The principles of biochemistry are employed to preserve or enhance natural or desired flavor, color, or physical characteristics of foods.

For example, enzymes tenderize meat and clarify juices. Natural anti-oxidants prevent deterioration in flavor. Chemical reactions are manipulated to enhance the ripening and produce the desired color of fruit.The food scientist uses controlled microbiological fermentations in the production of cheese, beverages, and other fermented foods.

On the other hand, he prevents microbial spoilage and minimizes loss by judicious application of appropriate physical principles such as refrigeration, heat, drying, salting, smoking, and atmosphere control.Proper incorporation of additives in the form of minerals, vitamins, and amino acids improves the nutritive quality of foods. Conversely, exclusion of objectionable materials such as pesticide residues, hormones, antibiotics, and toxins protect the safety of our foods.

## AGRICULTURAL ECONOMICS

Agricultural economics is a social science, a specialized branch of economics. Its subject matter is concerned with the principles of economics as related to the problems of agriculture.The work of agricultural economists encompasses problems involving production economics, agricultural business management, land tenure, credit, appraisal, resource conservation and development, marketing and pricing, interregional and international trade, industrial structure and competition, and agricultural policy. Agricultural economists integrate technology, public policy, and business principles in the development of information to assist producers, consumers, and policy makers in making decisions.Agricultural economists are particularly concerned today with:

- The continuing disequilibrium between agricultural and nonagricultural industries, manifested by low returns for resources employed in farming. Agricultural economists are identifying optimum methods of producing commodities and are identifying regions with greatest competitive advantage in producing various commodities.

  Cost and return analyses generate information which is helpful to farmers and managers of marketing firms in deciding what and how much to produce, what combination of resources to use in production, and the location and scale of production units.

  Increasingly, the efforts of many agricultural economists are concerned with economics of land use and development of natural and biological resources in an ever-broadening context. How can our land resources-agricultural land not needed at this time for traditional agricultural use, and forest and wildlands which are not being utilized most efficiently-be used best for recreation, wildlife, conservation, and forestry? How can human resources be developed to provide managerial proficiency for modern farms and yet provide flexibility for nonfarm employment?

- A second major problem concerns the processes of change in agriculture and impediments to change. Why are changes in the number and size of farms and marketing firms necessary? What factors determine the rate of change? What is the relationship of size to efficiency? How is competition affected by ownership patterns and industrial structure?
- A third major area is public policy. Many studies currently under way seek to identify the effects of alternative policies and programs upon resource use and incomes in agriculture. Other studies are concerned with policies designed to conserve and develop resources.

The basis for many of these studies is research which seeks to determine how and why the economy functions as it does, and what the implications are for resource use and income distribution to and within agriculture.

This research seeks to provide a factual basis for public policy decisions by determining the consequences of changes in economic and institutional conditions. The structure of international competition and the possible role of U.S. agriculture in trade with and development of foreign countries is receiving new emphasis by agricultural economists.

## Chapter 19

# Agricultural Engineering

Agricultural engineering recognizes and solves engineering problems connected with agriculture. Since agriculture is founded on biological systems, agricultural engineering is, in effect, the engineering of biological systems-the application of the laws of physical science to the control of biological reactions. Its practice involves solution of problems in plant and animal environmental control, planting, harvesting, transport, processing, and storage of food and natural fiber, and the creation and optimization of automated agricultural systems.

A distinguishing feature of the agricultural engineer's education is the combination of basic biology and agriculture studies with his engineering studies, so that he can understand the scientific foundations involved in the biological systems of agriculture and can put them to work.

This includes the engineering phases of soil and water management, sources of energy for agriculture, the functional requirements of plant life as related to agricultural machines, the functional requirements of farm animals as related to housing and environmental control, engineering factors involved in transport, processing, and storage of biological products, and integration of agricultural machine systems through mathematical interpretations and analogies.

Agricultural engineers specify power requirements for the machines and processes of agriculture and develop functional and mechanical requirements of the machines; design water systems to meet the needs of plants, animals, and rural people; develop engineering methods for increasing the value of food and fiber through processing; and design transport methods and materials-handling procedures.

The principal challenge to the agricultural engineer today is to create more powerful and efficient agricultural production systems. Tomorrow's challenge will be similar, but the techniques employed will be more sophisticated.

## THE CHALLENGE OF COMING DECADES

The challenge to our agricultural sciences for the future is no less dramatic than it has been in the past. Our U.S. population is growing at the rate of about 2% per year. At the moment, this is an increase of one person every 11 seconds, 8000 more mouths to feed each day. By 1980, our population may reach 245 million persons. World population presently numbers about 3.1 billion. It is expected to double by the year 2000. This means that there will be as many additional people on earth at the end of each 3-year period between now and the end of this century as presently live in the United States.

Moving from a world in which scarcity is the predominate pattern to a world in which freedom from hunger may at long last prevail is the challenge which faces this and succeeding generations. Dr. J. George Harrar, President of the Rockefeller Foundation, has stated: "We range in our agricultural practices throughout the world from those which approach automation on the one hand to the other extreme which reflects only that there has been little or no change from antiquity.

There are situations which require only two man-days to produce an acre of rice under total or semi-total automation versus 259 to 400 days in areas where work is all done by untrained manpower; there is tremendous power in scientific agriculture applied gradually and as rapidly as is compatible with situations. It can contribute enormously to the future food supply of the world even in the face of a rapidly increasing population.

## TREE FERNS

Beautiful though they are, the lowly ferns that decorate our temperate woodlands give little idea of the size that some of their giant cousins attain. The tree ferns of the tropics stand head and shoulders above all other members of the great fern group. Some of them raise their slender trunks to a height of 60 feet—about the equivalent of a 6-story building.

The 300 or more species of living tree ferns inhabit the cooler temperature situations of the tropics and subtropies. They thrive especially well on lush, rain-swept mountain slopes where the torrid climate is tempered by elevation. There they may form small forests, particularly on abandoned agricultural land.

Tree ferns may be found growing as high as timber line, which in the tropics may lie above 10,000 feet. Seemingly the rainier the climate the better these giant ferns like it. They are at their best in that montane belt where the cumulus clouds sailing in from the surrounding lands, impinge upon the mountains, saturating them with mist or rain and forming what have been fittingly called the "weeping woods." It makes little difference whether the habitat be in Hawaii or the West Indies, in the Andes or the distant Himalayas, for the conditions are essentially the same the tropical world around. In these areas, the traveler will see tree ferns in all their grandeur, silhouetted as often as not in the fog, their delicate crowns conspicuous among the surrounding mass of vegetation.

Some tree ferns have wandered out of the confines of the tropics into the temperate zone. In the mild oceanic climate of New Zealand, for instance, they grow even in sight of glacier ice. From that temperate land has come *Cyathea medullaris*, a tree fern that can stand even a few degrees of frost. This enables it to be grown in certain favored sections of the United States, such as California and Florida. On our Pacific Coast it can be seen in San Francisco's Golden Gate Park, where the Pacific fogs, as they roll in, apparently give these alien ferns a welcome taste of their native atmosphere.

## RESEMBLE SMALLER COUSINS

Tree ferns resemble their more familiar lowly cousins of our woodlands except for their over-size dimensions and certain technical differences known only to fern specialists. The stems of most ferns are short underground structures called rhizomes, usually lying in a horizontal plane. However, the stems or rhizomes of tree ferns grow stiffly erect to form a tough, fibrous trunk.

Tree fern trunks are not to be compared to tree trunks. They lack the solid woody cylinder and growth layer (cambium) typical of the trunks of most trees. The tree fern trunk is incapable of growth in girth and grows slowly only in length from a single terminal bud. The giant leaves or fronds are at first tightly coiled in the bud in the form of elegantly designed "fiddleheads" or croziers, as the reader may have observed among our common woodland ferns. The "fiddleheads" uncoil progressively, producing new whorls of leaves, and with each successive season, the gigantic green crown of the tree fern rises higher and higher.

Curiously, multitudes of tiny, aerial roots often grow down over parts of the trunk, covering it with an interlacing network. This serves as a wonderful footing for the growth of tropical air plants. As a result, tree fern trunks often look like green ferneries, cloaked as they are with mosses, tiny ferns, and similar plants.

In their arborescent form, at least, our existing tree ferns are perhaps our closest approach to those distant ancestors which, in the age of coal, first gave the earth a green cloak of forests. Most of the arborescent fernlike plants of those ancient times were the so-called "seed-ferns" (Pteridospermae). Where coal was being formed, these were more abundant than the tree ferns. Some of these fern allies of long ago reached a height of 100 feet. There were, at the same time, true ferns of arborescent habit in those days, though none of them can be included in any modern family.

## KINDS AND USES

The family that includes most of the living tree ferns is the Cyatheaceae. Another family, the Dicksoniaceae, includes most of the neartemperate species, all of the southern hemisphere ones of Australasian origin, and also the genus

*Cibotium*, which includes two Mexican species. These latter are commonly grown in the United States as display plants for hotels and the like. They have pale green, much divided leaves, which are several feet long. They are sold before the trunks have begun to assume tree form.

Orchid fanciers in the tropics have long recognized the value of tree fern trunks and often cut them to length and hang them up as supports on which to grow their exotic specimens. Fern trunks are termite resistant and make practically indestructible posts. Many a simple tropical dwelling stands on them. Long ago, the fierce Carib Indians of the Lesser Antilles found another use. They noted that dried tree fern trunks made fine tinder and used it as their standard material for carrying fire from place to place. Dried fern trunk will burn for hours, very conveniently without smoke or visible flame. The Caribs called it *wâtu hakuiyâ*, meaning "voracious fire."

But of all the uses of the giant ferns, the one that seems most fitting is their utilization as ornamentals in horticulture. No botanical garden would consider its conservatory complete without at least one arborescent fern. Many a home owner in the tropics and subtropics feels the same way about these lovely plants. Certainly few species can offer anything comparable to the lacey tracery of their magnificent leaves.

Next time you see one of these splendid plants, imagine a whole forest of them, on one of those breathtaking mountain slopes in the tropics, amid swirling mists and tumbling torrents. If you visit such a region on your next vacation trip—perhaps the Luquillo Mountains of Puerto Rico, Jamaica's Blue Mountain Peak, or the knife-edged ridges of Hawaii—you will be able to conjure up a good impression of what a good pteridophyte forest would have looked like eons ago, when the coal beds were being laid down for races yet unborn to burn in their hearths in climates too cold for tree ferns.

## MOLDS AND MEN

Osiris looked down from the sky and saw Man bent under the weight of his troubles. With a smile of compassion the god handed down the gift of fermentation.

That is a Fourth Dynasty Egyptian myth. But primitive man did lighten his load with fermentation. He fermented everything that could be fermented. He made beer and wine from grasses, roots, trees, berries, honeys, pine cones. He even fermented fish.

Then he applied the heavenly skill to the brewing of other beverages: beers from cereal grains, wines from grapes, meads from honeys. Spiced date wine and Tequila-like spirits of cactus became passé. Every man began to insist on "brand names" like *Latium* of the House of Lucius Titus-a full-bodied red wine that was shipped all over the world.

In the 1950's, Captain Y. P. Cousteau and his crew from the French Navy's Undersea Research Group sampled some Titus Latium (vintage 230 B.C.) which they recovered from an argosy in the harbor of Marseilles. Their comment was dry: "The Third Century Before Christ was not a good vintage century." Apparently there was more method than madness in the ancient practice of diluting wine with water.

The ancients cherished the gift of fermentation and even fought wars for the best vineyards. Religious cults sprang from the grape.

Somewhere along the line early man realized that fermentation did more than make his head whirl. The chance contamination of wine by bacteria pointed up the possibilities of other products. Vinegar, for instance. This sour liquid could be carried by the legionaires and mixed with the water of foreign lands. Undiluted, it could be used as an astringent. According to Pliny, vinegar was also used to convert lead into a white pigment.

Other fermentations proved extremely useful. Linen was made from flax fibers retted by bacteria in stagnant water. Meats, fish and certain vegetables were preserved by the action of one kind of bacteria against the growth and activity of harmful bacteria. Dye was made by fermenting the indigo plant. Coffee and cocoa are still fermented to separate the bean from the pulp. Arrow poisons, hypnotics and soporifics were extracted from fermented decoctions of herbs, roots and mushrooms. And cheeses were made from the fermented milk of camels, ewes, goats, llamas, yaks, mares, caribous, cows, water buffalos, and a long list of other mammals.

Crusaders returning from the seraglios of the Holy Land told wild tales of dancing girls, wines, medicines and riches beyond expectationhoping to kindle interest in more crusades and foment an escape from medieval monotony. Tales of the "chalice" endowed with miraculous medical cures, the "Philosopher's stone" that commuted baser metals to gold, the "Mithradaticum," a universal antidote, sent alchemists scurrying to their cells to mix, distill, percolate, extract and ferment everything within reach. Some of their formularies make the witches brew of Macbeth look like a housewife's recipe for pink lemonade.

Reports on the potency of his "ferments" boosted the alchemist's standing in society. He reached his zenith about the time the Bubonic Plague was decimating Europe's populations in the Fourteenth Century. Promulgating such theories as "poisons counteract poisons" and attributing disease to a "malevolent admixture of the elements," the alchemist did neither medicine nor fermentation much good.

Over the next three hundred years, man gained greater knowledge of the world around him. But knowledge and understanding are not synonymous. Even after the little Delft microscopist van Leeuwenhoek demonstrated "animalcules" in his teeth in 1687, no one suspected their nature or importance until 1732, when Boerhaave examined the "schmutzdecke" of vinegar. He postulated that living organisms changed it into sour wine (vinaigre).

One hundred five years later Kützing confirmed Boerhaave's observations by actual studies of the minute blobs of life responsible for vinegar formation. But Kützing's revelation was ignored by most scientists for almost 40 years. The world was not ready to acknowledge the role of microbes in fermentative change.

It took a chemist with a flair for public relations to hammer home the message of the microbes. Louis Pasteur not only developed the bacterial concept of disease, but established the principle that fermentation is a process brought about by living cells. Though his definition of fermentation ("life without air") is not correct, he got the essential idea: fermentation is life.

In 1878, Lister diluted sour milk to isolate lactic acid bacteria and prove that they caused the milk to sour. Three years later, Charles E. Avery ventured into industrial fermentation to manufacture lactic acid.

Although yeast, vinegar and alcoholic beverages had been produced on an industrial scale for decades, Avery's enterprise made him a father of industrial fermentation.

Alcohol had been distilled on a small scale from wines and beers since the Middle Ages. This was uneconomical—especially for industry. *The Industrial Alcohol Act of 1906* legalized the quantity production of *denatured* alcohol for industrial purposes. The act also declared such alcohol tax free. As a result of this legislation, more than 100 million gallons of tax-exempt alcohol go to market each year—mostly for industrial consumption. *

From 1906 on, industrial fermentation grew like Topsy. World War I gave it new impetus. Britain wanted to blast the Kaiser out of Europe with an explosive called cordite. The Kaiser, in turn, sought to widen the English Channel from Calais to the Irish Sea with nitroglycerine.

Both lacked the necessary means. Britain needed acetone to manufacture cordite; Germany lacked glycerine to make nitroglycerine. Almost simultaneously, both found fermentation the key to their production problems. Glycerol could be made by fermenting molasses; acetone by fermenting corn. The British process proved more successful.

Acetone, a commercial solvent, was also used as dope to stretch and tighten the canvas on the wings of German Fokkers and British flying jennies. The Germans were hamstrung by embargoes and blockades and concentrated their war efforts on developing a synthetic rubber. The production snag in the German program was butadiene, the starting material. There wasn't any.

So Kaiser Wilhelm's scientists fermented butanol which, in turn, was converted to the diene. Here was a picture in irony: the British were making acetone by fermentation and throwing away the by-product butanol desperately needed by the enemy. The Germans were using the same fermentation and discarding acetone!

Henry Ford's mass production of flivvers brought butanol back into business after World War I. The fermentation-derived chemical was used its a solvent in the red, black, green and yellow lacquers on the automobiles of the era.

Fermentation was pushed further along when Italy slowed the flow of concentrated lemon juice and citrate of lime shortly before the war. These products were used in America for the commercial production of citric acid. A Brooklyn firm of manufacturing chemists, Chas. Pfizer & Co., Inc., was one of the larger importers of this raw material. The firm thought about making its own lemon juice. Lemons, however, do not grow in Brooklyn. Pfizer chemists tried fermenting sugar.

Hairbreadth controls were developed; intricate steps and ingenious methods were devised. Fermentation, long an art, became a science overnight. Sugar was converted by man and by mold into citric acid.

By 1923 Pfizer was producing citric acid commercially, by fermentation. A constant source of supply was assured. A saving of more than 75 per cent was realized on the price of citric. Then came Prohibition—and a nation-wide preoccupation with bathtub gin. For some reason, industrial fermentation sat out the next decade.

In 1934, Reichstein and Grüssner synthesized vitamin C. Sorbose, a necessary starting material, was not readily available. Industrial research teams then developed a method of converting sorbitol into sorbose by fermentation. This was the crucial hurdle in vitamin C production. The method was a boon to humanity in general and to the British Navy in particular. His Majesty's seamen had sucked lemons since Nelson's time to ward off scurvy, a vitamin C deficiency disease.

The citric acid process later paved the way for riboflavin (vitamin B-2), vitamin B-12, and as-yet-undefined growth factors like Vigo-fac, a farm animal growth stimulant. In terms of health and well-being, these achievements are overshadowed by fermentation developments since 1941. Work was begun that year by the Northern Regional Research Laboratories in Peoria, Illinois, and by Merck, Squibb and Pfizer on mass-producing Sir Alexander Fleming's discovery, penicillin. Eventually, the entire U. S. drug industry was enlisted. Research on the first "miracle drug" had the unlimited backing of the governments of the United States and England. The lives of their soldiers were at stake.

The *Penicillium* mold, resembling the exotic microorganism that makes Roquefort cheese, was grown in flasks that looked like milk bottles. Production, however, was slow and uneconomical. Large volume production was the only answer. Giant tanks were built to hold the fermentation liquor. Ship-sized propellers were designed to chum the broth into a frothy maelstrom. Sterile air was sent bubbling through the brew. This was submerged fermentation. It worked.

But there were other problems. The mold originally used for the production of the first penicillin didn't grow too well in tanks. It preferred the undisturbed surface growth in bottles. So the search began for a mold that could live and reproduce in the turbulent tanks.

In 1943, three government scientists, Raper, Alexander and Coghill, obtained an cantaloupe from a Peoria market. From it they isolated a mold called *Penicillium chrysogenum*, which proved adaptable to the submerged-culture techniques. The new mold stepped up penicillin production.

But even this was not enough. Demerec, working at the Carnegie Institute in Cold Spring Harbor, Long Island, bombarded the *P. chrysogenum* mold with X-rays, hoping to mutate it so it would produce greater quantities of the drug. This worked. Still Demerec wasn't satisfied. He tried chemicals, heat, more X-rays, and ultra violet rays. Out of this electro-chemical whirligig eventually came a culture that produced more than a thousand times as much antibiotic as the original mold. Penicillin got to Normandy by D-day.

The Demerec method has been used by industry ever since. Industrial scientists then began looking for other microscopic forms of life to make more antibiotics. Waksman found Streptomycin. Bartz isolated Chloromycetin. Duggar discovered Aureomycin. Finlay and other Pfizer researchers turned up Terramycin.

Strange new names were added to the lexicon and the man on the street began talking about Terramycin, Tetracyn, Sigmamycin *, Magnamycin, Viocin, Ilotycin and the like. The era of antibotics was here.

At the same time, the medical profession began paying closer attention to hormones from the adrenal cortex-regulatory chemicals from glands no bigger than a pullet egg. Merck and Co. developed a method for synthesizing cortisone by a thirty-odd step process, proof that science was not dependent on the animal body for its regulatory chemicals.

Upjohn, Pfizer, and others developed methods for converting a readily available chemical, Reichstein's Compound "S", into one of these hormones by means of molds. This was hydrocortisone. Prednisolone was another giant step in steroid research. Also made by fermentation, prednisolone is from four to five times more potent than natural compounds.

Many antibiotics and hormonelike steroids have gone to market since 1952. Fermentation-produced chemicals like itaconic and kojic acids have been made available to industry. Microbes have been coaxed to make the basic amino acid, l-lysine, an essential "building block" of proteins....

Cultivation of plant tissue by submerged fermentation is still a laboratory curiosity. But some day, submerged culture techniques may give us-at lower cost, on a year-round basis—plant alkaloids, botanical extracts, and medicinal and agricultural chemicals heretofore dependent upon climatic whims and seasonal changes.

Three basic types of fermentation processes are in use today.

In many parts of the world, flax, hemp, jute and ramie are still retted —the past fibres are separated from the surrounding stem tissues by uncontrolled or natural fermentation. The plants are simply exposed to the weather (dew-retting) or placed in ponds or streams (water-retting), where bacteria ferment them. Natural fermentation is also used, with some refinements, in the commercial production of preserved foods like pickles, sauerkraut and olives.

Beers, wines and liquors, on the other hand, are made by a semicontrolled fermentation process. Braumeisters, * wine makers and fermentologists in the alcohol, whisky, solvent, yeast and related industries would probably not take kindly to this label. The term "semi-controlled," is used here strictly as a comparative. These processes *are* controlled—but the controls generally are not so rigid as those used in the chemical and pharmaceutical industries.

Habitués of Morey's know there are many kinds of beer. Lager, pilsner, bock, ale, stout and porter are domestic favorites. Each has a different character. However, all are made by essentially the same process. The differences in flavor are determined by the strains of yeast used, type of water, fermentation temperature, quantity and quality of hops, methods of malting, and so on.

In controlled fermentation, the tanks may be jacketed, glass lined, and equipped with spargers, impellers, heating and cooling coils, electronic foam control apparatus, heat regulators, rotometers, exhaust valves, baffles, sight glasses and sampling tubes. The media in which selected organisms are grown to produce antibiotics and other fermentation chemicals is compounded to insure the optimum balance of essential inorganic ions, trace elements, buffering salts, growth factors, precursors and nontoxic defoaming agents.

Care is taken to maintain the proper ratio of available nitrogen and carbon in the raw materials used as nutrients. The inoculum, although preformed, must contain a specific quantity of actively proliferating cells at a definite stage in their growth cycle.

The sequence of events applies in nearly all controlled fermentations. The variables, of course, are the organisms, media, fermentation time and extrinsic factors such as heat, air supply, pH.

A specially-selected pure-bred microorganism is aseptically grown on a solid agar medium containing elements which favor the production of spores, the "seeds" of the mold. At maturity, these spores are washed from the mold with a sterile detergent-containing solution and microscopically counted. Appropriately diluted aliquots of the spore suspension are inoculated into small flasks containing sterile media. This encourages active growth and reproduction of the mold.

The inoculated flasks are then shaken for a predetermined length of time so the mold can grow. This vegetative growth, called mycelium, assumes the form of small balls. It reproduces itself by fragmentation. The mycelial balls are aseptically transferred to small containers called "seed tanks," in which the process is repeated—but on a much larger scale. This is staged fermentation.

From the time the inoculum is introduced into the seed tanks until the end of the fermentation, the operation is rigidly controlled. When the growth of mold is optimum in the seed tanks, their contents are pumped into production fermenters containing sterile production medium.

The vast quantities of production broth must be "cooked" enough to kill contaminating organisms without affecting the nutrients in the medium. Sterile air, or air mixed with other gases, is introduced into the fermenters at a carefully-controlled volume. Fermentation temperatures are also rigorously controlled. As metabolism progresses heat is evolved—and heat often destroys the end product. Likewise, pH (hydrogen ion concentration) must be kept in check. pH values far below or much above the optimum tend to cut yields.

How the fermentation medium is agitated also directly influences the yield. Agitation must uniformly distribute air and nutrients to the growing mold. It must not allow toxic metabolic products to accumulate and reduce the yields by "poisoning" the ferment.

Progress is measured periodically by sampling. This involves pH, mycelial volume and sugar or protein concentration tests. Nutrients, defoaming agents, buffering compounds, precursors and other agents are added as tests indicate, and elaborate precautions are taken to maintain sterility.

When it is no longer economically feasible to continue the fermentation, the liquor containing the product is separated from the vegetative growth and insoluble ingredients. The extract is crystallized and made into the various forms in which it is packaged.

There are other types of fermentations: production of enzymes, chemical transformations by microbes, manufacture of cheese, industrial and human waste disposal, and many others. They add weight to a truism: That fermentation chemistry is indispensable to human well-being.

Without highly-developed controlled fermentation techniques, antibiotics would still be laboratry curiosities. Citric, kojic, gluconic, itaconic, oxalic acids; vitamins, hormones, steroids and a long list of chemical intermediates used by industry would be rare and costly. The methods for producing chemical "building blocks" economically would not exist. In the years ahead, the gift of Osiris may be expected to loom even larger as the key to better living.

## Chapter 20

# Nutrition Health and Disease

### SOME MISINFORMATION ABOUT FOODS, DIETS

"Food cooked in aluminum is poisonous and causes cancer."

The American Medical Association and other authorities have proved that there is no danger involved in using aluminum cooking utensils. Many common plant foods contain aluminum which is absorbed from the soil. This fact does not stop us from eating these foods. Physicians use aluminum compounds in the treatment of ulcers.

"A good way to reduce is to omit breakfast."

Reliable studies show that a reducing plan including a well-balanced breakfast is more successful than one which omits breakfast. The individual feels better, has less fatigue, is more alert, less hungry and less tempted to "nibble" before lunch time. Everyone, including "weight-reducers," needs to replenish the body with vitamins, minerals and protein, as well as energy, after a 10- to 12-hour fast. Surplus body fat can supply some of the energy but not the other nutrients.

"Overweight is due entirely to heredity."

Some authorities believe that in certain cases there may be inherited body disorders which may account for a family tendency to put on weight. Regardless of the cause, the overweight person is eating more than his body needs—he must eat less, or foods lower in calories, to lose weight.

In some instances there may be common family traits of unusually good appetites, easy-going temperaments, perhaps combined with a preference for sweets and high-fat foods.

### Dark Bread has Fewer Calories than white Bread

Generally speaking there is very little difference in the caloric value of regular white bread and dark bread. Some dark breads containing raisins or made with molasses may have a few more calories than plain white bread.

### Eat all the 'Reducing Bread' you Want

If the so-called "reducing breads" have any worthwhile nutritive value, they will contain calories. Unfortunately, the composition of many of these breads is not given on the wrapper; therefore, the caloric value is unknown. In cases where the composition is given, the caloric value is usually only slightly less than regular bread.

### Water is Fattening

Water has caloric value. There is nothing in water that can change to fat and be stored in the body. A person will weigh more if weighed immediately after drinking a large quantity of water, but if the body does not need all this water, the excess will be excreted shortly. In certain diseases of the kidney and heart, excess water may be retained in the body. If you feel that you aren't overeating, but are gaining weight, you should see your physician.

### Toast has fewer Calories than Untoasted Bread

Toasting does not reduce the number of calories. Toasting does slightly reduce the water content, and so weight for weight, a pound of toast actually has a few more calories than a pound of untoasted bread.

"Homogenized milk is 'richer' in fat than unhomogenized milk." The fat content is not changed when milk is homogenized. Homogenization divides the fat into fine particles which remain equally distributed throughout the milk. This gives the impression of "richer milk."

### Margarine has Fewer Calories than Butter

Margarine and butter have the same caloric value. Vegetable oils from which margarine is made have the same calorie content as animal fat.

### Grapefruit Aids Reducing

Grapefruit contains sugar, vitamins, minerals, and citric acid. None of these in any way removes or helps to remove fat from the body. Although grapefruit is not as high in caloric value as some fruits, it does contain calories.

### Sugar is not as Fattening as Starch

One ounce of sugar has essentially the same number of calories as one ounce of starch.

## HORIZONS IN DENTISTRY

The primary task of dentistry is not to extract and replace teeth, but to preserve them. Consequently, dental research centers around the saving of teeth. Of course, research on prevention of the loss of diseased teeth and surrounding tissues is necessary, but what is even more important is the research on preventing dental diseases. This means that dental research should be concentrated on the treatment and prevention of:

- Dental caries—a breaking down of the hard tooth tissues,
- Periodontal disturbances—loss of the tissues that support the teeth, and
- Malformations—deformities of the oral tissues, and
- Conditions which are the aftermaths.

There is no inkling, on the scientific horizon today, of major changes in these public responsibilities of dentistry. How is dental research accomplishing the foregoing tasks, and how can it become more effective?

## RESEARCH ON FLUORINE AND DENTAL CARIES

The discovery that proper amounts of a trace element, fluorine, in the diet can influence the development of a body organ to make it diseaseresistant—in this instance, the tooth—is an unparalleled achievement in disease prevention. Heretofore, practically all of the really preventive measures for the control of disease have been useful on *acute* diseases only.

The *chronic* diseases have been, and are, strikingly resistant to preventive measures. The reduction of dental decay by proper amounts of fluorine, which do no harm to the body, has been repeatedly demonstrated. Yet no one knows precisely the manner in which fluorine acts. This must still be explained, and that explanation will probably come from intensified research on calcified tissues.

Some hints have been uncovered that ions other than fluoride, such as ammonium, sodium, and stannous (tin), may play a potent role in caries prevention. These leads should be relentlessly pursued.

## CALCULUS RESEARCH

Much of the loss of teeth among older people is caused by the wasting away of the bone and the gums that support the teeth. Much of this loss of the investing structures is caused by the deposition of unwanted, hard deposits called dental calculus, mostly on the necks of the teeth. Today, the only treatment available is removal of the calculus. What is needed is intensive research in how calculus forms so that preventive measures may be designed.

## BACTERIAL PLAQUE RESEARCH

The soft deposits on teeth—the *plaques*—are as undesirable as the hard deposits because they harbor bacteria that are thought to help break down the tissues which hold the teeth in place. What are these plaques? Why do some harden to form calculus while tooth decay occurs under others? Why do some individuals have extensive plaques while others have small plaques or relatively harmless ones? These are just examples of some of the problems that you yourself may solve sometime should you decide to make a career of dentistry.

## RESEARCH ON SALIVA

The teeth are constantly bathed in saliva—why don't they dissolve in it? Because the saliva contains dissolved salts that keep the teeth from going into solution. Generally there is a balance in the solubility relationship in the enamel-saliva system with salts being replaced by the saliva at just about the same rate as they are being dissolved into it.

However, imbalances occur all too frequently, causing enamel dissolution or the precipitation of calculus. If balance could be maintained at the proper level, much of what we know as dental disease would not exist. Who is going to discover the how and whys of such dental disharmonies—perhaps you, a potential dental scientist?

## RESEARCH ON ORAL HYGIENE

What evidence is there that brushing teeth keeps them from decaying or stops the formation of calculus? Dentists have recommended oral hygiene as a preventive measure for decades, yet what few data there are indicate that the loss of teeth, at least the loss in young people, has been on the increase.

How can this trend toward increase in dental caries be reversed and be made to go downhill? Progress can be made principally by research on the treatment and the prevention of dental caries in the younger age groups; also, perhaps, by better oral hygiene on the part of the individual. Some data are available showing a direct relation between oral hygiene and tooth decay. Here again, field research and analysis have proved very effective, just as they did in the discovery of the relationship between fluorine and dental decay.

## RESEARCH IN DENTAL MATERIALS

Most of the tissues of the body can restore themselves or fill in gaps caused by injury. When tooth tissues—enamel and dentin—are lost, they are not repaired or replaced. So much of dental therapy is unique.

Why? Because much of modern dentistry involves the replacement of hard and soft oral tissues, lost by disease, with relatively inert materials. This surgical and *prosthetic* therapy must be biologically safe, mechanically efficient, and lifelike. Some of this treatment is largely reparative, such as artificial dentures; but much of dental treatment is preventive. The treatment of small cavities in teeth arrests decay and retards and prevents tooth loss.

Since much of dental therapy involves the making of restorations of lost tooth structures (operative dentistry) and replacements of missing teeth and tissues (prosthetic dentistry), research on the materials used in the treatment is imperative. The preparation of standards for these dental materials—standards that define the materials by physical and chemical tests—is an important part of the science of dental materials.

Another important aspect of dental therapy is relating laboratory research on dental materials with clinical research....

Among the suggestions for needed research in dental materials is the development of insoluble, decay-preventing, adhesive cements and filling materials with the necessary physical and esthetic qualities.

The need for dental research is evident. It is startlingly significant that there are almost one-half as many dentists as there are physicians and surgeons for the rest of the body. This is just an "around-the-corner" way of saying that dental disease is nearly universal.

It appears that the incidence of dental disease is increasing and that dental defects are developing faster than dentists can correct them now or in the future. Since the incidence of dental defects seems to be pyramiding, it is neither economically feasible nor practically sound to try to cope with the situation merely by increasing the number of dentists. This indicates that only by the additional control and preventive measures that research may provide can dentistry hope to cope with dental disease.

## RATS, BATS AND HUMAN DISEASES

The recent occurrence of six cases of human plague—one of them fatal—in two of our western states is a timely reminder of the significance of the zoonoses to outbreaks of human disease. Zoonoses—singular, zoonosis—are diseases of animals secondarily transmitted to man. Two of the most deadly zoonoses and perhaps most widely feared, are the plague, and rabies. Not since 1924, when there were 41 cases of plague with 36 deaths in Los Angeles, California, have so many cases been reported in one year in the United States. Throughout the world, the number of plague cases recorded for 1964—almost 1500—far exceeded the number recorded in 1963. Although an epidemic in Tanzania accounted for much of the increase, there were also more cases in Vietnam. Other countries seriously afflicted with plague in 1964 include Ecuador, Peru, Brazil, and India.

## WILD RODENT PLAGUE

The six Americans stricken all had what is known as sylvatic plague, because they acquired their infection from woodland or prairie animals. Prairie dogs, or rather their fleas, were deemed responsible for transmitting the disease to five of the victims. These American burrowing rodents are related to marmots and live in colonies on our western prairies.

In the woods and prairies of 14 of our western states, fleas convey the plague bacillus from one wild rodent to another. Bitten by one of these infected insects, a human being usually acquires plague in its bubonic form—so-called because of the swelling, or bubo, which appears. Sometimes the plague bacilli escape from the enlarged lymph node underlying the bubo and reach the lungs. In this way, bubonic plague is converted into pneumonic plague, a fulminating form which can be transmitted directly from one human being to another.

When large numbers of the bacilli are present in the blood—septicemic plague—fleas can spread the disease from man to man. The dark spots on the skin, due to hemorrhages, just beneath the surface, gave to this malady its dread-ridden name, the Black Death.

## URBAN PLAGUE

Outbreaks of human plague acquired from wild rodents, occurring as they usually do in sparsely populated regions, do not readily grow to epidemic proportions. However, different species of rodents occupy overlapping territories.

The fleas of wild rodents dwelling in areas remote from human habitations can infect only a few domestic rats and wild rodents living in rural and suburban areas, or can initiate an epizootic outbreak. Their fleas, in turn, can transmit the disease to urban rats, and theirs, in turn, to urban dwellers of the species *Homo sapiens*.

It is possible for a single case of pneumonic plague in a crowded section of a city to initiate an epidemic of major importance. In the Los Angeles epidemic of 1924, 17 of the victims were members of a single family. Other victims included neighbors, nurses, an ambulance driver, and a priest who had visited at the home of one of the patients. During the 18 months following the epidemic, health officials carried out a "campaign of quarantine, disinfection, rat and flea eradication, and ratproofing buildings" at a total cost of about four million dollars.

A larger plague epidemic ravaged San Francisco during the worldwide outbreak which lasted from 1895 to 1905. In four years, 121 people were stricken with the disease and 118 lost their lives.

## FURRED AND FEATHERED HOSTS

The great San Francisco earthquake and fire of 1906 which followed on the heels of the epidemic, drove the plague-harboring urban rats out of the city along with the exodus of people. Soon ground squirrels nearby became plague-ridden. The enzootic area gradually spread northward and eastward, and invasion of the plains states east of the Rocky Mountains began a quarter of a century ago.

Even if the enzootic were limited to ground squirrels, its eradication would be formidable. But the problem is one of far greater scope, for it is now known that plague occurs naturally not only in over 200 species of rodents, but also in rabbits and hares as well.

Wild-rodent plague is remarkably persistent in certain areas. Persistent plague foci are perpetuated, not by the highly susceptible hosts, but by hosts which are rather resistant to the disease. Animals which hibernate may acquire infection just before they turn in for their long winter's sleep. During this time, plague takes on a peculiar latent form and does not flare up until the animal has emerged from the dormant state.

In this way, plague can persist over the winter in a hibernating species.

Other animals known to be instrumental in transmitting plague to human beings are camels, cats and shrews. By carrying infected rodents and their fleas from one place to another, birds of prey may spread the disease. These birds, although resistant to the disease, are sometimes infested with fowl fleas containing plague bacilli.

## THE DEADLY FLEA

Rodents are not really transmitters of plague to man; they are, in a sense, its victims. The real vector is the flea. These tiny arthropods commute from host, where they partake of several blood meals daily, to the animal's nest or burrow, where eggs are laid and larvae develop. The proportionate time spent at each location varies with the species and climatic conditions.

Plague bacilli have been discovered in about 100 different species of fleas throughout the world. In the rat flea, *Xenopsylla cheopis,* and the ground squirrel flea, *Hoplopsyllus anomalus,* multiplication of the bacilli occurs in the mid-gut which becomes "blocked" with blood and bacilli. In desperation, the flea goes from host to host in rapid succession, regurgitating part of the deadly mass into each new puncture wound. "Blocked" fleas are believed to play a key role in enzootics, although ordinary fleas may keep epizootics going.

In rodent burrows, fleas escape the rigors of climatic change. And even though the normal inhabitants of such burrows are destroyed, the fleas in and around them can remain infective for many months. The flea is truly a reservoir of plague.

Although suspect for some 15 years, insectivorous bats of the United States are now known to be a natural reservoir for the rabies virus, and may harbor other zoonotic viruses as well. Perhaps it was not without good cause that, throughout recorded history, bats have been regarded as revolting creatures of ill omen. Five people in the United States died of rabies following actual bites or other close contact with bats during the 1950's. By 1960, rabies virus had been found in naturally infected bats in 31 states. Twenty-four species of bat native to the United States have been incriminated.

For many years rabies enzootic in the vampire bats of South America has taken its toll in lives of both human beings and livestock. This is the first species of animal known to have the capacity to harbor rabies virus in its saliva in the absence of other signs of the disease. There is support for the view that this symptomless carrier state may also occur in other species of bats and in lemmings, ground squirrels, fowl and other animals as well.

This concept of inapparent rabies infections has spilled over into the field of human medicine. Human rabies was formerly believed to be consistently fatal. In recent years there is some suggestive evidence that "man may be subject to subclinical rabies or recover from an overt clinical.

## RESERVOIRING MECHANISM

The predilection of certain human viruses for the fat tissue of animals was a suggestion to investigators that the interscapular brown fat, so prominent in hibernators, might serve as a reservoir for the rabies virus in hibernating species of bats. Some bats hibernate in winter while others escape its rigors by migrating to warmer climates.

The hunch was a good one, for rabies virus was actually found in the brown fat of native American wild bats. In the laboratory, strains of rabies virus of bat origin often seemed to prefer the brown fat, multiplying in it at a much faster rate than in brain or salivary gland. In other words, some strains of bat virus are lipotropic rather than neurotropic.

There is still another "reservoiring mechanism." When bats hibernate the virus stops proliferating. Hibernation seems to put the virus into "cold storage." The virus, like the bat, is in a dormant state and does not actively proliferate until after arousal of the animal.

Rabies epizootics among wild animals are prone to occur where bat caves are numerous. Vampire bats are not hesitant about biting various animals and man. But no one has observed insectivorous bats in the act of biting foxes or other wild animals in nature or in the laboratory. The route of transmission of the infection from this kind of bat to wild animals was a riddle.

Then it was discovered that some species of bats excrete the virus either in urine or guano. It is also kown that the virus can enter the body through breaks in mucous membrane as well as through breaks in the skin. In fact, "men can contract rabies merely by walking into a cave inhabited by infected bats".

Recently, animals in screened cages have been placed in bat caves and they, too, have contracted the disease. One fox became infected after only 10 hours' caged sojourn in a bat cave. So attempts are being made to gather the virus in caves directly from the air where it is believed to be suspended in an aerosol form.

## YOUR HEART

The healthy heart is a very strong organ. It is only about the size of a large fist, but most of its bulk is muscle. Its one job is to pump into the arteries the blood returned to it by the veins. All the cells in the body depend upon the rapidly circulating blood stream for oxygen and nourishment and for removal of wastes.Your blood makes up only about 8 per cent of your body weight. But to keep that blood in circulation through miles of blood vessels during an ordinary day of work, play, and rest, the healthy heart pumps many, many hundreds of gallons of blood at an average daily rate of 70 to 80 strokes a minute.

Its normal pumping action is a continuous series of regular contractions and relaxations—beat-rest, beat-rest, beat-rest.Your heart normally rests twice as much as it works. But during periods of strenuous physical activity or emotional strain, it may beat twice as fast as usual and pump out twice as much blood. The faster the heart beats, the harder it works, and the less time it has to rest. However, most hearts can endure great physical exertion without difficulty. Strenuous physical exercise and work do not primarily cause heart disease, although they may aggravate difficulties already present.It's a fact that heart disease is widespread in this country.

More than 10 million Americans of all ages have some form of cardiovascular (heart and blood vessel) ailment. More than half of all deaths each year are due to heart and circulatory diseases. About 35 per cent of these occur among people 45 to 64 years of age.But while it is true that heart disease affects many people in middle and later life, it is also true that more people are living longer than ever before. Fewer lives are cut short by infectious diseases of childhood and youth which medical science now prevents and controls.

Thanks to modern knowledge and more accurate methods of diagnosing heart and circulatory disturbances, some ailments once incorrectly attributed to other causes are now properly classified as heart disease. This accounts, in part, for the statistics.Great strides have been made in research, diagnosis, treatment and control of heart disease. Progress has made it possible for thousands of people to survive and recover from heart and blood vessel ailments. Brilliant surgical techniques for correcting heart defects have been developed in recent years. Scientists foresee even greater advances in the near future. Right now:

- Most people who have heart attacks recover.
- High blood pressure (hypertension) can be controlled.
- Recurrent attacks of rheumatic fever, which damage the heart, can be prevented.
- Many heart defects can be repaired.
- Medical science can do a great deal for people with circulatory disorders.

Today's scientific knowledge is being applied to help people with heart disease live and work happily and productively. In addition, research has brought to light knowledge which can help people reduce the risk of acquiring heart ailments. So, although heart disease has by no means been conquered, there is every reason for optimism.

Sometimes the term "heart disease" is confused with "heart attack." But they are not the same. Heart disease includes many different conditions and disorders affecting the heart and the circulatory system. The word *cardiovascular,* which means heart and blood vessel, is therefore more apt and more descriptive. Here, we describe several common cardiovascular ailments:

## ATHEROSCLEROSIS

This is the most important form of "hardening of the arteries." It is a complex process which is believed to begin when the body's mechanism for handling certain fatty particles such as cholesterol—either eaten or developed within the body—breaks down. These fatty substances start to collect in the inner lining of the walls of the arteries. As more and more particles are deposited in the arterial wall, the channel through which the blood flows gradually narrows.

Atherosclerosis may begin to develop at a relatively early age. People may have the condition in a mild form without noticeable difficulty, throughout a long life. When symptoms do occur, they usually show up at middle age or later.

Intensive scientific studies go on all the time to find ways to prevent and cure atherosclerosis. Among the factors being explored are diet, smoking, exercise, stress, hormones and heredity. There is growing evidence that diet is an important factor contributing to heart attack risk. The substitution, whenever possible, of vegetable and fish oils for animal fats is now encouraged by many authorities as a precautionary measure.

This substitution tends to lower the level of cholesterol in the blood. It is known that populations with lower levels of cholesterol and other fatty substances in the blood have a lower incidence of coronary heart disease than those with higher levels. Any special, fat-controlled diet however should be prescribed by a physician.

## CORONARY ARTERY DISEASE

When you hear that someone has "heart trouble," it often means coronary artery disease. Today this condition is being seen more frequently than it once was among young and middle-aged adults. However, it occurs most often among people between the ages of 40 and 60, and affects more men than women.

The coronary arteries supply the heart muscle itself with blood. The chief cause of coronary artery disease is thickening of the coronary arteries (coronary atherosclerosis). When these arteries become narrowed, due to the process of atherosclerosis, the heart muscle receives a smaller amount of blood. This sometimes results in what is known as coronary insufficiency —which means that the heart muscle itself is less efficient due to the lack of proper nourishment.

Many people are able to live quite comfortably with coronary artery disease if they are careful not to place too great a strain on their hearts.

## HIGH BLOOD PRESSURE

In medical language, this is hypertension—a condition which affects about 5 million Americans. The popular notion that high blood pressure is caused by aging is not based on fact. There is no reason for anyone to feel that because he is getting along in years, he must be developing high blood pressure.

To understand high blood pressure, it is first necessary to know the meaning of "blood pressure." Everyone has "blood pressure," and everyone's pressure normally goes up and down. Blood pressure is simply the pressure of the blood against the walls of the arteries. It is highest during systole—the period when the heart pumps a fresh supply of blood into the elastic-walled arteries which stretch to accommodate it. It is lowest during diastole—the period when the heart pauses between beats to fill with blood. *Diastolic* pressure is more significant than *systolic* pressure. That is because diastolic pressure represents the basic pressure exerted on the arterial walls without the additional pressure due to the contraction of the heart.

The nervous system influences blood pressure. For example, it is natural for pressure to go up when a person is excited. The smallest branches of the arteries (arterioles) are controlled by nerves which cause them to constrict when you are keyed up with strong feelings such as joy, fear, or anger. When the arterioles tighten up in this way; the resistance to the flow of blood is greater, and this increases pressure in the arteries. When excitement subsides, the arterioles dilate, opening wider, and the pressure goes down.

When a person has high blood pressure, the arterioles remain in a more or less constantly constricted state. Other influences, besides the functioning of the nervous system, cause this constriction. It is generally accepted by physicians that certain chemical substances set free into the blood stream from various glands and organs also cause the arterioles to constrict. Why this happens in some people and not in others, what the relationship is between these substances and the influence of the nervous system, are still questions which scientists are exploring.

Sometimes high blood pressure occurs with other diseases, such as kidney trouble or disturbances of the endocrine glands. In these cases, when the disease is cured, usually the hypertension is, too. When the cause is unknown, as it is in the majority of cases, the condition is called essential hypertension. It seems to run in families and occur more often among people who are overweight. Nervous tension apparently aggravates the condition in some people.

## HYPERTENSION AND THE HEART

How does hypertension affect the heart? First of all, the heart must work harder, although not necessarily more rapidly, to pump blood through the arteries. Pressure in the arteries must increase to keep the blood flowing through the constricted arterioles at nearly the normal rate. As a result of this continual, excessive work over a long period of time, the heart muscle enlarges. This is an important step in the development of hypertensive heart disease. Often, but not always, the walls of the arteries become scarred, thickened, and lose their elasticity—that is, they become atherosclerotic.

What are the symptoms of hypertension? In some cases, there are none. In others, there may be headaches, dizziness, fatigue. Any or all of these symptoms do not definitely indicate high blood pressure, but they do mean that something is wrong and a medical checkup is in order.

Sometimes high blood pressure clears up by itself. Even if it does not, modern medicine can do a great deal to keep it under control. Very often high blood pressure can be lowered to a safe level by new and effective drugs, by changes in diet and living habits. Occasionally, surgery corrects the condition. The important thing in managing hypertension is to follow the physician's advice faithfully.

## LOW BLOOD PRESSURE

Many healthy people have this condition. "Low-normal" blood pressure, as it is sometimes called, actually involves less work for the heart and blood vessels to keep the blood in normal circulation. Although rarely associated with serious illness, low blood pressure is sometimes a symptom of disease. A physician can tell if it is significant.

## HEART ATTACK

This term usually means the sudden closing (occlusion) of a coronary artery by a blood clot (thrombus). Although it happens suddenly, this is often the result of a gradual process of atherosclerosis in the coronary arteries. When the narrowed artery becomes clogged, a portion of the heart is deprived of its blood supply. This condition usually causes a severely painful feeling of pressure in the chest, often accompanied by sweating, weakness, shortness of breath, and possibly loss of consciousness. Sometimes there is nausea and vomiting.

This, plus the pain, may be mistaken for acute indigestion. Only a physician can tell the difference, and he should be called at once when these symptoms occur.

The doctor may prescribe various treatments, but most of the healing is done by the body itself. When a coronary artery closes, other arteries nearby increase in size and new branches develop to deliver blood to the part of the heart that needs it. This is called collateral circulation. At the same time, scar tissue begins to form in the damaged part of the heart.

Time required for recovery depends on the extent of the damage, rate of healing, and whether or not there are complications. Treatment is aimed at giving the heart a chance to heal. In some cases, the physician prescribes anticoagulant drugs to prevent formation of new blood clots.

Heart attacks are more common among men than women. Most patients recover from a first attack. In fact, healing is usually so successful that most people can return to work and resume usual activities after recovery. Of course, there may have to be some changes made in living habits. If so, the physician, taking into consideration the patient's particular problems and circumstances, will determine what changes are indicated.

Medical terms for heart attack are: coronary thrombosis, coronary occlusion, myocardial infarction.

## ANGINA PECTORIS

This means cardiac pain. Even when there has been no thrombosis, pain may be experienced if the coronary arteries are diseased. It may be a sensation of pressure in the chest or it can be a severe pressing or strangling pain under the breast bone and radiating down the arm. Such discomfort is usually a brief warning that the blood supply to a part of the heart muscle is insufficient at the moment.

Exertion, excitement, or digesting of a heavy meal may produce the pain in a susceptible person. Anyone who experiences it should see a doctor, who will give medication that will ordinarily relieve the pain almost at once. The physician will also instruct the patient in how to regulate his life and use medication to prevent or lessen the severity of such attacks.

## STROKE

This circulatory disorder affects many people, but not only the aged, as is commonly believed. What is a stroke? If the blood supply to part of the brain is reduced or completely blocked, a stroke occurs. Nerve cells, in that part of the brain deprived of blood, cannot function. As a result, part of the body which these nerve cells control also cannot function normally. When this happens, a person may become partially paralyzed, and may have difficulty speaking.

Stroke is commonly caused by a blood clot (thrombus) which forms in an artery narrowed by atherosclerosis. Sometimes it is caused by a clot which is formed elsewhere in the body and is carried to the brain by the blood stream. This is called an embolus. Another cause of stroke is hemorrhage, or bleeding from an artery in the brain.

Some people recover completely. Others may require a long time to make even a partial recovery. It depends upon the extent and type of damage. Immediate physiotherapy is highly important in helping a patient to regain lost or impaired muscular control—and prompt attention should be given to the prescribed exercises and massages, as well as positioning in bed.

The sooner rehabilitation can begin, the more effective it is. Surgery has been successful in treating some types of stroke—for example, those caused by blockage of the arteries in the neck. To prevent a possible recurrence, physicians can prescribe anticoagulant drugs for some patients, depending on the type of stroke.

## RHEUMATIC HEART TROUBLE

Heart trouble can result from repeated attacks of rheumatic fever, although, today, this can usually be prevented. Rheumatic fever is not so prevalent as it once was, but it still occurs among some children and young adults. It strikes most often between the ages of 5 and 15. Grownups may have heart disease which stems from childhood attacks of rheumatic fever.

Rheumatic fever can be precipitated by a streptococcus infection such as "strep" sore throat (sore throat accompanied by fever), "strep" ear infection, or scarlet fever. It attacks the connective tissues of the body, and may injure the covering and lining of the heart, the heart muscle, and the valves. Signs and symptoms which may indicate rheumatic fever are persistent, slight fever; pain in the joints and muscles; poor appetite; failure to gain weight; unexplained fatigue; repeated nosebleeds. Remember that rheumatic fever itself is not contagious. Nobody can contract it by being near a person who has it. But "strep" infections are contagious. Today, if a youngster contracts a streptococcal infection, adequate treatment usually prevents subsequent development of rheumatic fever. Prompt medical treatment of sore throats and other infections protects against rheumatic fever.

A first bout with the disease may not harm the heart. Many patients recover without any damage whatsoever. But once a person has had rheumatic fever, he is susceptible to recurrent attacks. To prevent this, the physician prescribes regular doses of sulfa drugs or antibiotics over a long period of time. This preventive medication is usually successful.

## CONGENITAL DEFECTS

Every year, about 30- to 40-thousand babies are born with defective hearts. An inborn or congenital heart defect means that the heart, or a blood vessel near the heart, failed to develop normally during the prenatal period of growth. What causes most of these malformations is not yet known, although researchers are delving into the question. Inborn heart defects are sometimes mild and sometimes serious. If the trouble is mild, surgery may never be necessary, and the person may lead a normal life. More serious defects can often be helped by surgery, and each year research paves the way for more defective hearts to be helped by surgery.

Even when this is not possible, good medical care is beneficial. When surgery is used, it may correct the defect completely or partially so that circulation becomes normal or at least nearly normal. Youngsters with congenital heart defects need close medical supervision during their growing years so as to avoid any infections which might cause difficulties. Standard immunizations are a "must."

## BACTERIAL ENDOCARDITIS

This is an infection of the heart lining (endocardium) usually caused by bacteria of the streptococcus family. People with congenital heart defects or rheumatic heart disease are the most susceptible. Today this dis ease can usually be treated successfully. Modern medicine, using antibiotics, is able to cure most cases. To prevent the possibility of bacterial endocarditis, doctors prescribe antibiotics before performing certain surgical procedures—even tooth extractions—which might allow the infection to gain a foothold in a susceptible person.

## WALTER REED AND THE CONQUEST OF YELLOW FEVER

The story you are about to read is one of the great adventure stories of medical science. Before the mystery of yellow fever was finally solved, couragious American soldiers had braved hardships and danger; a gallant young scientist had given his life; and his comrades in the great adventure had gone on to complete without him one of the most brilliant scientific achievements in the annals of medical history. The leader of the little band of scientists and volunteers who solved the mystery of yellow fever was Walter Reed, and with him the story begins.

In 1898 the Spanish-American war broke out. Sanitary conditions in the army camps were deplorable, and it was not long before epidemics of typhoid fever appeared among the volunteer troops. Major Reed and two distinguished civilian physicians were appointed to a commission to study the origin of these epidemics and how to stop them. The year-long painstaking investigation made by this commission, of which Major Reed was chairman, added greatly to the store of scientific knowledge needed to fight this dangerous disease.

The work of the typhoid-fever commission was no sooner completed than another menace to the health of American soldiers appeared in Cuba where the war was being fought. This menace was yellow fever—long a challenge to medical research.

Before telling the story of how Walter Reed and his co-workers met this challenge, let's pause to review its history so that we shall have a better understanding of what their work has meant both to our country and to the world.

## YELLOW FEVER ON THE MARCH

Yellow fever is an acute infectious disease of the tropics and subtropics, where its nickname is yellow jack. It took its name from the lemonyellow skin tint seen in victims with severe cases. In the worst epidemics of yellow fever as many as 80 out of every 100 persons who caught it died. Those who survived were usually immune for life.

The earliest known occurrence of yellow fever in the Western hemisphere was in Central America in 1596. The disease spread rapidly and made large areas of the tropical Americas almost uninhabitable for the white man. It invaded New York for the first time in 1668, Boston in 1691, and Philadelphia in 1695. In the 200 years from 1705 to 1905 more than 160 epidemics of yellow fever occurred in the United States. In New Orleans, Philadelphia, Memphis, Charleston, Norfolk, Galveston, New York, Baltimore, and many other cities, it took a heavy toll of life while doctors looked on, helpless.

But mere dates and statistics do not tell the full story of the terror of yellow fever. That can be made clear by looking at the experience of just one city during a yellow-fever invasion.

In 1793, six years after the adoption of the Constitution of the United States, yellow fever appeared on a street near the busy wharves of Philadelphia. As it spread, fear gripped the people and all the streets and roads leading from Philadelphia were crowded with families fleeing to the country for safety. So many doctors were sick or had died of yellow fever that "at one time there were only three physicians who were able to visit patients, and at this time there were probably not less than 6,000 persons ill with the fever."

Dr. Benjamin Rush, famous in yellow-fever history and then a physician in Philadelphia, relates that a cheerful contenance was scarcely to be seen in the city for six weeks. About to enter a patient's house he met a 2-year-old child on the street who looked up at him and smiled. "I was strangely affected by this sight," Dr. Rush tells us. "Few persons were met in the streets except those who were in quest of a physician, a nurse, or the men who buried the dead. The hearse alone kept up the remembrance of the noise of carriages or carts in the street."

## EARLY CLUES

What causes yellow fever? How is it spread? For more than 200 years learned men had searched for the answers to these questions. Probably most of you know these answers. The cause of yellow fever is a specific virus, that is, a germ so small that it can pass through the finest filters. This virus is carried from one person to another by the bites of female *Aëdes aegypti* mosquitoes.

From what you know now, it is easy to see how closely the mosquito's habits tally with the spread of yellow fever. See if you can tell why the following clues taken from old accounts of yellow fever might have put early investigators on the right track if they had even dreamed that a greatly feared and often deadly disease might be carried by so lowly a creature as a mosquito.

- Epidemics of yellow fever commonly started in the low wet regions or around the docks of cities near the sea coast.
- In the high and dry parts of a city the disease was not contagious. In many epidemics people from low-lying sections fled to the higher part of the city or to the country districts. Although many of these people came down with yellow fever after they had left their homes, the disease did not spread to other people in the new neighborhood. This observation made many intelligent investigators think that the disease must be present in the air of certain districts and not in others.
- Some observers noticed, also, that whenever the wind blew strongly in a certain direction, yellow fever broke out in its path. When the air was still, the infection was limited to the houses of an already infected neighborhood.
- Yellow fever flourished when the weather was hot, but it was stamped out by frost. This observation made

many investigators believe that "heat was a very common exciting cause of the disorder."

If you have interpreted these clues correctly, you know why they point to the mosquito as the guilty party in the spread of yellow fever. Mosquitoes breed in still water. Mosquitoes travel through the air. Mosquitoes are blown in the direction of the prevailing wind. Mosquitoes are active in hot weather and are killed by frost.

## THE MOSQUITO ACCUSED

The first suggestion that both yellow fever and malaria are carried by insects came from Dr. Josiah C. Nott, of Mobile, Ala., in 1848. But to Dr. Carlos J. Finlay, of Havana, Cuba, belongs the full credit for first advancing the theory that mosquitoes are the carriers of yellow fever. He outlined this theory in a paper read before the Royal Academy of Havana on August 11, 1881.

Dr. Finlay did more than suggest that mosquitoes in general were the logical suspects. He accused the exact species of mosquito—now known as *Aëdes aegypti*. For 19 years—from 1881 to 1900—Dr. Finlay labored diligently to gather evidence in support of his theory.

Unfortunately he failed to plan and carry out convincing experiments and made some understandable mistakes in interpreting his findings. As a result, not a single doctor or research scientist, except his assistant, believed Dr. Finlay. Up to 1900 his theory was treated with ridicule and neglect.

In the year 1900 the Spanish-American war had ended, but Cuba was still under American military control, with General Leonard Wood as Military Governor. Yellow fever had become epidemic in Havana and in the town of Quemados, 6 miles away.

Near Quemados was Columbia Barracks where American troops were stationed. Already a number of American soldiers in the Cuban army of occupation had caught yellow fever and some had died. No wonder that Dr. George Miller Sternberg, Surgeon General of the Army, and his medical officers were worried!

For many years, Dr. Sternberg had been one of the world's leading authorities on yellow fever. In Havana, in Brazil, in Mexico he had searched laboriously and unsuccessfully for its cause in the bodies of yellow-fever patients. He alone was not to solve the mystery of yellow fever, but this accomplishment was to be the crowning event of his long and brilliant administration.

In May 1900 General Sternberg appointed a Board of four medical officers to investigate acute infectious diseases, and especially yellow fever on the island of Cuba. The four doctors were Major Walter Reed, the leader, and acting assistant surgeons James Carroll, Jesse W. Lazear, and Aristides Agramonte. All four were highly skilled bacteriologists, thoroughly familiar with the mysterious disease which they were about to challenge with the courage, self-sacrifice, and scientific accuracy that characterized their work.

Dr. Lazear and Dr. Agramonte were already in Cuba making studies related to yellow fever at the time they were appointed to the Board. Major Reed and Dr. Carroll joined them at Columbia Barracks, Quemados, Cuba, on June 25. And so the great adventure began.

## THE EARLY STEPS

At first the Board gave its entire time to a thorough search for the causative agent—the germ—of yellow fever. Several investigators, including Sternberg and Finlay, had found bacteria of one kind or another in the bodies of yellow-fever patients. Then in 1897 an Italian scientist, Dr. Giuseppe Sanarelli, claimed that he had found the specific cause of yellow fever in the form of a certain bacillus.

As a yellow-fever epidemic was raging in Quemados, the Board was able to obtain plenty of material for study. After examining cultures made from the blood of 18 living patients and from the tissues of 11 patients who had died, the Board stated that in not a single one had they found Sanarelli's bacillus or any other bacteria whatever that might be the cause of yellow fever. As no germ could be found, Walter Reed and his co-workers decided to abandon their search for the cause of yellow fever for the time being and to go on to the next big question—how is yellow fever spread?

At that time almost everyone thought that yellow fever was carried by fomites—that is, by the articles of clothing, bedding, and other materials that had been contaminated by the excretions of yellow-fever patients. Careful reading of the reports of yellow-fever epidemics and his own personal observations had convinced Walter Reed that fomites had little, if anything, to do with the spread of yellow fever. On the other hand, he was greatly impressed by the peculiar way in which the infection jumped from one house to another in Quemados even when no contact existed between any persons or with any contaminated articles (fomites).

Another thing that especially impressed him and Dr. Lazear was the length of time it took to change a noninfected house into an infected one. In the report of a yellow-fever epidemic made in 1898, Surgeon Henry R. Carter of the Marine Hospital Service had shown that the first case of yellow fever in each one of 16 widely separated houses was not followed by new cases for about two weeks or more. Where were the germs responsible for the first case before they got around to claiming new victims?

Dr. Ronald Ross, an English army surgeon, had recently found the parasites of malaria in the stomach wall of *Anopheles* mosquitoes which had fed upon the blood of malaria patients. Was it possible that mosquitoes which had bitten yellow-fever patients had the germs in their bodies and could later pass on the germs to others by their bites?

The time had come to call on Dr. Finlay. Major Reed reported that the doctor received his visitors with great courtesy and gave them reprints of everything he had published relating to yellow fever. What is more, he presented them with eggs of the species of mosquito—*Aëdes aegypti*-which he had been using in his experiments. This precious gift consisted of a small basin of water to the edge of which, above the water line, a number of jet-black cigar-shaped eggs were clinging. Back at the laboratory the investigators raised the level of the water in the basin, and after a short period the eggs hatched into larvae, or wrigglers. Six or seven days later the wrigglers changed into pupae, or tumblers. After another 26 hours the adult mosquitoes took to the air. With these mosquitoes, hatched from the eggs furnished by Dr. Finlay, the first series of experiments to test the mosquito-transmission theory were made.

But before undertaking the experiments it was necessary to make a grave decision. As it was then believed that yellow fever could not be given to animals, the only way of investigating it was to experiment on human beings. It was a tremendous responsibility to expose men deliberately to a dangerous disease for which there was no specific cure.

After careful thought, however, the Board decided that a solution of the yellow-fever mystery, and all that it would mean in the saving of life, justified experimenting on human beings. Of course, those who volunteered for the experiments must give their free consent and be told of the risks they ran. Moreover, the members of the Board agreed that it was their duty to experiment on themselves as well as on the volunteers.

Before the first series of experiments began, Dr. Reed was called back to Washington to work on the report of the typhoid-fever commission. Dr. Carroll was put in charge during his absence, but it was Dr. Lazear who made the mosquito-biting experiments.

## THE FIRST EXPERIMENTS

In this first series of experiments 11 volunteers were bitten by female aegypti mosquitoes after these mosquitoes had already bitten patients with well-marked cases of yellow fever. Of these 11 persons, only two developed the disease. The first positive case was that of Dr. Carroll. He had a severe attack of yellow fever from which he recovered.

The second positive case was that of an Army private, William Dean, who had been bitten by the same mosquito that had bitten Dr. Carroll and also by three other contaminated mosquitoes. He came down with a mild, but wellmarked, case of yellow fever six days later. The case of William Dean had special importance because his was the first experimental case in which it could be proved that there had been no other source of infection except the bites of contaminated mosquitoes. Dr. Carroll's case was defective because he had gone places where he might have been infected in some other way between the time he was bitten and the time the disease developed.

Dr. Lazear had been the sixth person to be bitten in the early experiments. The results were negative. Later, on September 13, while collecting blood for study from yellow-fever patients in Las Animas Hospital, a mosquito settled on the back of his hand. He allowed it to stay there until it had drunk its fill.

Five days after the bite, on September 18, he came down with yellow fever. On the evening of September 25, 1900, he died at the age of 34 and "in dying," in Major Reed's own words, "added one more name to that imperishable roll of honor to which none others belong than martyrs to the cause of humanity."

Major Reed returned to Cuba early in October and Dr. Carroll handed Dr. Lazear's notes over to him. A careful study of these notes convinced Major Reed that "the mosquito serves as the intermediate host for the parasite of yellow fever." He said so in these words at the end of the famous preliminary note on the etiology (cause) of yellow fever which he read at the annual meeting of the American Public Health Association in Indianapolis late in October.

But what of the uncertainty about Dr. Carroll's case? What about those nine negative cases? It is not difficult to imagine that there was more headshaking than handclapping at that meeting.

And rightly so, because it is the duty of medical men to withhold acceptance of any new theory until its truth has been proved beyond shadow of doubt. Major Reed's brilliant scientific achievement—what entitles him to be called "the conqueror of yellow fever"-is that he produced that proof.

## ®THE EXPERIMENTS AT CAMP LAZEAR

To demonstrate that the mosquito is the only carrier of yellow fever, it was necessary to conduct experiments in such a way as to make it impossible for the men experimented on to get yellow fever accidentally.

Major Reed and his associates took a piece of ground about 6 miles from Havana and built a camp there, which they named Camp Lazear after their dead comrade. The camp site was well drained and freely exposed to sunlight and winds. In this camp were quartered men who had never had yellow fever and who were therefore called nonimmunes. These men were the American soldiers and the Spanish immigrants who had volunteered for the experiments. When it became known that soldiers were wanted for yellow-fever experiments, the first volunteers were John R. Kissinger, a private in the Hospital Corps, and John J. Moran, a headquarters clerk. Major Reed talked the matter over with them, explaining the risk of suffering and even of death.

They held to their purpose. Kissinger volunteered, to use his own words, "solely in the interest of humanity and the cause of science." Major Reed's comment on this young man was: "In my opinion this exhibition of moral courage has never been surpassed in the annals of the Army of the United States."

The Board proposed to attempt the infection of nonimmunes with yellow fever in three ways:

- By the bites of mosquitoes that had previously bitten yellow-fever patients;
- By the injection of blood taken from yellow-fever patients; and
- By exposure to the most intimate contact with fomites. All three sets of experiments were carried on at approximately the same time.

Experience had shown that if a person is going to contract yellow fever, he develops it within six days after exposure. Therefore, if the men were kept in quarantine for two weeks without developing the disease, this fact would show they had not become infected before they entered camp.

Things were now so arranged that if any man developed yellow fever, the Board would know that the disease was the result of the experiment in which he had taken part, and nothing else.

## FOMITES PROVED INNOCENT

To test the fomites theory a small frame house called "the infectedclothing building" consisting of one room, 14 by 20 feet in size, was erected at Camp Lazear. It was tightly built, and the doors and windows were so placed as to admit as little sunlight and air as possible. The room was kept like the hold of a ship in the tropics—warm, dark, and moist.

The building was now ready for the experiment. Three large boxes filled with sheets, pillow slips, blankets, and clothing contaminated with the excretions of yellow-fever patients were placed inside; and on November 30, 1900, Dr. Robert P. Cooke, acting Assistant Surgeon, United States Army, and two privates of the Hospital Corps, all nonimmune young Americans, entered the building.

They unpacked the boxes, giving each article a thorough shaking in order to fill the air with the specific agent of yellow fever if it was contained in these fomites. They then made the beds with the soiled bed clothing and slept in them. Various contaminated articles were hung about the bed in which Dr. Cooke slept. For 20 nights this room was occupied by these nonimmunes. They packed up the soiled articles every morning and unpacked them at night, but not one of the men developed yellow fever. From December 21, 1900, to January 10, 1901, the room was again occupied by two nonimmune young Americans. These men slept every night in the soiled garments and on the bedding used by yellow-fever patients throughout their illness. They also remained perfectly well.

The experiment was repeated a third time with the same results. This series of experiments blasted the fomites theory. Remember that the young men who took part in them shared the universal belief that contact with fomites was practically the same as catching yellow fever. Their bravery was as great as that of the volunteers who caught yellow fever as the result of mosquito bites.

## THE M0SQUITO PROVED GUILTY

The mosquito-biting experiments began on November 20. At first the results were discouraging. But on December 5,

Kissinger was bitten by mosquitoes which had bitten yellow-fever patients from 15 to 20 days before. Four days later he had a well-marked case of yellow fever, from which he recovered. Four more positive cases developed in one week. In all, 13 men at Camp Lazear were infected by means of the bites of contaminated mosquitoes, and the disease developed in 10. Fortunately, they recovered. No one else in the camp became ill.

As a result of these experiments it was found that yellow fever could be carried from one person to another by the bite of a female *Aëdes aegypti* mosquito that had bitten a yellow-fever patient in the first three days of his illness, and had then been kept for at least 12 days before it was allowed to bite a human being who had never had yellow fever.

If that plan were followed, the person bitten would generally come down with the disease within six days. It now became clear as to why it took so long for a case of yellow fever to infect a house.

Mosquitoes had to bite the patient during the first three days of his illness. Then 12 days had to go by before they could pass on the disease by biting another person in the household. But after that interval of 12 days they were a menace to every nonimmune who entered the house or its immediate neighborhood.

## HOW A HOUSE IS INFECTED WITH YELLOW FEVER

Since it was proved that a house could not be infected with yellow fever by fomites, the next question to be settled once and for all was: "How does a house become infected?" The experiment designed to answer this question was made in the "infected-mosquito building."

It was similar to the infected-clothing building except that it was well ventilated. It was divided into two parts by a wire screen that extended from top to bottom and allowed air but not mosquitoes to pass freely from one side to the other. "Now," said Major Reed, "I am going to infect one side of this building with yellow fever and not the other side." On December 21 he set free in the larger room 15 mosquitoes that had previously bitten yellow-fever patients. Five minutes later John J.Moran

Moran entered the room containing the mosquitoes. In the room on the opposite side of the screen were two nonimmunes who acted as controls. Moran paid three short visits to the mosquito room within two days and all 15 mosquitoes bit him. On Christmas morning this "plucky Ohio boy," as Major Reed called him, came down with yellow fever and had a sharp attack which he bore without a murmur.

The two controls who had slept each night in the mosquito-free room and continued to sleep there for 13 additional nights remained in perfect health. Therefore Reed concluded that as the air on both sides of the wire screen partition was exactly the same, it must have been the presence of contaminated mosquitoes that infected the side in which Moran contracted yellow fever, and the absence of mosquitoes that made the other side noninfectious.

By the end of the year 1900 the experiments at Camp Lazear had left no grounds for doubting that one of the greatest medical mysteries of all time had been solved. Mosquitoes—and mosquitoes only—were the carriers of yellow fever.

## THE VIRUS OF YELLOW FEVER

You will remember that the third way in which the Board proposed to produce experimental yellow fever was by inoculating nonimmunes with blood drawn from yellow-fever patients. They succeeded in doing this in three out of four cases. This served to show that the specific agent, or germ, of yellow fever is present in the blood of its victims. After the completion of the work at Camp Lazear, Dr. Carroll remained in Cuba to continue further blood-inoculation experiments planned by the Board. In October 1901 he was able to produce unmistakable yellow fever in two American soldiers by inoculating them with filtered blood serum derived from a yellow-fever patient.

Major Reed and Dr. Carroll recognized that the production of yellow fever by the injection of blood serum that had previously been passed through a laboratory filter capable of removing all bacteria was a matter of extreme importance. It seemed logical to them to conclude that "the specific agent of yellow fever is of such minute size that it passes readily through the pores of a Berkefeld filter." They had, in other words, discovered that the cause of yellow fever is a filterable virus.

## THE FRUITS OF VICTORY

Walter Reed passed sentence on the mosquito carriers of yellow fever in these words: "The spread of yellow fever can be most effectually controlled by measures directed to the destruction of mosquitoes and the protection of the sick against these insects."

The task of executing this sentence in Havana fell upon William C. Gorgas, then a Major in the Medical Corps, United States Army, and Chief Sanitary Officer of that city. Practically no one thought that it could be done! For the guilty mosquitoes were legion—and they were everywhere—at times in such numbers as to cloud the sky. To tell Gorgas that a thing was impossible was not the right way to keep him from trying it.

The experiments at Camp Lazear had not convinced him that the mosquito is the only way, or even the ordinary way, of spreading yellow fever. But "if it is the mosquito," he said, "I am going to get rid of the mosquito." He was aided in his gigantic undertaking by the yellow-fever mosquitoes themselves. These mosquitoes never stray far from the human habitations where the females can get the meals of human blood which they must have in order to lay their eggs.

Moreover, they prefer to lay their eggs in clean standing water in artificial containers—rain barrels, pails, pitchers, tin cans, broken bottles, flower vases, pans, and other household utensils. Gorgas had a record made of every habitation and every family in Havana—then a city of 300,000 population—with a list of every possible water receptacle in and near each house and a note as to its location. He ordered his inspectors on their monthly visits to check each one to make sure that it was kept free of standing water and that rain barrels and other water-storage receptacles were kept screened. The end of yellow fever in Havana was as dramatic as old "yellow jack" itself had always been. For 140 years this disease had not been absent from Havana for a single day.

On September 28, 1901, Gorgas recorded the last case to occur there for several years. This spectacular achievement awed the scientific world and converted Gorgas himself from a doubter to an enthusiastic disciple of "that great man Walter Reed." Later Gorgas freed the Isthmus of Panama of yellow-fever mosquitoes, with the result that the United States was able to build the Panama Canal.

## THE END OF THE ADVENTURE FOR WALTER REED

After the Yellow Fever Board broke ranks at Camp Lazear in February 1901, Major Reed was called back to Washington. During the brief remainder of his life he devoted most of his time to teaching at the Army Medical School and at Columbian (now George Washington) University.

On November 23, 1902, he died following an operation for appendicitis. It is good to know that before he died he saw the great city of Havana delivered from her ancient foe, and the way made clear for the saving of his own beloved country from a great plague.

As leader of the heroic undertaking in which the mystery of yellow fever was solved, Major Reed received many honors during his lifetime. But it was not until after his death that his country began to appreciate the greatness of what he had done. The one tribute that he would perhaps have valued most came in 1909 when his comrades of the Army Medical Corps named the great Army hospital in Washington after him. Today, Walter Reed General Hospital symbolizes military medicine at its best in treatment, research, and teaching. In 1945 he was elected to the Hall of Fame at New York University, the first physician to be so honored, and in 1951 the Walter Reed Society was founded to perpetuate his memory.

## YELLOW FEVER TODAY

The heroism of Walter Reed's little band of scientists and volunteers has run like a golden thread through all the later work done in connection with this disease. Later investigators proved that Reed and Carroll had been right in thinking that the specific cause is an agent so minute that it can pass through a laboratory filter. This agent is now called the virus of yellow fever. You will get some idea of its smallness from the fact that the head of a pin could accommodate 25 million particles of the virus with room to spare.

In 1927 it was found that certain kinds of monkeys are susceptible to the virus, and so it became unnecessary for experimenters to use human beings as subjects. In 1937 a vaccine against yellow fever was perfected and has since been used to immunize millions of people living in, or traveling to, parts of the world where yellow fever is still a threat.

Vigorous campaigns against *Aëdes aegypti* mosquitoes have banished yellow fever from the centers where it had once been a major cause of death. In the two Americas as a whole, the last case of yellow fever known to have been transmitted by these house-dwelling mosquitoes was in Trinidad in 1954. Before that, however, it was discovered that certain forest-dwelling mosquitoes transmit yellow-fever virus to men and susceptible animals—particularly monkeys—living in the tropical and subtropical forests of Africa and Central and South America. Vaccination is the only way of controlling jungle yellow fever, as it is called, because it is impossible to get rid of its forest mosquito carriers by any known antimosquito measures.

## PLANT PATHOLOGY AND HUMAN WELFARE

Diseased plants have always been with us and influenced our wellbeing. The history of mankind through the ages has been the story of hungry men in search of food. Tribes that tended sheep or cattle for their own food needed forage for their animals as well. The desire for food and forage has brought on many conflicts, small and large, from the beginning of time.

Diseases of plants have causes similar to those among animals and man. In early times, physicians using medical terms wrote about plant diseases—even though plants do not feel ease or disease as men do. But plants with disease may be so impaired that their usefulness to themselves or to mankind is seriously reduced. With the dawn of history we find reports of famine resulting from mildew and rust, induced by fungi. Certainly, plant diseases were troublesome even in the earliest historic times. Doubtless they were present in evolutionary times, as fossils show. Many of the extinct plants probably had diseases that hastened their disappearance.

How plant diseases developed is not definitely known. Probably when weakened plants died, various microorganisms assisted in decomposing them and in returning the materials to the soil to enrich it. Presumably most of these microorganisms worked only on dead plants and animals, clearing away debris and building humus. However, some microorganisms apparently were able to cause decay while the plants were still alive. This was a parasitic mode of life. By further adaptation, other microorganisms, such as certain rust and mildew fungi, were able to attack and grow on living tissue only. Certain common names for plant diseases came easily enough. The rust fungi often had the color of iron rust. Terms like soft rot, dry rot, wilt, and leaf spot are descriptive.

## CAUSE OF DISEASE

Diseases are not always caused by microorganisms. Injuries that occur because of accidents, poisons, or defects in the environment often result in diseased tissue. Fire, flood, frost, higher animals, insects, and machinery, as well as poor agricultural techniques, may also produce damage. Many chemicals are beneficial, but when improperly used they too can cause losses.

Important environmental factors may be damaging; for example, the weather can be too hot or too cold, too wet or too dry. The light can be too intense or too weak; or the length of daylight may be too short or too long. There can be too little or too much fertilizer. In addition to the common mineral elements, such as potassium, phosphorus, and nitrogen, many plants require traces of boron, zinc, copper, manganese, or other nutrients.

Between the nonparasitic diseases and those caused by pathogens (literally, disease-causers) is a large additional group caused by viruses. There is no agreement about whether or not viruses are living. But this depends on what we regard as life, rather than on our ignorance about these viruses. Much of our basic information about the viruses that cause serious diseases of animals and men has come from the study of plant viruses, especially tobacco mosaic virus.

Most groups of living microorganisms include some that may be pathogenic. The concept of pathogenicity was established by Prevost in 1807 but went unrecognized until the middle of the last century. The research was done with covered smut, a disease in which the grains of wheat were replaced by fungus growth, looking mostly like black dust but being in reality a black powdery mass of spores. The disease was always present, but was not epidemic.

Sometimes it caused heavy loss in one field and little in another. When Prevost put some of the black dust in water and watched it under a microscope, the black dustlike bodies grew.

He also observed that traces (about 1 part per million) of soluble copper from a dirty copper kettle would prevent growth. These two keys were basically important. However, they were not turned in the lock to open the way for the control of bunt until later in the century.

The idea that specific living microorganisms could cause disease was accepted quite slowly. Many received such ideas only with chuckles of amusement—whether disease of plant, animal, or man made no difference. Many preferred to generalize about disease as a principle. It was to be overcome by another principle, health and natural good. Comparable ideas persist today.

But the means of securing adequate evidence was soon at hand. For plants we use the rules of proof for pathogenicity, known as Koch's postulates, adopted from medicine. Briefly, the pathogen must be consistently associated with the diseased tissue, it must repeatedly be isolated and characterized in culture, it must reproduce the disease symptoms when inoculated onto or into the host plant, and it must again be isolated and proved to be identical with the first isolate.

Most of the pathogens causing plant diseases are bacteria or fungi. One has only to mention a few examples, such as those causing blight of potatoes (fungus), fire blight of apples and pears (bacterium), chestnut blight (fungus), and Dutch elm disease (fungus). Such major crops as wheat, corn, rice, potato, apple, and tomato are subject to 50 or more diseases. In addition, certain flowering plants prey on other plants and cause disease. Examples are mistletoe, dodder, and witchweed. By common consent, the eelworms (nematodes) appear among the plant pathogens. Only in recent years have we recognized their great importance.

On the other hand, some microorganisms are beneficial. Certain bacteria induce the root nodules of legumes, such as alfalfa, clover, and peas, and thus make possible the fixation of atmospheric nitrogen. Various fungi aid germination of seed and normal growth of plants such as orchids, heaths, and trees by association with the roots.

Fortunately, the pathogens of plants rarely produce diseases in animals or man. However, some of the fungi are poisonous when eaten. Ergot of rye has been notorious. This fungus grows as long, hard, purple masses replacing some of the grains in a maturing head of rye. It was often included in low-cost bread until its poisonous nature was discovered. A toxic alkaloid in the fungus induced abortion and gangrene. Epidemics of ergot poisoning swept over France nine times in the 17th century and eight times in the 18th century, and thousands died. Various other fungi, including certain mushrooms, are poisonous when eaten.

## SOME HISTORIC LOSSES

Enormous losses of food, history-making losses, have been frequent in the story of mankind. Before epidemics were understood, they were considered as special kinds of plagues. Many factors have combined for the final result—the plant pathogens often need suitable weather, especially warm temperature and high moisture.

Sometimes insect vectors, in addition to the damage they themselves do, add further injury by spreading or injecting disease-inducing microorganisms or viruses. But the importance of microorganisms was slow to be recognized. All could see the devastations of swarms of locusts—they descended on fields, ate every green leaf or twig, and left fields bare—but the stealthy work of fungi seemed obscure to the farmers.

Various famines, and even the outcomes of wars, have been influenced by the shortage of food caused by plant disease microorganisms. Speculation about the supernatural was common. After the Irish Famine, or potato murrain, which began in 1845-46, the real cause was worked out. Potatoes had originated in the Andes in America and, no doubt, the blight fungus did too. However, blight in Europe corresponds in time to the advent of steamships. Men have argued the possibility that diseased potatoes, which decayed on slow sailing vessels, could have landed from fast steamships and started the potato murrain. In any case, this is doubtless an early example of a way in which growers have moved dangerous microorganisms from one country to another.

The potato famine continued for several years. Starvation occurred in Ireland and in western Europe. Rebellions, looting of landowners' property, even murders, took place. During this time, America benefited by the immigration of many of the western European people. They were a select group. A few were criminals, but usually only those with initiative and an adventurous spirit would say "good-by" to relatives and tradition for a try in the New World.

The proof that the potato famine was caused by a fungus helped explain the riddle of the common barberry bushes and rust. Stem rust on wheat was proved to be caused by a fungus that appeared first red, then black, and had the barberry as an alternate host. The rust was worse near barberries, and laws had been passed against growing them.

Rusts on stems or leaves of wheat had been a plague since the dawn of history. The Romans had a god called Robigus whom they, tried to appease for a good harvest. During the

1st century, annually on 25 April, they held a procession to his sacred grove, poured red wine over the altar, and sacrificed a red dog. This was also the time that rust first appeared generally, and that the dog star was thought to be malign. Robigus was thus cajoled into accepting a cheap red substitute for the red on the wheat, into calling off the dog star, into saving the wheat, and thus into letting his mortal followers have their bread.

Rusts of stems and leaves probably prevented wheat from becoming a major food crop in the southeast of the United States—hence the wide use of corn products there. Rust was serious during World War I in both Europe and North America; in the single year 1916, it was estimated that 300 million bushels of wheat were lost in the United States and Canada. Such slogans as "Food will win the war" were common. The food losses from potato blight and other diseases in Germany doubtless helped to shorten World War I.

Wheat rusts, wheat smut, and potato blight have been carried all over the world, but they are only several examples of the many diseases on all our crops. The world's important food crops include bananas, beans, cassava, corn, peanuts, potatoes, rice, soybeans, sugar beets, sugar cane, wheat, and many others. Plants yielding fiber, lumber, drugs, and so on, are valuable in our daily lives. We must not forget rubber and tobacco. Each has many diseases. Merely to list the plants and diseases by name occupies a large book. Ignorance about them often causes hardship in developing countries. Let us consider two examples.

Coffee rust put an end to a thriving coffee industry in Ceylon. In 1879 the Ceylon government sent an appeal for help to the botanical institution at Kew in England. It was a decade too late. The rust took the leaves off the trees, and after several defoliations the trees died. After much economic hardship, Englishmen planted tea and the English became confirmed tea drinkers.

Coffee rust has spread widely; it occurs in Malaya, India, Java, Sumatra, the Philippines, and the east and west coast of Africa. Big growers doubtless can handle it, but the small growers cannot afford control measures. With planes going so fast and frequently, how long will it take the rust to reach South and Central America, where over 99 percent of the coffee trees are susceptible? And how much more in aid money will be needed to help coffee-growing countries? A seldom-considered Damocles' sword is hanging over American coffee growers.

But American fungi also present some threats to the Far East. Rubber trees growing there were native to South America. But in America, a fungus causes a defoliating leaf spot, making rubber-growing expensive. By lucky circumstances, the fungus was left behind when Englishmen took the rubber trees to Malaya, and rubber plantations have spread in the Far East. Here again, how long will it be before the leaf spot moves to and attacks the eastern rubber trees?

These are only two diseases. Many more could be discussed. Their control or prevention is in the hands of a relatively few scientists throughout the world. Their chief means of operating is mentioned very briefly.

## CONTROL

The control of plant pathogens is ordinarily based upon an understanding of the pathogens' points of vulnerability. Since, with occasional exceptions, individual plants are of relatively small value, the plant pathologist is concerned with the control of epidemics. An epidemic occurs when there is a rapid repetition of the cycle whereby the pathogen enters the plant, becomes established and reproduces, comes to the surface again, produces spores and so forth, and is carried somehow to another host. Under favorable conditions a new generation can occur in from 8 to 14 days.

Each infection produces hundreds, sometimes many thousands, of spores. These phenomenal "population explosions" can cause devastation of a crop. Plant pathologists secured the ideas of cycles from medical workers. Control measures can be applied at various points in this cycle where weak links occur in the chain of events. For example, the use of quarantines, inhibiting agents, special cultural practices, repulsion of disseminating insects, and the development of disease-resistant plant varieties, all have successful applications.

Most of the other control measures, in the minds of some people, are merely stopgap procedures to be employed until disease-resistant varieties can be developed. However, the development of new varieties must be more or less continuous. For as smart men breed a variety that resists disease, the microorganisms may have or evolve a different race or strain of the pathogen or even a different pathogen to attack it. So the plant breeder has a problem to keep one jump ahead.

For breeding, undisturbed or natural areas are most important. Here the breeder searches for native plants having desirable characteristics. When natural areas are destroyed, his problem becomes serious. Every effort must be made to preserve a suitable number of natural areas over the world. Every year of delay means that potentially important plants, bearing genes for resistance or other valuable characteristics, become extinct.

In Europe and North America, research has made possible the reduction of plant diseases so that in some cases we have an abundance or a surplus in storage of products such as wheat, corn, or cotton. (Fungi also may attack food in storage).

This supply is possible only as a result of continued research of the pathologists, along with their colleagues in agronomy, horticulture, entomology, soils, genetics, engineering, and so forth. After the War Between the States, research stations and substations were established frequently in the areas to be served, because what was suitable for one place might not be right for another. The large supply of crops was produced by the applications of many years of research.

Although this supply in North America may cause problems, one should not overlook the enormous asset it provides in case of armed conflict. Better still, it serves as a strong deterrent for men in any nation who might try to overcome any country in North America. At the same time, it may provide food for developing countries while they work on their own problems.

## DEVELOPING COUNTRIES

Many people in Euorpe and North America do not realize that much of the world's population goes to bed hungry every night. Also, it has often been erroneously thought that if some of the South American, African, and Asian countries would use our seed and our methods, there would be plenty of food to go around. Unfortunately, the procedures that work in Europe and North America are quite apt to fail in other places. For that matter, procedures that are suitable for southern Wisconsin may not be suitable for northern Wisconsin.

Research has been necessary in order to learn the application of basic knowledge for particular localities. Consequently, these questions are raised: Without research by plant scientists, including pathologists and their colleagues, can the developing countries be wholly or partly self-sustaining? Likewise, can they be encouraged to produce much of their own food rather than depend upon purchases or gifts? Do they need a philosophy of responsibility and activity along lines that would make them largely or entirely self-sustaining?

The 2-year assignments of experts to developing countries too often have proved inadequate. What is needed mostly is long-term research, comparable to that sustaining North America and western Europe. This cannot be emphasized too strongly.

Among the primary problems are these: What can we do with the arid lands? They occupy about one-third of the world's land area.

What can we do with fertilizers? This is important almost everywhere, including the apparently fertile wet tropics, because many such areas are badly leached. What can we do about weeds? Can chemicals wisely replace cultivation?

Some have argued that an increase in food production is self-defeating, especially when the products of arable land are consumed by the exploding populations. However, populations must be fed until the explosion can be controlled. Meantime, the prevention of plant diseases holds great promise. But, the measurement of losses points up the magnitude of the problem.

## DISEASE LOSSES IN MODERN TIMES

Plant disease losses in developing countries are difficult to determine at present. Vallega and Chiarappa have given a few examples. In 1947-48 in New South Wales, Australia, stem rust caused a loss of wheat estimated at 270,000 tons. This represented food for about 3 million people for 1 year. In South Africa, disease losses in sorghum have recently averaged about 1 million tons. This is enough food for 5.5 million people for 1 year. A small "Irish Famine" occurred during 1951 in Chiloe, southern Chile. The mean loss there of potatoes from the late blight area was 40 percent, with 23 percent for the entire country.

Ghana and Nigeria have suffered severe losses in cocoa. Coffee has never come back as a crop after the coffee-rust epidemic in Ceylon in 1880-90. In South India, the loss from coffee rust sometimes may run to 70 percent; in East Africa, 30-percent loss from coffee rust is frequent. In the Philippines, of an original population of 250,000 coconut trees on San Miguel Island prior to the appearance of cadang-cadang in 1961, only 80 trees were still bearing fruit in 1965. In the state of São Paulo, Brazil, 75 percent of the 6 million citrus trees were infected and killed by tristesa in 1949. And so goes the story of disease. Many times, in a single-crop area, great economic disturbances occur that result in much suffering because of crop losses.

In the United States, the figures on losses are much more accurate. LeClerg has given many details, some of which are summarized in Table 1. Here the average annual loss from pathogens for 1951-60 was $3,251,114,000. To this total must be added the average annual loss from nematodes of $372,335,000, and the average annual expenditures of $2,255,000 for suppression of plant diseases and nematodes. In round figures, this is an annual loss caused by plant disease in the United States, held down from an unknown very large sum, to just over $3% billion.

There is an excellent report by Ogawa of losses in California, and this serves as an example of the situation in a leading agricultural state. A summary appears as Table 2, which shows losses in landscape plantings, apparently omitted by LeClerg.

After total destruction no loss appears on the record. The foregoing discussions and tables of losses do not take into account the losses from crops that have already been wiped out. For example, nothing is noted about the loss from fire blight on commercial Bartlett pears in central or eastern United States.

Fire blight killed the trees. Nothing is said about losses from chestnut blight. The chestnut trees were killed by the blight fungus. So it goes with Vicland oats, stone fruits in the Salt River Valley, Arizona, hops for beer near Milwaukee, certain varieties of corn, citrus, cotton, flax, and many other crops. This is a continuing process. In some places the Dutch elm disease alone, or together with phloem necrosis, is now destroying the American elms. Without the use of control measures, the crop losses from plant diseases would be really serious.

Doubtless harvests would often be reduced greatly and some crops would be completely destroyed. Prices would be enormously increased, and for some things we would go hungry, even in the United States. And our souls would miss much of nature's beauty.

## PATTERN FOR CONTROL

To control plant diseases, the time-honored means is as follows. Basic research is done on the host, on the pathogen, and on their interaction, as well as on the environment. The most promising control measures are tried in fields that represent the extremes of environment likely to be encountered by the plant in question. After successful control measures are found, they are imparted to the agricultural workers via the extension procedures. They are also given to students. The success achieved in the United States has often followed this pattern.

This is a procedure that is promising for developing countries as well. But it takes years of trial and error in each new location to insure success. Although the basic principles may be known, their application must be modified in most cases for a new location and usually for a new environment.

To make such environmental studies, crude facilities were built years ago. Now, some institutions provide batteries of growth chambers with excellent control of temperature, moisture, light, and nutrition, suitable for growing small plants. For larger plants, many possibilities are provided by mountains having warm bases and a cool top, wet and dry sides, ridges and valleys for variations in light, and so on. But suitable precautions are necessary to prevent dangerous pathogens from spreading.

The introduction of foreign disease organisms is one of the major problems for all countries, but especially for developing countries. Few of them have adequate quarantine laws and services. Still fewer have enough competent men and facilities for the operation of such services. Thus, when men import promising crops, at the same time they too often import disease organisms and injurious insects that make their plant growth difficult or impossible. Quarantine officials complain particularly against diplomats and their wives, who claim diplomatic immunity against examination of their luggage as they move about.

## THE OUTLOOK

No gift of prophecy is needed to say we shall always need food. We eat plants and certain animals, but these animals also eat plants. Wherever crop plants can grow there are diseases that need control.

Dedicated men will take the basic work already done, will try it on the land, and will pass on to the growers the results of successful trials. Likewise, these or other men will educate the students.

This is particularly important for developing countries. But here the basic information developed elsewhere must be tried under local conditions. Since planting, growing, and harvesting usually occur only once a year, a competent man usually needs a year in an area new to him to learn what problems are important. Then the second year he can make experiments. Many times, his first experiments fail because of some unanticipated circumstance.

So several years of undisturbed continuity are necessary-as any Agricultural Experiment Station can testify. Then taking the results of successful experiments to growers requires much skill. In developing countries, this is often best done by dedicated natives. For best results, they should devote their lifetimes to plant pathology.

One can hardly overemphasize the importance of taking tried and true successful local results to the growers. Neither can one overemphasize the importance of educating the native students who face the challenges of the future.

Further progress depends on many factors. They are similar oftentimes to the needs in sister sciences. For example, we need a better means of classifying knowledge so that we can go to the libraries and find what is known about a given subject. Also, in the United States and, to a lesser extent, in Europe, the subjects plant pathology, entomology, soils, genetics, and so on are separated for convenience in administration and teaching. But in nature, they are seldom separated—all work together. Correspondingly, men who are trained in these subjects often have a better chance of success if they work together as a team. A team should be able to continue its function, however, with or without a substitute, if for some reason a member drops out.

Additional basic information also will foster progress. An indication of some of the numerous fundamental problems is given in various recent publications. Among the subjects needing further research are: improved control with chemicals (including chemotherapy); effect of temperature, moisture, and other environmental factors on the host and the pathogen (and their interaction).

Genetics of the pathogens, genetic development of the host plant to increase resistance; chemicals in the host or parasite that influence either resistance or pathogenicity, and chemicals (phytoalexins) produced by the interaction of host and parasite that affect resistance; ecology, especially concerning the effects of crop rotations on soil-borne diseases; biological control;

hidden disease agents and inhibiting agents that reduce crop yields; nematodes and their direct damage, as well as their introduction of other pathogens; tissue culture, the comparison of the metabolism of host and of pathogen; microscopic observation of the responses of host cells to parasitic attack under various conditions; mycorrhiza and pathogens near the roots; and so on and on.

**Table: Average Annual Loss in Crop Groups Due to Disease.**

| *Average annual loss* | *Commodity 1961-2001 inclusive* ($1000) |
|---|---|
| Field crops | 1,890,836 |
| Forage crops and pastures and ranges | 808,701 |
| Fruit and nut crops | 223,505 |
| Ornamental plants and shade trees | 14,099 |
| Forage seed crops | 23,584 |
| Vegetable crops | 290,389 |
| Total | 3,251,114 |

These are merely a dozen among many possible examples. The order in which they appear and the omission of others have no significance. The field is so broad that a plant pathologist must be able to work with other colleagues who understand related subjects from agronomy through to zoology.

**Table: Summary of Crop Group Losses and Disease-Control Costs**

| *Losses and control* | *Costs* |
|---|---|
| Commodity | |
| *($1000)* | |
| Citrus and subtropicals | 23,088 |
| Field crops | 79,749 |
| Forest crops and products | 18,015 |
| Fruit and nut crops* | 105,467 |
| Ornamental crops† | 136,339 |
| Small fruit crops | 9,564 |
| Vegetable crops | 290,389 |
| Total | 6,62,611 |

*Mostly grapes, peaches, plums, and prunes.

†Mostly landscape plantings.

No doubt future crops will be protected by future dedicated plant pathologists, who will find great personal satisfaction in recognizing and accepting the great challenge and opportunity for service. They, along with other colleagues, will be helping provide mankind with some of the basic needs-including food, clothing, and shelter which we, in the United States, too often take for granted.

Diseases have always been present in plants and have caused many changes in the lives of men. They have influenced what men ate, what they wore, the shelter they used, and so on. Industries have flourished and have disappeared. Famines have occurred; populations have moved; even the outcomes of wars have been influenced. Important plants are frequently attacked. Losses each year in the United States are held down from unknown enormous sums by various control measures.

The control of disease is one of the promising means by which developing countries can maintain and advance themselves. Dedicated scientific men, no doubt, will continue serving mankind by their research, extension, and teaching about plant diseases, if given the opportunity. They will need to cooperate with colleagues in related fields.

## MEDICAL UTOPIAS

In medicine, even more than in other fields of science, theories and practices have always been under the sway of *a priori* philosophical attitudes and rationalized beliefs. The social forces that have influenced medical history range from the primitive fear of demons to the current wave of faith healing, from Rousseau's assertion that "hygiene is less a science than a virtue" to the modern illusion that diseases can be conquered by drugs. Among all the medical utopias that have flourished in the course of time, none has blossomed so constantly and in so many forms as the belief that disease can be entirely eliminated from the earth. At the present time this illusion is based on an uncritical faith in the magic power of experimental science.

But fundamentally it arises from the mystical belief in the existence of what Thomas De Quincey, in "*Confessions of an English OpiumEater,*" called "that sort of vital warmth which would probably always accompany a bodily constitution of primeval and antediluvian health."

Many thinkers of classical Greece certainly believed that reasonable men could achieve the millennium of health by the exercise of wisdom. Witness the cult of Hygeia, which was an expression of the faith that men could enjoy mens sana in corpore sano if they lived according to reason. Carrying this doctrine to its logical conclusion, Plato wrote that the need for many hospitals and doctors was the mark of a bad city; there would be little use for them in his ideal Republic. In Imperial Rome, Tiberius asserted in a similar vein that anyone who consulted a doctor after the age of thirty was a fool for not having yet learned to regulate his life properly without outside help.

The widespread conviction that health is purchasable, not only in limited areas but also on a world-wide scale, seems to be justified by the advances made during the past half-century in the fields of nutrition and infection. In reality, however, it has not yet been shown that these achievements justify the wide extrapolations made from them.

There is overwhelming historical evidence that the evolution of diseases is influenced by many determining factors which are not as yet amenable to social or medical control, and may never be. The changes that have occurred without benefit of conscious human intervention in the prevalence of various diseases during the past few centuries should serve as a warning that it is unwise to predict the future from the short perspective of the past decades.

Granted the lack of precise information, it is clear that there have been spontaneous ebbs and flows in the prevalence and severity of many diseases. Plague invaded the Roman world during the Justinian era; leprosy was prevalent in western Europe until the sixteen century; plague again reached catastrophic proportions during the Renaissance; several outbreaks of the sweating sickness terrorized England during Tudor times; syphilis spread like wildfire shortly after 1500; smallpox was the scourge of the seventeenth and eighteenth centuries; tuberculosis, scarlet fever, diphtheria, and measles took over when smallpox began to recede; today virus infections occupy the focus of attention in our medical communities; and long before viruses had become scientifically fashionable, pandemics of influenza at times added a note of still greater unpredictability to the pattern of infection.

Coming now to our times, who could have dreamed a generation ago that hypervitaminoses would become a common form of nutritional disease in the Western World; that the cigarette industry, air pollutants, and the use of radiations would be held responsible for the increase in certain types of cancer; that the introduction of detergents and various synthetics would increase the incidence of allergies; that advances in chemotherapy and other therapeutic procedures would create a new staphylococcus pathology; that alcoholics and patients with various forms of iatrogenic diseases would occupy such a large number of beds in the modern hospital?

Despite the dark spots on the health picture of the modern world, many facts seem to provide support for those who claim that we are approaching medical utopia. Largely because of the control of microbial diseases and the improvements in general nutrition, the immense majority of children survive. Moreover, even those suffering from diabetes, congenital heart abnormalities, and many other formerly fatal diseases can now have an almost normal span of life. True enough, progress has been much less dramatic with regard to the diseases which affect adulthood.

Yet even in this age group several types of ailments and accidents which once were uniformly fatal can now be treated so effectively that they have been all but eliminated as causes of death. As a result, more and more persons in our communities exceed the life span of threescore and ten to which so few could aspire in the past.

The achievements of modern medical science are, indeed, almost miraculous. Surgery restores to function broken limbs and damaged hearts with amazing safety and little suffering; sanitation removes from our environment many of the germs of disease; new drugs are constantly being developed to relieve physical pain, to help us sleep if we are restless, to keep us awake if we feel sleepy, and to make us oblivious of worries. Within limits it is true that we can control many of the diseases that "our grandparents and parents took for granted." But this does not prove that we know how to control the disease problems that will be encountered in the future, or even, for that matter, those of the present. It is seldom recognized that each type of society has diseases peculiar to itself-indeed, that each civilization creates its own diseases. Furthermore, there is no evidence that the techniques developed for dealing with the disease problems of one generation can cope with the problems of another.

Progressively mankind managed to cope with this problem by developing certain kinds of natural immunity and by introducing sanitary practices. Today the situation is different. Horses have been displaced by automobiles, and the diseases they brought about have therefore disappeared. But unsought results of this advance have been the pollution of the atmosphere by automobile exhausts, and forty thousand fatalities each year on our highways. We are beginning to think constructively about these new problems and may learn to do something about them, but we may expect that new difficulties will soon arise from further improvements in methods of transportation.

Among the new problems arising from the partial control of man's ancient plagues, some have an economic basis and are apparent chiefly among the underprivileged peoples of the world. In the past, microbial diseases acted as one of the checks in population size by killing large numbers of children and keeping down the number of persons who reached old age.

To put it crudely, nature balanced the books for us. Today, public health measures, supplemented by the use of insecticides and antimicrobial drugs, greatly reduce the number of early deaths. As the birth rates show as yet no sign of decreasing, the world population is growing in an unprecedented manner. In Ceylon, for example, the partial control of malaria resulting from the use of DDT has resulted in a sudden increase of the population during the past few years.

Unfortunately, the results of infection control are not an unmixed blessing, especially as some national economies are not capable of making orderly adjustments to the new state of affairs. It is of particular importance in this regard that food production and especially the supply of well-balanced proteins cannot keep pace with the increase in world population.

Advances in agriculture and industrial technology will, of course, improve food supplies, but not fast enough for the needs. While the world population increased by eight per cent during the past ten years, food production increased by only five per cent, and in consequence more people go hungry today than did a decade ago. It is to be feared that food deficiencies will in the long run cause more physiological misery and suffering than have been prevented by the partial control of infection.

Less apparent as yet, but also constituting a threat for the future, are the economic and biological consequences of the survival of persons suffering from various types of physical and mental deficiency. The availability of techniques capable of postponing death in every age group and arresting almost any type of disease will increasingly present to the medical conscience difficult alternatives.

For example, to save the life of a child suffering from some hereditary defect is a humane act and the source of professional gratification, but the long-range consequences of this achievement mean magnified medical problems for the following generations. Likewise, prolonging the life of an aged and ailing person must be weighed against the consequences for the individual himself and also for the community of which he is a part. These ethical difficulties are not new, of course, but in the past they rarely presented issues to the medical conscience because the physician's power of action was so limited.

Soon, however, ethical difficulties are bound to become larger as the physician becomes better able to prolong biological life in individuals who cannot derive either profit or pleasure from existence, and whose survival creates painful burdens for the community. Increasing numbers of these persons cannot pull their full weight in society and require constant medical supervision and economic assistance. They constitute a social burden which is likely to grow heavier with time, precisely as a result of medical advances.

Even more important than these economic considerations, however, is the fact that many of the biologically defective individuals who are saved from death transfer to their progeny the genetic basis of their deficiences. Speaking of our "load of mutations," Professor H. J. Muller has repeatedly emphasized that, as medical science becomes more effective in prolonging survival, there will be an increase in the frequency of detrimental genes allowed to accumulate in our communities.

Whatever the theories of physicians, laboratory scientists, and sociologists, it is of course society that must decide on the types of threats it is most anxious to avoid and on the kind of health it wants—whether it prizes security more than adventure, whether it is willing to jeopardize the future for the sake of present-day comfort. But the decision might be and should be influenced by knowledge derived from a study of the manner in which different ways of life can affect the future of the individual and of society.

## Chapter 21

# Ecology

The fundamental and basic concepts of animal ecology are also the fundamental and basic concepts of human ecology. Primitive man had few, if any, characteristics not also found in animals. Modern man is certainly more highy developed and specialized psychologically than any animal, but specialization of one sort or another is found in every animal species.

Specialization is a product of evolution, and those characteristics or traits of man often considered to be unique can all be traced back to primitive characteristics, traits, or potentialities found also in animals. Many analogous traits not in direct line of descent to man have been evolved in other organisms. Let us enumerate a few.

Erect bipedal locomotion occurs in birds. A squirrel will use its hands to manipulate a nut, or a monkey a banana, as expertly as man. Laughter in man appears as an exaggeration of facial expressions and specific behavior indicating pleasure found among mammals at play. Sticks are used as a tool by a Galapagos finch to pry insect larvae out of holes in dead wood; spiders set a web net to capture food from the air as a fisherman does from the water; and hermit crabs use empty shells as homes and shields against predation by enemies.

Many tropical species maintain sexual and reproductive activities throughout the year as does man who comes from tropical ancestors. Parental care of young is perhaps more complex and highly developed in birds than in any other animal, including man. Intercommunication between individuals is well evolved within animal species, not by use of a vocabulary or the written word it is true but by a great variety of scent stimuli, color displays, sounds, call-notes, and songs, each conveying a special meaning. It may not be entirely co- incidental that the songs of birds have an esthetic appeal to man; perhaps they have also to birds themselves who have evolved specific song patterns, melodies, and qualities of tone through a long period of natural and sexual selection.

Social hierarchies exist among many gregarious species of animals and territorial behavior is evidenced among solitary nesting ones, equivalent to similar behavior in the human species. The leaf-cutting ants of the tropics practice agriculture when they bring leaves into their nestmounds on which they cultivate a mold used as food. They likewise practice sanitation by systematically removing and depositing feces and wastes at some distance from their nests. The organized industry of ants, termites, and some birds is well shown in the construction of colonial nests.

Division of labor is expressed at various levels in the animal kingdom, beginning with separation of reproductive and somatic cells in the colonial protozoans and the differentiation of tissues and organs in the lower invertebrates and attaining complex dimensions among the social insects where different individuals are adapted for special functions. Division of labor is also evident in the biotic community where various species play special roles in the cycling of nutrients, the capture and transmission of energy, the regulation of population balances, the creation of microclimates, etc., analogous to the role of different trades or vocations in the human community. Animals may even have a primitive type of culture and ethics, if we mean a mores or behavior that is beneficial and traditional for the species and which is handed down to succeeding generations in large part by learning of the young.

Man's chief claim to uniqueness lies in the highly specialized development and functioning of his brain. His morphological and physiological capacities to do things are very limited, likewise his innate tolerance of even moderately rugged climatic conditions. However, his highly evolved intelligence has enabled him to circumvent many of his limitations: he has invented complex tools to do his tasks, he uses fire and air-conditioning to keep his shelter comfortable, his manufactured clothing allows him to invade microclimates for which otherwise he is not physiologically adapted, and his industrial skill has enabled him more completely to exploit natural resources.

Conceptual thought, capacity for abstraction and synthesis, the use of symbols in speech and writing, ability to reason and anticipate events, self-consciousness, and the spiritual and esthetic values seen in religion and art were doubtless latent in primitive man and had to be developed through effort and practice to reach the fruition seen in modern man. This cultural evolution has gone on mostly independently of biological evolution, and the mind of modern civilized man is the latest step in an evolutionary process that began only about 40,000 to 5000 years ago and reached explosive proportions only within the last 5000 years.

This specialization, although unique to man, is actually no more unique in evolution than specializations of other sorts found in many other animals. Man still requires food, water, and space; a favorable microclimate; protection from enemies; and reproduction. To obtain these requirements for life he must fit into an ecosystem as do other animal species, even though the ecosystem is a highly modified one of his own creation.

If one looks at the geological time table, one sees that during the last billion or so years life has evolved from single celled plants and animals living in the sea to more complex invertebrates and primitive fish, invasion of freshwater and land habitats, and evolution of insects, amphibians, reptiles, birds, and mammals. Man did not come physically into this succession until a million or so years ago. There can be no prediction nor certainty but that in the next million years—or billion years —many new types of animal life will appear; and man, as we know him today, will either be exterminated or greatly altered in appearance, form, and function.

Man probably originated, as did most groups of vertebrate animals, in the Old World tropics and dispersed as shifting climates, vegetation, and land bridges permitted, into Europe, northern Asia, North and South America, Australia, and southern Africa. In these different areas, populations of man became isolated for many thousands of years and were prevented from interbreeding with other populations by physiographic, climatic, or biotic barriers.

As is common with animals, this geographic isolation allowed genetic variations to occur and become established, involving color of skin, texture of hair and other characteristics. Variations in tribal customs evolved. Occupancy of new environments and contacts with different biota exposed populations in different areas to different selection pressures and they came to occupy somewhat different ecological niches.

If this geographic isolation had persisted a few more thousand years, doubtlessly reproductive isolating mechanisms, already partially formed, would have become fully established, and *Homo sapiens* would have differentiated into a number of new species. The evolution of civilization, however, interrupted this process of speciation. Artificial means of transportation encouraged world-wide travel, and this brought the bypassing of geographical barriers, the breaking down of ecological and ethological evolving reproductive isolating mechanisms, and the fusing of the races and their cultures again into one, so that the species was preserved in its biological unity.

The population density and distribution of primitive man varied in relation to natural conditions according to the normal or Gaussian curve, reaching a peak under optimum conditions and tailing off toward both maximum and minimum limits of physiological tolerance for environmental extremes. Shelford's law of tolerance is a generalization of these relations between abundance, physiological comfort, and environmental factors applicable to all organisms. Modern man has skewed the curve somewhat by this fabrication of artificial home and working environments but the law still applies.

Liebig's law of the minimum and Blackman's limiting factors state that growth, activity, or even existence of an organism is determined or regulated by that essential factor or condition in shortest supply. Modern man has developed methods of distributing essential elements or products to alleviate local deficiencies, but when his transport system breaks down or in areas where it has not been perfected, the working of this law is apparent.

Primitive man was certainly a constituent member of the biotic community. He was dependent upon the vegetation for shelter from the weather, as protection against enemies, as a source of food, and for use in other ways. His niche in the community has been compared with that of one of the larger carnivores. He had to compete with other carnivores and the larger herbivores for the essentials of life, and his role in the community was determined by his successes and failures in establishing favorable interrelations.

When primitive man changed into modern man with the perfection of his intelligence and his invention of new tools, he changed from an ordinary member of the community to an ecological dominant. By building a home of his own that gave shelter from both weather and enemy and the development of agriculture, he was no longer dependent on the community as it was originally constituted. He modified the community to suit his own needs, determined to a large extent what other species he would permit to associate with him, and established the rules and conditions for their doing so.

During geological time there have been a long series of organisms that have assumed dominance for periods of time, and often by no less drastic changes in the habitat and community. Man, in fact, has not yet obtained complete dominance in the arctic tundra, the desert, the tropical rain forest, or the ocean. Even in temperate regions, whenever man relaxes his dominance or mismanages the habitat, there is reversion of the region to dominance by other native species.

Civilized man, as primitive man, is still dependent on plants for nutrition and energy. Civilized man, however, has created an ecosystem where the kinds of nutrients that he likes and the energy that he needs can be obtained more efficiently than in ecosystems which he does not dominate. Crops have been developed through selective breeding and have been cultivated by special methods to give high yields. Competition for food from other species has been greatly reduced by use of fences, guns, pesticides, and the hoe.

Plant crops are either consumed directly or fed to domesticated animals in a higher trophic level, so that meat as well as vegetables are eaten. In removing this harvest from his ecosystem he removes nutrient elements that in natural ecosystems circulate continuously, first through plants, then animals, then to the transformers in the soil that render them available for reabsorption by plants. Man acknowledges the important role of nutrient cycles in the ecosystem when he returns the nutrients that he has taken for food in the form of fertilizers unsuitable for his direct assimilation.

In natural ecosystems, food chains start with green plants because of their unique ability to capture solar energy and to synthesize nutrients from raw materials. With each additional link in the food chain through which energy and nutrients are transmitted there is a loss of some 80 or 90%. Each trophic level uses energy for maintaining its own existence and activities and there is wastage through prey not eaten, nonpredatory deaths, excreta, and heat. Consequently, with addition of links in the food chain, there is decrease in number of individuals or biomass that can be supported and in variety of species.

The biomass of herbivores is always greater than the biomass of carnivores. In densely populated parts of the human world, man has largely given up the luxury of eating beef, pork, fowl, and lamb and depends instead on wheat, rice, or corn, thereby eliminating one trophic level and taking fuller advantage of the productivity of the land. If man could reduce the length of the food chain still further through bypassing plants and using solar radiation directly for the manufacture of food and for other uses, he would have a supply of energy several hundred times greater than is now captured by plants.

The sigmoid growth curve represents approximately the manner and rate by which cells, organs, individuals, populations, and communities increase in size and complexity. In the lower part of the curve, conditions for growth are very favorable and growth proceeds at an ever accelerating pace. However, inhibiting factors come to exert an increasingly important role and beyond the point of inflection they gradually bring the growth curve to an asymptote.

With animals, the size of the asymptotic population is determined by available space, suitable food, and favorable climate. Factors that may ultimately stabilize the population at the asymptote are density-dependent, that is, they vary in intensity of their action with the density of the population. These factors consist of predation, disease, emigration, competition, and fecundity. Some species, certain insects and perhaps Arctic rodents and grouse among others, never attain stabilized populations. Their populations continue to increase to the limit of space or food supply or until unfavorable weather occurs. Then there is a crash and the few scattered survivors start the growth process over again.

Modern man has largely eliminated predation as a mortality factor, he has conquered many diseases and may conceivably subdue the rest, and he has emigrated to all favorable parts of the world even though he has not settled in large numbers as yet in some parts of it. With the harnessing of solar energy and intelligent use of minerals, man may be able to produce a super-abundance of food through a sort of artificial photosynthesis as well as apply this energy industrially.

Likewise, he may obtain control over the weather. Space is limited, however, at least on earth, and it is difficult to see what he can do about it aside from building megalopoli, skyscrapers, underground burrows, sea platforms, or vast air chambers underwater. Space may be the limiting factor ultimately setting the limit to the size of the human population on earth. The amount of space reduction that the human animal will tolerate may not be just space to stand on but space in which he can carry on a comfortable existence.

Competition will continue to be a potent factor, as it is in regulating all animal populations. War may someday be eliminated and the world someday may function as a single economic unit, but it is difficult to conceive how competition between individuals can be significantly reduced as the limits of space are approached; more likely it will become accentuated in the final struggles for existence.

With the advent of civilization and modern industrialization, all efforts to date have been to reduce mortality rates and expand available resources. This has thrown the stabilized populations of primitive man out of balance and brought on a new accelerating phase in the growth curve. Populations will not again be brought into stabilization until birth rate and mortality rates again balance. This means that the birth rate must be reduced in proportion to the reduction of mortality rate.

This occurs in populations of wild animals more or less automatically through changes in physiology and behavior. Man's biggest challenge in the present age is whether his great intellect will give him the self-discipline and skill to regulate his own population growth to the level best suited for the perpetuation of the culture that has evolved as well as for his physical existence.

The laws of Nature apply to man as they do to animals. There are no exceptions. If he can come to understand what they are and how they work, he will know better what to anticipate concerning their effects on himself. He cannot ignore the dynamic forces of the environment with impunity, but being blessed with an intelligence far above that of other animals, he can guard against them or alleviate their effects to his own advantage.

## BIOLOGY, SOCIETY, AND CULTURE IN HUMAN ECOLOGY

Human ecology seeks to understand man and his problems by studying individuals and populations as biological entities profoundly modified by human society and culture and by studying the effects of physical, biological, and cultural environments upon man and those of man upon his environments.

The viewpoint of ecology is at once statistical and typological. The observational reality is always a set of tolerances pertinent to the physics and chemistry of an individuals' metabolism and behavior and to those of his behavioral products. Yet this reality persists only through genetic and social mechanisms that transcend individual life sequences to provide the continuities, feedback stabilities, and progressive adaptations of biological and cultural evolution.

## GENERAL CONCEPTS

Biologically, man has both generic and unique characteristics as an organism. Among the former, the phenomena of distinctiveness and of persistence over time, through physiological and morphological changes, are central.

By what processes is this constancy maintained? To answer this question we must have some concept of a free-living organism in its natural life-situations. Many biologists prefer to view organisms as systems of selfregulatory processes. By this view alone one cannot really understand organisms. Equilibrium and constancy are important, but a more fundamental organismic characteristic among free-living animals is direction.

Direction arises both from the organism and the environment, for the survival of the organism depends upon its finding an environment which is adequate for it. To succeed in this search the organism must select, from all sorts of environmental stimuli that act upon it, only those events that are pertinent to it. This milieu is certainly not "definite and static"; rather it forms continuously as the active organism grows, matures, and ages. Thus organism and environment are in essence inseparable. It is this bond that gives rise to direction in organismic behavior and to selection in evolution.

In finding this adequate environment, the free-living animal organism exhibits preferred behavior. To determine whether a particular phenomenon which seems to be preferred behavior is essential and genuine, one must consider the entire organism. As Goldstein states it, "We are dealing with genuine attributes or constants if we find, by examining as many fields as possible, order and 'adequate performance' in the rest of the organism. In this way we apprehend certain characteristics of the organism with which we are dealing, certain norms and constants of its nature."

Among these "constants" of the nature of the organism are ways of behaving; sensory and motor thresholds; intellectual characteristics; affectivity; psychic or mental and physical traits; physiological attributes such as temperature, respiration, pulse, blood pressure; chemical qualities of the blood; blood types; and reactivity to noxious and stressful circumstances. The living organism tends to approach these relative constants or the "average mean." Thus, Goldstein remarks, "We are only in position to speak of one and the same organism, if, in spite of temporary changes, these constants become manifest."

There are two groups of "constants": those which express "the essential nature of the species" and the "individual constants." Goldstein emphasizes the fundamental significance of the "individual constants": "On the basis of the constants of the species, the life of the normal and especially of the defective individual cannot be sufficiently comprehended, notwithstanding certain congruencies between the individuals of the same species. For that objective, an acquaintance with the nature of the individual, that is with the *individual's normal constants*, is prerequisite."

## SPECIES CONSTANTS

Man is a species that has proved unique in its fitness to survive and multiply under extremely varying selective pressures. In the Pliocene, the protohominids appear to have been brachiating dwellers in tropical forests, a habitat which favored depth perception, color vision, sound production and hearing, and truly prehensile hands. Increasing drought in the late Pliocene and early Pleistocene forced adaption to a forest-edge and grasslands environment, in which both escape from predation and increasingly successful predation were basic elements of survival.

During this period, the precursors of man (Australopithecenes) become effective runners, in consequence of mutations and selection leading to the human heel, the reduced jaw, and the S-shaped spine. In general, less robust and more juvenile body builds sacrificing strength for endurance were evidently favored. These factors, jointly with the direct advantages of better perception, memory, symbolization, cerebral inhibition, learning and communication, fostered the absolute and relative growth of the brain.

Over the past half-million years, man has adapted to a wide range of nontropical climates, as a consequence both of migrations and climatic fluctations during the ice ages. These events must have been associated with periods of population growth and dispersal alternating with others of decline and concentration. Mechanisms such as these account, in fact, for the extraordinary combination of long-persistent geographical variability and full reproductive unity characterizing man.

Even more important have been the selective advantages of variability, role specialization, and cooperation in human society. These and other expressions of polygenesis even in small populations of man are undoubtedly related to the propensities toward out-breeding and mating stability ubiquitous in man.

The prolonged mobility and self-sufficiency, in primitive conditions, of pre-adolescent children, in combination with taboos on parent-child and sibling intercourse and the continuous sexual receptivity of the human female, seem to have been the behaviors crucial to expressing these propensities.

The increasing range and effectiveness of taught behavior—culture-have vastly altered selective pressures during the past half-million years. It limited the restrictions of cold through fire, shelter, and clothing. Effective cooking ended the limitations of unaided human digestion. Thus, with seed-gathering and, later, agriculture, hunger became less of a control on human numbers. But sedentary life, land clearance, the elimination of other potential pathogenic hosts, and symbiosis with domestic animals made disease immunity and behavioral safeguards against infection basic selection mechanisms. The significance of this later selection has been insufficiently stressed.

Larger populations made more mutations possible, while group survival became far less random than earlier. On one hand, new technology in some areas radically increased population-bearing capacities. On the other, endemic diseases (parasitism, malaria, tuberculosis) systematically induced child-birth and infant mortality, while epidemics carried off entire susceptible populations (e.g., smallpox and measles in North America and Oceania). Thus, intense growth in favored areas—the Near East, India, China, Middle America—and stagnation or decline elsewhere, have predominated since the rise of agriculture.

Associated with the control of food and the increasing size of human populations has been the increasing importance of management: the determination of goals, the discovery of means, the organization of efforts, and the distribution of gains—in human societies. An allied process has been the increasingly high evaluation of play and symbolic expression as ever more structured societies have repressed wider areas of biological desire.

At the same time, the change from simple face-to-face communities to stratified states brought with it lessened capacity to identify and harness individual variation. Social criteria, such as the accidents of conquest or defeat, urban or rural residence, white skin or black, have dominated role choices. The significance of complex individual variations in human adaptation has thus been diminished through feedbacks from history. This diminution has also lowered man's evolutionary adaptability.

At his present stage of evolution, man as a species has many distinctive characteristics. He matures slowly; some 25% of his life span is occupied with maturing. This long period of maturing is, however, merely an exaggeration of a primate phenomenon. Bolk has called the slowing "foetalization." Another term is neoteny.

Neoteny has had several important consequences. First, it is advantageous for learning and social cooperation to be taught while the organism is docile and sexually nonaggressive. Second, neoteny may be the cause of nakedness in man. According to Huxley, "The distribution of hair on man is extremely similar to that on a late foetus of a chimpanzee, and there can be little doubt that it represents an extension of this temporary anthropoid phase into permanence." Hairlessness is, except for desert inhabitants, unique among terrestrial animals. The hairless state, Huxley speculates must have encouraged humans to protect themselves against their enemies and the elements and thus have been a spur to intelligence.

Perhaps this hairless state has played a role in man's dependence upon evaporative cooling as a method of heat economy. This physiological regulation, of course, is not exclusively human, for many terrestrial animals employ evaporative cooling as a defense against overheating. What is unique about man in this respect is that he is an "eccrine animal." He is probably the only animal utilizing active sweat secreted by the eccrine sweat gland as a process of evaporative cooling. Other animals sweat but their sweat derives from apocrine glands.

One might speculate that the long period of maturing of the young has been genetically related to the post-mature longevity characteristic of man. The survival of animals beyond the reproductive period is unusually brief. In man, long survival allows the species to utilize social benefits accruing from speech and tradition. The menopause of the human female is unusually complete and sudden, and her long postreproductive period is exceptional among mammals. Williams suggests that these features may have developed in man because of the selective advantages offered to the slowly maturing offspring.

Man exhibits great reproductive variability. The differential fertility among men is enormous, ranging from infertility to families of 1 to 12 and even 20. This differential fertility is of greater selective advantage than differential mortality; it provides a basis for rapid changes in human gene pools.

The rapid and profound evolution of man's brain (primarily encephalization) allowed for plasticity and potential variety. This plasticity has had several different consequences. With encephalization and growing cortical dominance, man's power of attention has shifted to self-stimulative phenomena. As a result he experienced not only tensions, e.g., memory tensions of inhibition, but also symbolic satisfactions. While experimental neurosis can be produced, under natural conditions avoidance of conflict is the general rule.

"Only the peculiarities of the human mind," according to Huxley, "have forced its partial abandonment upon man." Association mechanisms bring into relation knowing, feeling, and willing. Both unified mental life and mental illness result. Repression is a device for resolving conflicting impulses, but this repressed input may become harmful and lead to mental disturbance. Taraxis or disorder has come to dominate the lives of men. Events appear to be overwhelming and incomprehensible; everything appears to be in disorder. We suffer confusion, defeat, frustration, failure, remorse, humiliation, grief, and so on. Richards writes:

If one looks at biology and the biological species, it is curious that the human being should be so sensitive and subject to external events and tormented by things that do not physically hurt; that it cannot accept adversity with the same philosophy that rabbits, rats, and guinea pigs can. To be sure, experiments can be set to torture the rat, the dog, possibly even the rabbit in taraxic patterns, but they have to be most deviously contrived. With man these things happen daily by themselves.

Thus creativity and disorder, distress and sublimation, are complementary properties of man in society and culture.

## CONSEQUENCES AND IMPLICATIONS

Man's unique endosomatic and particularly his exosomatic, evolution have brought him to a position as the dominant animal. In almost all environments, he dominates the ecosystem and holds much of his own evolution in his own hands. Cultural evolution—so inexorably interwoven with man's unique psychoneurological traits—has brought him to this position of power. He not only dominates but also consciously manipulates ecosystems. By virtue of this power he has created a biological revolution directed primarily toward the immediate enhancement of human welfare at the expense of competing species and often with reckless disregard for long-range consequences. Only in recent decades has there developed a broad awareness of the need for a continuing strategy of resource management.

A few examples will illustrate the problems man now faces. He has created what Rostlund appropriately terms "the domesticated landscape," that panorama of domesticated plants and animals, cultivated fields and pastures, and monosystem reforestations. This domesticated landscape (*anthroposere* may be suggested as a technical term) comprises a series of artificial ecosystems designed to assure man of plentiful food and raw materials to meet his domestic and industrial requirements.

The domesticated plants and animals selectively bred to inhabit these ecosystems depend upon the continuing intervention of man for their survival. Because of competition with other organisms which continually invade these ecosystems man has been forced to institute control measures. These measures nave been chemical more often than biological. The use and misuse of the biocides, moreover, have often complicated the problems of management, for not only do these chemicals have their intended effects but also unintended ones, ofttimes remote. Through his actions man tends to create a negative balance in the nutrient cycles of ecosystems. He extracts more than he returns.

For example, he has caused a definite species reduction. Furthermore, since his artificial ecosystems are most susceptible to erosion through the agencies of water, wind, and fire, he must treat his fields and pastures with concentrated chemicals—the fertilizers. These fertilizers must be applied with care else more undesirable reactions accrue. There has resulted what in essence is a continuing managerial struggle to maintain the artificiality of the domesticated landscape.

The capture of energy is another problem for resource management strategy. It can be argued, for instance, that cultural advance is directly dependent upon an increased ability to capture energy. In the United States today, man's consumption of energy outside of the food chain exceeds that used directly and indirectly in food by some 25 times. While this energy has been vital in the support of man's technological establishment, there have resulted two profound consequences.

In the first place, the disposal of technological wastes has become an increasing problem. Water and air have provided ready vehicles for discharge. The exponential increase in man's domestic and industrial use of energy has now demonstrated that the vital natural resources are not unlimited in their capacity to dilute and degrade his wastes. Water pollution and air pollution have become central problems of our times and there is ample evidence accumulating that these pollutants exert diverse and profound disturbances throughout ecosystems.

In the second place, this capture of energy has brought to him enormous power, power with which man can destroy himself as well as all other forms of life. This power will necessitate a drastic revision of his mode of thinking. He must turn from traditional views of ingroup survival to concern for the survival of mankind. Chisholm states the matter precisely:

The whole method of survival by groups in competition to the death with other groups has broken down. The survival group, for the first time in human experience, has become the human race itself. From now on we will survive as members of the human race or not at all, but we have no previous experience of this situation and no traditional concern or education for survival of the human race.

The occasion for such concern had not arisen until about fifteen years ago and was not foreseen or provided for by our parents or ancestors. Now we are all threatened with extinction by our own traditional survival patterns, a position which most of us still find impossible to accept as real, because we have been taught from infancy to depend upon our "conscious" values, and even to consider changes in them is commonly felt to be immoral and disloyal.

This same view was eloquently expressed by Lord Brain in a recent article entitled "Science and Antiscience":

The evolution of the human race is now threatened by a failure of integration. [By "integration" Lord Brain means bringing man into *adequate* relation with the ecosystem.] That integration is a social function, necessary both within individual national societies and, in the interest of our common humanity, between those societies.

The magnitude, speed, and imbalance of cultural advances over the past century have created other problems. New levels of anticipation have spread more rapidly than productive skill and technology. The control of epidemic diseases has had insufficient returns, too often in work opportunities for a growing, healthier labor force. More unemployment and more pressure on resources have been the tragic results. In addition, the world's population is already so large and the rates of growth are seemingly so rapid that the prospect of Malthusian limits has become real.

Vast increases in productivity, better systems of international exchange, and drastic increases in birth control are essential. These problems involve complex moral as well as technical issues. In a world of prejudice, the problems of eugenics, except for overwhelming diseases, can scarcely be broached. This is especially true since traits of selective value in some environments, such as the sickle-cell gene and perhaps diabetes, are very deleterious in others.

## INDIVIDUAL CONSTANTS

A basic question in human ecology is thus a clarification of relationships and values. What is "health?" How does one measure health? No one really knows the answer to either question. Why must we understand health so that we can measure it?

As man more and more rapidly disturbs his ecosystem, he makes it increasingly difficult for each individual to find an adequate environment. The rapidity of environmental changes may soon exceed the norm of reaction, the adaptability, of the human organism. Thus it becomes urgent that we gain some insight into the limits of the adaptability of individual organisms.

The disturbances which man has created in the ecosystem are manifold. He has polluted the water and the air and has contaminated his food. His gregariousness has led him to live in congested urban agglomerations where, if he is well off, his private quarters and place of work are air-conditioned. His physician administers to him an increasing array of chemicals designed fundamentally to assist him in coping with illness but which take their toll in iatrogenic diseases.

Problems such as these have led to an increasing concern about "environmental health." Attention is now turning to controlling the environment so that it will not become inadequate.

Questions that now confront those working on problems of environmental health include:

- What are the long-term effects on human health of low levels of toxic chemicals such as biocides and water and air pollutants?
- What are the long-term effects of living in atmospheres with closely regulated temperatures and humidities?
- What effects accrue from such labor-saving devices as the automobile which eliminate the necessity of walking?
- What impact do these man-made changes have on the ecosystem and how do these impacts affect the health of man?

To provide answers to these questions we must be able to measure health with sufficient precision to detect deviations from the healthy state before there is morbidity and mortality. To do this means a radical reorientation of our thinking, for health has heretofore been defined in terms of lack of evident morbidity and mortality.

To arrive at a productive model of health we must begin with the individual's normal constants, needs, and requirements. Several features of individuality must be integrated in this model. One particularly characteristic constant of individuality is "the constant in the temporal course of processes." Goldstein writes:

Every human being has a rhythm of his own, which manifests itself in various performances. A performance is normal when an individual can accomplish it in a rhythm which is his adequate rhythm for this performance. Just as for physiological processes, like heartbeating and respiration, this is valid for physicochemical processes. The time constant indicates a particular characteristic of the personality.

In addition, there is a relation between biochemical and physiological individuality and sensitivity to noxious environmental influences and susceptibility to illness. Individuality is largely quantitative rather than qualtative; each individual mobilizes different configurations or patterns of component homeostatic processes in making adjustments to environmental change. This diversity is probably polymorphic. Through systematic study of the diversity will emerge knowledge of the individual's adaptive capacity, which is required to deal constructively with the problems of environmental health.

Some examples may be given to illustrate these general points. Without a doubt, air pollution is a serious problem. The acute disasters serve to emphasize the grim prospects. Low levels of air pollution are now present in many urban centers. How do these low levels affect the health of the population? The individuals we should be most concerned about are those most sensitive to air pollution. If attention is devoted only to the "individual at the mean," many persons will be severely incapacitated or dead before the group within the "normal range" are affected. The most susceptible persons must be identified, for their individuality is a matter of fundamental importance in establishing criteria for air quality and other environmental standards.

The human being has a propensity to store excess energy as fat. This propensity creates significant health problems for a segment of the population identified as obese. Obesity predisposes to a variety of illnesses—for instance, diabetes mellitus, hypertension, and coronary artery disease. There is an individuality in the matter of fat storage. The ectomorph stores fat more readily than the endomorph. This fact suggests that the configuration of the component physiological regulations of the ectomorph differs from that of the endomorph. The needs and requirements of these types are not the same.

That these types exist suggest that there are probably polymorphisms of physiological regulations in the population. These polymorphisms can only be elucidated through systematic studies of individuality. Such studies should contribute to an explanation of why some people have a high frequency of illness and others have a low frequency, why some persons are successful in finding an adequate environment and others are not.

## EPILOGUE

Human ecology is, above all, a way of asking questions that may be productive in understanding man's evolution, nature, and problems of adaptation. It recognizes that, in man, biology, society, and culture are deeply interrelated. Culture and society have transformed man's environment but, in doing so, have changed rather than eliminated the selective pressures to which he must adapt effectively if he is to survive.

The problems posed by human ecology are very difficult. In fact, until the theoretical structure of operations research and the modern armament of instrument and data processing had become available, few of these problems could be effectively attacked.

Today, both the necessity for, and possibility of, effective human ecological research present great challenges for the biological and behavioral sciences.

## Chapter 22

# The Ecology of Disease

The study of the incidence and transmission of disease is the province of the special science of epidemiology, although this study could equally well be called the ecology or the natural history of disease. Epidemiology stems from *epidemic*, an unusual or severe outbreak of a disease. Epidemics have played an important role in all through human history, and many minds have been devoted to analyzing and fighting them. The epidemic, the unusual situation, however, cannot be understood without a study of the endemic, or the more usual condition, in which the disease is scarcely noticed because it is a steady and continuing part of the community environment.

The word epidemiology can be applied very broadly, to cover such subjects as the causes and incidence of mental disorders, of accidents, or indeed of any sort of disease situation, but in the context of this book, we are concerned chiefly with the epidemiology of infectious disease, since this is basically a biological problem that is dependent on the interactions of hosts, pathogens, and environment. Incidence means numbers of cases, and epidemiology is in large part a statistical science, although the statistics are meaningless except in terms of the biology of the situation.

We are only beginning to understand the complicated host-parasite interactions in infectious disease. Some parasites are highly host-specific, that is, they attack only one species of animal, or perhaps only a single kind of tissue within the animal. Parasites which require two or more different hosts to complete their life cycle may be highly specific for both of the alternating hosts, though one might be an insect and one a mammal. Other parasites are less specific in their host relations.

In general, related animals are apt to have related parasites, and particular parasites often seem to have a long evolutionary history of association with their hosts. But there are many exceptions; the only way we have of predicting the specificity of a parasite is to test it in the laboratory on numerous possible hosts. Many human pathogens are highly host-specific for man. This is true, for instance, of the viruses associated with the common cold, and it has greatly handicapped their study, since they cannot be tested on experimental animals.

Pathogens vary greatly in virulence, in the extent of damage they cause to the host. The same strain of pathogen may be highly virulent in one host and not in another. A given strain of yellow fever, for instance, may infect and multiply in a species of opossum with no apparent injury to the opossum, and at the same time be highly fatal for certain monkey species. Indeed, it may be fatal for one kind of monkey and harmless for another kind. Individuals of a given host species also may vary greatly in their reaction to a particular strain of pathogen, ranging from relatively resistant to susceptible. Differences in virulence of human pathogens at different times, or in different populations, are a commonplace in the history of disease.

There is always some sort of a serological (biochemical) reaction on the part of a host in which a parasite has become successfully established. These reactions are investigated by the sciences of serology and immunology. In many cases the parasite acts as an *antigen* which provokes the development of neutralizing *antibodies*. This is particularly clear with the viruses, and the course of the disease can be visualized as a race between virus multiplication and antibody production.

With a pathogen such as the yellow fever virus, the host may be killed before the virus is controlled by antibody production, or the antibodies may win and the virus be eliminated. Some pathogens, especially many viruses like those causing yellow fever, smallpox, and measles, produce in the host, after one infection, the ability to continue manufacturing antibodies as long as it lives, and it thus remains *immune* to a second invasion of the same virus.

We do not really understand this process: one theory holds that traces of the virus remain established in some part of the body and continue to stimulate antibody production. The principle of vaccination, with smallpox and yellow fever, is to infect the host with an innocuous strain of virus, making it immune to subsequent infection with virulent strains. With other viruses, like those of influenza, antibody production may be only temporary, so that the host presently is again susceptible to infection.

The self-limiting infections—smallpox and measles, for instance—must obviously have a continuing fresh supply of new hosts if they are to continue; such pathogens, then, are normally endemic in dense host populations and, they reach scattered populations only as occasional epidemics.

A great many pathogens show an alternation of hosts, and are not normally directly contagious from one individual to another of a given host species. Four human diseases—schistosomaisis, malaria, yellow fever, and plague—will serve as examples of different kinds of transmission mechanisms.

## SCHISTOSOMAISIS

The species of blood flukes of the genus *Schistosoma* are pathogenic for man and are important agents of disease in some parts of the world. The life histories vary in detail, but the cycle generally starts with the eggs being passed out in human excreta; the miracidial larvae that hatch from the eggs must find certain species of aquatic snails to serve as intermediate hosts.

After a period of development in the snail, forms called *cercariae* escape from the snails into the water in large numbers. If they encounter a mammal host in the two or three days that they are able to live, they bore into the skin (infection may also occur from drinking), develop into the adult phase, and start producing eggs again. The schistosomes that infect man may also infect other mammals with habits that bring them into snail-infested water.

## MALARIA

Four species of the protozoan genus *Plasmodium* cause various forms of malaria in man. In the human host, the parasites live within the red blood cells; the parasites normally multiply asexually, but from time to time sexual forms, *gametocytes*, are developed. If these are picked up from the blood by an appropriate mosquito (of the genus *Anopheles*), the male forms (*microgametes*) burst from the host blood cell in the mosquito stomach and fuse there with the female *macrogametes*. The resulting zygote penetrates the stomach wall and forms a cyst of multiplying cells that eventually burst out (in the form of *sporozoites*) and make their way through the mosquito body to the salivary glands.

When the mosquito next bites, they are injected into a new vertebrate host. The human *Plasmodia* are all host-specific for man; other species attack primates besides man, as well as birds, lizards, and a few other vertebrates. All, as far as is known, have mosquito intermediate hosts, but different *Plasmodia* infect different kinds of mosquitoes.

## YELLOW FEVER

Yellow fever is unusual because it shows two epidemiological patterns, one urban and one sylvatic. The pathogen is a virus which causes a self-limited infection in man (the human host either dies within around 12 days after infection or recovers with a lifetime immunity to re-infection). In the urban form, the intermediate host is a mosquito, *Aëdes aegypti*, which in tropical America, at least, is always closely associated with man, breeding in rain barrels, till cans, vases, and so forth.

When the mosquito bites a host with virus circulating in the blood stream, it becomes infective for new hosts after a period of about 12 days (the incubation period depends on the temperature) and remains infective for life (which, in the case of a mosquito, is usually short, at most 30 days or so). The virus was long thought to be host-specific for man, and since there is no other known host in urban areas, the virus requires a large human population for survival.

Under these circumstances, it seemed possible to eliminate yellow fever from the Western hemisphere by eliminating it from the large cities (like Rio de Janeiro) that served as foci of infection.

When a campaign with this purpose was started, however, it was discovered that the disease also exists in remote parts of the upper Amazon and Orinoco river systems, where the *Aëdes* mosquito does not live, and where the human population is small and scattered. It turned out that in these inland regions various species of monkeys served as the mammal host of the virus, and that forest mosquitoes (*Haemagogus*) are the intermediate hosts; man enters the cycle only accidentally, though a man, infected in the forest, could return to a town or city and set off an urban epidemic.

## PLAGUE

The Black Death, which caused such devastating epidemics in Europe in the Middle Ages, is still an important disease in some parts of the world, especially in India. It is now endemic in the western United States in wild rodents and causes occasional human deaths, providing a constant worry for public health officials. The pathogen is a bacterium, *Pasteurella pestis*. Plague shows several epidemiological patterns.

Normally it is a disease of rodents, and the great epidemics of history involved both rat and human populations. The bacillus is ordinarily transmitted from rat to rat by fleas, and fleas may also carry it from rat to man and from man to man. In a severe human epidemic, a "pneumonic" form of the disease develops, which is directly contagious from man to man.

In the course of a plague outbreak in San Francisco in 1900, ground squirrels outside the city became infected, and a sylvatic form of the disease has persisted in the western United States despite all efforts to exterminate it. Some 18 species of wild rodents have been found to serve as hosts, and at least 30 species of fleas associated with these rodents are able to transmit the disease under laboratory conditions.

## WASTEBASKET OF THE EARTH

Contamination in the food we eat, the water we drink, and the air we breathe suggests close connection with the soil. Usage such as "your hands are dirty," serves as a reminder that the thin surface layer of the soil is the wastebasket of the earth, the collector for the disposal of all matter that has once lived and moved.

### ROLE OF THE SOIL

The thin layer of the earth's surface is an intensive transformer of all the waste it collects. In that shallow stratum elements are separated out of combinations and reunited into other compounds, effecting vast changes in every kind of matter. These activities include transformations of energy by such processes as oxidation, reduction, hydration, hydrolysis, and molecular rearrangement. Oxidation, like combustion, dissipates energy in the form of heat, which escapes from the earth. Reduction concentrates it in compounds of high heat and fuel values. This is illustrated most significantly by the plant's reduction of carbon dioxide by water, storing the sun's energy in the resulting carbohydrates.

These transformations of matter and energy create, nourish, protect, and maintain all living creatures. These creatures vary in physiological complexity from the simplest microbial cell to man, whose high state of evolution has equipped him with a mind to comprehend—and to modify-his environment, even as far as unwittingly contaminating it.

The term *contamination* has many meanings. The wastes of one population may be either the poison or the life-support of another. We consider populations of different forms of life to see whether their coexistence is cooperative or competitive, and whether their respective wastes may be of benefit or harm to each other.

A high degree of cooperation through evolution and adaptation characterizes nature's management of environment. But when man manages environment, competition stands out-with frequent examples of contamination. As managers, we are biased toward aims and benefits for man alone. We have concerned ourselves little with how, in modifying the environment for ourselves, we may be disrupting it for all the other populations of the biotic pyramid—microbes, plants, animals, and man—all supported by the one creative foundation—the soil.

Instead of speaking of the soil as "dirt" to emphasize its contamination of ourselves, the converse might be more appropriate. Man is upsetting much of the natural environment by his contamination of the air, the water, and particularly the soil, reducing its ability to dispose of wastes safely, and to nourish healthy populations of men and all other living things. It is a dangerous boldness to believe that we can manage environments completely by technologies designed for our economic advantage.

## THE TRUE AGRARIAN

Given to a belief in the homocentric purpose of the earth, we have come to take our soil for granted. This view is quite the opposite of that of the pioneer—living mainly by agriculture—who respected and studied his environment and struggled to be naturally fit for his evolutionary survival there. He considered the seasons, the annual amount and distribution of the rainfall, the degrees of heat and cold, the winds and storms.

He did not consider land as a commodity. The pioneer appreciated the fact that the soil had been built by nature during the ages pre-dating him, brought about by the climatic forces breaking down the rocks, growing the microbes, the plants, the many other kinds of life. For him, those were the natural forces of soil construction, and he knew he must maintain soil productivity if he were to survive. The pioneers were truly agrarian people. For them the soil was holy ground. Too, it was living soil.

Because of the scientific organization of our recently increased knowledge about the soil, we forget that the decomposition of rocks, the growth of vegetation, and the complete return of that organic matter in place—all under nature's management—are what brought about productive soils. At first, these soils were not contaminated against the healthy coexistence of a specific, but limited, set of species. The early balance represented an evolutionary set, each form unique in relation to the others, to the soil's particular geologic—climatic setting, and its degree of development in a given area. Examples of this limited balance would include the particular plant and wildlife forms of virgin forests, or the American plains or prairies. All of these conditions were major determinants of the coexistent species; the survival of each was dependent on the survival of the others.

Man's management of agricultural crops and livestock has not been directed by knowledge of the limitations of soil fertility, nor by knowledge of the required climatic-geologic setting for crops and livestock with each in its natural ecological climax. Instead, we have been given to transplanting any species from anywhere to everywhere for economic gain, ignoring biological benefits or dangers to the species involved. We have brought in higher (or lower) species while depleting the very soil fertility support required to grow them in health.

We have thrown natural evolution into reverse. We struggle to nurture species we have made unfit for the environment because creative forces there cannot offer the required quality of food and energy support. We now need to view the pampered species as contamination against all other lower species which would otherwise arrive at their natural climax. We must accept the fact that the soil, with its dynamics of producing and accumulating organic substances through plants with the support of climatic forces, is still the only energy supply on which all kinds of life depend. Transformation is the major role played by the soil.

Granting that the entire biosphere is dependent on the inflow of that bit of the sun's energy fixed in organic substances by photosynthesis, it is helpful to note that as Nelson G. Harston, Frederick E. Smith, and Lawrence B. Slobodkin said in their article, "Community Structure", "Population Control and Composition", in *The American Naturalist* in 1960, populations divide themselves appropriately into three trophic levels: the decomposers, the producers, and the predators.

## THE DECOMPOSERS

The decomposers are represented by the varied kinds of microbial life which live by degrading organic debris through processes giving the microbes their energy and growth substances. The slow accumulation (at great depths below the earth's surface) of these remnants as fossil fuels of organic origin and a high degree of oxygen removal indicates that this heterotrophic group did not have much in the way of energy-giving "leftovers."

Nor can these remains be considered contamination when they are far beneath the soil surface. But when they are brought into the atmosphere, and into the highly aerobic surface soils, after laboratory work has turned them into products such as antiseptics, pesticides, and herbicides, they are the most extensive and powerful biochemical contaminations we have yet known. Their range of disturbances covers the entire biotic pyramid.

Hydrocarbons, which can disrupt the transformations in the soil's surface zone, were buried by nature at great soil depths. These high concentrations of energy are now our industrial fuels. But even such deep burial of the wastes resulting from the use of atomic fuels will not serve as safe removal. Atomic fuels, with their lingering rays for a lingering death of all living cells, are not respecters of the beneficial portions of the soil's microbial flora.

Nutritional requirements of microbial decomposers are met by the contents of the debris and the soil. The essential elements remain very much in the cycle of use and reuse, since sulfur, phosphorus, nitrogen, and carbon occur as major elements in the leftovers. Oxygen is almost absent there, but carbon, linked to hydrogen, occurs in high concentration.

Ever since the work of Pasteur, our fear of microbes has singled them out as our environment's major contamination. We have made them the victims of vengeance and we boil them under steam pressure at every opportunity. Their disrepute has been shared recently by dusts, fungus spores, pollen grains of trees, grasses, weeds, and other particles. Today the professional allergist devotes his attention to atmosphere contaminants and to other substances disturbing the mucous membranes and similar tissues of the human body.

But when microbes are viewed as disposers and transformers they become major benefactors for other populations. Microbes in surface soil serve as wrecking crews and salvage agencies. Simplifying the residue of past populations for energy release and reuse, these decomposers make the upper stratum of soil the real living foundation of the entire biotic pyramid. They keep open the sewage disposal systems of all the population levels.

Microbes are uniquely equipped to maintain their own populations. They reproduce at the rate of one generation per hour or even faster. They synthesize their own extra- and intra-cellular compounds—some of which we term antibiotics—for protection against competitors for their environment. The antibiotic quality may be merely an evolutionary accident that served to bring death to competitor cells. Nature's conservation practices use wastes from some life forms to make the environment serve its own survival more completely.

Benzene rings characterize the chemical structure of the antibiotic terramycin, for example. Modified ring structures with substitutions of nitrogen and sulfur for some of the carbon are found in penicillin, aureothrycin and other microbial products developed for their bacteriocidal effects on the human body. That ring structure represents highly reduced organic compounds such as those found in crude oil and coal.

The chemical structures and biochemical energy potential represent the opposite of that of natural organic wastes dumped on the surface soil. While natural organic wastes offer much as energy through microbial oxidations, crude oil and coal are too stable for biochemical transformation and energy release, even though they rank high as industrial fuels. They are seldom broken down by digestion. They overload the liver, the chemical censor of the human body. They are leftovers from anaerobic microbial populations, and are well removed as serious contaminants by their natural placement far below the surface soil and by disposal well beyond the entire biotic pyramid.

The benzene ring, in a simpler compound distilled from coal tar-carbolic acid—was an early antiseptic. But now, long-chain compounds and ring structures of carbon, sulfur, or nitrogen substitutions, or in chlorinated, nitrated, and sulfonated forms as synthetics from the industrial chemistry laboratory, are being distributed extensively, acting as deadly poisons against the populations in the biotic pyramid. They come into the atmosphere in "smog" and carbon monoxide, and in herbicides, pesticides, and the like.

Those microbial wastes welling upward from the depths of the earth, in their natural forms and in our more poisonous alterations of them, must be considered contamination by man of his own environment and of the environments of all populations that support him. His efforts to so completely destroy microbes are contributing to his gradual destruction by his own hand. Decomposers are not respecters of man when they release their own wastes as contaminants.

## THE PRODUCERS

The second group among the populations, or trophic levels, of the earth are the plants, which produce organic compounds carrying the chemical and other energy transferred from the sun. They are the only means of storing and distributing that supply. Energy is collected by photosynthesis, the unique process whereby the chlorophyll of the leaves binds it into compounds of carbon, hydrogen, and oxygen in the molecular arrangement of carbohydrates.

Plants are the source of energy for all biochemical processes, and of their own starter compounds into which they synthesize nitrogen, sulfur, and phosphorus to yield the different amino acids of proteins and living tissues which grow, protect, and reproduce. Plants are the only producers since they are the sources of food energy and growth potential synthesized directly from the chemical elements and flowing through all the other trophic levels.

Plants root themselves first into the soil and then extend their tops into the atmosphere. They may be vulnerable to contaminants from both directions. Their unique position and special processes bring both inorganic and organic decompositions from the soil into biochemical union with water, and carbon dioxide, and nitrogen. Plants, representing that limited zone where earth and atmosphere meet, act as a kind of interface for concentrations of different kinds of matter, becoming the significant stratum for the creation of all that lives.

All life is possible because the synthetic power of sunlight operates through the plant fluid, chlorophyll. It is a mobilizer, or chelator, of the chemical elements, unique because the plant itself creates it. Its chemical structure consists of the inorganic element magnesium as the core, combined with nitrogen, linked with carbon, and all three connected with hydrogen. Even this chelator's composition represents a chemical union between the soil, which furnishes the magnesium, and the atmosphere, which yields the nitrogen, carbons, and hydrogen.

Because plants support themselves by their own capacity for combining elements and using solar energy to create compounds, they surpass all other populations in the struggle for survival.

Microbes can use the elements in synthesis but must decompose the organic compounds synthesized by plants to provide the necessary energy. Plants, in turn, profit in their extended survival because the microbes simplify the accumulated organic matter to keep the soil's inorganic elements and the atmosphere's organic elements—carbon and nitrogen—in cycles of reuse. Otherwise the accumulated products of plants would contaminate their own environment.

Plants and microbes may be considered in a symbiosis for survival independent of more complex populations. But even that symbiosis may be disrupted by one or the other symbiont acting as a competitor, a parasite, or a predator. Either may even produce contamination by its waste products: plant compounds may be poisons for microbes and microbes may be poisons for plants.

By competing for essential inorganic elements—calcium, magnesium, phosphorus, potassium, nitrogen, sulfur—and the several "trace" elements, one may limit the other via the soil.

All trophic levels above the plants must live by the compounds of the latter's synthesis—carbohydrates for energy; proteins for growth of tissue, protection, and reproduction; and inorgano-organic combinations associated with the proteins. Proteins in plants result, not from photosynthesis, but from the plant's biochemical processes, which require expenditures of stored energy and assembly of inorganic and organic requirements by the roots.

The roots penetrate only a limited volume of soil, living there largely through activities of the decomposers. Plants, like microbes, use carbon dioxide waste—from the roots—to produce active hydrogen in the resulting carbonic acid to mobilize the soil fertility elements for plant survival. These two populations, the producers and the decomposers, as contaminants or as transformers, determine the environmental support of all other populations.

## THE PREDATORS

All the heterotrophic populations crowd each other to get to the food delivered to them by that mundane team of autotrophs—the symbiotic team of microbes and plants. That crowding, under the deficiencies provoking it, makes the wastes of each a contamination for the other whenever the natural processes of the producers and decomposers are disrupted. All populations above plants and microbes are included in the third category: the predators. They prey upon the plant population, and upon each other to an increasing degree, since the plants are not nourished by the living soil completely enough to be ample prey for man and all the other predators and parasites.

Predators—other than man—do not destroy their prey completely before the numbers of predators drop down so far that the numbers of prey mount to domination again. Increase in prey favors an increase in predators, and then, in turn, a decrease in prey, to yield naturally alternating dominations but not the extinction of either. But man is not merely the predator of one trophic level; he also aims his technology at complete extinction of many populations. Thus he breaks the law of predator-prey relations and of survival. When he causes extinction, he reduces the number of basic segments by which the biotic pyramid supports man.

Interpopulation predations increase as the soil becomes less able to grow the vegetation of nutrition in required quality and quantity. When the soil's inorganic elements have been depleted, then the exploited soils must eventually register their damage on all trophic levels. The most serious effects fall first on man, raising the question of whether the baffling degeneration of our bodily health, now increasing our concern about what was once commonly called "disease," may not suggest patterns of hidden causes connected in some way with the climatic-fertility pattern of the soil. Thus, human ecology may develop into the most important science.

## HUMAN LIFE CAME LATE AND MUST GO EARLY

When the numbers in any population are charted as a graph against a base of time, they give a sigmoid curve. Its shape suggests the top of the letter S pulled to the right while the bottom remains attached. The introduction of a single living microbe into a given volume of medium results in a slow increase in numbers with the curve moving along the near-horizontal. But soon the move turns toward the vertical, suggesting a population explosion. Then the increase lessens, ceases, and, finally, there is a population decrease followed by eventual extinction. Contamination by its accumulated non-transformed wastes, coupled with exhaustion of nutritional and other environmental supports, eliminates the population from its limited setting.

Populations of older countries, similarly charted, suggest the final third of the biotic curve. France, Scandinavia, Italy, and Spain, starting as far back as 1900, are illustrations. The United States illustrates the lower two thirds, or the beginning of the chart. Predictions from 1920 onward pointed to the failing percentages of increase, if immigrations were limited.

Since the multiplication of man still conforms to the natural laws of biological phenomena, and since human multiplication is by no means managed to give biological advantages for improved survival, we must characterize man, at this stage, as the main biological liability, not only to himself, but to the other populations supporting him. *He is the contamination in the environment.*

Man's disregard of his ecological limitation to the temperate zone—by the required supply of proteins alone—should be cited as only one powerful factor. As one moves from the temperate to the torrid and humid tropical zone, the desperate struggle for proteins at any trophic level is evident in the carnivorousness and cannibalism that are common. Here, life forms are mainly predators. The same is true as one moves from the temperate to the frigid zone.

The migrations into the frigid zone, into climatic-soil-fertility settings of little or no soil construction, are badly handicapped by the paucity of producers and decomposers. The predators win support by their carnivorousness or through lifelines reaching back into the temperate zone. In the tropics, high rainfall and temperatures have caused excessive decomposition of the rocks to the point of destruction of soils which then fail to fully nourish plant life.

Even proteins are there replaced by poisonous compounds of many producers. Carnivorousness and cannibalism characterize the survival of decidedly limited populations. The dry tropics are in the same category. Populations at climatic extremes are limited because the producers, operating in combination with the decomposers, are not providing the necessary proteins in complete array, nor with their natural accompaniments.

## CAN MAN MANAGE HIS OWN SPECIES

The epoch of man is but a minute segment in the paleontological column of the earth's populations as they have come, gone, or remained. Man has shifted away from the rugged individualism of open country and its diversity of agriculture according to the laws of nature. He has collected himself into congested cities which have been said to require monocultures and chemicalized agricultures; he has controlled the environment of those cities according to the dictates of technology and economics, disregarding nature's laws and even his own biochemistry.

During the latest part of the brief epoch of man's existence, his technologies and their political complications have served to harvest the natural living resources, to exploit soil fertility and to compel the human march from east to west. The march has rolled on until the resources of most recent possession—those of the western hemisphere—are dwindling rapidly under the political demand for coexistence of the Western world with those bringing up the rear in the march—a total world population approaching three billion predators.

Because they must behave according to the natural laws controlling biological bodies, each population below man has—through evolution—exhibited itself as a climax crop for a limited time in its limited ecological setting. But man, with mental capacity transcending that of the others, has used most of the other levels to his advantage and their disadvantage so that, in general, he does not conform to the pattern of evolution.

His technological powers disrupt the pattern and destroy the very conformers that support him. Unwittingly, his development and management of environmental control over materials and energy have increased the contamination of the soil and the atmosphere. He has destroyed the decomposer populations to the extent that the natural basic support of the producers is weakening; the entire biotic pyramid is tumbling because of man's dominance at the top.

In managing her contaminations, nature either transforms them through biotic disposers in the surface soil, or buries them safely at greater depths. Man, managing his contaminations as a helpless novice, seems to be on his way to his own destruction by and amongst them. In all probability, nature, not he, will determine the final outcome.

## SPACE TRACKS

In the few years that have passed since the mid- Twentieth Century, we have come to the realization that technology has developed both the power and the tools to enable us to explore more completely earth, space, our moon, and even nearby planets. These explorations, which at present fall almost entirely in the realm of the physical sciences and engineering, are proceeding at a rate few can grasp.

Meanwhile, biologists, and especially those concerned with the studies of the natural environment and its faunal and floral elements, have noted the marvelous accomplishments with mixed feelings. Particularly they have envied the new tools for making numerous measurements, which in turn, have led to the accumulation of vast amounts of information. The field biologist, unfortunately, still depends largely on sight and hearing for his information and, as a result, often obtains data too discontinuous and incomplete for reliable statistical analyses and interpretation.

Yet the same biologist also recognizes that the information and measurements that have been collected by the physical scientist, who have always been considered to be in a discipline apart, have been but little concerned with life of the earth on which we live. Interdisciplinary research between biologists, physical scientists, and engineers in studies of our natural environment has been slow and pursued with reluctance.

More than thirty years ago the conflict was expressed by the pioneer ecologist Dr. Royal Chapman who, in 1929, wrote: "Ecology is bound to become quantitative. Many of us are observing this inevitable tendency with regret. There is a feeling that the wonders of observational natural history are to be brushed aside by the cold, dry calculations of a mechanistic mathematics" But he added: "The urgent needs are, first, more accurate measurements of environmental factors and the populations which make up the natural associations, and, in the second place, better methods of evaluating the measurements of the factors."

A brief review of one aspect of man's efforts in these directions may help to put into better perspective the need for expanding interdisciplinary research and the real potential of modern instrumentation as applied to the biological sciences.

Watching and listening to the passage of animals in their seasonal migrations has been a necessary and traditional behavior pattern of man since he began to evolve as a flesh eater. Increasing curiosity about bird migrations led to the first markings of individual birds several hundred years ago, and the usefulness of this technique was well recognized by the early 1900's. As a result, several million birds and many other animals have been tagged with numbered metal bands or with various color markers.

But this method has told us only a little of an animal's movements, for band recoveries represent a relatively small percentage of the marked animals. Usually, all we have available as an end result are two points on a map—the banding and recovery sites—with a known time interval, but an unknown distance of travel, between. This information has served many purposes, but has contributed little or nothing to understanding the animal's orientation, navigation, or relationship and responses to its environment.

Now, observations of an animal's motility are of utmost significance. Since the environment apparently furnishes or modifies the stimuli that trigger basic responses, the ability to follow an animal's movements continuously and to attempt to correlate these movements with environmental factors seems essential to advances in ecology. Nearly all animals are at some time motile, and since the causal factors of this motivation are not adequately understood, major emphasis must be given to a continuous recording of animal movements correlated with other environmental events.

For instance, various researchers have gathered data indicating a relationship between the distribution and abundance of animal populations and the biotic and physical forces of their natural environments. However, only a few quantitative studies on an individual animal's motile responses to environmental factors have been attempted *under natural conditions*. In addition, relatively few studies have been concerned with accurate measurements of the effects animals have on their immediate microclimates and the relation of these alterations to the animal itself and to its habitat.

In order to describe a total energy balance of an environment, the following physical properties must be measured: absorbed and reflected solar and infrared radiations, air and soil temperatures, relative and absolute humidities, precipitation, and barometric pressures. Current advances in instrumentation give hope for truly accurate measurements of these and many other variables. Movement studies have previously been based almost entirely on visual observation, recapture of previously trapped and marked individuals, and on the appearance in study areas of species intermittently captured by various sampling methods.

From these studies have come current interpretations of an animal's response to environmental change, the area of its "home range," its migration and dispersal, population structure, and even conclusions on systematics, evolution, and biogeography. Under conditions of adversity, an animal adjusts physiologically (for example, by changing heart, respiratory, and metabolic rates).

It may shift its position in its normal home area; it may migrate; it may die. If we use the problem of animal movements and environmental stimuli as a basis for interdisciplinary research among engineers, physical scientists, and biologists, what can we present as a feasible technique to give us results of an importance equal to those being obtained by researchers in the physical sciences?

Of course, the development of radar during World War II enabled us to "see" movements of birds both day and night on the radarscope and on film, and also to see weather and the responses of the birds to it. It now allows us to measure, within the radar's range, the magnitude and speed of the migration, altitudes of flight, and direction.

But we cannot very accurately identify the birds as species nor can we distinguish between individuals and flocks. The use of radar by biologists evolved slowly—partly as a result of wartime security measures and partly because of the reluctance of many scientists to recognize that nearly all "angels" seen on radar were really birds.

Within the last three or four years, the availability of transistors and other miniature and now microminiature electronic components has given us another major breakthrough in the study of animal movements in their natural environment. Tiny radio transmitters, which, together with batteries, weigh less than an ounce and which can transmit a signal for weeks, have been placed on or in animals. To date, ruffed grouse, cottontail rabbits, woodchucks, porcupines, skunks, and other animals have been located intermittently by portable receivers carried in the field.

Some of the results obtained have been both surprising and dramatic. The transmitted signal was sometimes altered by the behavior of the animal and by one or more body functions. Changes in the signal have distinguished between a resting and a flying bird, and analyses of signals recorded on paper have given accurate measurements of rates of respiration and wingbeats of a mallard duck.

Information on relatively sedentary animals and their environments may become known by telemetering data to ground stations over short distances. But radio signals do not curve, as does the earth's surface, so signals traveling along the ground are largely lost over any considerable distance. Thus, tracking long migratory movements would require either numerous receiving antennas on ground towers or planes equipped with receivers—a tremendous expenditure in equipment and manpower.

Is there any way we could obtain line-of-sight radio signals from numerous animals that may be moving over vast distances in all three of the earth's mediums? How can we find out where the penguins of Antarctica go after mating season; the routes of the wandering albatross or the Caribbean turtle; the forces governing caribou movements; the track of the Canada goose? One tool in common current use by physical scientists and engineers could answer all these and many more. That is the artificial satellite.

Satellites carrying various intruments have already relayed back to earth vast amounts of data in the fields of geophysics, radiation, and other aspects of space studies. Most instruments for making these measurements and the radio transmitters for sending the data back to earth are very small in size and light in weight. What equipment would be needed to locate animals on various parts of the globe and to transmit the information back to earth? To obtain the necessary calculations, I went to my colleague, William W. Cochran, Director of the Bioelectronics Laboratory in the Minnesota Museum of Natural History. His experience in designing and building both satellite- and animal-borne radio transmitters and receivers places the following instrumentation plan beyond the realm of mere speculation.

A satellite could be instrumented to receive and relay to ground stations the signals obtained from many transmitter-carrying animals in various parts of the world. Initially, we might seek only the migration track of each animal. Cloud patterns and perhaps other meteorological and geophysical data that might affect animal movements might be available from measurements made by other projects (such as the Tiros satellites) for areas of the earth over which birds would fly or swim.

These data could provide information on such phenomena as disorientation of birds under conditions of overcast. Successes in tracking animals on a world-wide basis by the use of a satellite would, it is hoped, provide guides for further instrumentation to telemeter one or two physiological or environmental factors concurrently with migratory movements.

Species that could be tracked by this method, and for which prototype transmitters have already been designed and tested, are larger birds such as geese, swans, cranes, penguins, and albatrosses. Among mammals the caribou could be tracked in its migrations. Species in these groups are large enough to carry currently available transmitters and power supply.

Their modes and regions of travel and their physical environments are so different that numerous tests of the instrumentation could be provided effectively. For example, Canada, blue, and snow geese, which could be tagged with radio transmitters, pass over much of continental North America, moving mostly in long, direct flights. The longest nonstop flight is probably from James Bay, the southern extension of Hudson Bay, to the Gulf Coast of Louisiana and east Texas.

Instrumentation to track these and other animals on a global basis could be relatively simple. Each animal could carry a transmitter and power supply weighing approximately 20 to 60 grams. The satellite, traveling 18,000 miles an hour (approximately 5 miles per second), at an altitude of 200 miles above the earth in a polar orbit, would cross over both North and South Poles once about every 103 minutes.

In the satellite would be a receiver and transmitter weighing a few pounds—or possibly even less—and occupying perhaps half of one cubic foot of space. Power supply, of course, would require additional weight and space, but might be supplied by solar cells, which would extend the period of tracking and would weigh less than batteries.

In order to receive the signals from the satellite at all times in its flight, twenty-four optimum-located receiving stations would cover the earth. A tape recorder at each of the ground stations would record the signal that had been received in the satellite from an animal-borne transmitter and then retransmitted to earth. This tape would then be run through an audiospectrometer. The spectrograph would be displayed and used—together with the known positions of the satellite in its passes within the line-of-sight range of each animal—to locate the animals at various times of day and night.

Some of the details of this tracking method might be of interest. The tiny radio transmitters, each with its battery pack, would be no larger than the first joint of a thumb. It would be attached to a simple harness with the antenna forming one of the harness loops around the body of the animal. It has already been found that animals adjust relatively easily to harnesses that do not inconvenience their habits. The entire unit would weigh some two ounces or less, depending on the number of batteries. The transmitter signals would be intermittent pulses, each at a slightly different frequency and, therefore, distinguishable from others in the same part of the earth.

To help determine just how far a signal from these small instruments could be received, we attached half-ounce transmitters to high-altitude balloons used for research by the University of Minnesota Department of Physics. With radiated power of much less than one microwatt—far less than that anticipated for satellite tracking—we received the signal from 270 miles away when the balloon was 25 miles above the earth. And this was in a region of considerable radio interference. The test adequately demonstrated the feasibility of long-range line-of-sight transmission.

The proposed satellite would contain a receiver and an antenna tuned to that band of frequencies occupied by the transmitters. Satellite transmitter power should be at 500 microwatts to assure a satisfactory signal (distinguishable from all types of interference) at the ground receiving sites. A lower power could be used at the expense of adding or utilizing more ground receiving sites.

Signals from the satellite could be received by a ground station at a distance of about 1,200 miles. The satellite could receive signals from the animals up to about 800 miles. Thus, the effective coverage area for one receiving station for tracking animals is about 4,000 miles across. Therefore, one station near the center of North America could conceivably cover most of the United States and Canada. A few stations in the Pacific islands could cover the whole Pacific area.

Each orbit of the satellite "scans" a swath about 1,600 miles across. The overlap in successive orbits depends upon the latitude. No overlap would occur at the Equator and a 100 per cent overlap would occur at the Poles. Thus, in equatorial zones, two locations, or fixes, a day are possible. In temperate latitudes, four fixes daily are possible, and in arctic latitudes about 14 (the number of orbits per day). Of course, to get a fix the satellite must be in range of both the receiver and the tag on the animal, so the maximum number of fixes is obtained only if ideal receiver distribution is used—that is, at least one receiver must be so located as to cover any point in the area concerned.

This means that receivers can be spaced somewhat less than 1,200 miles apart, depending on the area covered, whereas for computing coverage, a radius of about 2,000 miles from the ground station can be used. For example, on Pass No. 1 of the satelllite both ground station receiver and tag transmitter are in range of the satellite; but on Pass No. 2 only the tag is in its range. Thus, if we consider the number of fixes as the number of times the satellite is in tag range, we must provide receiver coverage to insure satellite contact.

This means an overlap of receiver circles, each 1,200 miles across. But if we are only interested in complete coverage and not in the maximum number of fixes possible each day, we can use overlapping 2,000-mile circles. This is of importance at low latitudes, where tagged animals could be missed.

We have said that if a satellite is 200 miles high it will cross over both the North and South Poles once every 103 minutes, or every one hour and 43 minutes. (Its orbit can be thought of as fixed in space, for although precession occurs, it is very gradual.) If the satellite passes northward over point A on the Equator at noon, it will pass southward over the Equator 51½ minutes later, approximately on the opposite side of the earth.

About 103 minutes later it will return, but this time it will pass over the Equator about 1,600 miles *east* of point A, which is the distance earth (and point A) has rotated during the satellite's orbit. The next orbit will be about 3,200 miles east of point A, and seven orbits after crossing A the first time, the satellite will pass over the Equator about 11,500 miles east of A, going north. Since 11,500 miles is about one half the circumference of the earth, point A will see the satellite coming south on its seventh orbit.

In each ground receiving system would be one FM receiver and associated antenna tuned to the frequency of the satellite transmitter, one two-channel tape recorder, and one highly accurate clock to emit sound signals that could be recorded as time intervals.

Equipment of this type is available and relatively simple to operate. Each time the satellite passed within the 1,200-mile maximum range, a recording would be made of the modulation of the satellite transmitter and, on the tape's other recorder channel, of time tones accurate to one second (the satellite travels at five miles per second). These recordings are preserved for analysis.

The signals from the transmitters may or may not be heard through the interference, depending on the distance between the satellite and the tagged animals. Much of the time they will be inaudible. The tapes will be run through an audiospectrometer with a resolution of 10 cycles per second, or less; the narrow band width of the spectrometer will reduce the interference to a point at which the signals are discernible.

When these signals are plotted on paper as radio frequency versus time, they form an S-shaped curve. This curve is a result of the Doppler effect between the satellite receiver and the signals coming to it from the transmitter on the animal. This curve, plus accurate time, plus the satellite's path would enable us to locate the animal in a circle 50 miles or less across. Paradoxically, extracting these data from a single tape may take some 1,500 times as long (about four continuous days) as the four minutes or so required to record it initially, although the time could be reduced by using audiospectrometers with higher capacity.

The complexity of analyzing the tapes suggests the creation and use of analysis centers. And here still another field is introduced into this cooperative scientific venture—the adequate utilization of computers whose programing could be jointly undertaken by physical and biological scientists. The information that could conceivably be transmitted by, say, a goose equipped with a transmitter, is almost limitless, at least theoretically.

The speed, altitude, and direction of flight are only a few examples. Wing-beat and respiration rates could be sent to the satellite, as could temperatures of the individual animals. These factors could be examined by one of the new biological computers, which have been specially designed to extract desired information from tapes that might be cluttered with various kinds of interference. Latitude and longitude calculations to locate the animals could also be determined to accuracies of about 20 miles, depending on the particular electronic system utilized.

With knowledge of long-distance movements of only a few kinds of animals in various parts of the world, we could begin to obtain data on environmental and physiological factors, and with these to work with, answers could be found to such questions as: does the wandering albatross really wander, or does it follow routes prescribed by environmental factors?

To plan ways in which to use natural resources properly and to expand basic knowledge of the earth's ecologies, more extensive information on the biological and physical factors of the natural environment is obviously essential. It is also essential that such information be acquired at a rapid enough pace to bridge the gap that now exists in data acquisition between the physical and the biological sciences. If biologists are ever to participate in the application, to their own disciplines, of measurement programs such as those demonstrated during the I.G.Y., they must be ready to understand and to use advanced instrumentation whenever it is expedient.

On the last day of October, 1962, seven Canada geese, each wearing one of our transmitters, were monitored at their mid-continent stopover before the last leg of their fall migration flight from the Arctic to Texas. For twelve days previously they had done little flying—they ate, preened, walked, or swam. At that time we could determine the location of each goose and what each was doing by the signals from their transmitters. During those twelve days we learned much about goose behavior at a stopover area.

When they left on that last long flight, and disappeared over the horizon, their signals were lost to us on the ground. But those signals were still being sent out into space where, if a satellite had been "listening," it might have been able to tell us infinitely more than we know now about the mysteries of migration.

## BIRD MIGRATION

There is little to attract visitors to the sparsely-settled Baraba steppe of western Siberia; the landscape is uninspiring, the summers are dry, and winter comes early and is harsh. Yet this otherwise uninviting land is the site of Lake Chany, which covers 1300 square miles, and creates one of the finest wildfowl nesting areas in the world.

Much of the marsh land near the lake has been reclaimed and is farmed by the Ural-Yenisei Agricultural Collective; but the men and machines who work the rich, black soil disturb the tranquility of the lake region only twice each year, first to sow and then to harvest. By early August, the harvesting crews have already come and gone, taking with them the crop of early ripening wheat. Thus there is seldom a witness, unless by a naturalist come for that purpose, to the yearly gathering of the Graylag Geese on the marshy south shore of Lake Chany.

The geese have mated during the spring, and have passed the summer incubating the eggs and rearing the six or so goslings born to each couple (life-long monogamy and familial togetherness is the rule among geese). One evening in early autumn—August is early autumn in Siberia—the geese cease feeding and splashing among the heavy growth of reeds that mask much of the lake.

A day or so of general restlessness follows, and ther. the birds begin to gravitate to one particular point on the lake shore. As if drawn by a magnet, they waddle and swir , from points all around Chany. Gathering by tens of thousands, they wait briefly until all the families have arrived, and silently set off on one of the least understood events in the natural history of birds, the yearly migration.

## AN ARMY ON THE MARCH

Many types of birds nest along the shore of Lake Chany, and in the fall migrate to various destinations. In the case of the Graylags, the goal is the great Ganges Valley about 2000 miles away. In making the passage, some of the most forbidding terrain in the world will be crossed. Often flying at altitudes in excess of 16,000 feet, the birds will clear range after range of mountains, skim above desolate high plateaus, and battle fierce Himalayan gales. They will complete the trip in about three weeks.

Off hand, three weeks does not seem like a particularly speedy journey for these wildfowl, who can cruise at 45 miles per hour all night long, and who are capable of speeds up to 70 m.p.h. But allowances must be made, because when they begin their journey, the Graylags cannot fly. The geese start southward almost as one, a silent army some five miles wide; in the next ten days they will cover somewhat over 100 miles—on foot.

In *The Great Migrations* Georges Blond describes this marching exodus of the Graylags. The birds must walk at first, he points out, because the overpowering urge to migrate comes upon them while they are still molting. The old wing feathers have been shed and the new growth is not complete.

Yet if the geese would only wait another 10 or 12 days at Lake Chany, they would be able to fly over their walking route in less than three hours. Why then do they strike south at such an unpropitious time? Or, to raise the larger question that ornithologists have yet to fully answer, how does any migratory species know when the time has arrived to be moving on?

## THE PUZZLING URGE TO MIGRATE

Broadly speaking, of course, migratory birds travel for self-preservation. Unlike the sedentary species which are able to cope with changing conditions of weather, food, shelter and the like, the migrants must either seek more favorable locales or perish.

It appears unlikely, however, that changes in temperature, food supply, or cover vegetation are, as biologists once thought, solely responsible for launching the birds on their way. There are many known cases when a premature change in the weather has wiped out entire flocks, because the time to migrate had not yet arrived.

The birds simply sat and froze to death. Another theory has had it that sex hormones, which are produced in greater and lesser quantities at certain times of the year, were chiefly responsible for initiating migration. Injection with male sex hormones, for example, causes an increase in fatty tissue in many migratory species. It is from fat deposits that birds derive the energy needed for their arduous flights. But as with changes in the environment, the level of sex hormones cannot be the sole determining factor of when to migrate. It is known that castrated birds will also migrate at the appropriate time.

Currently, one of the most favorably received theories is that a change in the activity of the hypophysis gland, which stimulates the gonads and performs other physiological functions, is the triggering mechanism.

This glandular change is thought by some ornithologists to be influenced by the seasonal increase or decrease of daylight hours. Yet like all previous theories, there are exceptions to this one also. Birds in the tropics begin to migrate although changes in the photoperiod are so slight as to be nearly imperceptible. Thus, it is likely that different species respond to different clues, and many biologists are of the opinion that some of these clues are so subtle that, to date, only the birds are aware of them.

Many of the Graylags are lost during the walking phase of their migration. Villagers, and their dogs, waylay and slaughter thousands of the birds for food. But the loss is proportionately small, and the tenth night's march brings the survivors to Lake Kulunda, southeast of their summer home. Here the birds rest for three days.

On the fourth day, there is considerable pre-dawn activity. The wing feathers are now grown in, and every family makes short test flights. Then, with the first light of dawn, the entire army of geese takes to the air. For as long as their route keeps them over the steppes, the geese fly in broad waves, one parallel line after another, each line stretching over several miles. But at the edge of the steppes lie the mountains, and to fly over this terrain a more flexible formation is needed. Thus, an observer on the ground hears a sudden increase in gabbling, or "talk"; the long lines break into fragments, and each fragment smoothly adjusts itself into the familiar "V" formation of migrating wild geese.

Once the geese take wing, they complete their trip in good time. Flying only while it is dark, they cover from 160 to 220 miles each night, with one night's flight extended to 300 miles so as to avoid landing on a spur of the Gobi Desert. Storms, treacherous down drafts, and birds of prey will take a heavy toll before the geese reach the Ganges Valley. In the early spring, the Graylags will retrace their route and return to the breeding grounds at Lake Chany.

## NAVIGATING BY DAYLIGHT

Whether the migratory route is 9300 miles long, as that travelled by the American Golden Plover which flies from north of Newfoundland to the Argentine pampas, or the more modest 2000-mile trip of the Graylags, perhaps the most puzzling aspects of these yearly flights is the navigation involved.

When the birds are near their breeding grounds or wintering area, it is probable that they orient themselves by sighting familiar landmarks. It is known that migratory birds possess a remarkable visual memory. But what about the vast stretches between summer and winter homes? It is most unlikely that any bird could remember enough landmarks, and their relationship to other landmarks, to navigate thousands of miles. In any case, landmarks are of little value to night-flying birds. Moreover, it is known that some, and perhaps most, migratory species can orient themselves and head for home even when released in distant and completely alien lands.

Considerable experimental evidence suggests that some birds may look upward, rather than to the ground, for orientation. German researchers detained birds whose restlessness indicated they were ready to migrate. The birds were kept in a closed room; periodically sunlight was allowed to enter from one or another opening in the walls and ceiling. The birds always maintained the same general orientation to the sun, just as they would have had in flight. Diffusing the light, as if the day were overcast, tended to confuse the birds. Since the position of the sun shifts during each day, it appears that migrants must have some means, perhaps a "metabolic clock," to correct for this shift so as to remain flying on target.

## STEERING BY THE STARS

It seems that many night-flying species guide themselves by the stars. European warblers used in one series of tests were exposed to the night "sky" of a planetarium. The flight pattern of these birds in the wild was known, and as their apparent location was changed (by altering the position of the stars on the planetarium dome) the birds oriented themselves in their cages to correct for these changes, just as they would do in flight.

These and other experiments into bird navigation are explained further by Dr. Wesley E. Lanyon in his book *Biology of Birds*. As Dr. Lanyon made clear, although naturalists now have some idea about the nature of the signposts used by migratory birds, the actual mechanism of navigation still remains obscure.

But, however they manage it, it is certain that the Graylags will cease feeding one evening this August, and leaving behind the plants, grubs and seeds so prodigally provided at Lake Chany, will unhesitantly begin to shuffle across the Siberian steppes, right on course for the distant Ganges.

## THE FLOWER AND THE BEE

'The might of Britain depends on its old maids.' Old maids keep cats. Cats kill mice. Empty mouse holes form good nesting places for bumble bees. Bumble bees pollinate red clover, clover feeds beef—beef gives Britons their drive and energy! Even if this chain reaction appears somewhat fanciful, we have abundant evidence that our crop economy is geared to this amazing strong, yet delicate, link between the plant and insect kingdoms—between *the Flower and the Bee.*

I want to discuss some facets of this remarkable relationship and to indicate current lines of research in this field, which are biased, as you might expect, towards man's attempt to exploit it and profit by it.

## POLLINATION

All flowering plants are highly specialised organisms, and one result of this is that there is a separation in space between the male and female organs in the flower (stamens and ovaries), so it becomes necessary for the pollen to be carried in some way or other over this spatial barrier. This transference and placing of pollen grains on the receptive stigma of the ovary is what we term pollination—whether or not fertilisation and seed set follows is entirely due to whether the right kind of pollen has been placed on the stigma.

Wind, water, birds and insects may all act as pollinating agencies and bridge the gap, not only in individual flowers, but between one flower and another or between one plant and another effecting cross-pollination.

Many floral forms make self-pollination virtually impossible and establish an *outbreeding* mechanism for the species.

This appears to be a present-day evolutionary platform in many Angiosperm species. But will it remain so? The dandelions and hawkweeds have already evolved to another method of reproduction, *not using pollen at all.* They are *apomictic,* i.e., their seeds develop without fertilisation. Although their flowers still exhibit a beautiful mechanism for cross-pollination by insects backed up by a liquid inducement of concentrated nectar, it is obsolete. Try it yourselves by cutting off the stamens and styles of a dandelion head: it will still set good seed.

*Entomophilous* flowers, i.e. those pollinated by insects, exhibit many features which we find pleasing—colour, scent, form—but bees only visit flowers for food, pollen and nectar for rearing their grubs, for their daily needs, and, in the case of the *honey bee* for winter store as honey or packed pollen. An inhibitory substance in the pharyngeal glands of the bee prevents *germination* of pollen in cells which would possibly ferment the honey and ruin the winter store. But we now know, as a result of many fine researches on the senses of the bees, that these other attributes (colour, scent, form) have a meaning for the bees and therefore assist in attracting them as pollinators.

The delicacy and precision of the bees' senses can be judged from the following facts:

## SENSE OF SMELL

Bees can distinguish, recognise and remember flower scents. They can pick out one flower perfume from amidst thirty others. They are also very sensitive to the scent they themselves extrude from their abdominal scent organ.

## COLOUR SENSE

They can distinguish, recognise and remember three regions of the visible spectrum and ultra violet. This latter is a very real 'colour' for the bees and some flowers which appear self-coloured to us, may be exhibiting a colour contrast for the bees' eyes. For example, in *Mimulus* and the *Yel-low Flag Iris* low Flag Iris the opening to the flower tube is shown by strong ultra-violet radiation from the petals at this point. The bee will perceive this as a guide to the nectar of ultra violet on a yellow ground.

## SENSE OF TASTE

Bees can appreciate degrees of sweetness and even show preference for certain mixtures of sugars. Dr. Raw of Rothamsted reported that consistent preferences were shown for single sugars in the following descending order:

- Sucrose;
- Glucose;
- Maltose;
- Fructose;

But when a mixture was offered, an anomalous high preference for sucrose-glucose-fructose solution, which was quite different from predicted preference on the basis of an additive effect of the constituent sugars.

Recently it has been shown that the families which contain so many 'bee' flowers actually have this mixture of sugars in their nectar.

Also in several 'climax' groups of plants such as the Crucifers (cabbages), the nectar is composed of simple digestible monosaccharides-fructose and glucose—in virtually the same proportions as are found in honey. So, many of these highly evolved 'modern' plants are producing a nectar which is technically a 'honey' already. The digestive invertase of the Hymenopteran (bee) visitors is not required! The nectar can be assimilated immediately by the insect. Has this any biological significance?

To return to the bee senses. Von Frisch has proved that the bees can communicate their sensual experiences to their hive mates by dancing and hence recruit workers to forage a particular crop.

But this emphasises a difference between the hive and bumble bee. Honey bees do not connect the flowers with food instinctively and have to learn that their food is in the flower. Whereas a novice bumble bee worker on her first foraging flight will fly straight to a flower.

## TIME SENSE

This is highly developed in the honey bee. There is some evidence that this may be employed in pollen collecting. It has been found that different species 'present' their pollen at different times of day, e.g.:

| *Poppy* | *Early morning (peak 6* |
|---|---|
| | a.m. (5-10 a.m.) |
| Dandelion | morning (peak 10-11 a.m) |
| Raspberries and | all day |
| blackberries | afternoon |
| Broad beans | (12 noon-3 p.m) |
| Vegetable marrow | night (10 p.m.-3 a.m.) |

In a number of species the rhythm of pollen presentation shows a positive correlation with the rhythm of pollen collection by the bees. For example, in the Water Plantain pollen is presented from 11 a.m. to noon. Just before 11 a.m. bees will arrive and scout among the plants to see if any flowers are open (flower opening and anther dehiscence is simultaneous).

It is not only bees which have this time sense. In the early morning, hover flies have been seen clinging, head downwards, on the undehisced anthers of maize, lipping around the terminal pore of the anther and waiting for opening time. So this time sense can be employed for the mutual benefit of both plant and insect for in all cases of bees collecting pollen it was seen that pollination was also effected.

## RAPIDITY OF LEARNING

This is a striking feature of honey-bee psychology. To quote one, e.g., a 'Novice' bee on maize searched in between the glumes of the male flowers for 'something', then by chance passed under a group of stamens and dislodged some pollen on to herself and immediately turned on to her back and began to collect pollen.

## INDIVIDUAL TRAITS

Bees exhibit quite individual traits in collecting nectar and pollen. The pollen collectors are like women shopping, some neat and expeditious in their pollen packing, others much slower. One working the broom by shouldering between the standard and keel of the young flower, was able to remove all the pollen without springing the flower.

Then there are those bees which the American orchardists call 'stinkers', which take the nectar from apple-blossom while clinging to the outside of the flower, thus failing to pollinate them.

And here is a link between bee physiology and flower physiology. In early spring, fresh pollen is urgently required by the bee colonies for brood rearing. Flowers blooming in February, March and April require a temperature from 8·8-10° C. (50° F.) for free anthesis (free production of pollen). 10° C. has been stated to be the limiting temperature for (hive) bee flight; inaction of the wing muscles is said to occur below this. It is seen that the February-April flowers have pollen available at a slightly lower temperature than this.

So there is a pretty balance here, the bee food being available in the flowers at just about the same temperature level that the pollinating insect can fly. There is yet another parallel. Crocus and snowdrop flowers open at 4° and 5° C. respectively if in sunlight, and the bees were seen working these crops at 4° and 5·7° C. respectively if it was sunny. So direct sunlight compensates for the lower air temperature at a comparable level for both insect flight and flower opening.

There are other attributes of bee psychology which facilitate pollination. *Constancy* is one of the foremost of these. Coupled with the honey bee's sense of location it is immensely valuable to agriculturalist and orchardist. As long as the food—be it nectar or pollen—is plentiful, the bees will continue to visit a crop exclusively. But there is a danger in this very constancy, for, other things being equal, bees will be attracted to crops with the strongest nectar. They may even select individuals in one crop.

Take the case of the orchardist. Cross-pollination is essential for fruit set in certain 'self-sterile' varieties of apple (or plum), so orchardists frequently plant a mixture of varieties in the expectation that the bees will effect this.

But see what happened in an American plum orchard. Three varieties were planted, the sugar concentrations of whose nectar differed considerably:

| | |
|---|---|
| Maynard | 10·1 per cent sugar in nectar |
| Eldorado | 14·9 " " " " " |
| Milton | 28·4 " " " " " |

*Milton* attracted most bees and was well pollinated, and note that it would receive pollen from its own race. Now, if the variety is self-fertile, this is satisfactory, its 'pollination potential' is high if honey bees are the pollinators. If however, taking a hypothetical case, the variety was selfsterile its pollination potential would be very low.

This would also be so if it was self-sterile (or fertile) and had very weak nectar. It would only succeed in being cross-pollinated if it could compete on even terms, as regards nectar concentration with neighbouring varieties for the bees' services.

So a plant may miss the biological bus in this way. Another example of this is recorded by Vansell in Oregon.

Some plants have nectaries outside the flowers, e.g. on leaves, petioles, stipules, the so called extrafloral nectaries:

| Plant | Nectar | *Percentage* sugar |
|---|---|---|
| Common Vetch | blossom | 22·6 |
| " " | stipular | 56·5 |
| Hungarian Vetch | blossom | 25·2 |
| " " | stipular | 47·7 |

The bees work stipular nectars assiduously and neglect the flowers-these species have also missed the bus!

Another example of biological misfortune occurs in the loco weed (*Astragalus lentiginosus*), Nevada, and the buck eye (*Aesculus californica*). Both produce poisonous nectar, containing selenium and alkaloid, respectively. These are deadly to bees and *Aesculus* is particularly dangerous because the nectar is very concentrated (almost supersaturated). Even the movement of a flower may prejudice its pollination potential.

For example, when a bee alights on a legume flower, she 'trips' it. The wings and keel are pressed down and the stamens spring up and shower her with pollen. This latter action in Alfalfa (*Medicago sativa*), is so violent and deals the honey bee such a shrewd blow, that some workers say she is disconcerted and deterred from further foraging. Dr. Nielsen in Denmark has found a strain where the staminal column is nearly straight and does not strike the bee. It may have an economic future.

## INTERDEPENDENCE OF FLOWERS AND BEES

Bees will exploit any flower in which they find 'good' nectar (probably that containing over 18·0 per cent sugar) but some flowers with deepseated nectar in long-tubed flowers are dependent on bees for pollination. This link may be very close, e.g. in certain *Aconitum* spp. (Monkshood), where the geographical range of the species in Europe and the range of certain long-tongued *Bombi* is said to coincide. There are other close links.

There is sometimes a coincidence of the 'imago' stage of some wild bees and the flowering of certain plants, a classic example being the Pickerel Weed (*Pontederia cordata*) and the

Pickerel Weed bee (*Halictoides novae-angliae*). The flowering of the former coincides with appearance of bee, and it is believed that the bee does not visit any other plant for nectar or pollen. It is a monotropic bee.

Others are 'oligotropic', the Vernal Andrenas visiting only species of Willow and the Autumnal Andrenas visiting only species of Golden Rod.

Dr. Gorton Linsley reports that perhaps the most exquisitely precise bee-flower associations occur in the Californian desert between a species of *Oenothera* (Evening Primrose genus) and a black solitary bee (Anthophorid). Before dawn it is cold, dark and silent. As the first light comes, the flowers of the *Oenothera* open their petals and face the place where the sun will rise.

As the sun strikes them, the stamens, which stick out in front, release their pollen. At this very moment, the black bees arrive, land on the stiff stamens, and whip off the pollen (and pollinate the flowers).

The desert is bright with flowers and humming with bees. In an hour, the flowers have collapsed limp and dying in the hot sun, the bees have disappeared into cool crannies and all is still and dead once more.

## BEE BEHAVIOUR AND THE EVOLUTION OF PLANT SPECIES

The behaviour of bees may vitally affect the evolution of plant species. Related species growing wild may occupy the same range without apparently hybridising but when brought into breeding plots will hybridise freely! So the barriers to interbreeding cannot involve incompatability or hybrid sterility.

But it may be that interspecific pollination is not occurring in the wild because the pollinating systems of the plants are acting as isolating mechanisms. Verne Grant (U.S.A.) says this could be because the floral mechanisms of the two species may differ, so that a bee pollinating one, cannot pollinate the second—he calls this *mechanical isolation* the bees are 'specific', or constant to a species.

Here cross-pollination is mechanically possible but does not occur because of bees' flower constancy. He calls this *'ethological' isolation.*

This latter type does exist among three subspecies of *Gilia.* Two of these have lavender flowers and sweet scent and the third smells of creosote, and is deep violet.

| *Subspecies* | *Colour* | *Scent* |
|---|---|---|
| Gilia capitata capitata | lavender | sweet |
| G. capitata tomentosa | lavender | sweet |
| G. capitata chamissonis | deep violet | creosote |

Eighty plants of these three were randomised in a plot. If a bee visited the 'creosote' plant she remained constant to it (although investigating flowers of the other two). Others worked the two 'sweet' subspecies with free interchange. This isolation held during spring, later in summer some bees on 'sweet' subspecies occasionally visited the 'creosote' flowers and vice versa but this was not customary. These field observations were reinforced by sowing the seed of all the plants and classifying the progeny as subspecies or hybrids.

The two 'sweet' subspecies produced 58·7 per cent hybrids, but there was no contamination of these two by pollen from the 'creosote' subspecies. So here one subspecies was, by the bees' action, kept genetically pure, while heterozygosity was impressed on the other two. This kind of bee behaviour might possibly be a useful tool in the hands of the geneticists.

Two of the most important fodder crops in the world today are alfalfa and red clover. The demand for red clover seed exceeds the supply, and red clover is dependent on long-tongued bumble bees for its pollination. Realisation of this important fact has been slow, resulting in such mistakes as the introduction of red clover into New Zealand, where there were no indigenous *Bombi.*

The New Zealand story is particularly interesting. The error was remedied by the importation of four species of British bumble bees with phenomenal success. (But even this was faulty, because one of these was *Bombus terrestris* which has a short tongue and robs the flowers of their nectar by biting through the base of the corolla tube without effecting pollination.) Yields of 400 lb. of seed per acre was averaged on the plains and, in the valleys in the hills, a prize field reached 1100 lb. per acre. Then in the plains the yields gradually dropped to 170 lb. per acre. What had happened? The bees were still there, but they had become adapted to the New Zealand climate. When first exported, they produced colonies with plenty of workers at the time when the clover was in flower.

Then, gradually, their habits changed. The queens ceased to hibernate and remained active all the year, so that, to (quote Mr. Anderson of the New Zealand Grassland Division of the D.S.I.R., 'the colonial maxima are now much more widely spread over the year'. So the problem had to be tackled again. Dr. Elwood Montgomery suggested that the American species *B. impatiens,* which is specially adapted to life on the plains might be introduced. The Department of Agriculture endorsed this policy but the beekeepers opposed it, because they thought it would reduce the honey crop, and they also feared the introduction of acarine disease.

Now, the tube of the flower is too long for the honey bee to exploit red clover for nectar, although they may visit it for pollen. Nevertheless, the introduction was stopped, but an educative campaign initiated by Dr. Montgomery bore fruit, and reconciled the beekeepers to the government policy. But Mr. Anderson says the scheme has not yet been implemented.

What of the position in other countries? Today, in parts of America and in Scandinavia the natural population of bumble bees has been so diminished that the position is becoming critical.

Intensive methods of agriculture are responsible for this, involving, as they do, the destruction of nesting sites by the cutting down of woodlands, elimination of 'rough' ground and the use of fencing posts as boundaries instead of hedges. Other contributory causes are, the spraying of crops with insecticides when they are in bloom-particularly orchard trees, and the diminution of the natural flora by use of selective weed killers especially along roadsides and grass verges.

The majority of papers at the 'Symposium on pollination' held in Copenhagen last August dealt with the problem of establishing adequate numbers of the right species of bumble bees to pollinate alfalfa and red clover. Indeed, in the States they are experimenting in the domestication of bumble bees. As the 'fall' approaches, wooden boxes, filled with sphagnum or dried grass and baited with corn, are placed along fences and edges of woods. The mice are attracted to hiberate therein and they 'work up' the moss into 'desirable residences' for the queen bees next spring! Both Danish and American workers report a substantial increase in worker bees in the domesticated colonies.

## THE RAIN FOREST

My two favorite kinds of places in this world are coral reefs and rain forests. I don't know how I would vote if I had to choose between them, had to decide that I could go only to reefs and never to forests again, or vice versa. My idea has long been to live by a broad, sandy beach with a rain forest behind me and a coral reef offshore before me, with either open to exploration or contemplation. Maybe someday I'll achieve the ideal. There are places where it is possible-some of the islands of the South Seas, or Trinidad and Tobago in the West Indies, for instance.

Certainly they are different enough, the rain forest and the coral reef.

They have no inhabitants in common, nor even any general kinds of inhabitants in common. There is no way of comparing their appearance, either. The reef world is bright with color and movement. The forest is all green and brown, dim and still. The reef is Baroque, the forest, Gothic.

They have this in common: one is the product of the most favorable possible conditions for life in the sea, the other for life on land.

Sunlight, warmth, moisture, are always abundantly present, stable, and favorable throughout the year. Moreover, they have remained about the same over long stretches of geological time. As a result, there is a tremendous variety of different kinds of organisms in both environments—and these organisms, among themselves, have developed a tremendous variety of different kinds of relationships.

"Rain forests" and "jungle" are frequently taken to mean the same thing. But I have never liked the word jungle. It has all the wrong connotations. You hack your way painfully through the lush vegetation of the jungle, dripping sweat in the steam-bath atmosphere; snakes hang from trees and lurk under foot; leopards crouch on almost every branch and there is always a tiger just beyond the impenetrable screen of foliage. There are hordes of biting, stinging and burning things. The jungle is green hell.

I doubt that there is any place, outside of books and movies, where all these conditions are combined, though certainly there is plenty of nasty and difficult country, both in the tropics and out. The thickest tangle of vegetation is the second growth that springs up after rain forest has been cleared. Everywhere in the tropics, people follow a slash-and-burn type of agriculture. Trees are felled, allowed to dry, burned, and crops planted among the charred logs.

Sometimes crops are harvested for two or three years or more, but presently the land is abandoned and a new area cut. The abandoned clearing is taken over by a thick tangle of vegetation that for several years may be almost impossible to penetrate except by slow and painful cutting with a bush knife. Small mammals and rodents multiply in this vegetation, providing abundant food for the snakes that move in. Such places sometimes harbor swarms of mites, ticks, flies, mosquitoes and a wide variety of stinging things. It's about as close to a green hell as you can get.

The true rain forest, untouched, almost untrodden by man, is a very different sort of place. The forest floor is open, carpeted with the richly variegated browns of many different kinds of fallen leaves, sometimes brightly spotted with blue or red or yellow from flowers that have fallen from unseen heights above. The carpeting is thin, easily scuffed away to show the red lateritic clay soil so characteristic of the equatorial regions. There is no thick accumulation of leaf mold like that of northern forests, no rich accumulation of humus. The processes of decay are too fast to permit much organic accumulation in the soil.

There is little vegetation on the forest floor since the light is too dim for plants. There is a thin growth of tree seedlings (which have no chance to grow unless some catastrophe to a forest giant should open space), ferns, sometimes dwarf palms, or scattered thickets of huge-leaved aroids, the sort of plants that also grow well in the dim light of hotel lobbies. But basically, the forest floor is open, with vistas of a hundred feet or more, vistas framed and closed by the straight trunks of the trees that disappear into the vaulted green canopy that they support above.

The cliche often used for the forest is "cathedral-like." The comparison is inevitable: the cool, dim light, the utter stillness, the massive grandeur of the trunks of forest giants, often supported by great buttresses and interspersed with the straight, clean columns of palms and smaller trees; the gothic detail of the thick, richly carved, woody lianas plastered against the trunks or looping down from the canopy above. Awe and wonder come easily in the forest, sometimes exultation—sometimes, for a man alone there, fear. Man is out of scale: the forest is too vast, too impersonal, too variegated, too deeply shadowed. Here man needs his fellow man for reassurance. Alone, he has lost all significance.

The rain forest is perhaps more truly a silent world than the sea. The wind scarcely penetrates; it is not only silent, it is still. All sound then gains a curiously enhanced mystery. A sudden crack—what could have made it? An inexplicable gurgle. A single clear peal—that was a bird, probably a trogon. A whistle, impossible to identify. But mostly silence. The silence sometimes becomes infectious; I remember sometimes trying to blend into this world by moving along a trail without rustling a leaf with my feet or popping a twig.

In contrast with the reef, it is a monotonously-colored world. Everything is some shade of brown or gray or green. I have lugged cameras loaded with color film all day without finding anything that seemed to warrant color photography, and in despiration, got my companions to wear red kerchiefs and blue jeans so that they could provide color contrast as well as "human interest." But I always found color photography difficult in the forest where the dim light requires long exposures, and the light itself is greatly altered by being filtered through the thick, green canopy. Only by taking advantage of the margins of clearings, or by using a flash, can you be sure of results.

Perhaps I am making the forest sound too easy, too open, too cathedral-like, overdoing my rebellion against the idea of jungle. It is difficult to give an objective description, to convey an accurate impression of a landscape like the rain forest which may, in one person, arouse awe and wonder, and in another, fear and hatred. P. W. Richards, in his book, *TheTropical Rain Forest*

Tropical Rain Forest, has justly remarked that "tropical vegetation has a fatal tendency to produce rhetorical exuberance in those who describe it." The exuberance mostly tends toward the green hell side, but perhaps I have overdone the cathedral analogy.

I doubt whether the rain forest is anywhere easy to penetrate for any great distance. There are always obstructions: occasional fallen trunks, sudden tangled thickets, and above all, stretches of swamps and countless streams. Sometimes the streams are small, clear, shallow sandy brooks, looking no different from the forest brooks of New England, and easily negotiated.

But sometimes they are broad rivers, sometimes they move sluggishly over bottomless mud, sometimes they are choked with impenetrable masses of fantastic vegetation. The green hell analogy becomes vivid enough in these forest swamps. They are the reason that man has had so little success in making trails or roads through the forests; why he clings to the major rivers either for exploration or trade.

The world I have been describing bears no obvious resemblance to the world of the coral reef. This is partly, of course, because we have been looking at the forest from the bottom, from the point of view of walking on the forest floor, while we see the reef from above, floating over it. I loved the rain forest from my first encounter with it when I had just left college and was working at a research station of the United Fruit Company in Honduras, but I think I never really appreciated the forest until the years of yellow-fever study in Colombia, when I worked comfortably from platforms high in the trees—which gave a quite different perspective. But in any event, the similarities only become apparent when we examine the two habitats as biological systems, and for this we need a biological description of the forest.

Rain forest is the type of vegetation that occupies the lowland tropics in regions of high rainfall, where the rain is fairly evenly distributed throughout the year. The minimum rainfall to support such a forest is generally considered to be about eighty inches a year, though usually the rainfall is much higher, well over a hundred inches.

The seasonal distribution is as important as the total amount: where there is a pronounced dry season of several months with little or no rain, the forest changes considerably in character and is generally called a monsoon forest.

In such a forest the trees are not so tall, there is more undergrowth, and many plants drop their leaves during the dry season. The forest also changes with mountain altitude and various kinds of montane forests can be recognized. Heavy forests occur in some high rainfall areas outside of the tropics, in southern Chile, in New Zealand, and along the coast of the state of Washington in North America; but these again have a different character from the tropical forest and are best considered as a separate vegetation type.

There are three major areas of tropical rain forest: the American, the African and the Indo-Malayan. They cover all of the land masses crossed by the equator except the east coast of Africa.

The American forest is by far the largest and most continuous, covering most of the Amazon drainage in central South America and extending south on the inner side of the Andes in Bolivia into the drainage of the Plata, and north in Colombia into the drainage of the Orinoco. This is hardly separated by the northernmost ranges of the Andes from a Pacific strip that follows the coast from Ecuador to Panama and continues, in Central America, along the Caribbean coast almost to the line of the Tropic of Cancer in Mexico. There is an isolated stretch of rain forest along the southern coast of Brazil, and rain forest once covered much of the West Indies, though now there are only scattered remnants.

The African rain forest is the smallest of the three, and there is considerable debate both about its present limits and its former extension. It covers, essentially, the central drainage of the Congo, with a north and west extension along the gulf of Guinea to Liberia.

The Indo-Malayan rain forest is the most fragmented. It covers most of the large islands of the East Indies-Sumatra, Borneo, Celebes, New Guinea, the Philippines-and the Malay Peninsula, with outlying areas on the west coast of India, in Burma, on the coast of Indochina, and along the coast of northern Queensland in Australia.

In structure and appearance, the forests of the three areas are very much alike. The taller trees reach a height averaging about 150 feet, though individual trees more than 200 feet in height are not uncommon. The tallest reported rain forest trees are somewhat less than 300 feet. Rain forest trees are thus in general taller than trees in the temperate forests of Europe or North America, where the average for taller trees in the least disturbed forests is around 100 feet, with 150 feet an exceptional height. But trees in the tropical forest do not reach the gigantic proportions of the California redwoods or the Australian eucalyptuses. The tallest measured sequoias reach 364 feet, the tallest eucalyptuses, 350 feet.

The rain forest everywhere has a multi-storied canopy. The layers of the canopy, like the depth zones of the sea, are hard to define since they are not sharply separated, but it is customary to refer to the upper, the middle and the lower zones of the canopy.

This multiplicity of tree layers reflects the great variety of different kinds of trees that go to make up the rain forest. Most temperate zone forests are made up of one or two or a very few kinds of trees, with a maximum, in the Appalachian forests of the eastern United States, of perhaps 25 species of trees. Probably fifty species of trees is about the minimum for any rain forest, and in most places there are many hundreds of species.

Alfred Russel Wallace, who spent many years exploring both the Amazonian and Malayan regions wrote, in his book *Tropical Nature*: "If the traveller notices a particular species and wishes to find more like it, he may often turn his eyes in vain in every direction. Trees of varied forms, dimensions and colors are around him, but he rarely sees any of them repeated. Time after time he goes towards a tree which looks like the one he seeks, but a closer examination proves it to be distinct. He may at length, perhaps, meet with a second specimen a half a mile off, or he may fail altogether, till on another occasion he stumbles on one by accident."

Because of this immense variety, the catalog of rain forest trees is still far from complete, and anyone collecting botanical material in remote areas is liable to turn up species of trees unknown to science. And the trees belong to almost the whole range of plant families. Families that in the north are known only as herbs—the *Compositae*, or daisy family, for instance—are represented by trees in the rain forest. Even grass takes on a tree form in bamboo; and ferns, in the tree ferns.

There is a great development of woody vines—lianas. A considerable proportion of the foliage in the canopy, sometimes nearly half of it, is from the great vines that are supported by the trees, and these woody vines, like the trees themselves, belong to a great variety of species and to many different plant families.

Trees, woody vines and epiphytes are the characteristic life forms of plants in the tropical forest; and of these the epiphytes, the plants that perch on the branches and trunks of trees, are probably the strangest to the observer from the north. Lichens and mosses grow as epiphytes in northern forests, but in the rain forest the branches and trunks of the trees are covered with a bewildering variety of other plants: ferns, orchids, peppers, cactuses, bromeliads.

In all, something like 33 families of seed plants and ferns are represented by epiphytes in the rain forest flora.

Epiphytes grow in the rain forest, but without access to the soil they are faced with a water problem. Their niche on dry branches high in trees is in a way a sort of micro-desert. Many of them, the cactuses, for instance, are relatives of typically desert plants and have the fleshy look of desert plants. There is a good reason for this, since in both situations it is important for the plant to be able to conserve water, hence the frequency of succulent bulbs and leaves, as in the orchids. The bromeliads, as I mentioned in the last chapter, have solved the water problem by forming watertight tanks where they can collect their own supply, and where also they can get food from the rotting organic matter. The bromeliads are often very numarous, forming, as someone has remarked, a "marsh in the treetops.

The epiphytes are also faced with mineral problems. They must get the salts they need from the extremely dilute solutions in the rain that washes them, or from the humus and debris that collects in the cracks of the bark of the host trees or in the tangle of their own roots. Somehow they get sufficient minerals from these sources. Frequently the epiphytes live in a close symbiotic association with fungi, and the thin mycelial webs of the fungi help both partners in food collection. Fungi live in close relationships with the roots of many kinds of seed plants, forming associations that are called *mycorhizae.*

The roots of epiphytes also, with surprising frequency, are used as nesting sites by the ants that abound in the forest; and it has been suggested that the epiphytes have an additional source of food in the material that the ants are constantly hauling into their nests. This too would be a symbiotic relationship: the ants get a fine, well-protected home, built by the plant, and provide food by way of rent. Perhaps they also provide defense since many of these ants sting fiercely. They certainly defend the epiphytes valiantly against stray humans trying to make collections.

The rain forest crawls with ants of many different kinds, occupied with many sorts of business. Close associations between particular kinds of plants and particular species of ants are quite frequent. Sometimes the ants clearly serve to protect the plants that provide them with nesting sites in hollow stems. One learns to avoid brushing against the trunks or foliage of certain trees, like Cecropia, with the same care that one learns to avoid poison ivy or poison oak in more northern situations. With the ant-protected trees the fiery consequences of transgression are immediate as well as painful.

The ants are one example of the incredible abundance of insects in the rain forest. It is an abundance of kinds, rather than of individuals. Within a range of about ten miles of our laboratory in Colombia, we found 150 different species of mosquitoes (there are only 121 species known from all of the United States and Canada). But you may get more mosquito bites in northern woods than in tropical forests. In northern woods, the mosquitoes biting you are apt to be all the same kind; while in the rain forest, almost every bite will be from a different species of mosquito—if that is any comfort.

The task of collecting, naming and describing all of these insects is endless. Every collector brings back new things and no one really has any idea how many different kinds there are, which leads people, in guessing about how many species of insects there are in the world, to give figures that vary from one million to ten million. Whatever the figure, it is a big one, with a respectable proportion inhabiting the rain forest.

The biggest known insects are found there: for wingspread, butterflies and moths; for bulk, rhinocerous beetles; for length, walking sticks. But there are relative giants among almost all insect groups: cockroaches that look like small turtles, big flies, big wasps, monstrous grasshoppers.

There are also big mammals, elephants in the old world, tapirs in the new, and there is a gigantic frog in the Congo forest with a body approximately a foot long. But the woolly mammoths that roamed Europe, Siberia and North America until almost recent times were bigger than elephants, and in general with mammals, and birds, the forest representatives do not strike one as particularly big.

As a matter of fact, forest species—forest deer, for instance—tend to be somewhat smaller than their cousins in the savanna or outside the tropics. Perhaps it is only the cold-blooded animals —anacondas and boas and pythons, for example—that find a special opportunity for bigness in the rain-forest environment.

The color is mostly high in the canopy: the flowers of the trees, lianas, epiphytes; birds and butterflies. There is, it seems to me, a special tendency for animals, especially birds and butterflies, to be colored in metallic blues and greens. The wings of a great morpho butterfly, flashing in the sun, are especially famous, but many kinds of day-flying insects have metallic colors, even mosquitoes! Species that fly at night and species of the forest floor zone are, however, apt to be dull or at least softly colored. With the butterflies, as the with the birds, the brilliance tends to be a male characteristic.

The structure of the rain forest and the appearance of its inhabitants is much the same in America, Africa and Indonesia, but the three areas have had independent evolutionary histories, and are made up of quite different kinds of plants and animals. The only animal I can think of (besides man) common to all three regions is the leopard, which is hardly distinguishable from the jaguar of the New World. But though these big cats are at home in the rain forest, they are far from restricted to it, or to the tropics either for that matter. The leopard, in prehistoric times, ranged far into Europe, and the jaguar ranged all over both American continents until man managed to exterminate it in the settled regions.

The African and Indo-Malayan forests are more similar to each other in composition than either is to the American forest. They are separated now by the Indian Ocean within the tropics, and by the barriers of the Sahara and the Himalayas to the north. But the Himalayas are the youngest of mountains and the Sahara was once wooded, so that a few million years ago communication between Africa and Malaya was easier than now. But the Amazonian forest, all this while, was completely isolated by a sea barrier, only recently bridged by the isthmus of Panama.

Thus we find great apes in both the African and Indo-Malayan forests, gorillas and chimpanzees in the former and orangs and gibbons in the latter, but no great apes in tropical America. There are monkeys in all three regions, but the American monkeys belong to quite different families from the Old World monkeys, and have had a separate evolutionary history for a very long time.

Visitors to the tropics who expect to see monkeys everywhere are generally disappointed. In the rain forest, particularly, what one will see is very uncertain. The animals are there, all right, but they are not exhibitionists. They have learned, particularly, to be shy of possibly harmful human intruders. That this is learning, in part at least, is shown by animal behavior in places where the fauna has been long protected—in the park reservations and on the lands of Buddhist monasteries in the East.

One of the very few fully protected rain forest areas in the American tropics is a small island in Gatun Lake in Panama called Barro Colorado which was set aside in 1922 as a reserve to be used only for biological study. I only lived there for six months, and I probably saw more forest mammals in that period than in all my years of residence in other parts of the tropics.

Mostly you get only a fleeting glimpse of a band of monkeys as they take off rapidly through the treetops—they have spotted you first, and are having none of it. On Barro Colorado, the monkeys did not really become pals with the visiting scientists, but at least they didn't panic at the sight of a man. As a result, any time you went out on a trail, you were sure presently to come across a band of monkeys and could watch them for a while.

Coatis are longnosed tropical relatives of the raccoons—and they are even more charming, inquisitive and intelligent than their northern relatives. They are hated by hunters because, ganging up, they can kill the hunters' dogs, and they are not notably careful about poultry or gardens. Men and coatis are thus usually at odds, but on Barro Colorado they achieved a considerable mutual tolerance. If you stood quietly, the coatis would pay little attention to you while they went on about their business of rummaging among the leaf litter or scampering up the trees after insects and nuts.

Frank Chapman was there while I was, carrying on his life-long study of bird habits. He tried to devise a feeding station for birds that the coatis couldn't get at by hanging a cigar box from a wire trolley strung between a tree and the verandah of his little house, so that bananas, for birds only, could be suspended in midair.

It didn't take a neighboring coati long to figure that out. The coati, sniffing the banana from the ground, rapidly made for the tree, climbed it, did a tightrope act on the wire out to the cigar box, and got the banana. Chapman gave up and spent the rest of the time trying to fool the coati, with little success. He has told the story in his book, *My Tropical Air Castle*. As far as I know, coatis have not been investigated by psychologists, which seems a pity because they must have one of the highest I.Q.s in the animal kingdom.

Among the rain-forest animals, there are many survivors from the geological past. We tend to think of marsupials as primitive mammals that have survived and proliferated in the isolation of the Australian continent, except for the common and tough opossum of North America. But in the tropical American rain forest, there are dozens of kinds of marsupials. They are not as spectacular as the kangaroos and their relatives of Australia, but they are interesting enough.

There is a sleek water opossum (*Chironectes*) living on the margins of the forest streams; a bright-eyed woolly opossum (*Caluromys*) with lovely, thick fur; and many different species of tiny, mouse-like opossums (*Marmosa*) as well as several other genera. The sloths are another archaic group (and they look it) now confined to the trees of the American rain forest, though in the recent geological past there were also many kinds of ground sloths.

The catalog of ancient animal types surviving only in the rain forest is a long one, though most of the items on the list can be appreciated only by the zoological specialist. Because of this, the rain forest has sometimes been regarded as a sort of backwash, a refuge from the more strenuous and progressive parts of the biosphere. But the whole complex and wonderful system of adaptations within the forest is clearly a consequence of evolutionary forces operating within this environment. Thus, by another chain of reasoning one can come to the conclusion not that the rain forest is a backwash, but that it is the place where evolutionary change is most active.

It is in the rain forest that "jungle law" reigns supreme: the struggle for existence, nature red in tooth and claw. Here we find the most fantastic contrivances for catching food or for avoiding becoming food, the most perfect examples of animal camouflage. The struggle for existence is symbolized by the strangling fig which starts out as an epiphyte, a seedling growing high on some tree, sending down roots which reach the ground and grow until finally the host tree is smothered by the encircling fig, which then stands alone.

And I can think of nothing more devastatingly fierce, more irresistible, than a horde of army ants on the move, killing and dismembering every animal they encounter that cannot fly away or run fast enough. We had a snake pit in the laboratory where I worked in Honduras, where we kept many kinds of vipers for their venom, which was used for making antivenin. Once, I remember, we were invaded by army ants.

We tried everything we could think of to stop them or at least to make them change their course, but to no avail. The ants poured on in their tens of thousands, swept through our snake pit, and left us with a collection of bare skeletons. Life, then, can be grim enough for the forest inhabitants. Only the clever, the carefully protected, the extremely prolific, the most modern, seem to have any chance of survival.

Yet, scuttling through the leaves of the forest floor, and nestling in the debris collected around the roots of epiphytes high in the trees, are scores of kinds of cockroaches hardly different in any way that we can see from the fossils of their ancestors that lived three hundred million years ago in the forests of the Carboniferous period. Push over a rotting log and you may very well find a Peripatus: a soft, brown, delicate, multilegged, caterpillar-like thing that, on examination, turns out to be a very queer creature indeed. The ancestors of all the land arthropods, of the millipedes, centipedes, spiders, insects, the first animals to learn to live on land, must have been something like this. But this anachronism is still getting on very well in the warm, damp world of the forest, meeting (as far as we can see) unchanged the shifting hazards in hundreds of millions of years of forest life.

Of course, neither the rain forest nor the coral reef has any monopoly on anachronisms. The horseshoe crabs of the Atlantic coast of the United States might have crawled out of the most ancient seas; the tiny collembola that jump through the leaf mold of northern woods are hardly different from the earliest of insect fossils.

Every pond and every tidepool teems with survivors from the ancient past. I don't think we have any clear answer as to why some kinds of things have survived and other kinds perished as, with the onward flow of time, new kinds of organisms have come on the scene.

Our explanations are mostly circular: horeshoe crabs have survived because they were adapted or adaptable, that is, able to survive; trilobites became extinct because they weren't able to survive. One of the remarkable things is that very few of the major plant or animal types—phyla or classes—have become extinct. Proportions change—ferns and mosses give way to seed plants as trees, giving structure to the forests—but the ferns and mosses are still with us. Reptiles give way to mammals and birds as dominant animals in the forest scene, but reptiles are still with us. Change seems to be not so much a matter of absolute replacement as of diversification.

But with all these reservations and qualifications, the rain forest still seems to have more than its fair share of survivors from the past.

One can visualize the struggle for survival, the competition, the strenuousness of life in the rain forest. But then in the next instant, looking at this multitudinous accumulation of organisms one gets the feeling that there is so much warmth, so much light, so much moisture, so much food, that almost anything can survive and that almost everything does.

This is the paradox shared by the rain forest and the coral reef. Shallow, tropical seas and low, rain-drenched tropical lands with their associated swamps, lakes and rivers represent the most favorable conditions for the processes of life on the surface of our planet. The physical environment is not a problem except in special situations like that of the epiphytes of the forest where the collection of water or minerals may have particular difficulty. The problems of living things involve not so much the physical environment as other living things. When one moves away from the optimum of these two environments, the physical condition starts to become limiting.

The open sea has a less diversified biota than the reef because it has no solid substrate, no place where fixed organisms can grow; everything has to drift or swim. Descending from the surface of the sea, light and temperature become less favorable; going toward the poles in surface waters, temperature becomes less favorable.

Similarly, on land, as one moves from the rain forest to other types of habitat, various factors of the physical environment become limiting, present special problems which require special adaptations on the part of the organisms living in the habitat. Organisms must always cope with other organisms, but outside of the rain forest and the coral reef they must increasingly cope with climate too.

## Chapter 23

# Conservation and Economic Biology

### TURNING INSECTS AGAINST THEMSELVES

From that time in the distant past when a mosquito first dined on human blood, mankind has been engaged in a guerrilla war against insect pests. In recent years, the conflict has been escalated into all-out and total warfare. Famines caused by insect depredation of crops, and diseases transmitted by arthropod vectors, presently take millions of lives annually. Obviously, we dare not let up in our attack—but dare we continue? According to many conservationists and public health officials, we may win a Pyrrhic victory if we continue poisoning our wildlife and ourselves with the deadly insecticides now employed.

### OUTWITTING THE FOE

Clearly, new weapons are needed, and additions to the arsenal are being promised by scientists in many disciplines. Reports of synthetic hormones, aromatic chemicals and a variety of other agents, devices and techniques presently under investigation or being tested, indicate that the design for victory, as in conventional warfare, is based largely on outwitting the foe. For example, the old military maxim "Divide and conquer" has been given a new interpretation by the development of synthetic sex attractants. The strategy here is to divide the males from the females.

As is the case with many higher animals, at least some species of insects employ scent to attract mates. As is true of any good perfume, the fragrance is most effective when used sparingly. Male cockroaches, for example, are excited by as little as 30 molecules of the aromatic chemical produced by the female.

A perfumed female silkworm moth can seduce males from as far away as two and a half miles. It is interesting that in concentrated doses, the potent chemicals tend to confuse the males, or even to repel them—suggesting an obvious way that these attractants can be used as weapons. Indeed, a synthetic attractant has already been employed successfullly against the gypsy moth by spraying it over an area in such quantities that the males became confused, and many were unable to find the waiting females. Synthetic attractants also show promise as lures, drawing males to poisoned traps.

Not surprisingly, nearly every species uses a different scent as a sexual attractant, so the time-consuming process of extraction, isolation and analysis must be carried out for each attractant chemists want to synthesize. The "brute force" approach was taken by chemists determined to break the cockroach code. Clean dry air was blown through a container of female cockroaches, the air was collected as it emerged, and its volatile contents were condensed in a chilled container. After nine months of collecting, and with the unwitting aid of 10,000 virgin females, the experimenters were able to isolate only 12.2 milligrams of the pure attractant—enough, however, to identify the chemical constituants.

## INFILTRATING THE RANKS

An alternative approach to the strategy of destruction by sexual disruption is to release sterile males into a wild population. The sterility may be achieved by selective cross breeding, by the use of radiation, or by treating the insects with chemical sterilants. However the insects are made sterile, the tactic is to release enough of the males so that many of the fertile males will be deprived of mates.

Obviously, total eradication is not likely with this technique, but the release of sterile males might be employed as a holding action, keeping an insect population in check until the newer lethal weapons could be trained on the problem area. Of perhaps greater promise as a total control measure would be the release of males which carry a "daughterless gene." Such a gene would spread through a population and might eventually wipe it out.

## TURNING APPETITE TO ADVANTAGE

Just as an insect cannot help but respond to a sexual attractant, so too is it drawn unfailingly to food lures. These are chemicals which arouse the feeding response. One such chemical, combined with an insecticide, has already been used against the oriental fruit fly population on a Pacific island. The island was heavily infested with the flies until many small boards, coated with the lure and the poison, were dropped from the air. Within a short time, the flies were eradicated, and at a cost of only 50 cents per acre.

In the war against insects no holds are barred, and other control measures under investigation include the release of diseases lethal to these pests but harmless to higher animals, and the application of hormones which promote grotesque growth patterns and ultimate death. Several viruses, for example, are known which kill cabbage loopers, sawflies and other pests.

In application, diseased larvae of these pests are ground up, the mash mixed in a solution, and the solution sprayed over the crop. With this technique it takes only about ten infected caterpillars to protect an acre, and only one application is needed per season.

At one point in their life cycle, larvae respond to hormonal changes and pupate, to emerge later in the mature, winged form of the insect. A hormone has been synthesized which indefinitely delays pupating in many destructive forest insects. As a result of being sprayed with the hormone solution, the caterpillars simply continue feeding and growing larger. Eventually, they may grow so enormous that they pop, actually succumbing from a surfeit of tasty leaves.

## ALLIES IN THE INSECT RANKS

In his book *The Insects* Url Lanham notes that "Insects'

worst enemies are themselves." This fact has already been used to advantage as a means of controlling undesirable species.

Some years ago, the cottony cushion scale, an insect which attacks orchard trees, was brought under control by ladybird beetles imported in great numbers (they can be purchased by the quart) from Australia. Just recently, a species of parasitic beetles has been released in New Jersey to prey on gypsy moth eggs and larvae.

Gypsy moths have also been under attack in Pennsylvania, where their eggs are eaten by the larvae of tiny parasitic black flies, imported from Japan. Unfortunately, parasitic insects will seldom wipe out an undesirable species. Another approach to biological control, however, can be used to entirely rid an area of a pest. The ploy here is to import a species related to the pest which, because of competitive superiority, will drive the undesirable insects out. Such a program was successfully carried out in Sardinia, where the malaria-carrying *Anopheles labranchiae* mosquito was deliberately replaced by the hardier, and harmless *Anopheles hispaniola*.

It is apparent that many of the new weapons mentioned—and the listing is far from complete—are essentially tools and techniques for manipulating the unthinking responses "programmed" into insects. It is not likely that insects could evolve a resistance to a sexual attractant or a food lure without perishing in the wild state. Blind instinct may yet be the death of them.

## SCIENCE ATTACKS THE SCREWWORM

Today, more than ever before in its infant history, atomic energy is being utilized to advance the frontiers of science in the interests of peace and productivity. This is particularly true in the world of agriculture. Currently under way in this field is a revolutionary program—the use of nuclear power as a means of eradicating the screwworm, one of the worst menaces to livestock growers in the southern United States.

The deadly work of this pest represents a yearly loss of 20 million dollars to livestock raisers, particularly hard hit in the Southeast. To the cattleman especially, the fly is no longer simply an unpleasant menace—it has become a dangerous, uncompromising foe that threatens the industry.

This insect, technically known as *Cochliomyia hominovorax*, and identified as early as 1842 in Texas, has been a serious killer of herds since 1933. Severe screwworm outbreaks have been reported frequently in Florida, Georgia, Alabama, South Carolina, Mississippi and Tennessee. Surveys also have determined that from time to time infections were present in Kentucky and Virginia and as far north as New Jersey.

The screwworm's looks belie its destructive instincts. In appearance it resembles the common "blow fly," is bluish-green in color and has three dark stripes on its back.

Although mating only once, the female is capable of laying from 100 to 400 eggs at a time on the edges of wounds in warm-blooded mammals. Depositing eggs at four-day intervals, she is able to lay as many as 3000 whitish eggs. Within six to twelve hours the eggs hatch, the mature larvae leave the wound or lesion in about five days, and burrow in the soil. In approximately two weeks, the outer skin of the larva having hardened and formed a pupal case, the adult fly emerges. In two to five days the fly is ready to mate and continue the cycle. The average egg-to-egg cycle is twenty-one days, although cool weather may prolong it.

Predatory beetles and ants destroy some of the larvae and pupae, but not in sufficient numbers to effect any appreciable control. Nature's most efficient population regulator is temperature. If the daily average temperature is less than 55 degrees Fahrenheit for more than two months, screwworm pupae die in the soil. In winter, therefore, the "survival line" is considerably restricted, and the pest can usually over-winter only in Florida, south Texas and Mexico.

During the spring and summer, the screwworm fly is on the wing and at its most active. It can migrate thirty-five miles weekly, infesting all of Florida, the southern two-thirds of Georgia and the southeast corner of Alabama in the Southeast; most of Louisiana, Arkansas, Texas, southern New Mexico, southern Arizona and California in the Southwest; and

Oklahoma, Kansas and Missouri in the mid-West. However, the days of the screwworm seem to be numbered—at least in the Southeast.

In 1951, new techniques were undertaken to eradicate the insect in preliminary experiments. Using X-ray equipment, scientists and entomologists revealed that female screwworms, when sterilized by radiation, were infertile. Similarly, normal female flies, after mating with males made sterile by radiation, produced eggs that did not hatch. Could this be a means of control?

After achieving notable success in mass-rearing sterile flies under laboratory conditions, a small-scale field test was begun on Sanibel Island, off the coast of Florida.

The results of this exploratory study were not conclusive, since screwworm flies from the nearby mainland infiltrated the island and reinfested it. Yet researchers nevertheless gained first-hand experience in propagating, mass irradiation and liberation of treated screwworms. An important step forward had been taken. The battle lines were being drawn.

Backed by further experimentation in rearing techniques, sterilization by exposure to cobalt 60, and the behavior habits of the fly in the laboratory, the United States Department of Agriculture entomologists were ready for the first major offensive.

On March 26, 1954, seventeen thousand sterile male screwworm flies, the first batch to be exposed to the gamma rays of radioactive cobalt, were released by airplane over Curacao Island, in Netherlands West Indies. Systematic releases of flies were continued weekly.

Five months later, a ground survey showed that the screwworm population was decreasing, but not disappearing. The number of sterile flies was increased from 100 to 400 per week on each of the island's 170 squares miles. Eradication occurred rapidly.

Since December, 1954, Curacao, fifty miles from the nearest land and therefore safe from migratory flies, has been safe from the screwworm. Reviewing its findings, United States Department of Agriculture officials determined that the vexing problem in the Southeastern States could probably be handled as it had been during the Curacao test.

In Texas and the Southwest, however, the diagnosis was not as encouraging. The area's geographical location posed a knottier question, since normal flies can immigrate from Mexico with impunity.

Deciding on an all-out campaign in Florida, the Department was aware that many scientific problems had to be solved before such an important launching could be undertaken.

It was necessary to know how to blanket an area of 50,000 square miles rather than Curacao's 170. The weekly rearing and distribution of 50,000,000 sterile male and female flies presented its own problem. Many additional scientists had to be gathered. The Florida eradication scheme had to have more than the half-dozen men who had worked on the Caribbean project.

The job of raising and sterilizing a million flies a week was begun on a pilot plan basis. In the spring of 1957, fortified by more experimentation and study, entomologists decided to put their experiences and knowledge to use in the major operation.

Between April and August, in a zone forty by fifty miles southeast of Orlando, 2,000,000 sterile male screwworm flies were released from low-flying airplanes at a rate of 500 per square mile, each week for four months.

At first it was doubtful that the egg count would be reduced because of the infiltration of normal screwworms, as well as emigration of sterile flies, in the 2000 square mile area. But even with reinfestation from three sides, seventy percent sterility was obtained in twelve weeks. Valuable data also was compiled on the percentage reduction of fertile eggs at the center, as opposed to that beyond the area's boundaries.

At last, after years of testing, research and planning, a screwworm eradication program for the Southeastern States switched into "high gear." Last summer, with the completion of fly-rearing facilities on a gigantic scale at Sebring, Florida, extermination tactics were begun.

From first to last, the process requires intricate, exacting performance. When the screwworm pupae are five and a half days old, they are packed in canisters in lots of 18,000 and lowered into a cask of cobalt 60. They are exposed to this radioactive substance for about twelve minutes, or long enough to become saturated with 8000 roentgens of gamma rays. On removal, the screwworm pupae are placed in cartons resembling small egg boxes, and stored in an air-conditioned room at a constant temperature of 80 degrees F. until the flies emerge.

Then they are ready for distribution by twenty light planes that fly six hours a day, six days a week, from such central locations as Orlando, Gainesville, Miami, Sebring and Tallahassee, Florida, and Troy, Alabama. Besides the pilot, each plane carries a U.S.D.A. crewman who handles the systematic dispersal of 1000 boxes of flies a day.

Various kinds of release mechanisms had been tried before a Department scientist designed the most satisfactory one to date. It is an automatic dispersing apparatus that may be timed for various dispersal rates, and which opens and counts the cartons as they are dropped.

This method is a far cry from the one used at Curacao, where dispersal was a matter of tearing open fly-filled paper bags and throwing them out of the plane!

Every safety precaution is taken to safeguard all workers and scientists from overdoses of radioactivity. Technicians who work with the cobalt 60 casks are equipped with film badges and pocket chambers. In the event a cask malfunctions, it can be hoisted through the roof of the building and immersed in water before repairs begin.

Workers receive special attention so that they may not inadvertently carry any fertile female flies from the plant. Showers and fresh change of clothes are mandatory at the end of each shift.

By the middle of the past summer, approximately two and one-half billion flies had been reared for the 50,000 square-mile region encompassing the Florida Peninsula and the southeast corner of Georgia. As an added precaution during the two to three years it may take to eradicate the pest there, the average overwintering line has been extended 100 miles to the north.

Livestock inspectors, manning more than 100 fly traps, gather information weekly on the incidence and relative abundance of both wild and sterile flies throughout the treated area.

Although the project is far from completed, "Operation Screwworm" furnishes a fine example of how the power of the atom can be used to benefit man rather than to destroy him.

# Chapter 24

# Exobiology

## SIGNIFICANCE AND STATUS OF EXOBIOLOGY

The title of this paper, "Significance and Status of Exobiology," has been carefully selected. The *significance* of the term "exobiology" is in dispute and there are those who declare that the subject has no *status*. The term is of recent origin and is intended to denote the study of extraterrestrial life. The argument arises not only from those normally protesting the adulteration of the English language with newly-coined words but from those who literally interpret the prefix "exo" as meaning "out of, outside, or outer layer."

Some contend that a better term would be "xenobiology" in that the prefix "xeno" connotes "strange or foreign." Others feel that "biology" is sufficient, for it would encompass extraterrestrial creatures. These arguments may have semantic merit, but "exobiology" has been adopted by the National Aeronautics and Space Administration and is now a part of the literature. A dispute over nomenclature should not discourage biologists from pursuing the subject.

The more serious objection to the "science" of exobiology is that it has no status. Since there are no data on extraterrestrial life, how, some contend, can there be such a science? The subject matter is too important to permit such "sea-lawyer" rationalization to impede its investigation. I hope to demonstrate this.

The true significance of exobiology is best revealed by the questions it can help answer. The key question is "Is life limited to this planet?" Although we have no direct data, there is indirect evidence of two kinds. The first type consists of Earth-based physical observations of Mars which suggest the existence of life on that planet. Salisbury has summarized these. Perhaps the strongest evidence of this nature was the infrared spectroscopic analysis of Mars by Sinton which indicated the presence of aldehydes. Recently this interpretation has been questioned by Rea and Calvin, but they offer no better single fit for the data obtained.

The presence of water vapor on Mars has now been established, but in amounts which are considered too small by some to support life. Nonetheless, when all such data are considered, many scientists believe that the most likely explanation which generally accommodates them is that there is life on Mars.

The second type of data bearing on the question, "Is life limited to this planet?" is statistical. Although the size of the sample (one planet) is small, the statistical argument for life elsewhere is believed by many to be very strong. While Mars is generally considered to be the only other likely habitat of life in the solar system, Shapley has calculated that there are more than one hundred million stars which have planets sufficiently similar in composition and environment to Earth to support life. Our sample of one tells us that life does exist and, statistically, one would have to conclude that the probability of life on other planets is very high. Of course, yet unknown factors may operate to reduce significantly or even eliminate this probability. Nonetheless, the presently available facts require that biologists become seriously concerned with exobiology even though those facts do not include direct data on extraterrestrial life.

Let us, for the sake of argument, assume that somehow the search for life elsewhere in the universe had been completed and found to be negative. The chemical and environmental information gained from the search would, nonetheless, be invaluable in resolving the *biological* question of why life arose only on Earth.

On the other hand, if alien life is found, a parade of significant questions follows. "Is the life found biochemically similar to our own?" "Are 'they' and 'we' of common origin?" The most exciting prospect would be if the two types of life were different. On the other hand, if they were similar, the old question of panspermia would come to the fore. "Does life travel the void between planets or does it arise independently?"

The search for extraterrestrial life will certainly help resolve one of the most fundamental questions in biology: "Is life a distinct, discrete entity, or is it an inevitable manifestation of matter as chemical combinations become increasingly complex?"

After Darwin elucidated biological evolution, biologists appropriated the word for their exclusive use. However, the term "chemical evolution" is now used to describe the abiogenic production of life precursor compounds from inorganic chemicals under appropriate conditions. Only a few years have seen this area progress from the abiogenic production of simple organic compounds and amino acids to the abiogenic synthesis of fairly complex intermediates and the production of "proteinoids."

The theory of continuous generation of matter in the universe has gained considerable acceptance as has the evolution of the elements from neutrons or hydrogen gas. Is it possible, then, that life is just one segment of a vast continuum of evolution on a universal scale?

To speculate, did the following developmental sequence take place as part of this continuum: Space (as far back as we dare imagine), energy, elemental particles, hydrogen, the other elements, increasingly complex compounds, structured matter exhibiting the characteristics of life, consciousness, social evolution? Although there are many gaps in this theory of an evolutionary continuum, there is also much evidence for it.

In the portion of the spectrum of interest to biologists, we may not be too far away from achieving abiogenic synthesis of macromolecules which are identical to those functioning in living material. The largest gap facing the biologists in attempting to explain the development of a living cell from inorganic compounds is that of structure. Even if we could synthesize all of the chemicals of a cell, how can we assemble these materials into specialized components such as membranes, walls, ribosomes, mitochondria, and the like?

Here, too, however, there are now indications that at least part of the explanation lies in the natural geometric accommodation of very complex molecules. The finding of life elsewhere might shed light on the theory that the geometrically possible combinations of elements, as modified by chemical and physical constraints operating under a range of environmental conditions, inevitably produce self-replicating material subject to genetic mutation.

The interest in exobiology extends beyond possible simple alien forms of life. A question of utmost significance and fascination is "Is there *intelligent* life elsewhere?" The finding of even unicellular life on Mars would add greatly to the probability that the answer is "yes." Life on a second planet would double the size of our sample and would strongly indicate that life exists on many planets in the universe.

While prospects for intelligent life do not seem favorable in our solar system, it would then be exceedingly likely that other stellar systems contain planets favorably situated and endowed to permit the evolution of intelligent organisms. The consequences of such an assumption are awesome. If the Earth is but one of a great many planets supporting life, it becomes highly improbable that the Earth possesses the most advanced form of life.

The ramifications of a positive answer to the last question posed jar the imagination. The significance includes but transcends the realm of biology. Morrison has represented the growth of technology as a vertical line at some point on the galactic time scale. It would seem, thus, that, once having begun, the elucidation of all science occurs almost immediately.

Therefore, those planets possessing beings of greater intelligence than ours would be beyond this infinitesimally short blip on the time scale.

Contact with such beings could provide us with an almost instantaneous scientific denouement of the universe! While the time saved would be minute on the galactic scale, it would be very large compared to the life span of a man.

Such, then, is the scope of significance of exobiology. Now, what about its *status*? No one can deny that the significance greatly overshadows the status. While it is true that there are no generally accepted data, concerted research on the subject is in its infancy. We have not yet had the opportunity to deliver instruments to other planets. It now appears the possibility to do this is at hand, and it is hoped that the instrumented exploration of Mars will begin shortly.

However, there are other avenues of exobiological research including highly important experiments which can be conducted in Earth-based laboratories. Origin of life experiments, the abiogenic synthesis of macromolecules, and other empirical and theoretical investigations of life are potentially important to exobiology. If we could learn how life on Earth evolved from inorganic chemicals, we could probably answer the first question posed, "Is life limited to this planet?"

Pasteur was probably the first empirical exobiologist. He attempted to culture a piece of the Orgueil meteorite and found no growth. Today there is considerable dispute over the possible biological origin of some materials found in meteorites. Evidence was recently presented for the identification of exobiological material in the same Orgueil meteorite.

This evidence consists of a characteristic carbon chain length distribution of alkanes similar to that found in living material on Earth and isotopic ratios significantly different from those in terrestrial alkanes. On the other hand, the ease with which terrestrial microorganisms can contaminate meteorites has been demonstrated. The subject remains open, as does the possibility that the examination of meteorites can supply the first positive evidence of extraterrestrial life. However, the source of the meteorite would remain unknown.

Ground and rocket-based visual and infrared observations of Mars will be intensified during the 1965 and 1967 planetary oppositions. While information derived from these studies will not directly answer the question of life, the hospitality of the

Martian environment to life, as we know it, will be better revealed.

Instruments are now under development for the biological exploration of Mars. These instruments look for metabolism, growth, or chemical and physical properties peculiar to living matter. NASA is now considering the possibility of sending a package containing a number of such instruments to Mars in 1969. Plans are also being studied for landing much heavier instrument capsules in 1973.

The recent extraordinary photographs obtained from Ranger 7 indicate that the moon may contain a wealth of exobiological information. The entire surface of the moon seems to be composed of in-fall. It seems likely that, since its formation, the moon has been sweeping space, actively collecting vast quantities of materials, including planetary fragments from a great variety of near and distant sources. The galactic beachcombing that can be achieved through an examination of moon samples may produce fossils or other evidence of once living organisms. Such samples should become available within the next few years.

There are engineers who believe that manned exploration of Mars will be possible in the 1980's. This would permit a detailed exploration of Mars and, perhaps more important, the return of a sample to Earth laboratories for extensive studies. While we cannot say that planning for such a mission has yet begun, at least the subject is under discussion by the space agency. There is one other facet to the current status of exobiology. It concerns the means by which we will probe for life beyond the solar system.

Purcell presents what seems to be an irrefutable case against the possibility of manned space travel far beyond the solar system. He does this solely on the basis of the propellent mass required for a round trip journey in which the ship approaches the speed of light. Assuming the development of perfect nuclear fusion propulsion, he demonstrates that the ratio of the initial mass of the rocket to the final mass is $1.6 \times 109$. Thus, for every pound that made the round trip, a starting weight of 1.6 billion pounds would be required.

Even assuming that the ultimate is attained, the perfection of matter-antimatter propellant, the ratio of the initial mass to the final mass is 40,000. The tremendous energies thus shown to be required for manned interstellar space travel make such travel extraordinarily unlikely. It might also be pointed out that, at travel at or near the speed of light, tremendous radiation problems would result from collisions of the ship and occupants with atomic nuclei in space. In these difficulties, perhaps, lies the answer to the question raised by skeptics of intelligent extraterrestrial life, "Since they are smarter than we, why haven't they come here?"

Nonetheless, the solar system is not the limit of our exobiological exploration and indeed a brief, preliminary experiment probing beyond it has already been made. In contrast to the tremendous energies required for space travel, the electromagnetic energy required to traverse vast regions of space is quite manageable. Cocconi and Morrison show that our present capabilities are sufficient to permit radiocommunication over a distance of 10 light years.

Cameron, on the basis of the statistical frequency of environments similar to Earth, estimates that there are 2 × 106 advanced civilizations within our galaxy. He believes it highly probable that many of these contain inhabitants whose scientific achievements have surpassed our own. If so, they, too, are faced with the inefficiency of interstellar space travel. Accordingly, it seems possible that many distant societies have already contacted each other by electromagnetic communication systems.

During their existences, such societies would probably continually search for new civilizations which attain the technology required to join in this communication. Hence, we may now be the target of intelligent electromagnetic signals—signals that started their journey many years ago. The preliminary search for the signals which has been made, entitled *"Project Ozma,"* was conducted with the 85-foot radiotelescope of the National Radio Astronomy Observatory.

Two target stars, of the type thought likely to harbor Earth-type planets, were searched during May, June, and July,

1960. No signals of extraterrestrial intelligent origin were detected, but the prosecution of such a program will take many years of painstaking work. It will undoubtedly be undertaken at a serious level.

I have attempted to show that exobiology has considerable significance and that it is rapidly improving in status. In doing so, I fear that some of the possibilities discussed may not only have taxed your imagination, but may have exceeded your credulity. Perhaps some of these offenses may have done disservice to my goal of stimulating your interest and active participation in exobiology. However, the severest test I wish to impose on your imagination has been saved until last. It is that, in a universe containing on the order of one trillion galaxies, each consisting of hundreds of millions of stars, there is, nestling in an arm of one galaxy, one star which swings about it a planet completely covered with the only life to be found in all that infinite cosmic desert.

## THE BIOLOGY OF SPACE

To survive and to fulfill the missions assigned to him in space, man must learn to live there with a reasonable degree of comfort, efficiency, and freedom of movement. The greatest problem posed for man in this regard is the necessity of creating for himself in space an environment which will reasonably duplicate the one he is accustomed to on earth.

Man is a complex mechanism. He has developed through thousands of years of living under one set of general conditions. Although he can be trained and acclimatized, he cannot be reengineered. He must take his manner of living with him wherever he goes. This involves not only the air he breathes and the temperature and humidity he feels, but also his protection from heat, cold, weightlessness, and radiation. In addition to these, there are the related problems of food and water and the disposal of human wastes—plus the factors of fatigue and boredom.

Man uses in the neighborhood of 2 pounds of oxygen each day. This he gets, of course, from the atmosphere. In space there is no atmosphere. For human breathing purposes the atmosphere ends well below 50,000 feet. An aircraft pressurizes its cabin from surrounding atmosphere even at 40,000 feet, but because a spacecraft has no surrounding atmosphere, it must be pressurized from one of three sources carried along: compressed gaseous oxygen, liquid oxygen or chemical compounds which liberate oxygen.

Cosmic rays lose their original power and intensity when they enter the atmosphere. But, in space outside the atmosphere, any object is exposed to a powerful bombardment of cosmic ray particles.

Sunburn is caused by the ultraviolet rays of solar radiation. On earth the atmosphere gives lifesaving protection from solar radiation. But out in space its intensity is lethal.

Thousands of tiny meteoroids cross the regions of space. As they enter the atmosphere they normally are set afire by friction, become meteors, and burn up. But they constitute a possible hazard to any craft in space.

There are many psychological problems. Man is accustomed to a certain day-and-night cycle and performs best when he works, eats, and sleeps in his usual manner. He must be able to walk about and shift positions, to flex his arms, legs, and neck.

He must have some sort of visual orientation within the spacecraft and he must be able to make minor adjustments in his environment (like raising or lowering the lights) just to relieve the monotony. In space, man faces the problem of fatigue—the kind that comes from long commitment to one task and confinement within a relatively small compartment. This fatigue results in impaired judgment, decline in alertness, irritability, and indecision. His power of perception suffers.

## FOOD IS A MORALE FACTOR

Food is highly important to the space explorer both for nutrition and sustaining his morale. The cook, the engineer, the flight surgeon, and the nutrition expert have all combined in creating a menu the astronaut will look forward to, and of giving him a way to eat it. Three general feeding plans have been worked out for space flights. The first covers a short trip of 2 or 3 days. The second covers from 3 days to several months. The third or long-range covers a flight extending over months or years.

The first or short-range plan includes food bars, each of 250 calories, cut into bite-size pieces; water in plastic bags with plastic drinking tubes; both liquid and semisolid foods packaged in squeeze tubes (soup, meat, fruit, chocolate, milk, ham, turkey, and cheese); dehydrated foods and fruit juices to be reconstituted with water.

For longer range missions, ranging from 3 days to several months, the food experts assume that space scientists will have perfected a means of producing artificial gravity within the spacecraft. They also assume that water will be recovered from the atmosphere within the spacecraft and from body waste, purified and reused.

Where weight and space are crucial, the meals will consist of precooked, dehydrated food, plus ordinary beverages. Where weight is less of a problem, food would include canned, dehydrated, and instant types, none of which would require refrigeration. Spices, herbs, condiments, pickles, jellies, raisins, nuts, and candy would also be available.

Where weight and space are of no concern, the meals would be very much like those served today on the most modern commercial airliner.

For the long flights in the relatively distant future, those flights which may last for months or years, it is not feasible to carry expendable supplies. Such flights will require a completely regenerative, closed and balanced ecological system capable of producing food and oxygen (and absorbing carbon dioxide), and of recycling wastes to supply materials essential to the human body.

Since earth, itself, is a successful closed ecological system, man will probably take earth as his model in creating such a system for space.

The regenerative cycle is well established by Nature. Man and other animals eat the plants of earth, the products of the sea (and each other), return their wastes and even their bones to the earth in the recycling process.

## ENABLING MAN TO LIVE AND WORK IN SPACE

Steps have already been taken toward development of an artificial closed-cycle ecological system for space flight. Among those under development are prototype systems that can sustain five men in shirtsleeve comfort for 30 to 60 days; four men for 3 to 6 months; and eight men for a year. The systems will be designed to purify water and air for reuse and provide for waste disposal and for food.

Tests are being conducted to learn how man will perform and how he will react in the relative isolation of long space flights. One volunteer remained for about 5 months in a compartment approximately the size of an efficiency apartment. During the experiment, he performed directed tasks in sequence as if he were piloting a spacecraft but he saw no one else and had only voice contact with the outside. Tests with two or more volunteers are scheduled.

NASA is observing the effects of strong magnetic fields on organisms to determine any adverse effects from surrounding manned spacecraft with magnetic fields as a radiation shield. The artificial magnetic fields would function just like earth's magnetic field which diverts much lethal radiation hurtling to earth from outer space.

Biologists are considering induced hibernation for long space trips. An additional attribute of hibernation is that it appears to increase resistance to radiation.

NASA is studying algae and other plants as possible sources for a continued food supply. Moreover, studies are being carried out on development of compact, lightweight, nutritious, palatable, and morale-supporting meals that require neither refrigeration nor heating.

Some of these are in puree form to be taken from tubes; others are in solid form and must be chewed. (Foods that can be chewed are most desirable.) Some of the solid foods are dehydrated or freeze dried and must be reconstituted with water. In many cases, however, the moisture required can be supplied by the astronaut's salivary glands.

Space suits are being fabricated that will protect man from high vacuum temperature extremes, small meteoroids, and radiation in space or on the airless moon, yet permit reasonable freedom of movement.

A series of recoverable bioscience satellites will provide information about the effects of weightlessness for as long as a year on animals and plants. Scientists generally agree that lengthy periods of weightlessness can reduce muscle tone; i.e., weaken muscles. However, other biological effects are still to be determined.

Crewless instrumented spacecraft, such as Ranger, Surveyor, and Mariner, will precede man to the moon and planets. They will transmit vital information about the celestial body that he is to explore and about what he will face on the voyage through black uncharted space.

Orbiting Solar Observatories, Pioneer spacecraft, and interplanetary monitoring platforms (IMP) will furnish data on radiation and other space phenomena and contribute to development of reliable methods for forecasting large solar flares. Such flares flood space with lethal intensities of radiation.

## THE SEARCH FOR EXTRATERRESTRIAL LIFE

In a related program, NASA is investigating the possibility of life elsewhere than on earth. Life normally results where the conditions for it are appropriate. However, most scientists agree that the appropriate conditions need not necessarily resemble those on earth nor must all life be comparable to earth forms.

The conditions of space are simulated in earth-based laboratories. Micro-organisms are inserted in simulators to determine whether and how long organisms can survive the temperature extremes, high vacuum, and radiation of space. The results of these studies may have important implications relative to life on other planets.

Life scientists have simulated in the laboratory the kind of environment believed to exist on Mars. They have discovered that certain terrestrial lichens, mosses, and bacteria can survive for a limited time in this kind of environment. This does not necessarily mean that such organisms live on Mars but that organisms such as these could live on Mars. In the opinion of scientists, Mars is the best candidate in the solar system for some sort of life like that on earth.

Balloon-borne instruments have made infrared studies of Mars. These and other studies detected minute quantities of water vapor and considerable carbon dioxide in the atmosphere. Their existence suggests the possibility that specially adapted lower forms of life may exist on the Red Planet.

During its exploration of the moon, NASA plans to analyze lunar materials for signs of life. Such analysis may disclose remnants of extinct life or organic substances that might signify the presence of life. On the other hand, the study may bring to light prelife chemicals (chemicals identified with life) that have accumulated through natural processes over millions of years but have not been organized into living things. Such a find would be invaluable to the study of how life originated or evolved.

As part of this program, NASA is also supporting the examination of meteorites for organic compounds and studying the origin of compounds of living organisms under primeval geochemical conditions.

Several life detection devices are being readied for landing on other planets. These miniature laboratories are primarily designed to report on microbial life. They are quite small, and some weigh as little as 1 ½ pounds.

One is named *Gulliver* for Jonathan Swift's fictional discoverer of the tiny Lilliputians. After landing, Gulliver will fire adhesive cords outward and then reel them in. It is expected that dust and other surface substances will adhere to the cords. The cords will be immersed into or drenched by a nutrient solution containing radioactive carbon. If earthlike organisms are present, they will ingest and ferment the solution, creating radioactive carbon dioxide gas. This would be registered by a Geiger counter and the exciting results transmitted to earth.

Another device, which operates on a similar principle, is the *Wolf Trap*, named for its designer Dr. Wolf Vishniac. The

Wolf Trap will suck in samples of soil and air and immerse them in nutrient solutions. If the samples contain organisms, the solutions will undergo changes in acidity and turbidity which would be reported to earth.

Among the other life detection devices are the *Multivator* which is in effect a miniature laboratory that can make 24 biochemical tests for organisms, a microscope which can radio images back to earth, and a TV telescope that can survey the landscape for such possible signs of life as moving or waving objects.

Newer approaches include devices for determining the characteristic absorption of Polypeptides and DNA in the UV *range*; for determining *optical rotation* (typical of complex molecules of living or once living systems); *gas chromotography* for the detection and analysis of various life related materials; and *mass spectroscopy* for the identification of relatively large submolecular groups by means of density.

Another device may consist of a simple colorimeter which will measure the visible light absorption reaction (*"J"* bands) between certain dyes and proteins. Recent experiments have shown that by varying the pH and temperature, it may be possible to render this device "analytical" not only with respect to proteins but also with respect to other biologically significant molecules.

## TRAVELERS IN SPACE

During the conquest of Mexico in 1520 the Spanish invaders brought with them not only weapons and baggage, but also disease. The famous historian W. H. Prescott has told of "that terrible epidemic, the small-pox, which was now sweeping over the land like fire over the prairies leaving its path strewn with the dead bodies of the natives".

History is full of such incidents. Travelers to new countries have often spread infection. And just as often they have succumbed to unfamiliar ailments themselves or carried diseases back to their homes. A vital function of the U.S. Public Health Service is the quarantine at every port of entry to prevent travelers from spreading disease. The U.S. Department of Agriculture maintains even stricter controls against the entry ot plants and animals that might harbor infections.

The past has an important lesson to teach us in the space age. We face the possibility of finding unfamiliar forms of life elsewhere in the solar system or, perhaps, in the universe. Scientists recognize that organisms from the earth could pose a serious threat to life on other worlds, and extraterrestrial forms of life could represent an equally serious threat if brought back to our planet.

According to one estimate, common bacteria could completely occupy a planet the size of the earth in a few days or weeks if conditions were suitable for growth and spread. Who knows how fast an unfamiliar microbe could do the same here?

These problems may seem a long way off, since we will probably not make manned landings anywhere other than on the moon for many years. But they are really of immediate concern. Spacecraft have already crashed on the moon. These craft were not completely sterilized before takeoff, and we can no longer be sure that any bacteria that may be found on the moon originated there.

The scientific implications are very serious, for we would have liked to examine the moon's surface to find out what lifelike or prelife forms might have collected there. Some, but not all, scientists have seriously questioned whether organisms could survive under lunar conditions. Were we to detect one bacterial spore, it could have important meaning—if we knew for certain a spacecraft had not brought it there. The same questions will hold for the planets when we reach them.

With this in mind, researchers have conducted many laboratory tests to see whether earthly organisms could survive in an environment much harsher than ours.

## HOW ADAPTABLE IS LIFE

Several groups of scientists have tried to simulate certain characteristics of the environment of Mars to test some of the speculation about that planet. They wanted to find out what happens to living things when they are subjected to temperatures below freezing for long periods of time, to nearly waterless conditions, to alternate freezing and thawing, and to very low-oxygen atmospheres.

Many of the organisms die, of course. But what is remarkable is that some of them manage to survive, and a few even thrive under these difficult conditions. One investigation involved certain bacteria that grow without oxygen. The microbes were mixed with sterilized crushed volcanic lava to simulate soil conditions we might expect on Mars. They were dried and kept under a vacuum for eight months. Every eight hours the temperature was alternated between -23° and +25° C. (-10° and +77° F.) to parallei the changes due to the Martian day and night. In one strain 10 per cent of the bacteria survived. The results were similar for other species.

Other bacteria can survive and grow well under conditions that approximate a typical summer day at the Martian equator: 24° C. (+75° F.) at mid-day, -74° C. (-101° F.) at night, with only 4 ½ hours a day above the freezing temperature of water.

Single-celled organisms are not the only ones that can withstand extreme conditions. More than 30 species of multicellular plants have been tested in a low-oxygen atmosphere and have survived. The "air" in the test chamber had between.2 and.002 the amount of oxygen in the earth's air and was at a tenth the atmospheric pressure. The plants included the cucumber, which is very delicate, as well as peanut, barley, and turnip.

These plants did not merely survive; they did much better than they do in the earth's atmosphere. Their seedlings grew more rapidly and became taller. The plants ripened better and lived longer. This unusual atmosphere has also been shown to speed up biological processes in certain birds, amphibians, shellfish, and other marine animals. Seeds of winter rye will sprout in a nearly waterless atmosphere at less than a tenth the normal air pressure.

We should understand that such simulation experiments have definite limitations, ones the researchers are well aware of. The chambers in which the microbes and plants grow are designed to regulate temperature, pressure, patterns of light and dark, and other conditions in order to approximate as closely as possible the environment on Mars. But we do not have very precise information about the Martian environment. We do not know the nature of the soil on the planet, or the exact air pressure at the surface, or what radiation reaches the surface—to mention just a few of the many gaps in our knowledge. Actually all we know are a few "gaps." So the results of these experiments may prove only as useful as our guesses about Mars prove accurate.

Nevertheless, the experiments do teach us a couple of lessons. First, they indicate that living things are remarkably adaptable, far more so than biologists have thought. It appears that organisms can show tremendous flexibility in adapting to available atmosphere, water and warmth.

The experiments also remind us that we are still not very well informed about many life processes. It would be exceedingly risky to suggest that parallels might exist between life here and any other life that we might later learn exists. Perhaps we should not expect parallels, but we should not be surprised if we find them.

Second, the results re-emphasize the importance of sterilizing all space vehicles on lunar or planetary missions. If one organism is aboard and later colonizes the moon or a planet, we will have destroyed forever the original conditions there. We will have lost an invaluable scientific resource. This will be true even if, as some researchers believe, there is no life there beforehand.

## WHAT DO WE SEEK IN SPACE

Clearly, the journey will give us two unique rewards: a perspective on our own planet and a prospect on other worlds.

The first pioneering steps in space have already proved their value. Today orbiting satellites analyze the earth's atmosphere, reveal weather patterns, speed communications, and improve navigation. Tomorrow, more powerful craft will be developed, extending our reach and giving us new objectives.

Among these objectives, exobiology—the study of life beyond the earth—is the most subtle and demanding, for it insists "Know thyself." When we have learned as much about life as earthbound science can teach us, then we will be ready to lift our gaze to the planets and the stars.

In the past century, biology has experienced a remarkable development as a scientific discipline, particularly in our understanding of the biochemical mechanisms of life. Once Darwin had set forth the concept of evolution, this understanding could come about. For it is this magnificent concept that helps to explain how structures as diverse as the microscopic amoeba and the giant Sequoia could have developed.

Our understanding of life on earth has now progressed to the level of cells and molecules. Evolution, we now know, has a chemical basis.

Despite the outward differences of living organisms, their chemistry is the same. In all organisms the genetic material consists of nucleic acids. And in all organisms the cell structures are composed mainly of proteins. Indeed, on chemical inspection we find the composition of living cells so similar that it is difficult to tell the nucleus of a human nerve cell from the virus that might attack it.

In the world we know, nucleic acids and proteins come about only as copies of what has evolved before them. Their blueprints are handed down from parent to offspring. But how did these complex substances come about originally—without pre-existing cells or brains to guide their production?

Thirty years ago many people thought the answer to this question was beyond the reach of science. Today the question leads us to fundamental problems of exobiology:

Are nucleic acids the only substances that can link the generations of living things that might exist anywhere in the universe? Or are they merely the only links that earthly life has encountered? Are proteins the only means of building up cell structures? Or are they the accidental result of early chemical events on earth? It is the purpose of exobiology to probe for the answers—and for a larger one as well.

In the past, biology's domain has been limited to the thin shell of the earth, to the way in which one spark of life has illuminated one speck in the cosmos. By contrast, the basic laws of physics are derived from the motions of the stars; and we know the scope of chemistry from studies of the light emitted by stars at the boundaries of the observable universe.

As yet, biology has no such grand universal system. But there is a principle which we confidently expect will prove to be universal—the principle of evolution. Until we have evidence from the planets themselves, however, we can only speculate about evolution elsewhere.

The ultimate goal of the exobiologist, then, is to answer questions about *all* life, about *all* evolution. It would be difficult to point to any practical fruits of this research. Expeditions beyond the earth may be among the most costly experiments ever undertaken, but surely they count as one of the very aims of the human adventure. Is not the search for life beyond the earth the next stage of the great journey that has always led men into the unknown?

## GREGOR MENDEL AND HIS WORK

It is 145 years since, in a small village on the northern border of what was called Austria at that time, a boy was born in a farmer's house who was destined to influence human thoughts and science. Germans, Czechs and Poles had settled side by side in this part of the country, quarreling sometimes, but mixing their blood continually. During the Middle Ages the Mongolic Tatars invaded Europe just there. Thus, the place had been a melting pot of nations and races, and, like America, had brought up finally a splendid alloy.

The father's name was Anton Mendel; the boy was christened Johann. He grew up like other farmers' boys; he liked to help his father with his fruit trees and bees and retained from these early experiences his fondness for gardening and bee-keeping until his last years. Since his parents, although not poor compared with the neighbors, had no liquid resources, the young and gifted boy had to fight his way through high school and junior college (Gymnasium).

Finally he came to the conclusion, as he wrote in his autobiography, "That it has become impossible for him to continue such strenuous exertions. It was incumbent on him to enter a profession in which he would be spared perpetual anxiety about a means of livelihood. His private circumstances determined his choice of profession." So he entered as a novice the rich and beautiful monastery of the Augustinians of Bruenn in 1843 and assumed the monastic name of Gregor. Here he found the necessary means, leisure and good company. Here during the period from 1843 to 1865 he grew to become the great investigator whose name is known to every schoolboy to-day.

On a clear cold evening in February, 1865, several men were walking through the streets of Bruenn towards the modern school, a big building still new. One of these men, stocky and rather corpulent, friendly of countenance, with a high forehead and piercing blue eyes, wearing a tall hat, a long black coat and trousers tucked in top boots, was carrying a manuscript under his arm. This was Pater Gregor Mendel, a professor at the modern school, and with his friends he was going to a meeting of the Society of Natural Science where he was to read a paper on "Experiments in Plant Hybridization."

In the schoolroom, where the meeting was to be held, about forty persons had gathered, many of them able or even outstanding scientists. For about one hour Mendel read from his manuscript an account of the results of his experiments in hybridization of the edible pea, which had occupied him during the preceding eight years.

Mendel's predecessors failed in their experiments on heredity because they directed their attention to the behavior of the type of the species or races as a whole, instead of contenting themselves with one or two clearcut characters. The new thing about Mendel's method was that he had confined himself to studying the effects of hybridization upon single particular characters, and that he didn't take, as his predecessors had done, only a summary view upon a whole generation of hybrids, but examined each individual plant separately.

The experiments, the laws derived from these experiments, and the splendid explanation given to them by Mendel are to-day not only the base of the modern science of genetics, but belong to the fundamentals of biology taught to millions of students in all parts of the world.

Mendel had been since 1843 one of the brethren of the beautiful and wealthy monastery of the Augustinians of Bruenn, at that time in Austria, later in Czechoslovakia. His profession left him sufficient time, and the large garden of the monastery provided space enough, for his plant hybridizations. During the eight years from 1856 to 1864, he observed with a rare patience and perserverance more than 10,000 specimens.

In hybridization the pollen from the male plant is dusted on the pistils of the female plant through which it fertilizes the ovules. * Both the pollen and the ovules in the pistils carry hereditary characters which may be alike in the two parents or partly or entirely different.

The peas used by Mendel for hybridization differed in the simplest case only by one character or, better still, by a pair of characters; for instance, by the color of the flowers, which was red on one parental plant and white on the other; or by the shape of the seeds, which were smooth in one case and wrinkled in the other; or by the color of the cotyledons, which were yellow in one pea and green in the other, etc.

Mendel's experiments show in all cases the result that all individuals of the first generation of hybrids, the $F_1$ generation as it is called to-day, are uniform in appearance, and that moreover only one of the two parental characters, the stronger or the dominant one, is shown. That means, for instance, that the red color of the flowers, the smooth shape of the seeds or the yellow color of the cotyledons is in evidence while the other, or recessive, character seems to have disappeared. From the behavior of the hybrids of the $F_1$ generation, Mendel derived the first of the experimental laws, the so-called "Law of Uniformity," which is that all individuals of the first hybrid generation are equal or uniform. The special kind of inheritance shown by the prevalence of the dominant characters in the first hybrid generation is called alternative inheritance or the pea type of inheritance. In other instances, however, the hybrids show a mixture of the parental characteristics. Thus, crossing between a red-flowered and a white-flowered four o'clock (*Mirabilis*) gives a pinkflowered $F_1$ generation. This type of inheritance is called the intermediate, or *Mirabilis*, type of inheritance.

Now, Mendel self-pollinated the hybrids of the first generation, dusting the pistils of the flowers with their own pollen and obtained thus the second, or $F_2$ generation of hybrids. In this generation the recessive characters, which had seemingly disappeared, but, which were really only covered in the $F_1$ generation, reappeared again and in a characteristic and constant proportion.

Among the $F_2$ hybrids he found three red-flowered plants and one white-flowered plant, or three smooth-seeded and onewrinkled-seeded plant, or three plants with yellow cotyledons and one with green ones. In general, the hybrids of the $F_2$ generation showed a ratio of three dominant to one recessive plants. Mendel derived from the behavior of the $F_2$ generation his second experimental law, the so-called "Law of Segregation." Of course, the characteristic ratio of three dominant to one recessive may be expected only if the numbers of individuals are large, the Mendelian laws being so-called statistical laws or laws valid for large numbers only.

The third important experimental law Mendel discovered by crossing two plants which distinguished themselves not only by one but by two or more pairs of hereditary characters. He crossed, for instance, a pea plant with smooth and yellow seeds with another having green and wrinkled seeds. The first, or $F_1$, generation of hybrids was of course uniform, showing both smooth and yellow seeds, the dominant characters.

$F_1$ hybrids were then self-pollinated and the second hybrid, or $F_2$, generation was yielded in large numbers, showing all possible combinations of the parental characters in characteristic ratios and that there were nine smooth yellow to three smooth green to three wrinkled yellow to one wrinkled green. From these so-called polyhybrid crossings, Mendel derived the third and last of his exprimental laws, the "Law of Independent Assortment."

These experiments and observations Mendel reviewed in his lecture. Mendel's hearers, who were personally attached to the lecturer as well as appreciating him for his original observations in various fields of natural science, listened with respect but also with astonishment to his account of the invariable numerical ratios among the hybrids, unheard of in those days. Mendel concluded his first lecture and announced a second one at the next month's meeting and promised he would give them the theory he had elaborated in order to explain the behavior of the hybrids.

There was a goodly audience, once more, at the next month's meeting. It must be admitted, however, that the attention of most of the hearers was inclined to wander when the lecturer became engaged in a rather difficult algebraical deduction. And probably not a soul among the audience really understood what Mendel was driving at. His main idea was that the living individual might be regarded as composed of distinct hereditary characters, which are transmitted by distinct invisible heredity factors—to-day we call them genes.

In the hybrid the different parental genes are combined. But when the sex cells of the hybrids are formed the two parental genes separate again, remaining quite unchanged and pure, each sex cell containing only one of the two genes of one pair. We call this fundamental theoretical law the "Law of the Purity of the Gametes." Through combination of the different kinds of sex cells, which are pro duced by the hybrid, the law of segregation and the law of independent assortment can be easily explained.

Just as the chemist thinks of the most complicated compound as being built from a relatively small number of invariable atoms, so Mendel regarded the species as a mosaic of genes, the atoms of living organisms. It was no more nor less than an atomistic theory of the organic world which was developed before the astonished audience. The minutes of the meeting inform us that there were neither questions nor discussions. The audience dispersed and ceased to think about the matter—Mendel was disappointed but not discouraged. In all his modesty he knew that by his discoveries a new way into the unknown realm of science had been opened. "My time will come," he said to his friend Niessl.

Mendel's paper was published in the proceedings of the society for 1866. Mendel sent the separate prints to Carl Naegeli in Munich, one of the outstanding biologists of those days, who occupied himself with experiments on plant hybridization. A correspondence developed and letters and views were exchanged between the two men. But even Naegeli didn't appreciate the importance of Mendel's discovery. In not one of his books or papers dealing with heredity did he even mention Mendel's name. So, the man and the work were forgotten.

When Mendel died in 1884, hundreds of mourners, his pupils, who remembered their beloved teacher, and the poor, to whom he had been always kind, attended the funeral. But although hundreds realized that they had lost a good friend, and other hundreds attended the funeral of a high dignitary, not a single one of those present recognized that a great scientist and investigator had passed away.

The story of the rediscovery and the sudden resurrection of Mendel's work is a thrilling one. By a peculiar, but by no means an accidental, coincidence three investigators in three different places in Europe, DeVries in Amsterdam, Correns in Germany, Tschermak in Vienna, came almost at the same time across Mendel's paper and recognized at once its great importance.

Now the time has arrived for understanding, now "his time had come" and to an extent far beyond anything of which Mendel had dreamed. The little essay, published in the great volume of the Bruenn Society, has given stimulus to all branches of biology.

The progress of research since the beginning of the century has built for Mendel a monument more durable and more imposing than any monument of marble, because not only has "Mendelism" become the name of a whole vast province of investigation, but all living creatures which follow

"Mendelian" laws in the heredity transmission of their characters are said to "Mendelize."

As illustrations, I will explain the practical consequences of Mendelian research by two examples only. The Swede, Nilsson-Ehle, was one of the first investigators who tried to use Mendelistic methods to improve agricultural plants. In the cold climate of Sweden some wheat varieties, like the English square-hood wheat, were yielding well but were frozen easily. Other varieties, like the Swedish country wheat, were winter-hard but brought only a poor harvest.

Nilsson-Ehle knew that in accordance with the Mendelian law of independent assortment, the breeder is able to combine the desired characters of two different parents, like the chemist who combines the atoms to form various molecules or compounds. He crossed the late-ripening, well-yielding, square-hood wheat with the early-ripening, winter-hard, but poor-yielding Swedish country wheat. The resulting $F_1$ generation, however, was very discouraging. It was uniform, in accordance with Mendel's first law, all individuals being late-ripe and pooryielding, thus combining the two undesirable dominant characters.

In preMendelian times the breeder would have been discouraged and probably would have discontinued his efforts. Not so Nilsson-Ehle, who knew that the $F_1$ generation is hybrid, showing only the dominant traits, and that the independent assortment of all characters will appear only in the $F^2$ generation. Self-pollinating the $F_1$ plants he obtained an $F_2$ generation showing the ratio of nine late-ripe poor-yielding to three late-ripe well-yielding, to three early-ripe poor yielding, to one early-ripe, well-yielding wheat plants.

The desired combination of the two recessive characters, early-ripe, well-yielding, appeared only in the smallest ratio, one in sixteen—but because recessives are always true-breeding, or as it is called "homozygous," Nilsson-Ehle had only to isolate these plants and to destroy all others in order to obtain a new true breeding early-ripe and well-yielding variety which after a few years gave a crop large enough to be sold. Thus, by the work of the Mendelist, Nilsson-Ehle, culture of wheat was made possible even in the northern parts of Sweden and large amounts heretofore spent for imported wheat could be saved.

Another instance shows the importance of Mendelism for the understanding of human inheritance. Very soon after the rediscovery of Mendel's paper it became evident that the laws found by Mendel with his peas are valid also for animals and for human beings. Of course, the study of the laws of human heredity is limited and rendered more difficult by several obstacles. We can't make experiments with human beings. The laws of Mendel are statistical laws based upon large numbers of offspring, while the number of children in human families is generally small.

But in spite of these difficulties it was found very soon that human characters are inherited in the same manner as the characters of the pea. We know, for instance, that the dark color of the iris of the eye is dominant, the light blue color recessive. I remember a tragi-comic accident connected with this fact. At one of my lecture tours in a small town in Czechoslovakia, I spoke about the heredity of eye color in men and concluded that, while two dark-eyed parents may be hybrids in regard to eye color and thus may have children both with dark and blue eyes, the character blue-eyed, being recessive, is always pure.

Hence two blue-eyed parents will have only blue-eyed children. A few months later I learned that a divorce had taken place in the small town. I was surprised and resolved to be very careful even with scientifically proved statements in the future. Even more important is the Mendelian analysis of hereditary diseases.

If we learn that the predisposition to a certain disease is inherited through a dominant gene, as diabetes, for instance, then we know that all persons carrying the gene will be sick. In this case all carriers can be easily recognized. In the case of recessive diseases, feeblemindedness, * for instance, we know that the recessive gene may be covered by the dominant gene for health and that the person, seemingly healthy, may carry the disease and transmit it to his children.

With every year the influence of Mendel's modest work became more widespread. The theoretical explanation given by Mendel was based upon the hypothesis of a mechanism for the distribution and combination of the genes. To-day we know that exactly such a mechanism, as was seen by the prophetic eye of Mendel, exists in the chromosome apparatus of the nucleus of the cells.

The development of research on chromosomes, from the observations of the chromosomes and their distribution by mitosis to the discovery of the reduction of the number of chromosomes in building the sex cells and finally to the audacious attempt to locate the single genes within the chromosomes, is all a story, exciting as a novel and at the same time one of the most grandiose chapters in the history of science. A tiny animal, the fruit fly, *Drosophila,* was found to be the best object for genetical research. The parallelism between the behavior of the chromosomes and the mechanism of Mendelian inheritance was studied by hundreds of scientists, who were trying to determine even the location of the different genes within the different chromosomes and who started to devise so-called chromosome maps.

Correns, Baur and Goldschmidt in Germany; Bateson and his school in England; Devries in Holland; Nilsson-Ehle in Sweden, are the out-standing geneticists of the first decade after 1900. But soon the picture changed. The Carnegie Institution for Genetic Research in Long Island, under the leadership of Davenport and later under Blakeslee, became one of the world's centers of genetic research. In 1910, T. H. Morgan, then at Columbia University, later at the California Institute of Technology, started his investigations with the fruit fly, *Drosophila,* and founded the largest and most active school of geneticists.

The U.S. Department of Agriculture with its network of experimental stations connected with more than a hundred agricultural colleges became the most admirable organization for breeding of better crops and farm animals based upon the principles of Mendelism. The ideas developed by Mendel have found a new home here in the new world.

From 1905 to 1910, I tried by lectures and by articles to renew the memory of Mendel in my home country and to explain the importance of Mendelism to the people. This was not always an easy task. Once I happened to be standing beside two old citizens of Bruenn, who were chatting before a picture of Mendel in a book-seller's window. "Who is that chap, Mendel, they are always talking about now?" asked one of them. "Don't you know?" replied the second, "it's the fellow who left the town of Bruenn an inheritance!" In the brain of the worthy man the term "heredity" had no meaning, but he understood well enough the sense of an inheri. tance or bequest.

## THE LANGUAGE OF THE GENES

Our methods of communication with our fellow men take many forms. We share with other animals the ability to transmit information by such diverse means as the posture of our bodies; by the movements of our eyes, head, arms and hands; and by our utterances of non-specific sounds. But we go far beyond any other species on earth in that we have evolved sophisticated forms of pictorial representation, elaborate spoken and written languages, ingenious methods of recording music and language on discs, on magnetic tape and in a variety of other kinds of code.

Our remarkable skills in communication depend of course on our highly developed nervous systems with their provisions for transmitting information from various receptors to the central brain where it can be stored, integrated and reproduced. In this way it has been possible for us as a species to acquire knowledge and understanding in a cumulative way. Present-day cultures, including religion, art, music, language, literature, technology and science are the result. All these we pass on from one generation to the next, adding to the sum with each such passage.

In this lecture I am going to talk about the transmission of biological directions for development and function from generation to generation in organisms as varied as submicroscopic viruses and man.

## THE EGG CELL

Let me begin by reminding you that each of us begins development as a small almost microscopic single cell, the fertilised egg. It is about 1_!300 of an inch in diameter—about one-tenth the diameter of the head of a pin. A human egg does not look very different from the egg of a rabbit, a guinea pig, or a chimpanzee.

In fact it is not too unlike the egg cell from which spinach plant develops. Yet each of these cells obviously "knows" what its potential future is, for gross mistakes do not occur. The egg of a rabbit never gives rise to a guinea pig. Clearly the egg cell must contain information that says what it is going to do. A human egg carries specifications for growing into a person—not just any person but a very specific and truly unique one.

It is these specifications about which I wish to speak. I'll begin by posing five questions: First, how are these specifications or directions transmitted from one generation to the next? Second, how are they "written," that is, what is the "language" of heredity? Third, how are they replicated or "reprinted" as they must be with each of the many cell divisions that intervene between the egg of one generation and the egg and sperm cells of the next? Fourth, how are these directions "translated" during development? How are they used in the transformation of a tiny spherical egg into one of us? Finally, do errors or changes in the directions ever occur and, if so, what are the results?

Although it is by no means possible to give complete answers to these questions, we can come much closer to doing so than we could a few years ago. But before I go ahead, let me narrow my task considerably. The egg cell of an animal is, of course, complex, despite its small size. It consists of cell membranes enclosing cytoplasm, which constitutes the bulk of the cell, and a centrally located nucleus. I am going to confine my remarks largely to the nucleus.

In doing this I do not wish to leave the impression that I believe the cytoplasm to be unimportant. The fact is in all cellular forms it is indispensable. It is endowed with highly specific properties and in a real sense carries essential hereditary information. This is clear from the experiments of the kind Dr. Fischberg of Oxford University, Prof. John Moore of Columbia University and several other embryologists have made in which the nucleus of the egg of one species of animal is replaced with that of another. Thus, in the two species of frogs, *Rana pipiens* and *R. sylvatica*, the nucleus of the fertilised egg of one species can be removed and replaced by that of the other).

Whichever way the switch is made, the egg fails to develop, although the comparable transfer in which donor and recipient are of the same species leads to normal development. *Rana pipiens* cytoplasm is not compatible with a *R. sylvatica* nucleus and vice versa.

This is only one of several lines of evidence indicating that the egg cytoplasm has specific properties that cannot be easily modified by the nucleus. In confining my attention largely to the nucleus, I likewise do not wish to belittle the role of environmental factors.

They, too, are absolutely necessary. For a human egg cell to develop into a person, for example, a highly specific physical and chemical environment must be maintained, especially in the early stages of foetal development. Raw material of the proper kind must be available in the right amounts and at the right times.

Before and after birth the nervous system is being indelibly impressed with information. Especially after birth the information that is fed into the nervous system in massive amounts plays a large and important role in determining what we are. It includes a large input of cultural inheritance.

## THE NUCLEUS AND GENES

Getting back to the nucleus, there is an impressive body of evidence showing that in it is contained a very large part of the primary biological information that directs development and function. Mendel demonstrated in his garden peas that this information is particulate. Unfortunately, he was far ahead of his time and was quite unable to convince such persons as the German botanist Nägeli that what he found should be taken seriously. Had Darwin known of his work, the history of science might have been profoundly changed.

Mendel postulated the units of heredity that we now call genes. He did not know they were in the nucleus of the cell or that they were carried in chromosomes. But he did know that they determine whether pea seeds will be round or wrinkled, whether green or yellow, whether the flowers will be purple or white, whether the pea plant will be tall or short, and so on. He had a remarkable understanding about how genes are transmitted from one generation to the next. Although modified in important ways, his laws of inheritance still hold as first approximations to the ways of nature in communicating specifications from parent to offspring.

We now know that the most basic features of the process of transmission of biological directions for development are common to viruses, bacteria, algae, protozoa, higher plants, multicellular animals and man. Let me give you an example in man.

About seven out of every ten persons find the chemical substance phenylthiourea most unpleasantly bitter. The remaining three find it essentially tasteless. The two classes of people differ in the hereditary specifications. Individuals who taste it carry specifications in their cells that somehow say, "I can now taste phenylthiourea," whereas those who do not taste this substance carry the alternative form of this message, "I cannot taste phenylthiourea."

This message, in its two forms, is one unit of inheritance—a gene. Each fertilised egg, and all the cells descended from it contains two representatives of this message, one from the mother, via the egg, and one from the father, via the sperm. There are, therefore, four possible types of persons with respect to this particular bit of specification, viz.:

| *From the mother* | *From the father* |
|---|---|
| 1. Taste | Taste |
| 2. Taste | Not taste |
| 3. Not taste | Taste |
| 4. Not taste | Not taste |

The first is a pure taster. The next two are alike in having one message of each kind. They are "hybrid" for this message. They taste, which is a way of saying the positive taster form of the message is *dominant* to the non-taster form, or, vice versa, that the negative form is recessive.

When sperms or eggs are produced by individuals of these types, each carries one or the other of the two messages, either the maternal *or* the paternal one. Thus from any pair of parents it is easy to predict statistically the types of offspring, just as Mendel did with his peas.

How many such units or genes are contained in the nucleus of a human egg cell? We do not know with any degree of certainty. But we estimate that it is probably more than 10,000 kinds and less than 1,000,000, each present in duplicate except for sex-linked genes which are for the most part present singly in the male. What do these genes specify? Many characteristics: eye colour, hair colour, hair form, skin colour, and hundreds of other characters by which the species may differ.

Many biologists did in the early part of the century just after Mendel's work was rediscovered and while the young science of genetics was struggling for recognition, that these are all trivial characters. So they are, as are most of the characteristics responsible for the uniqueness of normal individuals. But we now have abundant evidence that most genes carry information of vital importance. Thus two genes in man carry the instructions for making the haemoglobin molecules that play such an essential role in carrying oxygen and carbon dioxide. Haemoglobin is *not* trivial.

Soon after the rediscovery of Mendel's classical paper, it was suspected that genes were carried in chromosomes. The clinching proof came in 1915 from the laboratory of the late Thomas Hunt Morgan. His student, Calvin Bridges, showed that errors in transmission of genes are exactly paralleled by errors in chromosome behaviour.

## Chapter 25

# DNA as Genetic Material

Chromosomes were known to be made up largely of protein and nucleic acid, the latter of the deoxyribose type. Deoxyribonucleic acid has now become widely known as DNA. Are the genes protein of DNA—or possibly both? At first their specifications were thought to reside in protein, for chemists knew that proteins are long polymer molecules consisting of linear sequences of amino-acid building blocks of some twenty kinds.

Since there are obviously almost unlimited possibilities in proportions and sequences of amino acids, it was easy to believe that the gene was a kind of coded message in protein. DNA, on the other hand, was believed by many to be a rather monotonous polymer built of four kinds of nucleotide units arranged in segments of four that were repeated many fold. If this were its structure, there would clearly be little opportunity for specificity; therefore it was not considered a serious candidate for the role of primary genetic material.

A series of investigations on pneumococcal bacteria, beginning in 1928 and leading up to the classical paper of Avery, MacLeod and McCarty, severely shook the faith in protein as the basic stuff of heredity. It became clear that the type of polysaccharide capsule produced by a pneumococcal strain could be altered experimentally by treating it with pure DNA from another strain.

Thus DNA from a type III pneumococcus could cause a type II recipient cell to be transformed permanently into type III, and henceforth to produce DNA specifying type III polysaccharide.

The experimental procedure by which this is accomplished is not quite as simple as the above account implies. Nevertheless it appeared that polysaccharide-specifying DNA from the donor strain might be entering the recipient and somehow replacing its homologous DNA. There were, however, alternative interpretations, which, though less straightforward, continued to be preferred by many geneticists and biochemists.

By 1953 more direct evidence had come from another source. By then the life-cycles of certain bacterial viruses—bacteriophage—had been worked out and sufficient genetic work done on them to make it clear that like higher organisms they exhibited particulate inheritance. In other words, these viruses, too, have genes.

Bacterial viruses consist largely of protein coats containing cores of DNA. Electron microscopy suggested that when a bacterial cell is infected by a virus particle, its coat does not enter the host cell. The matter was most elegantly settled by the use of radioactive tracers. Hershey and Chase (1952) infected bacteria with viruses whose protein coats were labelled with sulphur-35, a radioactive isotope. The labelling was done by growing a crop of viruses on bacteria that in turn had been grown on a medium containing radioactive sulphur in the form of sulphate.

After infection, the protein coats were removed from the bacterial hosts by shearing them off in a high speed blender. Separated by centrifugation, the virus coats and infected host cells could be separately examined for presence of radioactive sulphur. Almost all of it was found in the virus coats that did not enter host cells. Since proteins contain sulphur while DNA does not, this result suggests that only DNA enters the host on infection. Experiments in which DNA was labelled with radioactive phosphorus-32 led to the same conclusion.

Now the coats were largely unlabelled after infection had occurred, while the host cells contained most of the radioactivity. Since DNA contains phosphorus—one atom per nucleotide—and protein does not, it is clear that DNA and not much else enters the host cell. Since viruses have genes and only DNA enters the host cell, the viral genes must be DNA. Actually a small amount of protein does enter with the DNA, but labelling experiments in which the radioactivity entering a host cell in protein or DNA is followed to the next viral generation show that it is the DNA, not the small amount of protein, that is responsible for the transfer of genetic information from one generation of viruses to the next.

What about higher organisms—spinach plants and man? Are there primary genetic specifications likewise written in DNA? The evidence is not conclusive but since the most conservative hypothesis assumes they are, we proceed on that assumption until evidence to the contrary comes forth.

## STRUCTURE OF DNA

What is the structure of DNA that enables it to carry hereditary information? A tremendous step was taken toward answering this question in 1953 by a young American biologist, James D. Watson, and the English chemist, Francis H. C. Crick, working together at Cambridge University.

According to this model, DNA consists of a pair of antiparallel polynucleotide chains wound helically around a common axis and cross-linked through specific hydrogen bonding between purine and pyrimidine bases. Letting A, T, C, and G represent the four nucleotides characterised by adenine, thymine, cytosine and guanine, a four-unit segment of DNA can be represented in two dimensions as follows:

A-T-C-G
........
T-A-G-C

Paired dots represent hydrogen bonds. The two chains of the DNA molecule run in opposite directions. This is determined by the orientations of nucleotides in the two chains.

## INFORMATION IN DNA

Genetic information must somehow be determined by sequence of nucleotides. Since only A:T and C:G nucleotide pairs are possible in the normal structure, the two chains obviously carry complementary information. As we shall see this is of special significance in hypotheses of DNA replication.

One can think of information being carried either as a sequence of nucleotide pairs in the double molecule or as nucleotide sequences in the two single chain components.

How much DNA is in a single human cell? It is estimated that the total DNA of the 46 chromosomes of a fertilised human egg contains something like 5,000,000,000 nucleotide pairs. Since the genetic material is carried in duplicate in such a cell, one complete set of information is written in sequence of some 2,500,000,000 such pairs.

Prof. Crick has estimated that this amount of DNA is sufficient to, Making use of the information then available about DNA—nucleotide composition of DNA of various sources, X-ray diffraction observations of M. H. F. Wilkins and associates at King's College, London, general knowledge of the structural arrangements of atoms in nucleotide components, etc.—they succeeded in constructing a molecular model of DNA that seemed to satisfy essentially all requirements. Now, some seven years later, it is generally agreed that the Watson-Crick structure is essentially correct for the native DNAs of a number of organisms. It is now clear that their achievement is outstanding in modernday biology. This is so because their model goes so far in suggesting plausible answers to the questions posed at the beginning of this lecture.

According to this model, DNA consists of a pair of antiparallel polynucleotide chains wound helically around a common axis and cross-linked through specific hydrogen bonding between purine and pyrimidine bases. Letting A, T, C, and G represent the four nucleotides characterised by adenine, thymine, cytosine and guanine, a four-unit segment of DNA can be represented in two dimensions as follows:

A-T-C-G

........

T-A-G-C

Paired dots represent hydrogen bonds. The two chains of the DNA molecule run in opposite directions. This is determined by the orientations of nucleotides in the two chains.

## INFORMATION IN DNA

Genetic information must somehow be determined by sequence of nucleotides. Since only A:T and C:G nucleotide pairs are possible in the normal structure, the two chains obviously carry complementary information. As we shall see this is of special significance in hypotheses of DNA replication.

One can think of information being carried either as a sequence of nucleotide pairs in the double molecule or as nucleotide sequences in the two single chain components.

How much DNA is in a single human cell? It is estimated that the total DNA of the 46 chromosomes of a fertilised human egg contains something like 5,000,000,000 nucleotide pairs. Since the genetic material is carried in duplicate in such a cell, one complete set of information is written in sequence of some 2,500,000,000 such pairs.

Prof. Crick has estimated that this amount of DNA is sufficient to encode the contents of some 500 large library volumes. This is another way of saying that the genetic specifications for producing a person from an egg cell, given a proper environment, adequate food of the right kind, etc., might be written in English in this number of volumes.

What about the physical size of DNA molecules? The diameter of the double helix is 20 Ångström units. The base pairs are spaced at 3·4 Å intervals. Thus a continuous linear double helix of the 5,000,000,000 base pairs of a single human egg would be somewhat more than 5 ft. in length —only slightly less than the height of an average person. This is a convenient way to remember the amount of DNA in a single human cell. If the strands were packed side by side in a monolayer on the head of a pin 1 mm. in diameter, this amount of DNA would cover less than 1/200 of the surface.

## REPLICATION OF DNA

The Watson-Crick structure immediately suggested how DNA molecules might reproduce by separation of the double structure into single chains followed by sythesis of new partners against each of the old chains. Since the two originals are complementary, each carries the information necessary to direct the synthesis of its partner. It is presumed that single chains serve as templates against which free nucleotides are properly ordered by specific hydrogen pairing. Just how the problems of separation of paired chains helically coiled around a common axis is solved is not known. The forces required to "untwist" the double helix would not be great if it were to rotate around its axis in the manner of an automobile speedometer cable.

Is there any evidence that DNA replication in fact occurs in this way? Yes, two kinds of evidence indicate that it does.

One of these depends on labelling old and new chains so that they can be distinguished. This can and has been done with radioactive phosphorus-32, but there are difficulties in this method that have not been overcome in an entirely satisfactory manner.

A second method of labelling involves the use of the stable heavy isotope of nitrogen, N15. Bacteria grown on a culture medium containing only N15 eventually become fully labelled with this isotope. Since there are seven or eight nitrogen atoms per pair of nucleotides of molecular weight approximately 700, replacing all N14 with N15 increases the weight a bit over one per cent. Since the size does not change, the density increases by a corresponding amount. "Heavy" DNA molecules can be separated from the light variety in an analytical ultracentrifuge.

The method for doing this was first developed by Meselson Stahl and Vinograd and consists in centrifuging DNA in a cesium chloride solution of proper density. The cesium chloride molecules are thrown down in a high centrifugal field, but, being small, they diffuse sufficiently to establish an equilibrium density gradient in the centrifuge cell. If the range of density so established includes that of DNA, the DNA molecules in solution will form a band at a level exactly corresponding to their bouyant density. Being large they diffuse only slowly and hence form a narrow band, the position of which is easily established by means of an ultraviolet optical system. A mixture of N15 and N14 DNA molecules form two cleanly separated bands.

This method makes possible the following elegant experiment first carried out by Meselson and Stahl. Bacteria are grown on N15 medium until equilibrated, that is, until they become uniformly heavy. The bacteria are then transferred to a medium containing only nitrogen of atomic weight 14. After one cell generation, as determined by doubling of the total population, all DNA molecules should have one heavy old chain and one light new complement.

That is, they should be "hybrid" and hence intermediate in density between heavy and light DNA. They are. In a second round of replication—which doubles the population again-heavy and light chains should separate and each then directs synthesis of a light complement. Hence, half the DNA molecules after exactly two cell divisions should be hybrid and half light. Again they are.

If a population of hybrid DNA molecules—present after one replication of heavy molecules in an N14 medium—are heated under the right conditions, molecules about half the molecular weight are found and they are light and heavy in equal numbers. The evidence is fairly convincing that these are single chains of DNA.

This experiment does not prove that the Watson-Crick hypothesis of DNA replication is correct but it strongly suggests it. A perverse nature might have devised another way of giving the observed result.

## THE KORNBERG SYSTEM

An even more dramatic way of investigating the mechanism of DNA replication is that devised by Arthur Kornberg and his co-workers (1959). In a suitable buffer solution containing magnesium ions, the four nucleotides of DNA as triphosphates, and a DNA polymerising enzyme, DNA is rapidly synthesised if primer DNA molecules are added. Single stranded DNA, obtained by heating native DNA, is much more effective as a primer than is carefully prepared native material. Something like a twentyfold increase in DNA over that added as primer has been obtained.

That the primer is copied is indicated by the fact that the base and Vinograd (1957) and consists in centrifuging DNA in a cesium chloride solution of proper density. The cesium chloride molecules are thrown down in a high centrifugal field, but, being small, they diffuse sufficiently to establish an equilibrium density gradient in the centrifuge cell. If the range of density so established includes that of DNA, the DNA molecules in solution will form a band at a level exactly corresponding to their bouyant density.

Being large they diffuse only slowly and hence form a narrow band, the position of which is easily established by means of an ultraviolet optical system. A mixture of N15 and N14 DNA molecules form two cleanly separated bands.

This method makes possible the following elegant experiment first carried out by Meselson and Stahl (1958). Bacteria are grown on N15 medium until equilibrated, that is, until they become uniformly heavy. The bacteria are then transferred to a medium containing only nitrogen of atomic weight 14.

After one cell generation, as determined by doubling of the total population, all DNA molecules should have one heavy old chain and one light new complement. That is, they should be "hybrid" and hence intermediate in density between heavy and light DNA.

They are. In a second round of replication—which doubles the population again-heavy and light chains should separate and each then directs synthesis of a light complement. Hence, half the DNA molecules after exactly two cell divisions should be hybrid and half light.

If a population of hybrid DNA molecules—present after one replication of heavy molecules in an N14 medium—are heated under the right conditions, molecules about half the molecular weight are found and they are light and heavy in equal numbers. The evidence is fairly convincing that these are single chains of DNA. This experiment does not prove that the Watson-Crick hypothesis of DNA replication is correct but it strongly suggests it. A perverse nature might have devised another way of giving the observed result.

## THE KORNBERG SYSTEM

An even more dramatic way of investigating the mechanism of DNA replication is that devised by Arthur Kornberg and his co-workers (1959). In a suitable buffer solution containing magnesium ions, the four nucleotides of DNA as triphosphates, and a DNA polymerising enzyme, DNA is rapidly synthesised if primer DNA molecules are added. Single stranded DNA, obtained by heating native DNA, is much more effective as a primer than is carefully prepared native material. Something like a twentyfold increase in DNA over that added as primer has been obtained.

That the primer is copied is indicated by the fact that the base composition of the product is like that of primer. DNAs from different sources may have quite different rations of A:T to C:G base pairs, and it is therefore possible to determine whether the ratio of the primer is reproduced in the product. Without primer, DNA is spontaneously synthesised in the Kornberg system after a lag of two to four hours. But unlike natural DNA, this spontaneously synthesised material contains only A:T base pairs. The As and Ts alternating in sequence in each of the two chains as follows:

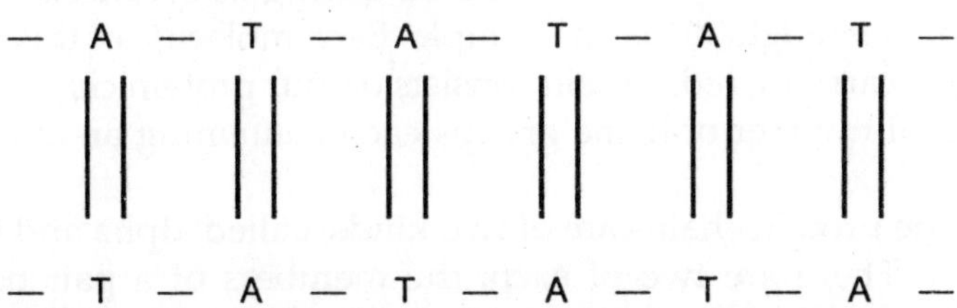


If this A:T copolymer is now used as a primer in a fresh system, more A:T polymer is formed without a lag period. C and G nucleotides are excluded although present in abundance in the system. Here too it appears that the primer is copied as postulated.

Again, the agreement between hypothesis and the facts observed in the Kornberg in vivo synthesis of DNA does not prove the hypothesis. But it enormously strengthens the case.

## USE OF DNA INFORMATION

How is information in the form of DNA made use of in the development of an organism like man? This is clearly a question of the most fundamental importance to biology. At the same time it is an enormously difficult one and we are a long way from knowing the complete answer.

All living systems contain DNA (or in some viruses a related form of nucleic acid, ribonucleic acid, called RNA) and protein. In cellular forms the proteins are of many kinds. We know that many genes, perhaps all, are somehow concerned with synthesis of specific proteins.

Many of these serve as organic catalysts—enzymes—or as components of these catalysts. Enzymes accelerate vital reactions that would otherwise proceed at rates too low to sustain life. In cellular organisms there are thousands of kinds of enzymes, each owing its specificity to a particular protein. We can therefore narrow the problem of gene action, at least in some cases, to that of protein synthesis.

An hypothesis at present widely used as a working basis visualises the process of protein synthesis in the following way: for each protein potentially capable of being formed, there is in the nucleus a specific segment of DNA that carries the information by which the twenty kinds of amino acid sub-units in that protein are properly ordered during its synthesis. Take human haemoglobin as an example. Each molecule of this vital oxygen-carrying red protein consists of four protein chains and an equal number of heme groups each containing an atom of iron.

The protein chains are of two kinds, called alpha and beta chains. There are two of each, the members of a pair being identical and each made up of about 150 amino-acid units arranged in a precisely determined sequence. For each of the two kinds of protein chains, there is presumed—with some evidence—to be a gene consisting of a segment of DNA. This segment is, by the definition I shall use, a gene. It may be something of the order of 1000 nucleotide pairs in length.

How is the information in the gene for the alpha haemoglobin chain made use of? The hypothesis assumes the following sequence of events: from the gene in the nucleus, information is transferred to RNA, presumably by the DNA somehow acting as a template in the ordering of the four kinds of ribose nucleotides in RNA.

This informational RNA then moves from the nucleus to the cytoplasm of the young red blood cell—before it loses its nucleus. There it is incorporated into microsomes, which are sub-microscopic bodies made of structural protein and structural RNA. Once in microsomes, informational RNA molecules serve as templates against which amino-acids are arranged in proper order to make alpha haemoglobin protein chains.

Prior to this the amino-acids are activated and attached to small carrier segments of RNA, each specific for its own amino-acid. Thus for each of the twenty amino-acids there is a corresponding carrier RNA segment. Carrier RNA molecules are somehow coded to specific sites on the informational RNA in the microsome. In this way each amino-acid is carried to its proper position on the template. They are then joined through peptide linkages to form alpha chains and are released from the template and microsome. Whether association of alpha and beta chains with their hemes, and with each other, occurs in the microsome or outside is not known.

Not all proteins for which genetic directions are available in the nucleus are synthesised in any one cell. There are ingenious control mechanisms in operation which determine whether the information in a given gene will be used and, if so, when and for how long. In some cases it is clear that the presence or absence of substrate determines whether a given enzyme will be synthesised. In this way it is assumed that enzymes are not made in quantity when there are no substrates on which they can work. Through studies on such control mechanisms we are beginning to understand how it is that cells with identical genetic information in their nuclei may do quite different things, depending on their environmental context or on their previous history.

There are now known many instances in a variety of organisms in which specific protein variation is known to be related to particular genes. I have already referred to haemoglobin protein in man. Similar situations are known in viruses, bacteria, algae, fungi, insects, higher plants, mammals, etc. The haemoglobin case in man is interesting in that it is known that gene changes may result in substitution of a single amino-acid in a chain of 150 units.

Thus sickle cell haemoglobin, synthesised under the direction of a modified form of the gene in control of the beta protein chain, differs from normal haemoglobin in that a single glutamic acid unit in the chain is replaced by a valine. In the presence of a third form of this same gene the same glutamic acid unit is replaced by the amino-acid lysine. In all, about a dozen modifications of human haemoglobin are known, many of them investigated genetically.

Another example in man is found in the genetic disease galactosaemia. Here is a specific enzyme, galactose-1-phosphate uridyl transferase, essential for the conversion of galactose into a usable form, which is partially or wholly inactive. The enzyme defect is referable to a gene change.

In bacteria, fungi and other micro-organisms correlations between gene change and specific protein modification are much easier to detect than they are in man with his several obvious disadvantages for biochemical and genetic investigations. As a result there are dozens of examples known in such forms as Escherichia, Salmonella and Neurospora.

## MUTATION

In terms of DNA structure gene mutation is believed to be the result of alterations in nucleotide sequence. Many mutations are thought to be the result of errors in DNA replication. In their original paper on DNA structure, Watson and Crick pointed out that if at the precise moment of partner selection a purine or pyrimidine were to exist in an improbable tautomeric form, it might form hydrogen bonds with a "wrong" nucleotide.

At the next round of replication such a "wrong" nucleotide would direct that its own complementary nucleotide be inserted in the paired chain. Thus one of the two daughter DNA molecules would differ from its sister in one nucleotide pair. Its informational content would be modified accordingly.

Experimentally mutations can be increased in frequency in a number of ways. High-energy radiations—ultraviolet, X-rays, etc.—are effective in doing this. Transmutation of artificially incorporated radioactive isotopes such as phosphorus-32 is known to produce mutations. Since radioactive isotopes occur naturally in low frequencies, they are no doubt responsible for a small fraction of spontaneously produced mutations.

Many chemical substances are known to be mutagenic. Perhaps the one best known from the standpoint of mechanism of action is nitrous acid. This specifically oxidises amino groups. In this way it may chemically change a natural DNA pyrimidine or purine in such a way that its hydrogen bonding specificity is modified.

The mutagenic properties have been especially well studied in a tobacco mosaic virus, an RNA virus. Here the natural pyrimidine, cytosine is converted to uracil, which is also a natural occurring pyrimidine in RNA. Adenine is oxidised to hypoxanthine, the latter not a normal constituent of RNA. In a similar way guanine is changed to xanthine. It appears that the oxidation of any one of 3000 of the 6000 nucleotides in a tobacco mosaic virus particle results in an inactivating mutation.

The purine and pyrimidine analogues, 2-aminopurine and 5-bromouracil, produce mutations, presumably by being incorporated in replicating DNA in place of their natural counterparts and thereby leading to errors in complement selection. Since these analogues presumably replace their natural counterparts, purine for purine and pyrimidine for pyrimidine, it might be expected that they would be effective in reversing the mutations they induce.

In a special class of bacterial virus mutants, investigated by Ernst Freese, this seems to be the case, whereas other classes of mutants, for example those induced by proflavine, are not so reversed by base analogues. It is presumed that the latter brings about replacement of purine by pyrimidine and vice versa. Investigations of this kind offer the hope of giving us a deeper understanding of the mutation process than we now have.

## FINE STRUCTURE OF GENES

If, as I have suggested, genes as functional genetic units of DNA are hundreds or thousands of base pairs long, it ought to be possible experimentally to demonstrate multiple mutational sites within a single such unit. Furthermore these should be arranged linearly within a unit. So called fine-structure studies on genes of bacterial viruses, bacteria, fungi and other organisms show that this is indeed the case (Beadle, 1960).

Clearly the unit of mutation is much smaller than the unit of function, for such sites within a functional unit undergo recombination in much the same way as do separate genes within a chromosome. On the basis of frequencies of such recombination it is possible to construct intragenic maps of mutational sites and thus to show that they are linearly arranged. The shortest distances that can be measured by such recombination are not far from those that are calculated to exist between adjacent nucleotides in DNA.

## THE CODING PROBLEM

How are sequences of nucleotides in DNA related to amino-acid sequences in proteins? DNA and protein can be thought of as four- and twenty-symbol systems. If they are both equivalent to simple linear codes, it is clear that it requires at least three nucleotides to specify an aminoacid, for the maximum number of two-letter words is only sixteen. Several possible coding systems have been investigated but so far the problem has not been solved.

Perhaps it will not be until nucleotide sequences in a gene can be compared with amino-acid sequences in the protein specified by that gene. As you know, complete amino-acid sequences have been worked out for only a very few proteins and unfortunately these are from organisms that are far from ideal for genetic study. There are now investigations under way in many laboratories designed to correlate genetic fine structure with protein fine structure. Virus proteins and bacterial enzymes such as alkaline phosphatase and tryptophane synthetase are among the systems that look especially favourable for such combined genetic and chemical studies.

## EVOLUTION

Presumably DNA replication bears no immediate relation to its informational content. Biologically useless DNA is replicated just as faithfully as that playing a vital role. Presumably, too, the mutation process is a random affair. If so, meaningful DNA sequences are much more likely to be made less useful through random mutation than they are to be converted into sequences more useful to the organism of which they are a part. In this sense they may be likened to random typographical errors in a useful message. Direct experiment verifies this expectation; most mutations are unfavourable in the context in which they occur.

If mutations are mostly unfavourable but are as faithfully reproduced as are their normal counterparts, why do they not accumulate with successive rounds of replication just as a typist would accumulate errors in the Lord's Prayer if she were to type copies in succession, each from the previous typing, in a purely mechanical way without proofreading or correcting errors! The answer is that biologically unfavourable mutations specify organisms with reduced reproductive fitness. The errors in their DNA are not corrected but the lines of descent carrying such errors are statistically bred out of existence over a shorter or longer number of generations depending on the degree of reduction in reproductive fitness. We call this natural selection.

Positive evolutionary progress depends on rare favourable mutations that increase reproductive fitness. They gradually replace the ancestral types from which they arose. Modern concepts of organic evolution hold that all living systems have evolved gradually, mutation by mutation, from pre-existing organisms. Sometimes, as in parasites, it is advantageous to become simpler and to count increasingly on the host for raw materials, a proper environment and protection against enemies. But from the beginning, the overall trend in many lines such as our own has evidently been toward greater complexity.

Speculation as to how organic evolution began in the first place leads to the conclusion that there is no clear qualitative break in the sequence of events that spans the advent of life on earth. This is only another way of saying that we cannot clearly distinguish between living and non-living systems. No matter what system one contemplates, it is possible to imagine a closely related one, only slightly simpler, that could have given rise to the former by a single mutation-like step.

The simplest living systems we know today are viruses. Tobacco mosaic virus particles are submicroscopic rods about 800 Å units in length. They consist of a cylindrical protein coat and an RNA core, the latter being made up of about 6000 nucleotides. In a living tobacco cell infected with a particle of this virus more viruses are synthesised.

The bacterial virus øX174 contains about the same amount of nucleic acid. Its units are almost spherical. They, too, have a protein coat and a nucleic acid core, but their nucleic acid is DNA, not RNA.

Some people say such viruses are not living—that they cannot carry on metabolism, synthesise their component parts, or do any of several other things that living organisms do. But they do reproduce their kind, given a proper environment. And they are mutable, and hence capable of organic evolution. Whether one calls them living or not depends on one's definition of life and that in the end must be purely arbitrary.

Let us imagine that virus-like systems were among the earliest "organisms" to evolve on earth. They were not like present day viruses, for there were no living cells from which they could derive their parts-nucleotides, amino-acids, perhaps a few enzymes, etc.

We can reasonably believe that at that stage of the earth's history there were myriads of spontaneously formed organic molecules around, including nucleotides, amino-acids, proteins, etc., and that these were the building blocks of the postulated virus-like creatures. After all we now have abundant experimental evidence that a variety of organic molecules are formed spontaneously under conditions assumed to have been characteristic of parts of the earth's crust a few thousand billion years ago.

If the building blocks of the postulated virus-like systems were around, erations depending on the degree of reduction in reproductive fitness. We call this natural selection. Positive evolutionary progress depends on rare favourable mutations that increase reproductive fitness. They gradually replace the ancestral types from which they arose.

Modern concepts of organic evolution hold that all living systems have evolved gradually, mutation by mutation, from pre-existing organisms. Sometimes, as in parasites, it is advantageous to become simpler and to count increasingly on the host for raw materials, a proper environment and protection against enemies. But from the beginning, the overall trend in many lines such as our own has evidently been toward greater complexity.

Speculation as to how organic evolution began in the first place leads to the conclusion that there is no clear qualitative break in the sequence of events that spans the advent of life on earth. This is only another way of saying that we cannot clearly distinguish between living and non-living systems. No matter what system one contemplates, it is possible to imagine a closely related one, only slightly simpler, that could have given rise to the former by a single mutation-like step.

The simplest living systems we know today are viruses. Tobacco mosaic virus particles are submicroscopic rods about 800 Å units in length. They consist of a cylindrical protein coat and an RNA core, the latter being made up of about 6000 nucleotides. In a living tobacco cell infected with a particle of this virus more viruses are synthesised. The bacterial virus øX174 contains about the same amount of nucleic acid. Its units are almost spherical. They, too, have a protein coat and a nucleic acid core, but their nucleic acid is DNA, not RNA.

Some people say such viruses are not living—that they cannot carry on metabolism, synthesise their component parts, or do any of several other things that living organisms do. But they do reproduce their kind, given a proper environment. And they are mutable, and hence capable of organic evolution. Whether one calls them living or not depends on one's definition of life and that in the end must be purely arbitrary.

Let us imagine that virus-like systems were among the earliest "organisms" to evolve on earth. They were not like present day viruses, for there were no living cells from which they could derive their parts-nucleotides, amino-acids, perhaps a few enzymes, etc.

We can reasonably believe that at that stage of the earth's history there were myriads of spontaneously formed organic molecules around, including nucleotides, amino-acids, proteins, etc., and that these were the building blocks of the postulated virus-like creatures. After all we now have abundant experimental evidence that a variety of organic molecules are formed spontaneously under conditions assumed to have been characteristic of parts of the earth's crust a few thousand billion years ago.

If the building blocks of the postulated virus-like systems were around, would they not have interacted, again spontaneously, to form nucleic acids, proteins, etc.? Nucleotides interact in the

Kornberg system to form DNA capable of replication. Since the role of man in this experiment is merely that of making the conditions more favourable for this particular reaction—that is, increasing its probability—would it not have occurred in his absence when the conditions became right?As every organic chemist knows, organic molecules are formed through the interaction of inorganic molecules—again when the conditions are right.

And as every inorganic chemist knows elements interact to produce inorganic molecules. Nuclear physicists tell us that elements themselves evolve from simpler elements as a result of processes that are both natural and inevitable, given the appropriate circumstances.Thus it is clear that the sequence:

hydrogen—helium—beryllium-8—carbon—oxygen—other elements—water--other inorganic molecules—simple organic molecules—more complex organic molecules like nucleotides, aminoacids, and small proteins—nucleic acids capable of replication—nucleic acids protected by protein coats—virus-like systems with protein coats serving catalytic functions—multigenic but subcellular organisms—simple cellular systems like bacteria—autonomous cellular forms like algae—protozoa-multicellular plants and animals and, in our line of descent, man himself—is a natural one that could have arisen by steps no one of which need have been larger than the individual mutational steps we know in today's living systems.

In the beginning there was a universe of hydrogen. How and when it was created—or whether it is and always has been in a steady state of continuous creation—science knows very little. In whatever way the universe began, there must have been built into it from the very beginning the potentiality of essentially unlimited orderly evolution.

## DNA AND THE CHEMISTRY OF INHERITANCE

Living organisms are unique, among the known forms of matter, because they are capable of creating their own specific, highly organized structure out of substances taken from far more disorganized surroundings, and can transmit this capability to their offspring.

Perhaps the oldest and most profound theoretical problem in biology is the effort to explain the curious paradox that, despite its unique capability for self-duplication and inheritance, a living organism is nevertheless a mixture of substances which are separately no more possessed of these properties than are the more prosaic molecules that never occur in living cells.

This question has been at the root of a long train of experiments, debates, and speculations that begins in classical times and continues unbroken through the development of present-day "molecular biology."

The basic issues are simply stated. If the component parts of a cell are not themselves alive, whence come the life-properties exhibited by the whole? Apart from the untenable notion of a mystic non-material "vital force" which supposedly animates the otherwise dead substance of the cell, the debate has elicited two main positions:

- There is, in fact, some special cellular component which possesses the fundamental attribute of self-duplication, and which is therefore a "living molecule" and the basic source of the life-properties of the cell.
- The unique properties of life are inherently connected with the very considerable complexity of living substance and arise from interactions among its separable constituents which are not exhibited unless these components occur together in the complex whole. In this view, only the entire living cell is capable of selfduplication.

There is at this time a widespread impression that this issue has now been resolved and that the cell does indeed contain a component—DNA— which, according to the theory of the "DNA code," possesses the basic attribute of life, self-duplication, and which guides the activities of all the inheritable processes of the cell.

The importance of this conclusion is self-evident, for it would answer, at last, the basic question of the origin of the unique properties of life, and, if correct, should lead to unprecedented technological control over these properties.

It is appropriate, therefore, to ask what criteria are required to establish that a molecule, such as DNA, is capable of self-duplication, to examine the degree to which the available evidence meets such criteria, and to determine whether the undoubted importance of DNA in the biology of inheritance may be due to some properties other than those attributed to it by current theory.

## SOME IMPLICATIONS

The chief conclusion to be derived from the foregoing considerations is that the unique capability of living organisms for self-duplication and inheritance arises from complex multimolecular interactions among at least several classes of cellular components.

Neither DNA nor any other cellular component is a "self-duplicating molecule" or the "master chemical of the cell"—terms which sometimes appear in current generalizations. There is no evidence from recent investigations of the biochemical aspects of genetics which requires abandonment of the conclusion, long established by biological data, that the least complex agent capable of selfduplication is the intact living cell.

The point of view developed here also suggests alternatives to a number of current views of phenomena related to inheritance:

- Various hypothetical schemes for regulating the "activity" of template genes are invoked in order to explain how the DNA code might operate in cellular differentiation and development. Alternatively, it may be suggested that differentiation is associated with the nucleotide sequestration system, tissue-specific changes in DNA snythesis (e.g., at chromosome "puffs") regulating nucleotide levels and thereby governing the size and metabolic character of the cell.
- The relationship between DNA and a species' characteristic sensitivity to ionizing radiation is often interpreted in terms of radiation-induced mutations in genetic templates. Alternatively, radiation-induced inhibition of DNA synthesis, which results in the accumulation of toxic concentrations of free nucleotides, may be regarded as the mediating process. This view is compatible with Sparrow's observation that in certain non-polyploid plants, radiation sensitivity is proportional to cellular DNA content, and is particularly affected by the relative proportions of heterochromatin.
- Sahasrabudhe has suggested that tumor cells may be characterized by marked changes in free nucleotide level resulting from nucleotide sequestration due to DNA synthesis. In this connection, it is also of interest

that tumor and other rapidly growing cells exhibit distinctive pathways of oxidative metabolism, which are in turn regulated by free nucleotide levels.

- The theory of the DNA code suggests a seemingly simple explanation for the emergence of the first forms of life from the prebiotic organic environment: that life began with the fortuitous appearance of a "self-duplicating nucleic acid molecule" which then organized the complex chemistry of life around itself. In contrast, the viewpoint developed here suggests that the primitive function of nucleic acid was to sequester free nucleotides, and thereby regulate overall metabolic activity. Thus, the nucleic acid template is to be regarded as a late development which improved the precision of the earlier modes of regulation of cellular activity, and which is, in any case, incapable of self-duplication.

The theory of the DNA code, if correct, also leads to important expectations regarding the feasibility of technological control over inheritance. If the nucleotide sequence of DNA were indeed a self-sufficient source of the inherited specificity of living things, then it might be possible, in the not too distant future, to synthesize artificial DNA molecules, which, on being introduced into living organisms, would artificially establish new inheritable characteristics quite outside the range of those normally observed.

However, if, as we have concluded, biological specificity is only partly due to DNA, then it is likely that the specificity represented by proteins may impose severe limits on the acceptability, to the cell, of abnormal DNA nucleotide sequences. This view suggests that technological control over biological inheritance may be far less plausible than indicated by inferences drawn from the code theory.

Finally, the viewpoint developed here bears on some fundamental questions regarding the relationship between physical theory and biology. The theory of the DNA code is often regarded as an example of the success with which "modern physics" can solve basic problems about living systems which have eluded the supposedly less critical analyses of classical biologists.

This view leads to the generalization that, despite its obvious complexity, the living cell must be governed by the "laws of physics"—as revealed by analysis of non-living systems—and that the most effective strategy for elucidating its unique properties is to study isolated parts which are sufficiently simple to permit their analyses in physico-chemical terms.

However, several developments in theoretical physics suggest that the fundamental properties of matter are in better harmony with the view that the unique properties of the cell are derived from interactions of its molecular parts and are inherently incapable of being elucidated by a simple summation of the observed properties of the isolated parts.

This viewpoint is a direct outgrowth of Bohr's theory of complementarity. It may also be closely related to the more recent theory of the "Bootstrap Universe.' This theory suggests that the unique properties of the atomic nucleus are inherently associated with its complexity, and are not accountable by the observed properties of fragments isolated from disrupted nuclei.

It may be suggested then, that, if biology is to be guided by the insights into the properties of matter that are afforded by modern physical theory, the role of DNA in the living cell must be viewed as subsumed under the complex properties of the system—the living cell—of which functional DNA is a part.

The theory of the DNA code is sometimes epitomized by the statement "DNA is the secret of life," an aphorism which appears increasingly to guide the course of current biological investigations. The viewpoint developed here suggests that biology might be more wisely guided by the aphorism, "Life is the secret of DNA."

# *Encyclopaedia*
# *of*
# BIOLOGICAL SCIENCE, TECHNOLOGY AND ENGINEERING

# *Encyclopaedia of* Biological Science, Technology and Engineering

## Volume 3

Dr. Albert Shawn

ANMOL PUBLICATIONS PVT. LTD.
NEW DELHI-110 002 (INDIA)

**ANMOL PUBLICATIONS PVT. LTD.**
**Regd. Office:** 4360/4, Ansari Road, Daryaganj,
New Delhi-110 002 (India)
Ph.: 23278000, 23261597
**Branch Office:** No. 1015, Ist Main Road, BSK IIIrd Stage
IIIrd Phase, IIIrd Block,
Bangalore-560 085 (India)
Tel.: 080-41723429
*Visit us at*: www.anmolpublications.com

*Encyclopaedia of Biological Science, Technology and Engineering*



First Edition, 2009

ISBN 978-81-261-4121-0 (Set)

PRINTED IN INDIA

Printed at Paras Offset Pvt. Ltd., New Delhi

# Contents

## Vol. 3

# Preface

Biology is the study of life and a branch of the Natural Sciences which studies living organisms and how they interact with each other and their Environment. It examines the *Structure, Function, Growth, Origin, Evolution, and Distribution* of living things. Also, it classifies and describes Organisms, their Functions, and how species come into existence. Four unifying principles form the foundation of *Modern Biology*: *Cell Theory, Evolution, Genetics and Homeostasis.*

Biology as a separate science was developed in the Nineteenth Century as scientists discovered that organisms shared fundamental characteristics. Biology is now a standard subject of instruction at schools and universities around the world, and over a million papers are published annually in a wide array of biology and medicine journals.

Most Biological Sciences are specialized disciplines. Traditionally, they are grouped by the type of Organism being studied: *Botany, Study of Plants; Zoology, Study of Animals; and Microbiology, Study of Microorganisms*. The fields within biology are further divided based on the scale at which organisms are studied and the methods used to study them: *Biochemistry* examines the fundamental chemistry of life; *Molecular Biology* studies the complex interactions of systems of *Biological Molecules; Cellular Biology* examines the basic building block of all life, *Cell*; Physiology examines the *Physical and Chemical Functions of the Tissues and Organ Systems of an Organism*; and *Ecology* examines how various organisms and their environment interrelate.

Author

## Chapter 26

# Heredity and Hiroshima

In the course of the past sixty years, we have learned much about heredity and its underlying chemical basis. In turn, genetic knowledge has contributed to man's welfare on many levels. The material rewards to society derived from such knowledge have already been large, and will undoubtedly prove larger in the future.

An exciting chapter in genetic history has been written in the production of high-yield strains of hybrid corn, of high-yield wheat strains resistant to attack by the fungal parasites, rust and smut, which formerly caused nationwide crop failure, and in the more recent, slightly incredible, use of X-ray sterilized male blow flies to help eradicate the pestilent screw worm in the South. These all attest to the value of genetic knowledge in eliminating pestilence and in breeding animals, plants, and microorganisms to give hardier, more productive strains and to increase the world's available resources of raw materials, antibiotics, and food.

Today a glance at the newspapers, with articles on population explosions, the drive of newly emerging nations for economic self-sufficiency, and live polio vaccines obtained from mutant virus strains, will again remind you of some of the consequences of man's curiosity about the socalled ivory-tower concepts of the gene, DNA, and messenger RNA.

However, let us turn our attention to still another topic of contemporary importance, caught up in recurrent ominous headlines, to a topic that now engages the attention of people throughout the world—the global genetic problem arising from the harnessing and exploitation of atomic energy. We know that a portion of the stupendous energy released from nuclear fission and fusion is released in the form of high-energy ionizing radiations, and in the form of radioactive isotopes of elements such as strontium and carbon. Ionizing radiations are highly mutagenic. Before the advent of atomic energy, the known mutagenic agents with which the human population had to contend were not of alarming concern.

For example, ultraviolet light has low tissue-penetrating properties and though it causes burning of skin and may even induce skin cancer, it has no marked mutagenic effect on other human cells. Chemical agents such as mustard gas are deleterious only when fools deliberately set them loose to be inhaled, and base analogues are ordinarily not present in the diet, although they may be used to treat or control various cancers. In general, these are mutagenic agents under our control. They can be used or forgotten at will.

Ionizing radiation, however, is a different kettle of fish. High-energy radiation is capable of deep tissue penetration and is massively destructive and mutagenic. Since our environment is bombarded by a certain amount of background radiation from naturally occurring radioactive elements and from outer space, ionizing radiation is a mutagen we cannot entirely control and against which we are never completely protected.

Our present concern is with the increasing amounts spewn into our environment from nuclear explosions. Each of us now receives more radiation than did our forefathers, and our children may receive more. Since the background radiation has shot upward in the past twenty years and since ionizing radiation is mutagenic, we must give thoughtful consideration to the possible genetic effects of this increase and of how great an exposure the human population can withstand.

Our concern with ionizing radiation is of two kinds, its effects on germinal tissue and succeeding generations, and its effect on somatic tissue, on us. Let us first consider somatic effects. Is there an association between somatic abnormalities, such as cancer, and radiation? The frequency of leukemia, a cancer characterized by an excess formation of white blood cells, can be increased by radiation. In the population of Hiroshima and Nagasaki that survived the atom bombs, there was a marked increase in leukemia during the subsequent five years. The association between leukemia and radiation is exhibited in other groups that have somehow been exposed to increased radiation. There is therefore a clear relationship between increasing amounts of radiation and increasing incidence of cancer.

We know that radiation results in somatic mutation. In view of the correlation found between leukemia and radiation, does this mean that cancer is mutational in origin? We do not know. We have experimental evidence that some animal leukemias can be induced by a virus; further, we know that interactions between bacteria and bacterial viruses are modified by radiation. We cannot, therefore, conclude that the increase in leukemia associated with an increase in radiation arises as a consequence of mutation, for it could result from the interaction of irradiation, virus, and host.

This is, perhaps, a fair assessment of the relation of radiation to cancer in general. Radiation could induce cancers for a number of different reasons. Since cancers have diverse origins, they may be caused by a virus, a mutation, or other factors. Regardless of the details, however, there is an association between an increasing amount of radiation and an increasing incidence of cancer. Radiation is clearly a potential somatic hazard.

We know of still other somatic effects of radiation, for a relationship between irradiation and aging has been observed. With increasing amounts of radiation, longevity is shortened. In fact, from animal experiments it appears that the shortening of human life may well be measured in days per unit dose. Again, increasing radiation must be viewed as a somatic hazard.

Uncontrolled release of nuclear energy also imperils the human population because it increases the abundance of certain radioactive elements. For instance, radioactive strontium 90 formed as a consequence of hydrogen-bomb explosions is extremely dangerous, since it has a long half life and can replace calcium in bone. It is carried into the ionosphere and is ultimately washed down and widely deposited over the earth's surface by rain and snow. From the soil, strontium 90 is absorbed by growing plants, the plants are ingested by cattle, and strontium 90 appears in milk and milk products.

When taken into the body, this radioactive isotope can take the place of calcium in the bone tissue, particularly in growing children. Strontium 90 in bone constitutes small centers of radioactivity and, if present in sufficient amounts, can give rise to bone cancer and a variety of other abnormalities. This is truly a new cycle of nature, a "twentiethcentury cycle." The amount of strontium 90 has already been markedly increased by nuclear explosions, and we have no clear-cut answers as to how much strontium 90 can be safely tolerated in the human diet or how big an increase in the total amount of strontium 90 can be absorbed without adverse effect on the population as a whole.

Safety levels in man are difficult to determine. They require long-term study, but our restless world denies us time. Decisions affecting the use of nuclear energy must be made, but, unfortunately, our information about the deleterious effect in man per unit increase of radiation is inadequate to permit us to offer unequivocal biological advice. The thoughtful person must agree that an increase in radiation is attendent with hazard, and that any political decision leading to an increase in radiation must be made on the basis of the gravest, most soul-searching considerations.

A second problem posed by bomb testing is the effect of ionizing radiation on germinal tissue. There is no question that mutation rates are enhanced by radiation. This was shown years ago by H. J. Muller with X-rays and Drosophila. By far the great majority of such mutations are lethal or deleterious. As background radiation increases, the average dose received by each of us increases, and presumably the total number of mutations present in our gametic cells also increases. If the total number of transmissible mutant genes is on the increase, the number of genes having a deleterious developmental effect obviously must also rise. For the sake of future generations, we desperately need information on how big an increase in the genetic burden a human population can withstand and survive!

In the populations of the cities of Hiroshima and Nagasaki that survived the atom bomb, despite the observed increase in frequency of leukemia, no substantial rise in the frequency of spontaneous abortions or in the number of stillborn infants has been noted. These findings, though fragmentary, are of interest, since the rate of spontaneous abortions and stillbirths perhaps should provide some clue about the increase in dominant mutant lethal genes in the irradiated population, and the increase in dominant lethals in turn might give at least a rough estimate of the increase in recessive lethal genes. But no increase was found. Perhaps dominant lethals in humans are rare compared to recessive lethals, or perhaps dominant lethals are quickly filtered out by cell death. Inbreeding experiments are necessary to answer this point, but such experiments are denied us.

Many indirect methods have been used to estimate the so-called "genetic load" (number of recessive lethal and deleterious genes carried in the human population). Analysis of the vital statistics of various small populations in which consanguineous marriages have occurred for many generations permits a rough estimate of genetic load.

The estimates have varied, but in general they have indicated a smaller load than would have been anticipated from studies of other organisms. For example, relatively large numbers of recessive lethals can accumulate and be maintained in laboratory populations of Drosophila, with a consequent reduction in the vitality of the population when inbred.

In human populations, the number of recessive lethals appears to be relatively small. It must be emphasized that the data on human populations are few and the analysis is uncertain. However, it is possible that lethals do not accumulate in man, but are eliminated within the first few divisions of the fertilized egg. If this is true and if the data obtained from the Nagasaki and Hiroshima populations illustrate what we can expect in human populations, then the evidence suggests that an increase in radiation may constitute a greater somatic than a germinal hazard; but irradiation and its effect on the genetic load of the human population remain virtual unknowns.

One fact does remain crystal clear. Increased radiation represents a serious human hazard, a hazard of recent origin and one that must be studied in detail. In the absence of overriding political considerations, background radiation obviously should not be increased without a sounder knowledge of the biological consequences of such an increase. How can more detailed information on human populations be obtained?

The answer, in part, must come from population studies of other organisms, and it also hinges on how successful we are in securing detailed information about the genetics of man. Techniques are now being developed that in time may enable us to grow differentiated cells in tissue culture, and we may soon be able to study differential human cells much as we now grow and study bacterial cells.—The problems are immediate and urgent and require the thoughtful attention of all of us.

# Chapter 27

# Origin of Life

## CHEMICAL ORIGIN OF LIFE

The problem of the origin of the universe is the greatest challenge the human intellect has ever faced. Indeed the origin of the universe, along with the origin of life and of intelligence, may be regarded as the fundamental question of our time. It is the purpose of this article to outline how modern science is searching for the secrets of life's beginnings.

While the very concept of the birth of the universe is staggering to the human mind, the unravelling of the riddle may come from a surprisingly simple observation or experiment. Abbé Lemaître's theory of an evolutionary universe arising from a primeval atom will stand or fall when an astronomer's penetrating gaze has compared the spatial density of galaxies 50 million light years away with those of 10,000 million light years away.

The rival cosmology of continuous creation demands the appearance of hydrogen at several billion trillion tons per second in the observable universe. This idea will be satisfactorily proved if nuclear physicists demonstrate the formation of hydrogen at the rate of one atom per year in a volume about the size of a New York skyscraper.

In our attempt to understand the origin of life, the enormity of the enigma is paralleled only by the complexity of the possible hypotheses. It is this difficulty which prompted the physicist J. D. Bernal to write in 1949: "It is probable that even a formulation of this problem is beyond the reach of any one scientist, for such a scientist would have to be at the same time a competent mathematician, physicist, and experienced organic chemist, he should have a very extensive knowledge of geology, geophysics and geochemistry and, besides all this, be absolutely at home in all biological disciplines. Sooner or later this task will have to be given to groups representing all these faculties and working closely together theoretically as well as experimentally."

Our ancestors thought that they had found a ready solution to the problem in the belief that life arose continually from the non-living. One had only to accept the evidence of the senses: worms from mud, maggots from decaying meat, mice from old linen and fireflies from the morning dew. Aristotle had propounded the doctrine of spontaneous generation in his "Metaphysics," and this teaching was accepted by the long line of European thinkers who had turned to him as the final authority in matters metaphysical and physical. The ancient Hindu scriptures described life as having evolved from non-living matter. The Rig Veda, for example, pointed to the beginning of life from the primary elements while the Atharva Veda postulated the oceans of the Earth as the cradle of all life.

The world's literature is full of allusions to the popular belief in spontaneous generation. Virgil, in his "Georgics," tells us how a swarm of bees arose from the carcass of a calf. Lucretius in "De Natura Rerum" refers to the Earth as the mother of all living things.

The Belgian chemist and physician, Van Helmont, even had a recipe for the making of mice. "If a dirty undergarment is squeezed into the mouth of a vessel containing wheat, within a few days, a ferment drained from the garments and transformed by the smell of the grain, encrust the wheat itself with its own skin and turns it into mice. And, what is more remarkable, the mice from corn and undergarments are neither weanlings nor sucklings nor premature but they jump out fully formed." Beliefs such as these could not long withstand the rigours of advancing scientific method.

Francesco Redi, a celebrated member of the Accademia del Cimento, demonstrated that the worms in putrefying flesh were larvae from the eggs of flies. His proofs were as simple as they were decisive. He showed that meat placed under a screen so that flies cannot lay their eggs on it never develops maggots.

But soon, with the [use of the microscope by Antọni van Leeuwenhoek to study natural sciences] in the second part of the 17th century, the doctrine of spontaneous generation began to reappear. Some investigators were unable to explain the origin of the varied organisms which the microscope showed in infusions from animal and vegetable matter. Among these organisms they discerned nothing that resembled the process of sexual generation. Thus they were led to assume that matter, once alive, keeps after death a special vitality.

About this time the English Jesuit, John Turberville Needham, published a work in which he described experiments to support the theory of spontaneous generation. The Italian Abbé Lazaro Spallanzani counteracted by showing that nutritive broths, sealed off from the air while boiling, never develop micro-organisms and hence never rot. Needham objected that by too much boiling Spallanzani had rendered the broth and the air above it incompatible with life.

In the middle of the last century Louis Pasteur demonstrated by a series of brilliant experiments that living organisms could not arise out of non-living material. Pasteur dealt the death blow to the theory of spontaneous generation which was based on incompetent observation and the willingness to accept the evidence of the senses. Unfortunately, his work also gave rise to the misconception that the problem of the origin of life could not be approached by scientific methods.

The question of life's beginning was therefore considered to be unworthy of the attention of any serious scientific investigator. But it is transparently clear that what Pasteur proved was that micro-organisms could not be grown from sterile starting material; his experiments had no bearing on the gradual formation of organic compounds leading to the emergence of biologically significant polymers. With the growth of organic chemistry a new approach to the problem appeared to be possible. J. J. Berzelius thought that organic compounds were produced from their elements by laws different from those governing the formation of inorganic compounds. This led him to believe that organic compounds were produced under the influence of a vital force and that they could not be prepared artificially. In 1828 Friedrich Wöhler converted ammonium cyanate into urea, a substance hitherto obtained only from animal sources. This synthesis weakened the distinction between organic and inorganic compounds.

The vitalists objected that Wöhler had obtained his ammonium cyanate from ammonia and cyanic acid, both of which were of animal origin. But already in 1781 Joseph Priestley had prepared ammonia by the reduction of nitric acid which was synthesized from its elements by Henry Cavendish in 1785. Potassium cyanide had also been obtained by K. W. Scheele by heating a mixture of potassium carbonate and carbon in the presence of nitrogen. Since potassium cyanide is readily converted into potassium cyanate, Wöhler's synthesis was truly one which started from inorganic materials. The experiments of these early synthetic chemists gradually removed the boundary between the inorganic and the organic world.

The basis of our approach to the problem is an evolutionary philosophy. The process may be traced forward from the primeval cloud of hydrogen gas or backward from the primordial germ which Charles Darwin placed at the beginning of life. The Darwinian theory of evolution has postulated the unity of the Earth's entire biosphere. According to Darwin, the higher forms of life evolved from the lower over a very extended period in the life of this planet. Fossil analysis has shown that the oldest known forms of living systems may be about 2000 million years old. Life, indeed, had a beginning on this planet. The consideration of biological evolution thus leads logically to another form of evolution, namely chemical evolution.

The evidence available from practically every field of science suggests that nature is a unity which has been divided into categories for human convenience. The division of matter into living and non-living is, perhaps, an artificial one which is convenient for distinguishing such extreme cases as a man and a rock, but would be quite inappropriate when describing a virus particle. Indeed, the crystallization of a virus by Wendell Stanley almost 30 years ago precipitated the need for revising our definition of the terms "life" and "living." These sentiments were powerfully expressed by N. W. Pirie in an essay entitled *"The Meaninglessness of the Terms 'Life' and 'Living'."* He compared the use of the terms living and non-living to the words acid and base as used in chemistry.

While sodium hydroxide is distinctly alkaline, sulphuric acid is a powerful acid. However, in between there is a whole variation in strength. The chemist has overcome the confusion arising from the use of these two terms by inventing the nomenclature of hydrogen ion concentration. He is thus able to describe all the observed phenomena in terms of one quantity; thus a fluid may have a pH of 4 or a pH of 8. According to Pirie, we may have to invent a similar quantity in order to avoid any vagueness that might arise in applying the term life to borderline cases such as the virus.

One of the first to speculate on the conditions necessary for the origin of life was Erasmus Darwin, the grandfather of Charles Darwin. In his "Temple of Nature" he had written: "All vegetables and animals now existing were originally derived from the smallest microscopic ones formed by spontaneous vitality."

Perhaps this idea had influenced Charles Darwin in his own thinking. Several years later he wrote to a friend about "Some warm little pond with all sorts of ammonia and phosphoric salts, light, heat, electricity, etc., present" in which he postulated that a protein compound was chemically formed ready to undergo still more complex changes.

About the same time, the physicist John Tyndall argued that every portion of a living organism can be reduced to inorganic matter. In his essay on vitality, published in 1866, he suggested that one could conceive of the reverse change from the inorganic to the organic, and that the special arrangement of elements in living bodies led to the phenomenon of life. In 1868 Thomas Huxley delivered a lecture in Edinburgh in which he pointed out that protoplasm was substantially the same over the whole range of living things. To him, the existence of life depended on certain molecules such as carbonic acid, water and nitrogenous compounds. These compounds were lifeless but, when brought together, gave rise to protoplasm.

To the Russian biochemist A. I. Oparin, more than to anyone else today, do we owe our present ideas on the scientific approach to the question of the origin of life. In clear and scientifically defensible terms, he pointed out in 1924 that there was no fundamental difference between a living organism and brute matter. The complex combination of manifestation and properties so characteristic of life must have arisen in the process of the evolution of matter.

The first English edition of his book *"The Origin of Life"* was published in 1938. A fourth edition of the book has recently appeared under the title of *"The Chemical Origin of Life."*

In 1928, the British biologist, J. B. S. Haldane, expressed his own ideas on the origin of life. He attributed the synthesis of organic compounds to the action of ultraviolet light on the Earth's primitive atmosphere. He suggested that the organic compounds accumulated till the primitive oceans had the consistency of a primordial "soup." Twenty years after the appearance of Haldane's paper in *The Rationalist Annual,* Bernal theorized before the British Physical Society in a lecture entitled *The Physical Basis of Life.*

Condensations and dehydrogenations are bound to lead to increasingly unsaturated substances, and ultimately to simple and possibly even to condensed ring structures almost certainly containing nitrogen, such as the pyrimidines and purines.

The appearance of such molecules makes possible still further synthesis. The primary difficulty, however, of imagining processes going thus far is the extreme dilution of the system if it is supposed to take place in the free ocean. The concentration of products is an absolute necessity for any further evolution.

The growth of biochemical and geochemical knowledge in the 20th century has stimulated a fresh interest in this age-old problem. The possibility of discovering extra-terrestrial life has given a further impetus to its experimental study. The first international symposium on the origin of life was held in Moscow in 1957. Ideas presented were grouped under headings which described five stages in the evolution of life from non-life: primary formation of primitive organic compounds on the Earth.

The transformation of primary organic compounds on the Earth; the origin of proteins, nucleoproteins and enzymes; the origin of structure and metabolism; and the evolution of metabolism. So much experimental work has since been done in laboratories all over the world that a second international symposium was convened in Wakulla Springs, Florida, in the autumn of 1963. The highlight of this meeting was the presence of both Oparin and Haldane, pioneer thinkers in the field.

Almost half a century ago, Jacques Loeb had studied the effect of electric discharges on mixtures of hydrocarbons and had suggested that a large number of organic compounds could be formed. However, the first experiment designed with the set purpose of testing some of the hypotheses on the origin of life was that of Melvin Calvin and his associates who, in 1951, treated water and carbon dioxide in the Berkeley cyclotron and obtained significant yields of formaldehyde and formic acid. In 1953, Stanley Miller, then a graduate student in Harold Urey's laboratory, assembled a sample of the assumed primeval terrestrial atmosphere consisting of methane, ammonia, water vapour and hydrogen, and exposed it to an electric discharge simulating lightning. Amino acids and other organic compounds found in living systems were formed.

Since this classic experiment, several investigators have entered the field. Notable among them are Sydney Fox of Florida State University and John Oro of the University of Houston. The majority of publications have dealt with the formation of amino acids and components of nucleic acid from a wide variety of conditions which may be considered prebiological.

Fox's work centered on the origin of proteins. A plausible answer seems to have begun to take shape. Proteinoids have been obtained by the thermal polymerization of 18 amino acids. These proteinoids have a distinct tendency to form microspheres having diameters in a bacterial range. Starting with ammonium cyanide, Oro has synthesized adenine and a number of biochemical intermediates of purines.

A starting point for any such experimental work must take into account cosmic abundances. Astronomical spectroscopy reveals that the most abundant elements in our galaxy are, in the order of rank: hydrogen, helium, oxygen, nitrogen and carbon.

With the exception of helium these are, indeed, the basic elementary constituents of all living organisms. We know from chemical equilibria that—in the presence of hydrogen—carbon, nitrogen and oxygen must exist in their reduced forms as methane, ammonia and water. The equilibrium constants for these reactions at 25° C are all of considerable magnitude. It is this atmosphere of methane, ammonia, water vapour and small amounts of hydrogen which we shall consider as the primitive atmosphere of the Earth.

The energies available for the synthesis of organic compounds under primitive Earth conditions were ultraviolet light from the Sun, electric discharges, ionizing radiation and heat. While it is evident that sunlight is the principal source of energy, only a small fraction of this is in the wavelength below 2000Å which could have been absorbed by the methane, ammonia and water (one Ångstrom is 10-8 cm). However, the photodissociation products of these molecules could absorb energy of higher wavelengths.

Next in importance as a source of energy are electric discharges, such as lightning and corona discharges from pointed objects. They occur close to the Earth's surface and, hence, would transfer the reaction products to the primitive oceans more efficiently. A certain amount of energy was also available from the disintegration of uranium, thorium, and potassium-40. While some of this energy may have been expended on the solid material, such as rocks, a certain proportion of it was available in the oceans and the atmosphere. Heat from volcanoes was another form of energy that may have been effective but, in comparison to the energy from the Sun, this was only a small portion and perhaps not widely distributed.

In our experiments with ionizing radiation, we have found that the electron beam from the linear accelerator at the Lawrence Radiation Laboratory of the University of California, Berkeley, provided us with a convenient source of electrons simulating potassium-40 on the primitive Earth. When a mixture of methane, ammonia and water was irradiated with electrons for about one hour, resulting in a total dose of approximately 1011 ergs, the largest single non-volatile compound formed was adenine. The production of adenine in this experiment was significant in the light of the multiple role played by this chemical in biological systems. Not only is it a constituent of both deoxyribonucleic acid (DNA) and ribonucleic acid (RNA), but it is also a unit of many important cofactors in living organisms.

In our experiments with electric discharges, tesla coils and luminous tube-transformers were used to simulate lightning on the primitive Earth. In a typical experiment, lasting 48 hours, most of the methane was converted into more complex organic compounds. In this experiment some of the constituents of the nucleic acid molecule have been identified.

The action of heat on the Earth's primitive atmosphere was duplicated by passing a mixture of methane and ammonia, in the presence of water vapour, through a heated Vycor tube at about 1000°C.

The effluent gases were absorbed in water. Analysis of the water-soluble material revealed the presence of several amino acids. This result has recently been reported by Fox, who identified 14 of the amino acids commonly present in protein. Analysis of the gas fraction has shown that a great portion of the methane was converted into higher hydrocarbons, including ring compounds such as benzene, toluene and anthracene.

Chemosynthesis by meteorite impact on planetary atmospheres has been suggested as a possible pathway for primordial organic synthesis. The reaction is probably a result of the intense heat generated momentarily in the wake of the shock wave following the impact. In a very preliminary experiment simulating these conditions, by firing a ballistic missile into a mixture of methane, ammonia and water vapour, we have been able to identify some amino acids and a few ultraviolet absorbing compounds which may be of biological significance.

In all the experiments just described, whether using ionizing radiation, electric discharges or heat, one of the primary products appears to be hydrogen cyanide. The second product is formaldehyde. S. L. Miller and Urey identified both these compounds in their early work. C. Palm and Calvin recorded similar results. Our own experiments point to the same conclusion. In subsequent experiments we have, therefore, used hydrogen cyanide and formaldehyde as our starting materials.

When a dilute aqueous solution of hydrogen cyanide is exposed to ultraviolet light, a wide variety of organic compounds can be formed. Among these have been identified adenine, guanine and urea. Adenine and guanine are the two purines in the nucleic acids. Urea is an important biochemical intermediate. The reaction with hydrogen cyanide appears to proceed even without an external source of energy. When a solution of hydrogen cyanide is left standing at -10°C, it is spontaneously converted into more complex organic compounds. In experiments with very dilute solutions of formaldehyde as starting material, the two sugars, ribose and deoxyribose, have been identified.

The same forms of energy that have been instrumental in the production of the sugars and of the purines have now been demonstrated to be responsible for the synthesis of nucleosides, nucleotides and peptides. When a solution of adenine and deoxyribose, in the presence of the cyanide ion, is exposed to heat or ultraviolet light, the nucleoside deoxyadenosine is rapidly formed.

This result confers a unique role on hydrogen cyanide. Not only is it a pathway for the purines but also, in the synthesis of nucleosides, it performs the function of a catalyst. Since the Earth's primitive atmosphere was reducing in nature, some ultraviolet light in the region between 2400Å and 2900Å must have reached the surface of the early oceans. The activation of dissolved purines, pyrimidines and their nucleosides which absorb ultraviolet light in this range may have been a step in the synthesis of nucleoside phosphates. To test this hypothesis, several experiments were performed.

The conversion of adenine to adenosine, adenosine to adenosine monophosphate, adenosine monophosphate to adenosine diphosphate, and the diphosphate to the triphosphate has been experimentally established. In these experiments, ethyl metaphosphate was used as a source of phosphorus.

It can be expected that this was not the most abundant source of phosphorus on the primitive Earth but it is likely that other, more abundant, phosphate salts can substitute for ethyl metaphosphate.

The synthesis of peptides under abiological aqueous conditions has been considered to be a matter of great difficulty. In our laboratory, however, we have recently found that when an aqueous solution of glycine and leucine was exposed to ultraviolet light in the presence of cyanamide, the dipeptides glycyl-glycine, glycyl-leucine, leucyl-glycine, leucyl-leucine were formed.

The action of cyanamide may be analogous to that of dicyclohexylcarbodiimide which has been extensively used by H. G. Khorana and his co-workers for the synthesis of nucleotides.These results show that, under simulated primitive Earth conditions, molecules of biological significance can be synthesized.

They lend support to the hypothesis of chemical evolution because the starting materials are the simple constituents of the primordial atmosphere; the sources of energy used are those that are most likely to have existed under primitive Earth conditions; the concentrations of materials used are very small; and the conditions are aqueous.Recent studies in quantum biochemistry by B. Pullman and A. Pullman of Paris have thrown new light on some very significant aspects of chemical evolution.

It is a striking fact that many of the molecules which are essential to living systems are conjugated systems exhibiting the phenomenon of electronic delocalization. In the nucleic acids, for example, the purines and pyrimidines are conjugated systems.

Although the proteins do not, at first sight, appear to enjoy this property, a closer look shows that the matrix of hydrogen bonding which exists in a protein molecule provides a certain measure of electronic delocalization. In the high energy phosphates, there is interaction between the mobile electrons of one phosphoryl group with those of another.

The porphyrins, such as chlorophyll and haem, which are of paramount importance in living systems, are highly conjugated molecules. This leads to the following conclusions:

- Evolutionary selection used the most stable compounds.
- On account of electron delocalization these compounds were best adapted for biological purposes.
- The possibility of life as we know it was made more probable by the appearance of the compounds.

There is no reason to doubt that we shall rediscover, one by one, the physical and chemical conditions which once determined and directed the course of chemical evolution. We may even reproduce the intermediate steps in the laboratory. Looking back upon the biochemical understanding gained during the span of one human generation, we have the right to be quite optimistic.

In contrast to unconscious nature which had to spend billions of years for the creation of life, conscious nature has a purpose and knows the outcome. Thus the time needed to solve our problem may not be long.

## CHEMICAL EVOLUTION

How did life come to be on the surface of the Earth? Darwin himself recognized that his basic idea of evolution by variation and natural selection must be a continuous process extending backward in time through that period in which the first living things arose and into the period of "chemical evolution" which preceeded it. We are approaching the examination of these events by two routes.

One is to seek for evidence in the ancient rocks of the Earth which were laid down before the time in which organisms capable of leaving their skeletons in the rocks to be fossilized were in existence. The Earth is believed to have taken its present form approximately 4700 My ago. We have found in rocks whose age is about 1000 My certain organic molecules which are closely related to the green pigments of plants, chlorophyll. This seems to establish that green plants were already flourishing before that time.

We have now found in rocks of still greater age, namely, 2500 My, the same kinds of molecules mentioned above which can be attributed to the presence of living organisms. If these molecules are as old as the rocks, we have thus shortened the time available for the generation of the complex biosynthetic sequences which give rise to these specific hydrocarbons (polyisoprenoids) to less than 2000 My.

The second approach is to attempt to reproduce in the laboratory those chemical processes induced by energy of various kinds—radiation from the sun, from radioactivity, from electrical storm, etc.—which could give rise to simple organic molecules and polymeric combinations of them, ultimately leading to systems which could be called alive. This attempt has also succeeded along a variety of lines, and many of the present-day biologically important molecules have been constructed abiogenically from the primeval atmosphere, thus providing laboratory evidence for the hypothetical processes. The latest step in this approach has been to demonstrate the formation of polypeptides in dilute aqueous solutions through the agency of molecules formed in the primitive atmosphere of the Earth.

Finally we must seek evidence for the same processes in material found elsewhere than on the Earth, such as other parts of our solar system, e.g. the Moon and Mars. We can expect to know whether such materials exist at all in the rocks of the Moon within this decade. We may even know something more definite about the botany of Mars during this same period.

## LIFE BEGINS

The planet earth was solitary and desolate four billion years ago. Oceans tossed breakers against lifeless shores. Often dark clouds blotted out the searing sunlight, and lightning stabbed from thunder clouds. Volcanoes erupted, pouring lava over the barren land and along the ocean bottom. You could not have breathed the atmosphere that surrounded the globe then. The air was made up of ammonia, methane, hydrogen, and water vapor.

Yet somehow, out of this seemingly inhospitable environment, life emerged. There are a number of explanations of how it might have happened. One of the most widely accepted assumes that the vast amounts of energy available acted on the simple gases in the atmosphere to produce the substances that led to life. Most of the energy came as ultraviolet radiation from the sun, but there was also energy in the form of heat from volcanoes and radioactive elements in the earth, and electricity from lightning. Once the right substances were produced, it was likely that life would emerge.

The substances that could have been produced included amino acids, sugars, proteins, and organic acids such as acetic acid (vinegar) or the lactic acid found in milk and muscles. These compounds rained down and collected in the oceans until they made a "hot thin soup," to use the phrase of the British biologist J. B. S. Haldane ( 1892-1964), who had many pioneering ideas about the origin of life. From this soup more complicated materials developed.

It may seem as though we are indulging in science fiction here. But the basis for this sort of thinking was worked out in great detail by the Russian biochemist A. I. Oparin ( 1894-) in his book *On the Origin of Life.* Oparin suggested several steps along the path to life that scientists have been able to confirm by laboratory experiments.

## A KEY EXPERIMENT

One of the most important of these experiments was conducted in 1953 by Stanley L. Miller, then a graduate student working in the laboratory of Harold C. Urey at the University of Chicago. Into a glass apparatus Miller put a mixture of the gases that probably made up earth's early atmosphere. For a week this mixture was subjected to continuous electrical discharge from a spark coil. When the products of the experiment were analyzed, Miller found several new materials, particularly amino acids. Since this pioneering investigation, other scientists have confirmed and extended Miller's experiments.

For instance, John Oró of the University of Houston in Texas heated a concentrated solution of hydrogen cyanide (H-Ca"N), and ammonia in water for several days at temperatures up to the boiling point of water. He used hydrogen cyanide as a starting material because it is formed in large amounts in Miller's procedure, and because it has been detected in the tails of comets. When Oró analyzed the mixture that had been heated, he found adenine in his solution. And adenine, as we know, is one of the two purines of DNA.

Cyril Ponnamperuma of the National Aeronautics and Space Administration's Ames Research Center in California demonstrated that adenine could be produced by another process. His technique involved exposing ammonia, methane, and water to high-energy electrons from a device known as a linear electron accelerator.

The first experiments of this sort were carried out by Melvin Calvin of the University of California. Further ingenious experiments were performed by Philip Abelson of the Carnegie Institution of Washington and others. These are only a few of the investigators in this field, and the number of researchers is increasing.

The important point about the research that has been completed is this: It is clear that an atmosphere rich in hydrogen and hydrogen compounds is essential for the synthesis of the substances that support life. So far no one has succeeded in making compounds of biological importance using gases or solutions to be found as part of an oxygen-rich atmosphere such as the earth has today.

## THE EFFECTS OF ULTRAVIOLET RADIATION

And so some of the pieces in the history of life are beginning to fall in place. We have not explained one important part of the story, however. How did the hydrogen-rich atmosphere of those early ages change to the atmosphere we breathe today?

The ultraviolet radiation that long ago penetrated to the earth's surface affected the water vapor in the atmosphere. It broke the bonds holding the atoms together so that hydrogen gas and oxygen gas were produced. Gradually the very light hydrogen escaped into space and the heavier oxygen stayed behind, making up a greater percentage of the atmosphere. Other reactions led to the breakup of ammonia and methane, too, and the escape of more hydrogen.

This mechanism did not provide all the oxygen in the air today, since much is produced by plants from carbon dioxide. But it was an important step in the earth's development. We not only breathe oxygen—it also protects us from the burning effects of ultraviolet radiation. This radiation breaks the bonds of oxygen molecules, $O_2$, to produce a much more active form of the gas, called ozone, which is written $O_3$.

There is a layer of ozone 14 to 15 miles up in the sky. This layer, known as the *ozonosphere,* might be thought of as a coating of suntan lotion for the earth, screening out the ultraviolet but allowing other radiation from the sun to pass through. Today enough ultraviolet still penetrates to the surface to make a sunburn painful or even deadly on occasion.

## CHEMICALS THAT HELP THEMSELVES

Even after producing the building blocks of life's chemicals, we are a long way from producing those substances that actually support life. How could proteins and nucleic acids have come into being? One series of experiments has demonstrated that artificial proteins can be made by heating amino acids under the proper laboratory conditions. These experiments were carried out by Sidney W. Fox of Florida State University.

Dr. Fox has been able to make microscopic spherules (sphere-shaped bodies) that appear to behave in water very much like the natural proteins of a living cell. This is not to say that natural proteins were first formed by such processes. But the studies do show that it is possible to synthesize these complicated substances outside the intricate chemical-manufacturing plant of the cell.

There are many ways life's chemicals may have been put together originally, and only a few of these processes have been satisfactorily duplicated in the laboratory. As you read in the next few paragraphs about one of the ways that has been suggested, you should bear in mind that all of these ideas are subject to revision as more information is acquired in the coming years. Furthermore, the steps described represent only one line of thinking; many other approaches are being explored by biologists in many parts of the world.

One step in the development of the molecular structure of proteins and nucleic acids has been suggested by Dr. Calvin. It would depend on substances known as *catalysts,* materials that can speed up or slow down the rate of a chemical reaction. There are catalysts that speed up the rate of their *own* formation.

These are known as *autocatalysts*. When an autocatalyst is produced, it tends to cause the formation of more molecules like itself. An autocatalyst would have had a great natural advantage over other substances, for it would have become more plentiful than substances that lacked this characteristic. This would have permitted a kind of chemical evolution.

Oparin has pointed out that certain proteins tend to attract each other when they are in a water solution. This attraction occurs because the proteins are electrically charged. These clumps of proteins are then encased by water molecules which form a sort of skin. The clumps with their water "skins" look very much like amoebas—the single-celled organisms that are among the simplest animals known. Oparin discovered that certain chemical reactions that occur in living cells will take place in such clumps. So it is conceivable that living cells might have been formed by a clumping of molecules. Another characteristic of large molecules is their tendency to settle out of solution in a flattened form, like stacks of molecular waffles.

## HOW DID CELLS DEVELOP?

Given molecules with more than one of the characteristics we have discussed, one can imagine what might happen. For instance, the key links in DNA are large, flat molecules that might have tended to settle in layers. If a very elementary nucleic acid were to be assembled this way, it does not seem unreasonable to speculate that eventually some duplication of the molecule might occur.

This seems particularly likely in view of the fact that a synthetic DNA that is autocatalytic has been manufactured in the laboratory. This DNA has only two bases instead of four—adenine and thymine without guanine and cytosine. When placed in a suitable mixture, this DNA stimulates production of more DNA.

Thus it might have been possible for substances manufactured from the raw materials in the earth's early atmosphere to have gathered in the oceans. They could have interacted and, by well-established natural processes, produced primitive cells having some of the characteristics we associate with life.

Whether this hypothetical pathway can be substantiated by experiment remains to be seen. Many of our ideas about the beginnings of life may have to be revised. Many scientists are confident, though, that they will be able to demonstrate how life's progression from nonliving forms might have occurred.

It must be clearly understood, however, that there is a huge gap in our knowledge about the way building-block compounds came together to form living organisms capable of reproduction, mutation, and response to their surroundings.

Although the ideas we have discussed here were put forward by highly qualified scientists, there is as yet no real evidence to help us fill the gap. All is speculation, and we will have to hope that as yet unperformed experiments will provide us with the clues we need to begin piecing together this part of the picture.

## LIFE WAS MORE THAN ACCIDENTAL

Nevertheless, the evidence about the formation of the building blocks has led scientists to become more and more convinced that life on earth did not come about by mere accident. Given the chemical conditions after the earth had coalesced, and given the inherent characteristics of these chemicals, it seems very likely that life should have come into existence.

That it took perhaps two or three billion years is no surprise. In the universe, time is not measured in the brief lifespan of man, but in eons. Sometime, somewhere, when all the conditions were right, life began. We probably will never be able to say "Life started then," and name a specific time. It may have started when the first molecule of some DNA-like substance was formed. Even though it would have been a small occurrence it would have been a profound one.

You might well ask at this point whether life is originating today, right here on earth. If it is so probable an occurrence, why can't we see it coming into existence now, under our very noses? This is a fair question, and there seems to be a good answer for it. In the distant past, the collections of molecules leading to life were able to accumulate in the first seas because there were no organisms about to "eat" them or use them up.

There were no complicated chemical life forms taking in simple foodstuffs and converting them to sugars and starches, proteins and nucleic acids. Today the chemicals leading to life are consumed before they can generate their own new forms. This is evolution in action again. The organisms already in existence are far better able to compete for food and vital materials than the simpler chemicals that once spawned these organisms.

Although we seem to have made a good guess as to which path life took in its meanderings through time, we still have many puzzles to solve. What we would like to know most of all is how specific proteins and nucleic acids first appeared. If we knew this, we would be better prepared to understand the further development of life. It is here that space exploration could help us complete our chart. We might find on other planets, or in the dust that swirls between them, some signs of life.

We have shown you something about the nature of life as a chemical process and from the standpoint of evolution. You have seen how living things might have developed, and you can follow future events in this area, events that are sure to make headlines for years to come.

## ON THE ORIGIN AND EVOLUTION OF LIVING MACHINES

If we wish to contrast the living and the non-living we may compare a living cell with a mixture of chemicals in a test tube. If we wish to find similarities we may do better to compare the living cell with a man-made machine, say an automobile, which may also be contrasted with the contents of the test tube.

Clearly we would not try to describe an automobile by grinding up its various parts and subjecting them to chemical analysis, and we would not expect to learn all about the living machine by following, exclusively, a similar attack. It seems hardly necessary to say this and yet we seem at times to go astray in just this direction, whether we are trying to study the nature of viruses, the growth of cancers, the mechanism of genetic inheritance, or, as in the present case, the origin and evolution of life.

A certain parallel may be drawn between the non-living machine and the living machine in that both have evolutionary histories even though these are of different kinds. In tracing the evolution of the nonliving machine we might begin a few hundred thousands of years ago when the first usable tool was chipped out of a piece of stone by an ancestor of modern Man.

Then trace the story through innumerable steps including the first smelting of metal in Neolithic times, and the development of the effective steam engine less than two centuries ago, to the modern automobile and other complex machines. Viewed in this way it is seen that the evolution of the non-living man-made machine pre-supposes the evolution of Man himself; and this takes us into the history of living machines, as one of which Man has evolved.

With the story of the non-living machine so entangled with that of the living machine, the definition of a point of origin for the former must be vague or quite arbitrary. And if we try to imagine the various functional components of a living machine emerging from a mixture of chemical compounds—which we think happened sometime within, say, the last four thousand million years—we meet similar difficulty in defining the moment of emergence; perhaps we do better to think of the origin of life as spread over a considerable interval of time, blurred to our view with regard to both the kind and order of the steps that occurred. How the chemical compounds from which the living machine could be constructed, themselves came into being may be regarded as another problem.

But again, separation of this problem from that of the origin of the living machine itself may be quite arbitrary. The analogy between living and nonliving machines probably should not be pressed much farther than to point out that neither could be expected to arise spontaneously from a mixture of their component molecules; that is, without some special conditions or events extraneous to the simple chemical mixture itself. It should hardly be necessary to point out that I do not mean by this statement that extraphysical factors or vitalistic concepts need be invoked—the present argument is surely a mechanistic one.

We could find many properties of living machines that do not have their counterparts in non-living ones; but perhaps that which is most striking and most important from the standpoint of origin and evolution is the property of self-replication. Up to now such a property has not been built into any non-living machine. Self-replication of parts, and of the machine as a whole, is of particular importance in the present discussion; for if we are to assume that evolution of living systems has taken place by a mechanism of variation and natural selection, using these terms in the sense of modern Darwinism, this property becomes of paramount importance.

The existence of a distinguishable biological species depends upon the accurate replication of a pattern which determines the characteristics of that species. And yet if this pattern were always copied exactly there could be no evolution by natural selection: for this, errors in copying must occur and be replicated in subsequent generations. If by such a copied error—or mutation— a characteristic is conferred that better fits the species to the environment in which it has to exist, the chances of survival of that species are increased, whereas species with disadvantageous mutations tend to be eliminated. This is the process of evolution by natural selection described in oversimplified fashion.

Our knowledge of some aspects of the process of replication is increasing rapidly at the moment—so rapidly that we may at times forget, in our enthusiasm, that the origin of a replicating machine is something different from the copying of such a machine once it exists.

The point was emphasized by the late John Von Neumann who, while conceiving a machine that could replicate itself, admitted his inability to imagine a machine that could create itself. The difference between replication and origin was pointed out by Sir William Hardy a good many years earlier in an essay entitled, *"To Remind"* (which seems to have reminded few), basing his argument on the difficulties of the origin of a machine that could manufacture one only of two possible optical isomers, as living organisms do.

But to be more specific, let us consider the process of self-replication as we see it in modern living machines. And for the sake of argument let us try to reduce this process to a minimum of functional components that have to do with replication, leaving aside many other intriguing properties of living systems. If we are to give any physical solidity to our ideas of replication it seems necessary to assume some sort of spatial pattern, which we may think of as a template composed of large molecules having specific configurations that provide necessary genetic information for the construction of the individual organism making up the species.

We have to think that not only are materials replicated by building against this template but that in some way the template itself is replicated in the course of self-duplication of the machine as a whole, so that the pattern can be passed along from generation to generation. Such replication entails the expenditure of free energy, since assembling the material against the template and taking the replica off cannot both be spontaneous acts; so there must be an energetic, or thermodynamic, component of the machine. And, as we shall see later, the mobilization of free energy in the living cell also entails a kinetic component.

Thus it is necessary to think of the replicating machine in terms of no less than three functional components; let us refer to them as spatial, energetic, and kinetic. So in thinking about the origin and evolution of such a machine we must take into account the origin of all three of these components rather than of any one of them alone. It is difficult to conceive how a machine that embodied all three could have come into being as a sudden act; and it would seem more likely that the different components were introduced separately.

At any rate, until the complete self-replicating machine had emerged, evolution by natural selection, in the sense in which this word may properly be used, was not possible; and here an error may creep into our thinking about the origin of living systems.

For natural selection is sometimes wrongly invoked, directly or tacitly, to explain the origin of things which had to be present before evolution by natural selection could take place, that is, to explain events leading up to the appearance of the self-replicating machine in terms of things that could have happened only after such a machine was already in existence. This incorrect application of the concept of natural selection may lead into the realm of teleology and finalism. Undoubtedly there was going on before living machines were perfected to the point where natural selection was possible, a process which I like to call "chemical evolution"; and from this evolution resulted many types of compounds that are indispensable components of living machines today. An outstanding example is the amino acids, which, combined in proteins, are found playing basic roles in all living species.

## SOME IMPLICATIONS

What are the implications of the argument that has just been presented? First, let us review the similarities and differences between choice of mutants by natural selection and choice of reaction pathways in nonliving systems. Perhaps the differences and the similarities one sees depend somewhat on where one stands. I think most biologists recognize intuitively a difference between the two processes, whereas chemists may stress the similarities.

Applying the same framework of physicochemical mechanism, as I have attempted here, may emphasize similarities, tending to subordinate the differences in evolutionary result. But the functional differences remain nevertheless real.

In chemical evolution in a non-living world the direction of the pathway taken is a function of thermodynamics and kinetics, even when the expected direction of reaction is upset by radiant energy or other physical factors. In such a system, the reaction pathway may respond immediately and directly to a change in the environment, say the introduction of a catalyst which speeds up a given reaction step. The catalyst may be a product of the reaction itself, which then is an autocatalytic reaction; but this does not alter the picture very much because it is difficult to distinguish between such a reaction system and its environment.

In the living, replicating system, on the other hand, this distinction is more clear-cut, there being little direct influence—and virtually no specific influence—of the external environment on replication of pattern. In chemical reaction, a given molecule which has reached a high enough energy level to react is for the moment different from its fellows to that extent; but, the reaction once achieved, that particular molecule is not distinguishable from other product molecules.

The molecule does not maintain its individuality with respect to its energy content beyond a fraction of a second—the product state represents a new average condition, so in the long run it is the average and not the individual that counts. The same argument applies to formation of free radicals or intermediates in chemical reactions, which are transitory steps in the reaction that do not replicate and so perpetuate their kind.

Presumably, mutation too involves a temporary increase in energy at some place in the molecule in order to get over a barrier, but this is a fleeting change. In the case of a "successful" mutation the resultant alteration in pattern is maintained by replication. The individuality of the mutant pattern may thus be preserved for a very long time, being duplicated over and over again and so persisting through many generations, long after the particular molecule that was originally altered has ceased to exist as such. There is a sort of "memory" in this process that does not enter into strictly chemical evolution.

Natural selection of a new pattern is a slow process in which many generations may be involved in accomplishing a single step; it amounts to picking out altered replicating patterns, that is, mutated genes. This is not accomplished directly, but by the overt changes the mutation produces in a larger system, the cell or multicellular organism, that contains the new pattern.

Thus the gene is isolated from direct impingement of the environment upon it; and if we wish to consider natural selection as a feed-back mechanism we must include the environment in our system, and the impingement of the organism and environment may be difficult to analyze.

We see from all this that living machines may, in a sense, arrest the ordinary time course of events to an extent not paralleled in any simple, cyclic, physical, or chemical system, at least any that I am able to picture. And this offers correspondingly greater opportunity for the development of complexity in the living world as compared to the non-living. Does the picture I have presented modify in any way our ideas about evolution by natural selection?

Basically I think it strengthens the concept by giving it a more intimate and plausible physical basis. It is a great tribute to Charles Darwin that we are still seeking today in the. light of a century of scientific progress to explain with greater exactitude the principle he laid down, and which we still accept.

But, on the other hand, consideration of the more intimate mechanism may tend to weaken our anticipation of perfection in the results of evolution by natural selection. It emphasizes to us that while mutations may be regarded as accidental events there are great limitations as to the kinds of accidents that can occur.

Fundamentally these limitations are set by properties of the molecules that make up the cell; but how much these properties have influenced the course of evolution by natural selection would be difficult to express in rigorous form. It is also difficult to know to just what extent one must attribute the direction taken by biological evolution to the direction taken by the chemical evolution that preceded it. We find reason to think that some substances are common to all living systems because they were present at the beginning as the result of previous chemical evolution. The amino acids, for example, seem already to be reasonably accounted for in this way, and no doubt other essential components of living material will be.

There are certain dominant themes in the biochemical tapestry, but there are many details that seem to have been woven with considerable vagary suggesting both the guiding hand of natural selection and incidental "accidents" of mutation, some of which may have little or no survival value. We are constrained to view the role of natural selection in terms of the fossil record since the beginning of the Cambrian, 500 million years ago; and one of the things this record tells us, when we combine the evidence with that of comparative biochemistry, is that the major molecular patterns of inheritance must have been pretty well-established before that time.

What has happened since would seem to have involved relatively small changes, considering the degree of biological organization already established then. But living systems had existed and evolution has been going on for a long time before the beginning of the Cambrian—possibly for two or three thousand million years. The major themes of organic evolution were established in the course of this earlier period, and it seems likely that strictly physico-chemical factors played a more prominent role in setting limits upon the direction natural selection could take then than they did later.

But however the limits were set they must have continued to restrict the possible directions evolution might take. The first replicable and mutable cell would seem to have had before it the possibility of vastly more mutations than has any modern cell of any existent species.

Of course, I do not mean by this that the primordial living system contained in a physical sense all these possible patterns, for surely more became possible by recombinations in the course of the many millions of years of evolution than were present in the first living machine. What I mean to say is that the limitations on the direction of evolution are much greater for any existing species than they were for the primordial system from which it ultimately stems.

The limitations imposed by physico-chemical factors upon the possible directions of mutation must constitute restrictions on the degree of perfection of adapation to environment. I think this is sometimes lost sight of. There seems to be widespread tendency to disregard such limitations and to assume complete randomness of mutation, which may encourage overemphasis of the perfection of adaptation of the organism to the environment, and encourage a kind of teleology that may be misleading—we may risk turning Darwinism into a cult. This may also encourage false social analogies.

As regards the idea of randomness in mutation, we must remember that this is something we use as a tool and should not let it confuse us in thinking about evolution in an over-all or global sense. What we mean by randomness of mutation is that at a given instant, given a particular set of conditions, there is an array of possibilities for mutation; and for purposes of analysis, we may proceed to treat this array as though it were a fixed thing. But thinking evolutionarily we see that the possibilities for mutation are continually changing, and the environment as well, and that used in this way the term, randomness, only applies to a given instant we have chosen to study.

The question whether life exists on other worlds and how it originated there has occupied thinkers for a long time but there has been recent increase in interest, particularly since Sputnik and the development of the so-called Space Age. With recognition of the vast extent of the universe, which seems only to grow larger as we extend our observations, there has been a tendency to people this universe with living organisms.

It may be estimated that there is a very large number of planets in our galaxy that are sufficiently like our earth to support life as we know it. But to assume on this basis that such life exists there, tacitly admits that, given the same kind of a chemical environment as once existed on our earth, living organisms will come into being as a matter of course.

If the conditions for origin of life are more exacting, the probabilities become less, however, and such estimates need to be revised downward. From what has just been said it seems that one should assume more specific conditions than just an appropriate mixture of molecular species in order for a living, replicating machine to arise. Just how detailed the specifications were to permit the origin of life on this planet, and just how many rarely probable events were involved, it would be very difficult to estimate.

When we think of the broad array of living species which exists, we realize that their evolution by natural selection has also involved a great many rare events. For each mutation represents a rare event and there must have been many mutations involved in the evolution of each species. As we look back, it appears that had things happened in slightly different ways our present array of species might be quite different.

The farther back we go the more critical the picture becomes, that is, a mutation happening sometime early in pre-Cambrian time might have had much more effect on the direction of evolution of species than a mutation that occurred, say only a few million years ago.

If we include in our thinking Man's cultural evolution, which has entailed many rare events of a still different kind, we see that the chances of finding beings very much like ourselves elsewhere in the galaxy may be small indeed, even though there may be many planets that could have accommodated them, had the hazards of history followed the same course.

That there may somewhere be replicating machines constituted from the same molecular species as our own seems somewhat more likely, but the conditions and sequence of events that were required to establish them may make this, too, a tenuous possibility. Such thinking should, perhaps, not dampen enthusiasm for the exploration of the possibility of life on other planets, but it might temper our expectations of what we may find, and affect planning as regards what to look for.

A great deal of what I have said in this paper has, of course, been in the nature of speculation, but I hope that I have paid reasonable respect to established observation and theory in the various fields of science it has been necessary to touch upon. No doubt I have stretched both the chemical and the biological picture but I hope I have not distorted them unduly in trying to get them into the same frame. It seems to me it is time to take that risk; for have we not been trying for too long to approach many problems of life from opposite ends, with only a vague hope that we may somehow come together in the middle.

## Chapter 28

# Spontaneous Generation

From antiquity up to the end of the nineteenth century, there existed a theory called "spontaneous generation." This entailed a belief in the origin of living forms, both plant and animal, from nonliving substances. Swallows and frogs, for example, were thought to spring from mud in the spring. Likewise, flies were generated from pieces of meat. Today, belief in spontaneous generation has largely disappeared from the minds of educated people.

The theory opposed to this is called "biogenesis" and it can be described as favoring the genesis of plants and animals from other living things; in other words, all life comes from life. So many modern concepts and practices depend upon the fallacy of the theory of spontaneous generation that life would be considerably changed were it true.

The belief that biogenesis is true enables biologists to control nature for the good of mankind and to predict what is likely to happen in nature in the light of our present knowledge. In the field of entomology, for example, we can predict that the eggs of a certain kind of insect came from the female of that same species and that the eggs will eventually develop into an insect like one of its parents. By destroying the eggs or larvae of this insect we can control its numbers on the earth, an especially desirable situation if the insect in question is carrying malaria or yellow fever organisms.

The practice of sterilization in the laboratory, the hospital, and elsewhere is grounded in the firm conviction that organisms do not arise without parents. The housewife heats her fruits and vegetables for a certain length of time at a certain temperature, seals the cans or jars, puts them away, and feels certain that, barring a failure in sealing, the cans or jars will remain unspoiled and wholesome. This means that the organisms in them have been killed and that others cannot arise from the contents by spontaneous generation. In the laboratory, culture media are subjected to heat and pressure for predetermined times and will remain sterile indefinitely.

Surgical instruments are likewise sterilized in the hospitals and remain in that condition until needed. Pasteurization, named after the great Louis Pasteur, is also an adaptation of our knowledge about biogenesis. In the pasteurization of milk, wine, or beer, for example, the harmful organisms are killed and others of the same kind can not arise as long as the mixture is kept unopened.

The germ theory of disease also rests upon the foundation of biogenesis. For example, one of the troublesome problems of surgery was to prevent infection during an operation. Hospitals, as late as the middle of the last century, were in a very deplorable state. To decide upon an operation was tantamount to a death sentence because almost all surgery was accompanied by infection and death. The medical profession was very conservative and refused, for a long time, to realize the presence and importance of bacteria.

Antisepsis, which is said to have started with the great physician Dr. Joseph Lister, Professor of Surgery at the University of Edinburgh, means the control of disease-causing organisms; in this case, on the operating table. If these organisms can be killed off completely, an infection will not occur and the patient has a much better chance of recovery. Dr. Lister's work, with phenol as the antiseptic agent, stems directly from the work of Louis Pasteur. Amputations which were almost universally fatal were made safer. Dr. Lister, for example, had only six fatalities out of a group of forty such cases, a remarkable achievement in those days.

It took several centuries to disprove the theory of the spontaneous generation of life. The following history of the struggle leading to the death of the theory illustrates many of the problems of scientific investigations.

## EARLY IDEAS

Science has been one of the most successful fields leading to the improvement of man's lot upon the earth because, while working in the present, it leans very heavily upon all verified data of the past. Let us take a look backward into history to learn what we can of the early trends of thought on the subject of the continuity of life.

It can be said that early man had no doubts as to his own immediate ancestry; the fact of human birth was self-evident. He also knew the origin of many of the common animals. He probably realized, too, that the way to obtain a desired plant for agricultural purposes was to gather seeds of that plant and sow them in the ground.

The well-known Greek, Aristotle, had some fairly modern ideas about reproduction. He said: The nightingale produces her young at the beginning of summer. She produces five or six eggs—insects copulate and produce their young during the winter —wild animals produce their young once a year, unless, like the hare, they breed while they are nursing their young—fish also generally breed once a year—the scorpius breeds twice—the crustaceans incubate upon their ova, which are placed beneath them.

In various places in Aristotle *History of Animals* we find mention of such things as ova, copulation, semen, uterus, and so forth. Naturally, there were comparatively few investigators in those days and large segments of knowledge remained undiscovered. Aristotle had a tremendous wealth of information but there were many things that he could not know. For example, he could not completely understand the developmental cycle of certain insects. He said:

Whatever are produced spontaneously in living creatures, in the earth, or in plants, or in any part of them, have a distinction in the sexes, and by a union of the sexes something is produced, not the same in any respect, but an imperfect animal, as nits are produced from lice and from flies and butterflies are produced egg-like worms, from which neither similar creatures are produced, not any other creature, but such things only."

Knowledge almost always proceeds from the obvious to the less obvious. Aristotle knew of sex and the union of sexes—he knew the facts of generation when the offspring resembled the parent, but he was somewhat at a loss to follow an insect egg through a caterpillar stage to a resting stage and back to the adult again.

From such cases and others dealing with less easily observable species, Aristotle and those who worked with him and after him sponsored the doctrine of spontaneous generation. When a life history was not clear, this doctrine was invoked. Some of his ideas on this follow:

Some (plants) are produced from the seed of other plants, and others are of spontaneous growth; some animals are produced from animals of a similar form, the origin of others is spontaneous and not from similar forms—those which spring from putrid matter, this is the case with many insects; others originate in the animals themselves, and from the excrementitious matter in their parts— (the productions of others) is spontaneous, for some of them (insects) spring from the dew which falls upon plants. Some originate in rotten mud and dung.†

Eels, which are fish, have always been somewhat mysterious creatures. Their life cycle was a puzzle for centuries and even at the present time many facts about them are not yet known. We cannot understand how adult eels find their way to the breeding grounds in the region of the Sargasso Sea. We do not know how it is possible for the young eels, hatched there, to find their way over many miles of ocean back to their parent's home. It is not surprising that, several hundred years before Christ, Aristotle had this to say about eels—

Eels are not produced from sexual intercourse, nor are they oviparous nor have they ever been detected with semen or ova—the originate in what are called the entrails of the earth, which are found spontaneously in mud and moist earth.

It is very interesting to remember the Bible teaching about the creation of life as recorded in Genesis. Those scientists who favored biogenesis and those who advocated spontaneous generation, at least for certain organisms, quoted the Bible as favoring their side of the question. The biogenesists said that God made man and all other organisms at one time and "set up the rules" of reproduction and lineal descent. Presumably there could be nothing new on the earth. The other school pointed to the statement by the author of Genesis that God made man out of the dust of the ground. He made organic things (man and other creatures) out of lifeless material, dust.

In the Bible there is also an interesting account of bees arising from the body of a lion. This is found in the book of Judges and reads as follows:

Then went Samson down—and, behold, a young lion roared against him. And the Spirit of the Lord came mightily upon him and he rent him as he would have rent a kid. And after a time—he turned aside to see the carcass of the lion; and, behold, there was a swarm of bees and honey in the carcass of the lion. We should hasten to add here that the above passage does not prove that the biblical writers believed in spontaneous generation.

## SEVENTEENTH-CENTURY INVESTIGATIONS

All that is required is to cork up a pot containing corn with a dirty shirt; after about twenty-one days a ferment coming from the dirty shirt combines with the effluvium from the wheat, the grains of which are turned into rats, not minute and puny, but vigorous and full of activity.

So may he doubt whether in cheese and timber, worms are generated; or if beetles and wasps in cow dung; or if butterflies, locusts, grasshoppers, shell-fish, snails, eels, and such like, be procreated of putrefied matters, which is apt to receive the form of that creature to which it is by formative power disposed.

To question this is to question reason, sense, and experience. If he doubts of this let him go to Egypt, and there he will find the fields swarming with mice begot of the mud of Nylus, to the great calamity of the inhabitants.

It seems almost incredible that the life cycle of mice and rats should not be fully known. It would have been a simple matter to put a male and female together and note that the young come from the union of the sexes and not from rags or mud. That would entail making an experiment and men at this stage in history still preferred speculation to experiment. In fact, they probably had only the haziest idea of experimentation.

That is why the simple experiments of Redi, to be recounted next, caused such surprise. People were so accustomed to saying "it is written" or "so says Aristotle" that they were startled when Redi, disregarding authority, proved that certain insects came from eggs and were not spontaneously generated in meat.

Francesco Redi, poet and physician to the Grand Dukes of Tuscany, became interested in the relationship between flies and meat. He was possessed of great curiosity and, apparently, a unique originality. His great hypothesis of 1668 is described below in his own words.

The Earth, after having brought forth the first plants and animals at the beginning by order of the Supreme and Omnipotent Creator, has never since produced any kinds of plants or animals, either perfect or imperfect; and everything which we know in past or present times that she has produced, came solely from the true seeds of the plants and animals themselves, which thus, through means of their own, preserve their species.

And, although it be a matter of daily observation that infinite numbers of worms are produced in dead bodies and decayed plants, I feel, I say, inclined to believe that these worms are all generated by insemination and that the putrefied matter in which they are found has no other office than that of serving as a place, or suitable nest, where animals deposit their eggs at the breeding season, and in which they also find nourishment; otherwise, I assert that nothing is ever generated therein.

Redi made a number of experiments to validate his belief. His most classical ones are recorded as follows:

I continued similar experiments with the raw and cooked flesh of the ox, the deer, the buffalo, the lion, the tiger, the dog, the lamb, the kid, the rabbit; and sometimes with the flesh of ducks, geese, hens, swallows, etc., and finally I experimented with different kinds of fish, such as sword-fish, tuna, eel, sole, etc. In every case, one or other of the above-mentioned kinds of flies were hatched, and sometimes all were found in a single animal. Besides these, there were to be seen many broods of small black flies, some of which were so minute as to be scarcely visible, and almost always I saw that the decaying flesh and the fissures in the boxes where it lay were covered not alone with worms, but with the eggs from which, as I have said, the worms were hatched.

These eggs made me think of those deposits dropped by flies on meats, that eventually became worms, a fact noted by the compilers of the dictionary of our Academy, and also well known to hunters and to butchers, who protect their meats in Summer from filth by covering them with white cloths. Hence great Homer, in the nineteenth book of the Illiad, has good reason to say that Achilles feared lest the flies would breed worms in the wounds of dead Patrocles, whilst he was preparing to take vengeance on Hector.

Having considered these things, I began to believe that all worms found in meat were derived directly from the droppings of flies, and not from the putrefaction of the meat, and I was still more confirmed in this belief by having observed that, before the meat grew wormy, flies had hovered over it, of the same kind as those that later bred in it.

Belief would be vain without the confirmation of experiment, hence in the middle of July I put a snake, some fish, some eels of the Arno, and a slice of milk-fed veal in four large, wide-mouthed flasks; having well closed and sealed them, I filled the same number of flasks in the same way, only leaving these open. It was not long before the meat and fish, in these second vessels, became wormy and flies were seen entering and leaving at will; but in the closed flasks I did not see a worm, though many days had passed since the dead flesh had been put in them.

Outside on the paper cover there was now and then a deposit, or a maggot that eagerly sought some crevice by which to enter and obtain nourishment. Meanwhile the different things placed in the flasks had become putrid and stinking; the fish, their bones excepted, had all been dissolved into a thick, turbid fluid, which on settling became clear, with a drop or so of liquid grease floating on the surface; but the snake kept its form intact, with the same color, as if it had been put in but yesterday; the eels, on the contrary, produced little liquid, though they had become very much swollen, and losing all shape, looked like a viscous mass of glue; the veal, after many weeks, became hard and dry.

Not content with these experiments, I tried many others in different vessels. In order to leave nothing undone, I even had pieces of meat put under ground, but though remaining buried for weeks, they never bred worms, as was always the case when flies had been allowed to light on the meat. One day a large number of worms, which had bred in some buffalo-meat, were killed by my order; having placed part in a closed dish, but in the second worms had hatched, which changing as usual into egg-shape balls (pupae), finally became flies of the common kind. In the same experiment tried with dead flies, I never saw anything breed in the closed vessel.

Redi's classical experiment is not only unusual in that it was an experiment but that he prepared two sets of vessels. One he covered and the other he left open. Both sets contained the same kinds of meat and fish.

Therefore, the results obtained could only be attributable to the covering, because this was the only part of the experiment that he varied. This type of experiment is called a controlled experiment and is the best kind from which to draw conclusions.

It is well to emphasize this point because the layman and many pseudo-scientists are continually trying out this or that idea with the desire to "prove something." Uncountable thousands of people assemble some facts, make up their minds and come to a conclusion, but their conclusions will have little or no value unless they run a "controlled" experiment and are careful to vary only one factor at a time. In the case of Redi, the only thing he varied was the use of gauze over one set of vessels.

It should be mentioned that Redi was interested in proving that certain flies came from the eggs of the flies. This idea was not original with Redi because he had read in the nineteenth book of Homer *Iliad*, as he mentions, that Achilles was afraid lest flies enter the wounds of the dead Patroclus and breed worms therein.

Redi noted that in both sets of jars the meat became putrid, but knowing nothing of the cause of decay or putrefaction, he knew nothing of the role of bacteria in these processes. The discovery of bacteria did not come until 1676, and the linking up of decay and bacteria had to wait until the discovery of Schwann in 1837, almost one hundred fifty years later. It is interesting to note that Redi still believed that gall-flies were spontaneously generated in galls. This, of course, was disproved later by other scientists.

The great furor over spontaneous generation subsided somewhat as a result of Redi's experiments, but in 1676, an amateur Dutch scientist by the name of Antony van Leeuwenhoek unintentionally started the controversy all over again. Working with pepperjuice in an attempt to find the cause for its acridity, he saw minute organisms which subsequently became known as bacteria. He said, "by chance observing this water on the 24th of April, 1676, I saw therein, with great wonder, incredibly many very little animalcules, of divers sorts."

The careful observations of Leeuwenhoek, communicated in a series of letters to the Royal Society in London, initiated the science of bacteriology and led to thousands upon thousands of research projects, the end of which is not yet in sight. Some of the important fruits of the work on bacteria are: the control of spoilage, water sanitation, Pasteurization, sterilization, lactic acid fermentations, vinegar formation, ensilage, crop rotation, and last but not least, the germ theory of disease.

The science of bacteriology might not have developed in the absence of suitable means for magnifying the organisms so that they could be seen. Leeuwenhoek, by tireless perseverence, succeeded in grinding lenses from tiny bits of glass so that he had finer hand-lenses than anyone else. They were not microscopes as we think of them today but they were at least sufficiently powerful to detect the tiny bacteria.

## EIGHTEENTH-CENTURY INVESTIGATIONS

From the discovery of bacteria by Leeuwenhoek in 1676, to about the middle of the next century, investigators worked successfully on various aspects of bacteriology but were unable to prove or disprove the spontaneous generation of the microscopic bacteria. Their abundance, rapid rate of reproduction and the ability of some of them to withstand unfavorable conditions made the work difficult.

He put some mutton broth in flasks, boiled them a short time, put corks in their mouths, and set them aside. After a suitable waiting period, he examined them and found that each held a good growth of bacteria. Reasoning from his data, Needham spoke out for the spontaneous generation of bacteria in his cultures.

Abbe Lazaro Spallanzani of Italy is said to have been the first man to witness the division of bacterial cells under the lens, the first to isolate a single organism of very small dimensions, and the first to use glass flasks. Spallanzani is noted most of all, however, for his famous controversy with Needham and others who supported the theory of spontaneous generation.

Needham's experiments had three serious flaws. First, he boiled his infusions for too short a time, second, he stoppered them with corks, which are porous enough to admit bacteria, and third, he failed to record exact information on the temperature and time of boiling. The recording of data is very important in scientific work. Spallanzani sealed his flasks shut by melting the glass and heated them longer than had Needham. Upon examination, he was able to say, "Nineteen vessels, containing infused substances, were hermetically sealed and kept an hour in boiling water. Being opened at a proper time, not a single animacule was to be seen."

Ordinarily this experiment would have settled the argument but Needham rejoined that the good priest had boiled his flasks so long that he had destroyed the "vegetative force" in the infusion and naturally the organisms could not grow. Let us see how Spallanzani met this challenge.

To this experiment of mine, it was objected that the long continuance of heat had perhaps entirely destroyed the vegetative power of the infused substances, or materially injured the elasticity of the air remaining included in the vessels; thus it was not surprising if *animacula* did not appear.

To estimate the weight of these objections, I conceived an experiment apparently decisive; which was, to make nineteen infusions, and boil some of them a short time, others longer, and the rest very long. If it was founded, the number of *animacula* would be less according to the boiling, if not, the number would be alike in all cases.... White kidney beans, vetches, buck-wheat, barley, maize, the seeds of mallows and beets were infused.... The yoke of an egg... was also infused.— A certain quantity of each infusion (was boiled) half an hour; another quantity, an hour, a third, an hour and a half; and a fourth, two hours. —Each of the four classes was marked with a different number, to avoid all

hazard of confusion or error: and because an equal temperature was most essential, all were deposited in the same place. The vessels, containing the infusions, were not hermetically sealed, but loosely stopped with corks; the only object of this examination being to discover, whether long protracted ebullition would prejudice or destroy the property of infused substances in producing *animacula*; if it did, there would be no difference whether the vessels were open or closed.

On the 15th of September, I made-infusions; and on the 23rd examined them for the first time. *Animacula* were in all; but the number and species different in each—From this it may seem, that although long continued heat had not prevented the production of *animacula*, it had contributed to diminish the number, or alter the kind.

Spallanzani thus clarified several important aspects of the problem. He proved that heating substances to kill whatever organisms they contained and then sealing them to prevent the entrance of any others, caused the infusions to remain clear and sterile. There was no spontaneous generation in his flasks. He also proved that long boiling did not destroy anything in the air that organisms required for their growth. Boiling did not alter the *mythical* "vegetative force" as charged by Needham.

Although Spallanzani's work seemed conclusive, there always was the lingering doubt that hermetically sealing the flasks deprived any organisms that might be present of the air they required. The work on oxygen by Antoine Lavosier and John Priestley in 1774 and 1775 had opened up a new field and showed the relationship between this gas and living organisms.

Joseph Gay-Lussac then showed that oxygen was lacking in flasks boiled and sealed. The enemies of biogenesis took heart from the discovery of oxygen for was not this gas necessary for cellular respiration? Apparently more clever experiments would have to be devised, experiments which showed the relationships among air, oxygen, and the origin of living organisms.

## NINETEENTH-CENTURY INVESTIGATIONS

Not until 1836 did anyone have a real contribution to make to the solution to the problem. An investigator by the name of Schultz made up infusions. He allowed the tubes to remain open to the air with this proviso—the entering air had to pass over sulphuric acid or potassium hydroxide. Although his flasks usually remained sterile, they sometimes spoiled. ( Tyndall found later that germs can sometimes pass unharmed through the acid). His fellow scientists, remembering Needham, protested that the acid deprived the air of the "vegetation force" in those cases where cultures remained sterile.

The emphasis now seemed to be on the entering air. Much work was being done on fermentations. This work was closely allied with spontaneous generation. Meanwhile Cagniard de la Tour made a great contribution when he found that yeasts were organisms and ventured the opinion that they caused fermentation. Theodor Schwann, co-founder of the cell theory, heated infusions and passed air into them through red-hot tubes and of course he observed no growth in his flasks. He also said that putrefaction and fermentation were due to living organisms. The great chemists of the day, however, ridiculed de la Tour and Schwann because they felt certain that fermentation was purely a chemical process caused by a ferment (or enzyme as we now call it).

Schwann's experiments could not always be duplicated, particularly with sugar infusions, and this was due, no doubt, to the fact that resistant bacteria entered the flasks at times. The chemists had sport, with Schwann in particular. As so often happens, both the chemist and the biologist were each partly right about fermentation but this knowledge came later. Both Oliver Wendell Holmes and Ignaz Semmelweis were of the opinion that puerperal or childbed fever was a disease caused by organisms and that it was spread from patient to patient by the unclean hands of the obstetrician. They were subjected to villanous attacks by colleagues but eventually proved that keeping the hands and operating rooms sterile with antiseptics controlled the disease.

In the middle of the nineteenth century, two men made another contribution to the solution of the mystery surrounding the genesis of bacteria. Their names were Schröder and von Dusch. They proceeded as had many other investigators but instead of sealing their flasks or passing the air through strong acid or through hot glass they simply plugged the flasks with ordinary cotton. After suitable boiling, the flasks remained sterile.

This is our procedure today in bacteriological work but it was argued by some that cotton, being of vegetable origin, required air for its own metabolism and hence no air or oxygen got into the flask and naturally the organisms could not grow. Sometimes, too, media with milk or meat without water, spoiled. Their experiments were not conclusive and were not universally accepted.

In 1859, Félix Archimede Pouchet, Director of the Museum at Rouen, published a treatise on spontaneous generation in which he said: "when after meditating on the question it became clear to me that spontaneous generation was still one of the means employed by nature to reproduce living beings, I set about finding means by which the phenomenon could be demonstrated." He added, "what then will these opponents say if I succeed in introducing the generation of living organisms, while substituting artificial air for that of the atmosphere?

Pouchet performed the following experiment. He sealed a flask of boiling water, inverted it over a mercury bath and thrust the neck of the bottle under the mercury. He broke the neck of the bottle under the mercury, and connected it with an apparatus which, when heated, produced oxygen. The gas bubbled in and displaced the water.

When half full of gas and half full of water, he took a piece of hay which had been heated and, with sterilized forceps, pushed it underneath the mercury and into the mouth of the bottle. The hay floated in the water. After a few days, the infusion was found to be full of small organisms. Triumphantly Pouchet asked where this evidence of life could have come from—not from the hot water—not from the heated hay—not from the oxygen which had been produced artificially?

Pouchet's experiment was the trigger mechanism which "set off" Louis Pasteur, Scientific Director of the Ecole Normale, and almost led to the ultimate solution of the problem of spontaneous generation. Pasteur's name is known to almost every educated person because of his great contributions from 1860-1864 on the question we are considering. He is revered, however, because of the great number of his contributions to the safety and welfare of mankind. Some that are worth remembering are his work on crystals, the improvement of the French wine industry, the diseases of the silk worms, his work on anthrax, and his famous work on hydrophobia or rabies.

When Pasteur began his work on the problem, he had before him the results obtained by many investigators. He probably admired the cleverness of Spallanzani, Needham, Schultz, Schwann, Schröder, and von Dusch, as well as some others. His problem, however, was to make a foolproof experiment, one to which no one could object. First of all, he repeated the work of his predecessors by sealing flasks shut with a flame, by stoppering others with cotton and still others with asbestos, and by passing air through heat.

In 1843, Helmholtz had given him another lead. This investigator showed that the air contained contaminants and that this contaminant was a solid since it could not pass through a moist animal membrane. Pasteur pursued this line since he had a "hunch" that this was the proper avenue of approach.

He sucked air through a tube fitted with a cotton plug. The plug caught the dust particles in the air and when he dissolved the plug and put the dust into sterile broth, the broth became filled with organisms. The relationship between dust and organisms was thus established. He was now ready to perform his most crucial experiment in the field of spontaneous generation.

Pasteur used a yeast broth in his flasks. He realized that he must not bring his air into the flasks over heat nor must he pass it through acid, nor must he stopper the flasks with corks or cotton. How then could he silence the critics who would be sure to arise? He simply drew out the necks of his heated flasks into the shape of a swan's neck, set the flasks aside and, after waiting until he could stand the strain no longer, he examined them and found that every one remained organism-free.

The air with its own oxygen entered the flask but the heavier dust particles (and the bacteria) settled in the low parts of the neck. Thus the flasks remained sterile. Bacteria did not appear if the dust was excluded. If he broke the neck of a flask and admitted air, the broth became alive with organisms or if he tilted an unbroken flask so that the medium picked up some of the dust in the neck, the same thing happened.

There remained, however, the successful and unexplained experiments of Pouchet. Before a large audience, Pasteur threw a powerful beam of light onto the mercury in Pouchet's apparatus and disclosed a film of dust on the mercury. It was easy to see that Pouchet had pushed some of the dust (and also bacteria) into his sterile flasks along with the hay. That was his error! However, even without this, Pouchet's flasks might have developed life. Pouchet had used an alkaline (basic) medium and hay. An alkaline medium is very favorable to bacterial growth and the hay, no doubt, contained spore-forming bacteria. This type is able to produce a very thick wall about itself and resist unfavorable conditions.

The boiling to which they were subjected by Pouchet was not enough to kill them in the spore form, and after the flasks cooled, they germinated in the favorable alkaline medium. No wonder Pouchet thought he had a good case of spontaneous generation. Pasteur, on the other hand, used an acid medium and, luckily enough, there were no spore-formers in his flasks. ( Spallanzani also had been lucky in regard to spore-formers). No wonder Pasteur thought he had disproved spontaneous generation. Today we know that if we wish to inhibit spoilage in milk, an alkaline fluid favorable to bacterial growth, we have to heat it to 62 degrees centigrade for thirty minutes.

Thus we see that although Pasteur's discovery was both glorious and simple, it was built up from the knowledge gained by a great many other earnest workers, plus a certain amount of "luck." We must now describe the decisive experiment on this subject, devised by the brilliant John Tyndall of England and reported upon in 1876, twelve years after Pasteur's researches. Pasteur's experiments had been good but were rather simple in nature; Tyndall's experiments were sophisticated and his results conclusive.

Tyndall decided to combine his knowledge of optically pure air with an experiment on spontaneous generation. His apparatus was in the form of a box. This box had four legs attached to it. The front of the box had a large glass window and each of the two sides had a small glass window. The back had an air-tight door.

The bottom of the box had holes in it through which Tyndall inserted test tubes in such a manner as to exclude any air. On top of the box he inserted some coiled glass tubes through which air, but no dust, could enter. Another hole in the top held a thistle tube. Tyndall now applied a sticky material to the inside of the box to hold the dust particles securely once they had settled.

He then directed a beam of light through one of the small side windows, through the box, and out the other side. By looking through the front window, Tyndall noted the light beam in the box as long as there were dust particles present; when none were in the air, the beam became invisible. At this point, Tyndall inserted the thistle tube in the top and filled the test tubes with various kinds of broth. Then the apparatus was moved so that the tubes could be boiled (by immersion in boiling oil). This procedure was to kill any organisms in the broth so that it would be "lifeless."

Tyndall found that the tubes of broth remained sterile. Some objected that the boiling had rendered the broth unfit to support life so Tyndall, taking a cue from the researches of Pasteur and others, opened the door in the back of the box for a short time. This exposure allowed some dust (with the adhering germs) to enter. The tubes spoiled in a few days. Boiling had not destroyed the "vegetable force" in the broth.

This clever experiment and the work of Pasteur, plus that of others, gave mankind a real understanding of the distribution and means of control of bacteria. The long trail which led to Pasteur and Tyndall stopped only momentarily. It continued on to Robert Koch, the discoverer of the cause of tuberculosis; to Schaudinn, the discoverer of the syphilis organism, and to dozens of others. It will continue on for a long, long time. There is still much to be learned, there are still diseases to be understood and conquered.

In 1935, a chemist by the name of W. M. Stanley opened up an entirely new field of speculation about the nature of life and its origin by crystallizing substances formerly thought to be living. These substances are known as viruses. Stanley worked with a virus causing a disease of tobacco plants called tobacco mosaic.

Now, as is well-known, living organisms do not enter a crystal stage; this is a property of non-living matter. If the virus is really non-living, then this would be the first instance of a non-living particle causing an infectious disease like tobacco mosaic. Stanley and his co-workers pursued the matter and eventually found another amazing aspect of viruses. The viruses were very large protein molecules and had the property of reproducing themselves. Here, then, we had a characteristic of living things. Were these viruses the "missing link" between the living and the dead? Were they generated spontaneously from inorganic matter? Was Pasteur's work all for naught?

Some scientists, impressed by the simplicity and ultramicroscopic nature of the viruses, cried out that here were the primeval forms of life. The boundary between the living and the non-living had been bridged. More cautious men pointed to the fact that the viruses were obligate parasites, depending upon living organisms for their perpetuation. This fact probably removes the viruses quite far from primitiveness.

Over thirty years have passed since Stanley's monumental work and we are not now sure as to the exact nature of viruses. We can not be sure that they are not spontaneously generated. Redi, Spallanzani, and Pasteur did not prove that the theory of spontaneous generation is false.

They proved, however, that certain organisms did not arise from inorganic matter under the conditions of their experiments. To say that spontaneous generation never occurred or never will occur is a universal negative and cannot be proven. Real scientists have ever been cautious about being too dogmatic. They remember too well how countless popular theories, seemingly backed by irrefutable proof, have been overthrown during the long history of science. Two quotations might be brought to bear upon this point:

Scientific research does not lead to a knowledge of nature which is free of error but it is an unending process in which the degree of truth is always increasing. We uncover errors in existing theories, we discover more and more things of significance, but the goal of absolute truth keeps on receding.

In no case does science tell us why natural phenomena occur in order-every scientific explanation rests upon something which is granted. The discovery that nature is orderly can throw no light on the origin of anything in nature.-

Order is not an explanation of anything but something that itself calls for explanation.

Those who couple the question of the origin of life with the theory of evolution have a feeling that because we have never witnessed an example of life arising from the inorganic world during several centuries of investigations, it has never occurred. However, they still feel that in the prehistoric past, when conditions upon the earth were different and more favorable for such an event, the first particle, endowed with at least some of the properties of life, arose.

Thus we have the extremely anomalous situation existing in which scientists believe in a physiochemical origin of life on the one hand and conduct experiments proving its unlikelihood on the other hand. We have finished with our story of spontaneous generation and have noted the long, sometimes bitter, struggle that attended the "defeat" of the theory.

We have seen that the pure speculation of Aristotle was replaced by the modern science of experimentation. This experimentation, carried on rather simply by Redi, culminated in the brilliant researches of Pasteur and Tyndall. It is to the credit of scientists that they tested one another's theories, that their experiments, if sometimes faulty, were honestly done, and that they constantly strove to discern the phantom light of ultimate truth.

## Chapter 29

# Manmade and Natural Selection

By changing what man knows about the world, he changes the world he knows; and by changing the world in which he lives, he changes himself. Herein lies a danger and a hope; a danger because random changes of the biological nature are likely to produce deterioration rather than improvement; a hope because changes resulting from knowledge can also be directed by knowledge.

The human species, *Homo sapiens*, mankind, is the unique and most successful product of biological evolution, so far. This has sometimes been questioned, I suspect without too much conviction on the part of the doubters, perhaps only to mock man's pretensions or to challenge his values. But man *is* the most successful product of evolution, by any reasonable definition of biological success.

Man began his career as a rare animal, living somewhere in the tropics or subtropics of the Old World, probably in Africa. From this obscure beginning, mankind multiplied to become one of the most numerous mammals, for there will soon be about three billion men living. Numbers may not be an unadulterated blessing, but they are one of the measures of biological success of a species.

Moreover, man has spread and occupied all the continents and most islands, except for the frozen wastes of Antarctica and of the interior of Greenland; he has learned to traverse seas and oceans and deserts; he is well on the way towards control or elimination of the predators and parasites which used to prey on him; he has subdued and domesticated many animal and plant species, made them serve his needs and his fancies, broadened enormously the range of utilizable food supplies, and learned to make use of a variety of energy sources.

Modern man lives no longer at the mercy of wild beasts and vagaries of the climate; he has reached a status where his continuation as a species is in no danger, except perhaps as a result of man's own folly or of a cosmic accident.

The evolutionary uniqueness of man lies in that in mankind the biological evolution has transcended itself. With man commences a new, superorganic, mode of evolution, which is the evolution of culture. Culture is a tremendously potent instrument for adaptation to the environment. A very large part of the evolutionary progress, both biologically and culturally, has come from adversity. Life faces environments which are more often niggardly than bountiful, more frequently inimical than benign.

For life to endure, it must develop defences and adaptations. Biological adaptation occurs through natural selection; new genes arise through mutation, sexual recombination creates new combinations of genes, and natural selection acts to multiply the successful genetic endowments and to reduce the frequencies of the unsuccessful ones. In man and in man alone, adaptation may occur also through alteration of culture.

Many species of mammals have become adapted to cold climates by growing warm fur; man alone has achieved the same end by donning fur coats. Birds have mastered the air by becoming flying machines; man has conquered the air by building flying machines.

Biological and cultural evolutions of man are not independent; they are interdependent. The superorganic has an organic basis. Formation and maintenance of culture presuppose a human genotype. Even the most clever ape cannot learn human culture.

Some writers have jumped to the conclusion that the genetic development of the human species was completed before culture appeared, and that the evolution of culture has replaced biological evolution. This is not true.

The two evolutions go together, interacting and usually mutually reinforcing each other. There is feedback between genetics and culture. Culture is an adaptive mechanism supplemental to, but not incompatible with, biological adaptation. To be sure, adaptation by culture proved to be more efficacious, and, before all else, more rapid than adaptation by genes. This is why the emergence of the genetic basis of culture was the master stroke of the biological evolution of the human species. The genetic basis of culture should be improved or at least maintained. It should not be allowed to deteriorate.

## NATURAL SELECTION, STRUGGLE, AND FITNESS

Man has not only evolved; for better or for worse, he is also evolving. Is he to become a superman or a demigod? Or will the fate in store for him be like that of so many successful species of the past, which eventually declined and became extinct? Long-term prophecies do not expose the prophets to the risk of being proved wrong too soon. I nevertheless do not wish to indulge here in prophecies or in designing utopias. I wish rather to investigate some of the evolutionary forces currently at work in the human species.

According to the theory set forth by Darwin and Wallace more than a century ago, adaptation occurs in biological evolution by way of the process of natural selection. Does natural selection operate in modern mankind? This has to be answered obviously in the negative, if by "natural" you mean a world uninhabited and uninfluenced by man. But it is here that we must proceed with the greatest caution. What, indeed, is selection, and when is a selection "natural" and when is it not? Darwin said that natural selection is the outcome of the survival of the fittest in the struggle for life. "Struggle" suggests strife, contention, competition. Darwin himself wrote that "from the war of nature, from famine and death, the most exalted object we are capable of conceiving, the higher animal, directly follows."

However, natural selection does not ineluctably depend on any of these things. Birch has defined competition thus:

"Competition occurs when a number of animals (of the same or different species) utilize common resources the supply of which is short; or if the resources are not in short supply, competition occurs when the animals seeking these resources nevertheless harm each other in the process." Natural selection may, however, take place when resources are not limiting, if the carriers of some genes possess greater reproductive potentials than others. Some cases have been observed in experiments with animals and plants when a genotype superior under competition is inferior in the absence of competition, or vice versa.

And who is the "fittest," whose survival results in natural selection? Does natural selection make us fit for life in the society of other men, or for wisdom, or for good will, or for unselfishness? It does not necessarily ordain any of these qualities. Darwinian fitness is a measure of the reproductive proficiency. Its guiding principle is "be fruitful, and multiply, and replenish the earth." It is quite indispensable to distinguish Darwinian fitness from excellence in human estimation. The two may go hand in hand, but sometimes they may be in opposition.

It is man and man alone who can probe, scrutinize, and question the wisdom of the evolutionary process which brought him into being. He may improve what nature hath wrought. When man chooses the individuals who are to become parents of the succeeding generation he practices artificial selection. According to Lerner, artificial selection is a process which has a goal that can be visualized. Any selection which is not artificial is natural selection.

## MUTATION AND THE NORMALIZING NATURAL SELECTION

There are several forms of natural selection operating in the human species, as well as in other organisms. Although the distinctions between these forms of selection are neither absolutely rigid nor always clear-cut, they are helpful for straight thinking about evolutionary problems. Stabilizing or normalizing natural selection is the simplest and most obvious.

It is a conservative force; it counteracts the spread in populations of detrimental mutants, hereditary diseases, and weaknesses of various kinds. Failure, or at least relaxation, of the normalizing selection in human populations leads to fears that an insidious process of genetic decay is at work in the human species. Is there a basis for these fears?

Mutations continue to arise in man, even as they have been arising since the dawn of life. Some mutants cause grave and even fatal hereditary diseases, such as retinoblastoma, hemophilia, epiloia, etc. Others cause malformations, such as achondroplasia, arachnodactyly, or brachydactyly. Still others, and probably a maj rity of mutants, cause small and unspectacular changes in the appearance of physiology or behavior of their carriers. Small mutations are difficult to study quantitatively. Not enough is known about them, even in Drosophila and in other organisms more favorable for experimentation than man.

Large mutations are certainly more often deleterious than useful, at least in the environments in which the species normally lives, and at least in homozygous condition. A useful major mutant is like a needle in a haystack; it may be there but it is notoriously hard to find among the many harmful ones. It is tempting to suppose that all or most small mutations will also be harmful, in various small and devious ways. This is what geneticists who are adherents of the "classical" hypothesis of the population structure believe to be the case. The evidence on which this belief rests is not fully convincing.

A brief consideration of some examples of the operation of the normalizing selection in man will be useful at this point. Achondroplasia or chondrodystrophy are formidable names for a rather common type of dwarfism in man, caused by the presence of a single dominant mutant gene. Achondroplasts have bodies and heads of about normal size, but their limbs are very short.

Some achondroplastic dwarfs are born in families in which at least one of the parents is a dwarf. Such dwarfs evidently inherit the genes for the abnormality from the affected parents. But some dwarfs are children of parents of normal stature;

these dwarfs carry the dominant gene for dwarfism newly arisen by mutation.

Mørch has recorded the birth of eight dwarf mutants among about 94 thousand children in Denmark; this indicates that about one sex-cell per 24,000 sex cells produced by normal parents transports a newly arisen gene for achondroplastic dwarfism, a mutation rate of the order of $4 \times 10^{-5}$ per generation.

The genes for achondroplastic dwarfism newly arisen by mutation are introduced into the gene pool of human populations; this happens relentlessly in every generation. Does it follow that the dwarfism will grow more and more widespread in the course of time? Not necessarily. The spread of the genes for dwarfism is opposed by natural selection. Mørch found that the achondroplastic dwarfs in Denmark produce only about twenty per cent as many children as do their nondwarf brothers and sisters.

This means that the adaptive value, or the Darwinian fitness, of the achondroplastic dwarfs is only 0.2 normal; it can also be said that the gene for the dwarfism is opposed by a normalizing natural selection of a magnitude of 0.8. How does natural selection operate on this gene? Why do the dwarfs produce so few children?

It turns out that this is chiefly because many of them remain unmarried. And their failure to find mates is not due to a weakened sexual drive but to their external appearance being not in accord with what is regarded in our society as handsomeness or attractiveness. Here, then, is an example of a natural selective process which operates not by causing early death of the carriers of certain genes, but acting instead via certain culturally conditioned forms of human behavior.

A person who remains childless may be described as genetically "dead"; but childlessness is a form of "genetic death" which does not immediately produce a cadaver. Some mutant genes are, to be sure, eliminated by a normalizing selection acting through "genetic death" which is also real death. Retinoblastoma is a form of cancer of the eye, which afflicts infants and children; unless the eyes are removed promptly by a surgical operation early death is inevitable. Neel and Falls found the mutation from the normal to the dominant mutant gene causing retinoblastoma to be about as freguent as that causing the achondroplastic dwarfism.

Before the invention of surgery which saves the lives of retinoblastomatic children, the normalizing selection was eliminating in every generation all the mutant genes for retinoblastoma which arose in that generation (i.e., in the sexcells which gave rise to that generation). The surgery may be said to frustrate this normalizing selection; the retinoblastomatics, though blind, grow up and become capable of being parents of children one-half of whom will now inherit the gene for retinoblastoma. The retinoblastomatics, like the achondroplasts, are now of two kinds—those due to new mutations and those having inherited the gene from their parents.

Genes which reduce the Darwinian fitness of their carriers less dramatically than do the genes discussed above are nevertheless also opposed by normalizing selection. Predisposition to diabetes mellitus, or at least to some of its forms, seems to be inherited through a recessive gene. The same seems to be true of some forms of myopia (short-sightedness). We do not know either the mutation rates which produce these genes, nor the selection rates which oppose their spread. What is of interest for our purposes is that these selection rates doubtless vary greatly depending on the environment.

Myopia is likely to be more incapacitating to people engaged in some occupations (i.e., hunters or automobile drivers) than in others (i.e., handicraftsmen or clerks), and besides myopia can be "cured" by wearing glasses. The onset of diabetes mellitus may be early or late in life; it may thus strike after the close of the reproductive age, and thus fail to reduce the number of the children produced and hence to reduce the Darwinian fitness. And some forms of diabetes may be "cured" by insulin therapy.

Now, to "cure" a hereditary disease obviously does not mean to change the genes which have caused it. Health and disease refer however to the states of well-being or infirmity of a person; a person consulting his doctor does not ask the latter to alter his genes, but only to advise how to change the environment in a way such that the genes with which the person has been born will react by producing a reasonable state of well-being.

A myopic wearing glasses, or a diabetic having received his insulin injections, is no longer incapacitated, or at least less so than both were before the "cure." The relief of the incapacitation may however increase their Darwinian fitness, and mitigate the severity of the normalizing natural selection.

## THE "NORMAL" MAN

The removal of a mutant from a population is called a genetic death or, less dramatically, genetic elimination. The death of a child with retinoblastoma eliminates from the population one mutant retinoblastomatic gene. When an achondroplastic dwarf fails to marry, or raises a family with fewer children than he would have if he were not a dwarf, a mutant gene for achondroplasia is eliminated.

The classical hypothesis of population structure assumes that there exists for every biological species, or may appear or be produced in the future, one normal, or best, optimal genetic endowment. This genetic endowment would give the ideal, the normal, the archetypical man, or the ideal Drosophila fly, or the ideal corn plant—the man, the Drosophila, the corn as they ought to be.

The way to produce this normal man is to let the normalizing natural selection remove, eliminate, purge from the population all the mutant genes. What then would remain would be the ideal constellation of genes. This hypothesis has the advantages of simplicity, though not necessarily of accuracy. It is a product of typological thinking, which has, to be sure venerable antecedents.

It goes back to Plato's eternal ideal types. Many scientists find the appeal of the typological mode of thought irresistible. Indeed, if the classical hypothesis were justified, the problem of the guidance of the biological evolution of the human species would be theoretically simple, however difficult it might still remain in practice. We would have to arrange for the normalizing natural selection to protect the ideal normal human genes from contamination by mutations. Or, to put it in positive rather than negative terms, we might strive to obtain a mankind in which everybody would be a carrier of the normal genetic endowment. The three billion humans, or whatever numbers this "normal" mankind might contain, would then be as similar genetically as identical twins. But they would presumably be strong and healthy and happy identical multitudes!

## DIVERSIFYING NATURAL SELECTION

Fortunately or unfortunately, depending upon one's point of view, the classical hypothesis of population structure is perhaps only a halftruth. That it is true to some extent is clear enough. Harmful mutants do arise, and some of them seem to be unconditionally harmful. It is hard to imagine an environment in which retinoblastoma could possibly be harmless, not to speak of usefulness.

Accumulation of such mutants in human populations can only augment human misery. To counteract this accumulation, one could, conceivably, reduce the mutation frequencies, or else eliminate the mutants as painlessly as possible. Unfortunately, for the time being, man has learned how to increase the mutability by exposure to high-energy radiations and by other means, but not how to decrease the mutation rates.

The concept of a single genetic endowment, ideal or optimal for each living species, is however a typological fiction. No living individual, and certainly no population and no species, exist in absolutely uniform environments. Environments vary in time and in space. How does life solve the problem of adaptation to a multiplicity of environments?

There are two possible solutions, and, with its habitual opportunism, evolution has used them both. The first is the genetically-conditioned physiological and developmental plasticity. The organism reacts to changes in the environment by adaptive modifications. For example, the human body has physiological mechanisms which enable it to conserve an almost constant internal temperature despite external temperature variations. In man, the most important kind of plasticity is his educability. Most humans can be trained for, and can acquire skills to perform passably in, any one of the many employments which most human societies have to offer.

Secondly, a population facing a diversity of environments may become genetically diversified. Natural selection favors different genetic endowments in different environments. This is the diversifying (also called disruptive) form of natural selection. Instead of one perfect genotype, diversifying selection favors many genotypes; it favors genetic polymorphism. A population which abounds in genetic variety has a better grip on a complex environment than a genetically uniform population; polymorphism is one way of exploiting the environment more fully.

## DIVERSITY AND EQUALITY

Adaptations by way of plasticity and by way of genetic diversity are not mutually exclusive. They are complementary. Culture is by far the most potent adaptive mechanism which has emerged in the evolution of life. Its potency is due to its being transmitted by teaching and learning, instead of by genes.

The ability to be trained, and to become competent in whatever occupation a person meets his opportunity, increases the social and usually also the biological fitness. Genetic specialization for one vocation may confer very high fitness for that particular vocation. But educability permits a choice of vocations, and thus confers fitness in complex and changeable environments. In an evolving culture new occupations constantly arise. Who needed aircraft pilots a century ago, but how many blacksmiths are now needed in technologically advanced countries?

Culture is, however, an adaptive contrivance to make people diversified, not to make them alike. If uniformity were advantageous, genetic fixity would most likely have emerged in evolution. Genetic diversity or polymorphism joins force with developmental plasticity. I could have probably learned many kinds of jobs other than that I actually fixed upon. But no amount of effort and training could make me a first-rate wrestler or sprinter or painter or concertmaster. I just do not possess the genetic wherewithal for these occupations. Other people do, and this genetic diversity enriches the store of man's capabilities. It is the leaven of cultural progress.

The facts of biology are compatible with the lofty vision of human equality. All men have been born equal; they certainly are not all alike. It is nonsense to think that only identical twins can be equals. Human equality is not predicated on identity, or even on identity of ability. It presupposes, however, something approaching an equality of opportunity to develop whatever socially useful gifts and aptitudes a person's genes have provided him with, and which he may choose to develop.

Culture fosters a multitude of employments and functions to be filled and served; equality of opportunity stimulates division of labor rather than sets it aside; it enables every person to choose among occupations for which he is qualified by his abilities. It is wrong to think that equality of opportunity makes genetic variation unimportant. It does precisely the opposite.

It makes the difference between people reflect meaningfully their genetic differences. Inequality of opportunity acts, on the contrary, to hide, to distort, and to stultify the genetic diversity.

## BALANCING NATURAL SELECTION

Far from being an unfortunate deviation from the ideal state, genetic diversity is an adaptive response of life to its environment. A gene useful in one environment may cause a genetic disease in another; a mutant harmful in combination with some genes may be useful in other combinations. Some adaptively ambivalent genes increase the fitness of their carriers when in single dose, in heterozygous state, but are harmful, or even lethal in homozygotes.

The sickle-cell anemia and the Mediterranean anemia, or thalassemia, are the best known human examples. These almost invariably fatal diseases are due to homozygosis for genes which in heterozygous condition seem to confer a relative immunity to certain kinds of malarial fevers. Heterozygous carriers consequently enjoy a superior fitness, hybrid vigor, or heterosis, in countries where these particular forms of malaria are prevalent. The genes for sickle-cells and for thalassemia are common in tropical and subtropical parts of the Old World, and they have been introduced in the New World as well.

Natural selection holds the "normal" and the sickle-cell and thalassemia genes in balanced polymorphism in populations of malarial countries. This is the balancing form of natural selection. How important is balancing selection in the human species is an open issue. It may well be that many of the genes responsible for the individual differences which we observe among so-called "normal" persons, differences in facial features, body build, health, intelligence, longevity, etc., are kept in balanced polymorphism.

Morton and Chung have presented evidence that the genes for the M and N blood types are maintained by balancing selection. The heterozygotes MN have a Darwinian fitness superior to MM and NN homozygotes. Just what is the advantage of the heterozygous state is unknown, nor is the nature of the drawback in homozygotes. The homozygotes are certainly viable; I am one of them. A very slight superiority in fitness in heterozygotes is, however, sufficient to maintain a balanced polymorphism.

There are grounds to suspect that the O-A-B blood types in man are also held in balanced polymorphism. The evidence for this is at present inconclusive, but recent works of Levene, Rosenfield, Morton, Crow, and others show that the genes involved are subject to a still different kind of selective pressure. This is a relative incompatability between the mother and the fetus when they differ in genetic constitution in certain ways. Such incompatibility has been known to arise because of the genes for a still different system of blood types, the

Rhesus, or Rh system. An Rh positive fetus in an Rh negative mother may suffer injury or even be aborted.

## DIRECTIONAL NATURAL SELECTION

The form of natural selection most important in the long run is directional selection. It was pivotal in Darwin's theory of evolution. Living species may respond to challenges of their environments by genetic changes. Genes once useful become inferior to others and are gradually eliminated, replaced by superior gene variants. The whole genetic system of the species is eventually rebuilt.

There is a fairly general agreement among biologists that the emergence of the human species was due in the main to directional selection. It has, however, been questioned whether directional selection, or indeed any form of natural selection, has continued to operate after the genetic basis of culture has taken shape. Now, let us admit that the success of mankind as a biological species was due precisely to man's culture being able to change ever so much faster than his genes can.

Man adapts his environments to his genes more often than he adapts his genes to his environments. But, as pointed out, the two methods of adaptation are complementary and not alternative. Far from making human environments stable and uniform, culture increases the tempo of change. Given an environmental flux, the necessary and sufficient condition for genetic change is availability in the populations of genetic variants, some of which are better and others less well adapted to shifting environments. Natural selection will multiply the favorable and depress or eliminate the unfavorable variants.

We do not know exactly how much genetic change has taken place in mankind at different stages of its history. Modern man might or might not be able to survive, even if properly trained, in the environments of his ancestors of 100,000, or even 10,000 years ago. Or, if he survived, he might not be as efficient or as happy in those environments as his ancestors were. We do not know for sure.

Neanderthal man may or may not have been capable of becoming a reasonably well adjusted citizen if raised in New

York or in New Orleans. Perhaps some Neanderthals might have been fit to become Ph.D.'s and to be elected members of the Society of Sigma Xi. But, on the other hand, they may have been unfit for any now existing education. Modern women are alleged to experience greater difficulties in childbirth than did their great-grandmothers. All this is conjectural and not rigorously proved. It is, however, a fallacy to assert that what is unproved did not occur. In point of fact, some of the above changes probably did occur.

## ARE CULTURE AND NATURAL SELECTION COMPATIBLE

We have seen that several forms of natural selection operate in modern mankind. But they certainly do not operate as they did during the Stone Age, or even as they did a century ago. Neither does natural selection operate always in the same way in wild and "natural" species, quite "unspoiled" by culture. This is inevitable. Natural selection depends on environments, and environments change. Human environments have changed a great deal in a century, not to speak of millennia.

The real problem is not whether natural selection in man is going on, but whether it is going on towards what we, humans, regard as betterment or deterioration. Natural selection tends to enhance the reproductive proficiency of the population in which it operates. Such proficiency is however not the only estimable quality with which we wish to see people endowed. And besides, a high reproductive fitness in one environment does not even insure the survival of the population or the species when the environment changes.

Normalizing selection is, as we have seen, not the only form of natural selection. The relaxation of some of its functions is, however, a cause of apprehension. Medicine, hygiene, civilized living save many lives which would otherwise be extinguished. This situation is here to stay; we would not want it to be otherwise, even if we could. Some of the lives thus saved will, however, engender lives that will stand in need of being saved in the generations to come. Can it be that we help the ailing, the lame, and the deformed only to make our descendants more ailing, more lame, and more deformed?

Suppose that we have learned how to save the lives of persons afflicted with a hereditary disease, such as retinoblastoma, which previously was incurably fatal. In genetic terms, this means that the Darwinian fitness of the victims of the disease has increased, and that the normalizing selection against this disease is relaxed. What will be the consequence?

The incidence of the disease in the population will increase from generation to generation. The increase is likely to be slow, generally no more than by one mutation rate per generation. It may take centuries or millennia to notice the difference for any one disease or malformation. However, the average health and welfare of the population are liable to show adverse effects of relaxed selection much sooner.

The process of mutation injects in every generation a certain number of harmful genes in the gene pool of the population; the process of normalizing selection eliminates a certain number of these genes. With environment reasonably stable, the situation tends to reach a state of equilibrium. At equilibrium, the mutation and the elimination are equal.

If mutation becomes more frequent (as it does in man because of exposure to high-energy radiations and perhaps to some chemicals), or if the elimination is lagging because of relaxation of normalizing selection, the incidence of harmful mutant genes in the population is bound to increase. And take note of this: If the classical theory of population structure were correct, all harmful mutations would be in a sense equivalent. For at equilibrium there is one elimination for every mutation, regardless of whether the mutation causes a lethal hereditary disease like retinoblastoma, or a malformation like achondroplasia, or a relatively mild defect such as myopia.

The problem is, however, more complex than this theory would suggest. It calls for research in what Wright describes neatly as "unfortunately the unpopular and scientifically somewhat unrewarding borderline fields of genetics and the social sciences." Although at equilibrium there may be one genetic elimination for every mutation, it is unrealistic to equate the human and social consequences of different mutations.

The elimination of a lethal mutant which causes death of an embryo before implantation in the uterus is scarcely noticed by the mother or by anyone else. Suffering accompanies the elimination of a mutant, such as retinoblastoma, which kills an infant apparently normal at birth.

Many mutants, such as hemophilia or Huntington's chorea, kill children, adolescents, or adults, cause misery to their victims, and disruption of the lives of their families. There is no way to measure precisely the amount of human anguish; yet one may surmise that the painful and slow death of the victims of so many hereditary diseases is torment greater than that involved in the elimination of a gene for anchondroplasia owing to the failure of an achondroplastic dwarf to beget children.

Looked at from the angle of the costs to the society, the nonequivalence of different mutants is no less evident. Myopia may be inherited as a recessive trait. Increases of the frequencies in populations of the gene for myopia are undesirable. Yet, it may become more and more common in future generations. However, only a fanatic might advocate sterilization of the myopics or other radical measures to prevent the spread of this gene.

One may hope that civilized societies can tolerate some more myopics; many of them are very useful citizens, and their defect can rather easily be corrected by a relatively inexpensive environmental change—wearing glasses. The effort needed to eradicate or to reduce the frequency of myopia genetically would exceed that requisite to rectify their defect environmentally, by manufacturing more pairs of glasses.

Diabetes mellitus is, given the present level of medicine, more difficult and expensive to correct than is myopia. Some diabetics may nevertheless be treated successfully by insulin therapy, helped to live to old age, and enabled to raise families as large as nondiabetics. The incidence of diabetes may therefore creep up slowly in the generations to come. Now, most people would probably agree that it is better to be free of diabetes than to have it under control, no matter how successfully, by insulin therapy or other means.

The prospect is not a pleasant one to contemplate. Insulin injections may perhaps be almost as common in some remote future as taking aspirin tablets is at present.

## TOWARDS GUIDANCE OF HUMAN EVOLUTION

Are we, then, faced with a dilemma—if we enable the weak and the deformed to live and to propagate their kinds, we face the prospect of a genetic twilight; but if we let them die or suffer when we can save them we face the certainty of a moral twilight. How to escape this dilemma?

I can well understand the impatience which some of my readers may feel if I refuse to provide an unambiguous answer to so pressing a problem: Let me however plead with you that infatuation with over-simple answers to very complex and difficult problems is one of the earmarks of intellectual mediocrity. I am afraid that the problem of guidance of human evolution has no simple solution.

At least I have not found one, nor has anybody else in my opinion. Each genetic condition will have to be considered on its own merits, and the solution which may be adopted for different conditions will probably be different. Suppose that everybody agrees that the genes causing myopia, achondroplasia, diabetes and retinoblastoma are undesirable. We shall nevertheless be forced to treat them differently. Some genetic defects will have to be put up with and managed environmentally; others will have to be treated genetically, by artificial selection, and the eugenic measures that may be needed can be effected without accepting any kind of biological Brave New World.

Let us face this fact: Our lives depend on civilization and technology, and the lives of our descendants will be even more dependent on civilized environments. I can imagine a wise old ape-man who deplored the softness of his contemporaries using stone knives to carve their meat instead of doing this with their teeth; or a solid conservative Peking man viewing with alarm the new fangled habit of using fire to make oneself warm. I have yet to hear anyone seriously proposing that we give up the use of knives and of fire now.

Nor does anyone in his right mind urge that we let people die of smallpox or tuberculosis, in order that genetic resistance to these diseases be maintained. The remedy for our genetic dependence on technology and medicine is more, not less, technology and medicine. You may, if you wish, feel nostalgic for the good old days of our cave-dwelling ancestors; the point of no return was passed in the evolution of our species many millenia before anyone could know what was happening.

Of course, not all genetic defects can be corrected by tools or remedies or medicines. Even though new and better tools and medicines will, one may hope, be invented in the future, this will not make all genetic equipments equally desirable. It is a relatively simple matter to correct for lack of genetic resistance to smallpox by vaccination, or for myopia by suitable glasses. It is not so simple with many other genetic defects.

Surgical removal of the eyes is called for in cases of retinoblastoma; this saves the lives of the victims, but leaves them blind. No remedies are known for countless other genetic defects. Human life is sacred; yet the social costs of some genetic variants are so great, and their social contributions are so small, that avoidance of their birth is ethically the most acceptable as well as the wisest solution. This does not necessarily call for enactment of Draconian eugenic laws; it is perhaps not overoptimistic to hope that spreading biological education and understanding may be a real help. Make persons whose progeny is likely to inherit a serious genetic defect aware of this fact; they may draw the conclusions themselves.

The strides accomplished by biochemical genetics in recent years have led some biologists to hope that methods will soon be discovered to induce specific changes in human genes of our choice. This would, indeed, be a radical solution of the problem of management of the evolution of our species, and of other species as well. We would simply change the genes which we do not like, in ways conforming to our desires. Now, if the history of science has any lesson to teach us, it is the unwisdom of declaring certain goals to be unattainable. The cavalier way in which the progress of science often treats such predictions should instill due humility even in the most doctrinaire prophets. The best that can be said about the possibility of changing specific genes in man in accordance with our desires is that, although such an invention would be a great boon, it is not within reach yet. And it cannot be assumed to be achievable.

Let us also not exaggerate the urgency of the problem of the genetic management of the evolution of our species. Another problem, that of the runaway overpopulation of our planet, is far more immediate and critical.

If mankind will prove unable to save itself from being choked by crowding it hardly needs to worry about its genetic quality. Although the problems of numbers and of quality are not one and the same, yet they may be closely connected in practice.

As steps towards regulation of the population size will begin to be taken, and this surely cannot be postponed for much longer, the genetic problem will inexorably obtrude itself before people's attention. The questions "how many people" and "what kind of people" will be solved together, if they will be solved at all.

Some people believe that all would be well with mankind, if only natural selection were permitted to operate without obstruction by medicine and technology. Let us not forget, however, that countless biological species of the past have become extinct, despite their evolution having been directed by natural selection unadulterated by culture. What we want is not simply natural selection, but selection, natural and artificial, directed towards humanly desirable goals. What are these goals?

This is the central problem of human ethics and of human evolution. Darwinian fitness is no guide here. If, in some human society, genetically duller people produce more progeny than the brighter ones, this simply means that, in the environment of that particular society, being a bit thick-headed increases the Darwinian fitness, and being too intelligent decreases it. Natural selection will act accordingly, and will not be any less "natural" on that account.

Human cultural evolution has resulted in the formation of a system of values, of *human* values. These are the values to which we wish human evolution to conform. These values are products of cultural evolution, conditioned of course by the biological evolution, yet not deducible from the latter.

Where do we find a criterion by which these values are to be judged? I know of no better one than that proposed by the ancient Chinese sage: "Every system of moral laws must be based upon man's own consciousness, verified by the common experience of mankind, tested by due sanction of historical experience and found without error, applied to the operations and processes of nature in the physical universe and found to be without contradiction, laid before the gods without question or fear, and able to wait a hundred generations and have it confirmed without a doubt by a Sage of posterity."

## THE TWO-MILLION-YEAR-OLD MAN

Discovery in East Africa of an entirely new species of primitive human being believed to be the direct ancestor of modern man was announced by Dr. Louis S. B. Leakey.

The British anthropologist and two other scientists who have studied the fossil remains have given the name *Homo habilis* to the new species within man's genus. The name means "able" or "having ability."

*Homo habilis* was smaller than a present-day pygmy, though manlike in most of his known characteristics. He first appeared nearly two million years ago in what is now Olduvai Gorge, Tanganyika.

Dr. Leakey and his wife Mary have found fossil parts of five individuals, including a young woman and child. The species was not alone. Existing at the same time in prehistoric East Africa were much more primitive types of Hominidae, the family of mammals to which man and his ancestors belong.

## FIRST PUBLIC ANNOUNCEMENTS

The dramatic discoveries were announced publicly for the first time here by Dr. Leakey at a press conference at the National Geographic Society, which supports his work.

Simultaneously, formal announcement was made in London in a scientific paper published in the April 4 issue of the British journal Nature. The authors were Dr. Leakey; Professor Phillip V. Tobias, of the Medical School of the University of Witswatersrand, Johannesburg; and Dr. John R. Napier, of London University's Royal Free Hospital Medical School.

Fossil bones of the five Homo habilis specimens were found over a period of four years in different deposits at Olduvai representing a geological time span estimated to cover more than a million years. Announcement was withheld pending evaluation by the Leakeys and scrutiny of the actual material by other scientists.

Professor Tobias went to East Africa to study the fossils with Dr. Leakey in his laboratory at Nairobi, Kenya. Dr. Napier made a study of skeletal material in London.

American anthropologists to whom Dr. Leakey has shown casts of specimens include Dr. T. Dale Stewart, Smithsonian Institution; Dr. William Straus, Jr., of Johns Hopkins University; Dr. W. W. Howells, Harvard University; and Dr. Clark Howell, Chicago University.

## JAVA MAN FOUND IN 1891

The discovery of *Homo habilis* may profoundly change present concepts of physical anthropology, the science that took a long leap forward in 1891 when Dubois found the skullcap of *Pithecanthropus erectus* (erect ape-man) in a Java stream bed.

In time, so-called Java Man (*c.* 500,000 years ago) and his close relative, Peking Man (*c.* 400,000 years ago), came to be regarded as very important landmarks in the search for early man—and not as primitive as originally thought. Lately they have been referred to as *Homo erectus*. "It now seems likely that the species of present-day man, *Homo sapiens,* is more likely to have evolved in Africa from *Homo habilis* than from some of the other hominids of the Pithecanthropine type in the Far East," Dr. Leakey says. This theory would take man's story very deeply into the prehistoric past. *Homo habilis* was living in Olduvai Gorge before, alongside, and after another type of hominid, *Zinjanthropus*, whose fossil bones were found by the Leakeys in 1959 sealed in by volcanic ash dated by the potassium-argon method as 1,750,000 years old.

Both *Homo habilis* and *Zinjanthropus* were tool makers. But *Zinjanthropus* dropped out of the picture, as did contemporary South African near-men (Australopithecines) to which he seems closely related. It would appear that nature experimented with these manlike creatures, found them wanting, and rejected them.

*Zinjanthropus*'s skull shows that he was low-browed and long-faced, with a relatively small brain. He certainly did not know the use of fire, though he probably shared the obvious ability of *Homo habilis* to plan and make tools. *Homo habilis* had a greater brain volume than *Zinjanthropus*. The former's brain was about three-fourths the size of that of a small modern man, according to Dr. Leakey. *Homo habilis* could use his hand with a precision grip, thus was well equipped to look after himself in the Early Pleistocene age of Africa.

## CHILD WAS FIRST EXAMPLE

The story of *Homo habilis* is closely interwined with that of *Zinjanthropus*. Evidence of *Zinjanthropus* came to light first at Olduvai in 1959 in the form of a shattered skull whose 400 fragments had to be put together like a three-dimension jigsaw puzzle.

During excavations in 1960 and 1961, Dr. and Mrs. Leakey found the lower jaw of a child and parts of its skull and hand, as well as the foot attributed to an elderly woman. These were in a deposit lower, thus older, than that of *Zinjanthropus*.

The new discoveries had the effect of a thunderbolt on the Leakeys. The pre-Zini child, though older than *Zinjanthropus*, was much more advanced. Strikingly different were the characteristics of the teeth. At the time, Dr. and Mrs.

Leakey were unwilling to give a scientific name to the new hominid, though they expressed the belief that it was related to present-day man. It was neither an Australopithecine like *Zinjanthropus* nor an aberrant of *Zinjanthropus*.

"During the closing months of 1963," Dr. Leakey said today, "Olduvai Gorge yielded very important new fossil hominid remains which fully establish that a species of the genus Homo-which is new to sciencewas living in East Africa side by side with *Zinjanthropus* in Lower Pleistocene times."

Dr. Leakey says the best of the new material consists of skull parts, a nearly complete lower jaw, and portions of the upper jaw of a young woman, 20 or 21 years old, whom he nicknamed "Cinderella." Additional fossil parts of at least two other individuals were there-consisting of teeth and skull fragments.

## MATERIAL CONFORMS

"The new 1963 material conforms exactly, in morphological character, to the earlier 1960-61 material," Dr. Leakey says. For scientific purposes, the fossils from Bed I have been made the basic type of the new species *Homo habilis*. The best of the material from Bed II becomes the first paratype. Dr. Leakey says, "The skull wall is very thin and has a rather small brain capacity, but in its morphological character it resembles a small skull of *Homo sapiens* more than anything else.

"Possibly the most important single feature of the brain case is to be seen in the occipital bone at the back of the skull," he adds. "This, in its morphological features, closely resembles *Homo sapiens* and is wholly unlike the form to be seen either in the Australopithecines or in *Homo erectus*."

The jaw and upper and lower teeth of the new species recall primitive *Homo sapiens* and are not like those of the near-man. Explains Dr. Leakey:

"In all of the Australopithecines, or near-men, the premolars are wide from side to side, and always wider than they are long from back to front. In the new species of *Homo* the opposite is the case."

"In *Homo habilis,* the canine teeth are large relative to the premolars. In the Australopithecines, including *Zinjanthropus,* the canine teeth are very small relative to the size of their premolars and the size of their large jaws."

The manner in which *Homo habilis'* teeth were worn is significant. He apparently had a softer diet which probably included the flesh of small game. The massive teeth of his contemporary *Zinjanthropus* were ground down by a coarse plant diet that probably included gritty roots.

## OVER-ALL PICTURE

"The over-all picture represented by the new species is one which has a number of resemblances to present-day man, rather than to the Pithecanthropines," Dr. Leakey said.

However: "In spite of a number of resemblances with *Homo sapiens,* both in the teeth and the skull of the new species, there are also a number of primitive pecularities to be seen. In particular there is the small size of the brain and the primitive characteristics of the hand. It is these things which made it necessary to set up a new species."

Recent excavations at Olduvai have been concentrated on Bed II. Prior to coming to the United States in February, Mrs. Leakey worked there continuously for 13 months with scarcely a break. Excavations at Fort Ternan, Kenya, another important fossil site being developed by the Leakeys, have slackened off due to unusually heavy rains last year and the new finds at Olduvai. Another important fossil discovery was made in Tanganyika in early 1964 by Richard Leakey, the Leakeys' 19-year-old son. The site was at Lake Natron, about 50 air miles from Olduvai.

The youth made a reconnaissance in a light plane over the rugged area, then led a small overland expedition by Land Rover and truck from Nairobi, Kenya. Members of the expedition crossed the lake on a pontoon boat and set up a base camp near steep escarpments where layers of volcanic ash and clay are exposed by erosion. Earlier, in 1959, Mrs. Leakey had spotted the inaccessible gorges from the south and thought them promising for excavation.

## JAWBONE DISCOVERED

The small party led by Richard Leakey and joined by Dr. Leakey's assistant, Glynn Isaac, made a systematic search of a part of the badlands. An African employee named Kamoya spied bleach-white bone gleaming from a wall of a gorge. The extraction of it began, and the specimen turned out to be the perfectly preserved lower jaw of a *Zinjanthropus* type of hominid.

To the Leakeys' surprise, the new jawbone fits almost precisely onto the *Zinjanthropus* skull found more than four years earlier at Olduvai. (The original *Zinjanthropus* skull still lacks a lower jaw). Geologically, the Natron deposits are more recent than the lower beds at Olduvai. Says Dr. Leakey: "The jawbone shows us that this type of Australopithecine continued to evolve practically unaltered for a very long period. It did not—as we once believed—gradually evolve in the direction of *Homo sapiens*.

"It also confirms the view that the *Zinjanthropus* type of hominid probably lived on a rough vegetarian diet. Both in the original *Zinjanthropus* palate and in the new màndible, the crowns of the premolars and the first and second molars show a considerable amount of wear to the enamel, in spite of the fact that the third molars, or wisdom teeth, were just coming into wear when the owner died. "In contrast to this, the premolars and molars of the paratype of the new species *Homo habilis*, whose age at death was comparable to the two Zinjanthropus specimens, are very much less worn."

## CONTINUITY AND CHANGE

If life comes from life, then we should be able to trace a chain of life back through the ages. And we can. It is possible to take an animal such as a horse and identify the early relatives of today's husky specimens. This can be done by carefully examining the buried remains of horselike animals found in many parts of the world.

As the record of millions and millions of years is studied, it becomes clear that life forms are not static. Several of the ancestors of today's horses were equipped with three toes on each front foot instead of a single hoof. A still earlier ancestor had four toes. There have been many other changes since the time, 55 million years ago, when that four-toed animal roamed the earth. That ancestor of the horse was no bigger than a terrier dog and was accustomed to eating soft foliage, not the harsh grasses that horses eat today. Only when they developed a harder tooth surface could horses begin to live off the grasses that were covering the earth between 13 and 25 million years ago. Only then were their teeth suitable for grinding the new food.

It is important to note, however, that there were horses in the past other than those in the main line of development, other than those that were direct ancestors of modern horses. Scientists who have studied fossils —ancient bones and other traces of life—have found that there were many more members of the family outside the main line than within it. These types have since become extinct and they have no present-day descendants.

In part, the concept of evolution explains the change and progression of living things over millions of years. The idea of such a pattern is actually very old. Some of the ancient Greek philosophers speculated about the nature of life and arrived at conclusions that occasionally come quite close to the theories we hold today. Empedocles of Agrigentum has been called the father of the idea of evolution. He believed that plants developed first and that animals came later, following a long series of trial efforts. Some scholars have seen in his thinking a hint of one of Darwin's theories: that the fittest organisms survive by chance instead of design. Empedocles argued that some forms of life were gradually replaced by others better suited to live in their surroundings.

Among the Greeks, Aristotle came closest in his thinking to our modern concept of evolution. He saw a single chain of life, an orderly progression from certain water animals to human beings. His ideas anticipated current theories in many ways. although he was entirely wrong in many of his beliefs. Since Aristotle's time, men have continued to puzzle over the pattern of evolution.

## EVOLUTION IN ACTION

Convincing evidence about whether evolution actually occurred, however, was not gathered together until the 19th century. The gatherer was Charles Darwin, who was then a young man in his twenties and much more a sportsman than a scientist. He had signed on aboard the H.M.S. *Beagle* as naturalist during a round-the-world surveying voyage.

His first clues to evolution came in South America. There he observed the remains of large, extinct animals that were clearly related to smaller ones still living on the continent. He found a huge fossil animal with armor resembling that of the existing armadillo. He also noted signs of gradual changes in animal forms as the ship progressed southward. Many forms seemed related, but had obvious differences too. How had this come about?

Darwin received an answer when he reached the rocky Galàpagos Islands off the west coast of South America. There he observed some finches, a common breed of bird. He found 14 species of finches, none of which was known elsewhere in the world.

True, they resembled a species found on the mainland, but three species that ate seeds depended on seeds different from those the mainland finches ate. The others fed not on seeds at all but on cactus plants or insects. One species even grasped a cactus spine in its beak to pry bugs from the bark of trees!

Eventually young Darwin reasoned that the birds must have originated from some mainland finches that had found their way to the islands. Over many, many years they must have developed widely varying feeding habits and in this way had evolved into separate species.

We now know that food supply is only one of many forces at work in shaping the nature of the world's populations. Predators or other natural enemies may bring about a change in living habits. During thousands or millions of years, the climate may change so that one living form may prove more successful than another, and thus multiply at the expense of a competitor.

## THE FOSSIL RECORD

Working with Darwin's concept, scientists have since filled in many parts of the story of evolution. One of the most productive lines of investigation has been the study of fossils. Fossils are traces of life long past. Such a trace may be a bone, an impression of a skeleton, seashell, or leaf in the hardened mud of ancient seas, or a whole animal, such as an ant preserved in jewel-like amber. The fossil record has been studied all over the world by hundreds of researchers investigating everything from the microscopic remains of protozoans to the bones of gigantic dinosaurs. Such fossils offer overwhelming evidence that evolution must have occurred.

A startling fact uncovered by these researchers into time is that all the basic types of animal structure are visible among fossils more than 350 million years old. This does not mean that the higher animals were developed then, but only that the pattern of bodily organization had been established. There were single-celled protozoans.

Their forms, preserved in stone, are visible because of the tiny skeletal structures some of them had developed. There were sponges, coral, worms, clamlike brachiopods, starfish and their relatives, mollusks of all sorts. There were arthropods equipped with horny skeletons outside the body, and chordates with skeletons of living bone inside the body. The various vertebrates (backboned animals) evolved from the early chordates, and include today's fishes, amphibians, reptiles, birds, and mammals. Invertebrates, animals without backbones, were more abundant than vertebrates. In fact, they still are.

## WHEN THE ANIMALS CAME ASHORE

Both vertebrates and invertebrates came ashore from the ancient seas in several forms. The ancestors of today's insects were primitive jointed arthropods related to the remote ancestors of crabs. But the most dramatic change in the earth's life forms began when backboned animals came ashore.

If you look at a goldfish or a trout, it is difficult to imagine how a fish could have evolved into an amphibian such as a frog or a toad, which seems to be perfectly at home on land. But there are fish alive today that show us the connection is not impossible. One species has both gills and lungs. There are others that breathe air through their stomachs or through accessory lungs, called air bladders. There is a mudskipper that chases insects, eats worms like a bird, and can use its fins to get up trees.

Certain primitive fish that roam the deeps of the Indian Ocean have fins very different from those of modern ones. These fins project from the body and can be turned about like the arm or leg of a land animal. The ancestors of amphibians might have had such fins and used them as flippers to squirm from pool to pool at the shoreline. And they had to have some way of gulping air when they were out of water. At first they must have changed so that land became as much a home as water. Then some of them became more at home on land than in water.

The step from amphibian to reptile represented a further adjustment to living ashore. Amphibians have a moist skin and lay eggs in water or moist places; their young usually begin life looking like fish and living in the water.

Reptiles, on the other hand, are dry-skinned, usually lay eggs on dry land or give birth to their young, and the offspring can immediately begin breathing in the air. Reptiles in their various forms dominated animal life on land for some 150 million years. One form even led to today's many types of birds.

## FROM MAMMALS TO MEN

Most of the reptile types failed to stand the test of time. It was an offshoot from the varied reptile stock that produced the next dominant group, the mammals. Mammals are characterized by hair or fur, and by suckling of their young. Only two of the transitional forms survive today, the duckbill platypus and the spiny anteater. These two animals have hair and somewhat better temperature control in their bodies than reptiles. The females produce milk, but they also lay eggs, like their reptile ancestors.

The mammals in their early evolution must have been small creatures, hardly worth the attention of the mighty reptiles. But within about 10 million years the majority of the large reptiles became extinct and the mammals came into their own. Their built-in temperature control made it possible for them to exploit all sorts of environments, from steaming jungles to chill mountain ranges. Because of this internal temperature control, mammals are called warm-blooded. Reptiles are said to be coldblooded; they cannot maintain warmth in cold weather. Birds also developed a warm-blooded system.

Over the ages the mammals expanded tremendously in type and number. They grew large, as the reptiles had done, but today's species are often smaller than their earlier counterparts. They also produced the ancestral line that led to man.

Of course, the evolution of animals could not have taken place without plant life abounding on the earth. Green plants supply food in the form of compounds that animals can turn into energy and tissues. Plants also remove carbon dioxide from the atmosphere and replenish the oxygen supply needed by animals. Plants too have adapted to life on land, with its fluctuations and changes in environment. More than a million and a half species survive today.

The results of evolution appear even more spectacular when you stop to think that a world now teeming with living things was once devoid of life.

## ANCIENT CHEMISTRY ALIVE TODAY

Perhaps the most important thing we learn from studying the concept of evolution is that all living things are related. Despite all the changes that have occurred, there is continuity—continuity more fundamental than the similarities between parent and child.

An interesting bit of chemical evidence in the fossil record bears this out. Proteins are substances involved in the vital processes of every living thing. They are made of simpler materials called amino acids. Recent studies of some fossils

300 million years old showed that the amino acids found in these ancient samples are the same proteins that serve life's purposes today.

This brings us back to the possibility of finding a way to detect life in space. Is there a chemistry common to living things on earth, a series of characteristics that we might look for to determine if life has evolved elsewhere? If so, our task may be greatly simplified. We should be able to send an automatic laboratory to a planet to identify a known substance. This would be easier than having to explore a planet's surface for all the possible forms of life.

## THE ROLE OF PALEONTOLOGY IN THE FORMULATION OF EVOLUTIONARY THOUGHT

Each of the disciplines of the biological sciences has played a dual role in the formulation of evolutionary theory: contributing data that have led to modern theory and providing a reservoir of fact and theory against which various facets of the evolutionary concept can be tested. A few of the disciplines have contributed directly to the development of the general idea of organic evolution, to the proposition that the history of life has been one of change with continuity through the agency of reproductive processes. Others, being recent in origin or in maturity, have been effective in contributing inductively only to the nuances of sophisticated modern versions of evolutionary theory.

Each of the biological disciplines is in some degree made more meaningful in itself and brought into logical relationship to others under the broad cover of the theory. Paleontology, a discipline separated from others by the temporal scope of its materials and by its methods, has to some extent fulfilled the two roles, as an agent of induction and a target of deduction, especially in the form of a tangible life history, and certainly takes on meaning when viewed in light of current evolutionary theory. In this essay an effort is made to assess these various roles of paleontology.

The theory of natural selection, which in one form or another is the most widely accepted today, and the broader and simpler idea that life has altered in an evolutionary way through time are quite different with respect to their relationships to origins in terms of paleontological data. We will deal with them separately and in their historical sequence.

The threads of the intellectual activities of the 16th and 17th centuries, and traces that go back even farther into the 15th, appear to have made emergence of a full-fledged concept of organic evolution virtually inevitable. Perhaps the most impelling base lay in the concept of evolutionary development of social history, formulated in particular by the philosopher Hegel.

But the astronomy of Galileo and Copernicus, the cosmology of Kant and Laplace, Descartes, Newtonian mechanics and the rise of materialism, recognition of uniformity and evolution in the physical history of the earth, travel and the expanding knowledge of zoogeography, and elucidations of the basic data of anatomy and embryology all begged for interpretation couched in a materialistic concept of the continuous change of life with time. First causes were removed, at least to a point beyond the scope of the immediate phenomena.

Speculations about an orderly change of life were certain, then, to emerge during these centuries, as indeed they did, just as they had appeared many times before when a philosophy of flux as a basic reality had arisen, contrasting to the concept of fundamental changelessness. There is no very clearly demonstrable relationship between the very early ideas of evolution, formulated for the most part in the pre-Christian eras, and our own, which finds its roots largely in events from the 15th century on. Thus in this discussion the evolutionary ideas of the earlier times, interesting as they may be, will be dismissed as only tangentially relevant.

It appears highly probable that some concept of organic evolution could and would have developed in the absence of any fossil record. Just what form it might have taken and how current theory might have been altered cannot, of course, be determined. It would have arisen, but it might have found the road to general acceptance much more difficult in the absence of the tangible evidence of life's history given to the layman by geology and the fossils in the rocks. The actual course of development of evolutionary theory, in any event, shows that the fossil record was very important in the thinking of those students who were the most influential in promulgating the idea of organic evolution.

*Vestiges of the Natural History of Creation* by Robert Chambers, published in 1844, while ridiculed for many reasons, presented the concept of evolution based on the factual evidence of the fossil record. The first one-third of the book is devoted in large part to the presentation of this evidence. The first edition was published during the time that Charles Darwin was developing his ideas and created considerable stir and interest in evolution and aided in establishing a climate in which the evolutionary philosophy of Herbert Spencer could flourish.

Long before, Erasmus Darwin, the grandfather of Charles, in his *The Temple of Nature* drew heavily if obscurely in poetic presentation upon the past history of life. His *Zoonomia* and other works also carried a transformationist doctrine. Buffon, struggling with the problem of origin and demise of species, similarly was stimulated by the existence in the rather recent past of animals now extinct. Lamarck, of course, was a knowledgeable paleontologist and was strongly influenced in his acceptance of the mutability of species by his understanding of the fossil record of life.

This list could be enlarged, but these are some of the persons who are justifiably well known for their part in the origin of the theory of evolution. Some evolutionists, on the other hand, seemed to pay little heed to the fossil record, or, perhaps, to be little aware of it. They were less influential. For the most part, those who gave evidence from the fossil record were the ones who stirred the interest that led to discussions and either recognition or rejection of the theory.

They touched upon ground common to many and thus comprehensible as material for debate. The more erudite and subtle presentations, such as those of Tennyson in his "In Memoriam" and, to some extent those of Erasmus Darwin, being literary or poetic in form, seem to have stirred less controversy, being read more for their poetic and literary than scientific content.

Today there are few serious students of biology who do not adhere to a broad concept of evolution of life in one form or another. By far the greatest number of these students subscribe to evolutionary theory that incorporates natural selection as a basic factor. In contrast to the importance of paleontology in the origin of the general concept of organic evolution is the very minor part that it appears to have played in the development of the theory of natural selection.

This is evident in the works of Charles Darwin, of Alfred Russel Wallace, and in studies and comments of those who had the germ of the concept in earlier times, for example, David Hume in his *Dialogues on Natural Religion,* published in 1739, or in the works of Charles Bonnet and Denis Diderot who first used the term evolution.

Darwin, in referring to the evidence of natural selection in the fossil record of the *Origin of Species,* succinctly stated that, "Geology assuredly does not reveal any such finely graded organic chain; and this, perhaps, is the most obvious and general objection that can be urged against my theory." In a broad sense, he pointed out, the record is in accord with the theory, and then proceeded to an excellent analysis of the reasons that there is no detailed correspondence. In no specific sense, however, can it be said that the idea of natural selection arose from the facts of the record of life.

It is notable in this regard, as well as with reference to the general idea of organic evolution, that many paleontologists were among the strongest critics of evolutionary theory. Cuvier and Owen are well known examples. Then as now, unless the record is viewed under the unifying evolutionary concept, its negative aspects are so pervasive that tremendous latitude in interpretation is possible.

This has often been seized upon by critics of evolutionary theory as a point of departure. Upon publication of a paper entitled "Morphology, Paleontology, and Evolution," in which some of the problems of interpretation of the fossil record were outlined, I received a great many papers and letters that presented "alternative hypotheses," some seriously conceived and some quite otherwise. Most took their beginnings in inferences based on the incompleteness of the record of life, relying heavily upon its negative aspects.

In summary of what has been said to this point, I will reiterate that paleontology appears to have played a very important role in the origin of the general concept of organic evolution, but it had only an insignificant and largely negative role in the development of the theory of natural selection.

We may now turn to the post-Darwinian phases of the development of theory, to examine what paleontology has contributed during the last century. The crux of the problem of natural selection after Darwin lay in the question of the origin of variations. This, along with the problems of preservation and dispersal, formed the special world of Mendel, de Vries, Weismann, Morgan, Fisher, Wright, Muller, Waddington, Dobzhansky, among many others, and more recently Beadle, Crick, Watson, and their colleagues. Other biologists, operating in many fields, were among the synthesizers of current evolutionary theory. Among paleontologists Simpson stands out as the one who more than any other successfully applied concepts of neobiology such as population genetics to the fossil record.

He emerged as a master of the synthetic approach and greatly enhanced the understanding of the fossil record and the integration of its data with those of other disciplines of biology. This era of synthesis, from about 1930 to 1955, probably marks the peak of both substantive and theoretical contributions of paleontology to evolutionary theory, at least up until now. Its history to the late 1940's has been written clearly and concisely by Simpson.

An important fact emerges from a study of what went on during this time. As significant as any substantive role that paleontology has played is another and subtler one, its effects upon the ways of thinking of those who have become thoroughly acquainted with the history of life. The synthesis brought this history to the attention of many who were not paleontologists proper and, in reciprocal fashion, enriched its impact upon paleontologists as they became more acutely aware of biology beyond the narrow limits of their own field.

Even some of the most sophisticated approaches to evolutionary theory today, in fields applying molecular biology or multivariate mathematics, show this impact as their new conclusions are often tested against the backdrop of phylogenies from the fossil record. The faith in these phylogenies is sometimes rather startling to the paleontologists who were instrumental in their formulations.

The all-over effect of this infiltration of the sense of history through neobiology, which is far from 100% effective, is hard to evaluate, but it surely has considerable significance. If evolutionary theory is considered in its broadest sense as a complex of theories, as argued by Beckner, then paleontology as a discipline with its special data must be thought of as one of the bulwarks of modern theory. It is tempting to launch into a discussion of the specifics, into considerations of matters of rates of change, of directional evolution, of chronofaunal evolution, community concepts, and other well-known areas with which paleontology has been intimately concerned.

These illustrate the roles that interpretation of the history of life can play in enriching, clarifying, and ramifying the concepts that contribute to the complex we call evolutionary theory. Such a diversion is not possible in the space available and not essential to the main purposes of this essay. That the fossil record plays a fundamental role in furnishing a history is self-evident. But it assumes this role only under the aegis of one or another governing concept to which it can itself contribute directly in only a minor way.

To attempt to visualize the importance of this point, it is amusing to speculate upon the conclusions that might be reached by some hypothetical observer totally unaware of life as we now know it, who saw only the evidence of the fossil record with its remnants of organisms arranged in rough temporal sequences. I will leave this speculation, the ground rules, and the conclusions to the readers to whom such "scientific fantasy" appeals.

As in the early days of evolutionary theory, (in the 18th and first part of the 19th century) there were many paleontologists who rejected the general theory of evolution, so later, after natural selection was proposed as a dominant aspect of the theory, many, accepting evolution, rejected natural selection as a critical element in it. So, of course, have many others who were not paleontologists or even scientists—Henri Bergson, for example. But among the biological scientists it was predominantly some of the paleontologists who failed to follow along the lines being revealed by developments of genetic theory.

There are good reasons for this. There was some degree of isolation between the various disciplines. More important, however, are the data revealed by the fossil record. There are great spatial and temporal gaps, sudden appearances of new major groups, equally sudden disappearances of old, including very rapid extinctions of groups that had flourished for long periods of time.

There were mass extinctions marked by essentially simultaneous death of several apparently little associated groups of organisms. At the time the record first is seen with any real clarity, the differentiation of phyla is virtually complete. As far as major groups are concerned, we see little clear evidence of time succession in differentiation with the simpler first and the more complex later.

Over shorter periods of time, but still far beyond man's capacity to observe directly, directional change with little fluctuation or diversion is a striking feature. "Bizarre" organisms seem to argue against a concept of nice adjustment through selection of the adapted. "Bizarreness" seems to be a prelude to extinction, often without tangible evidence of change of conditions of existence. The seemingly most superbly adjusted organisms sometimes were lost, while their awkward neighbors kept on.

Catastrophism, saltations in origins, innate directional forces, directional change in direct response to environmental stimulus, various finalistic, vitalistic, and teleological interpretations were the outcome. Cope, as a neo-Larmarckian,

Osborn with his concept of aristogenesis, Schindewolf and catastrophism or neocatastrophism, and Tielhard de Chardin with the Omega point will serve for illustration.

While it is true that all of the problems posed by the record can be rationalized under selection theory, the studies of students who followed "unorthodox" lines have made significant contributions to the knowledge and understanding of the history of life and, in one way or another, positively or negatively as may be the case, to evolutionary theory.

Much has come from efforts toward refuting what their work inspired, from the interest in evolution that it aroused (because often the conclusions were more palatable to laymen than the alternatives), and from insights into processes that were integrated into the broader theory of evolution developed later. Modern theory, however, depends very little upon any concepts developed *de novo* by these students. Most of their basic tenets, both those later accepted and those rejected, had been in existence, sometimes in less mature form, well before the exposition of the theory of natural selection by Darwin.

The metaphysical aspects of theories held by some paleontologists and by other evolutionists, while interesting, are beyond the scope of scientific treatment and will not concern us. But such things as catastrophism, parts of Lamarckian and neo-Lamarckian doctrines, or directional evolution at the level usually seen in the fossil record logically are proper subjects for scientific treatment. The concept of inheritance of environmentally induced characters has, of course, been tested many times and, at least at the gross level recognized by neo-Lamarckians, support has been consistently wanting.

Such phenomena as those covered by catastrophism and conceived as involving causality remain proper matters for study. We see no evidence of major catastrophic alteration of the direction of evolution of life in observations of modern organisms. Catastrophies do occur on various scales and a sub-species or species or even some ecological complex may sometimes be markedly affected. But this has not been clearly related to the sorts of things that seem to appear in the fossil record, with its instances of many kinds of events that may be widely spaced in time or are not *per se* repetetive and may occur only once. The lack of documentation over a short period of time cannot be taken as an indication that such events have not been important.

This, so it seems to me, points to a dilemma of the evolution-minded paleontologist and to a weakness of the fossil record as an agent of innovation as far as the fundamental, central core of evolutionary theory is concerned. First of all, the paleontologist cannot demonstrate that natural selection has or has not been a fundamental caustive factor in evolutionary events that he observes. The record can serve as a basis for testing and evaluating hypotheses based on modern materials and subsumed under the synthetic theory, or any other theory that may be considered acceptable on the basis of modern materials. Similarly viewed, it can supply data and hypotheses that can suggest pertinent studies by experimentation with living organisms.

In the realm of evolutionary events that emerge to view only with observations encompassing long periods of geologic time, an experimental interplay with living organisms is impossible. There are, however, many apparently long term processes and widely spaced nonrecurrent events that are not yet fully understood and that eventually may lead to new and important perspectives in evolutionary theory. It is in this area of research that the paleontologist can make unique contributions. He will have to rely primarily on induction, not because he is denied deduction but because he may have to make deductions that can be satisfied only by observation rather than by experiment, and he may not discover the fossil finds necessary to test his deductions.

This possibility of disappointment should not stifle a search for theoretical meaning. Two positive solutions are always possible. First, a theoretical breakthrough in an unrelated field may furnish the clue necessary to tie together the tag ends of the puzzle. Second, paleontologists themselves might sometime, somewhere, find fossil evidence giving their deductions a remarkably high degree of probability.

## FLOWERS, INSECTS, AND EVOLUTION

Any endeavor to unravel the evolutionary history of our fascinating flower-insect relationships reveals divergent interpretations among scientists. For such an undertaking a comprehensive knowledge of the natural sciences is required, especially of geology, botany, entomology, and above all, of paleobotany, paleoentomology, and evolution.

Many people realize the existence of interdependence between some plants and insects, but few are aware of the aeons of time involved during which this process reached its present state of development. Evolution has no definite end result and never a deliberate, predestined direction. In the case in point, it is expressed by a high degree of specialization and an intricate relationship that developed into such widely divergent biological entities as the flowering plants and the insects.

First, of course, we must consider the interdependency between plants and insects that exists in nature today; then we must apply the dictum, "The present is a key to the past." The geological record presents a chronicle of 400 million years of fossil life that can be traced through each subsequent younger formation.

The evolution of each living plant and insect has been a continuous chain of events, which, if broken, would have resulted in extinction. In turn, a complex and harmonious interdependency—or mutualism—exists between many plant and insect species. If that interdependency should be severely disrupted, mutual annihilation could result.

Early organization of flower structures appears to have been a slow, gradual process of adaptation and specialization without functional stimulus by insects. When the first insectlike animals appeared during Devonian time, the evolution of highly specialized reproductive structures among such plant groups as the seed ferns, the treelike club mosses (*Lepidodendron, Sigillaria*), and the giant horsetails or scouring rushes (*Calamites*) was already well on its way. Early "insects," according to fossil evidence, were doubtless primitive, wingless, and unspecialized. Their food requirements were not restricted, but may have comprised a varied diet of decaying debris.

The reasons for the sudden rise and expansion of the angiosperms-the higher plants, which have seeds enclosed in an ovary—are still shrouded in mystery, but a phylogenetic plasticity coupled with favorable ecological opportunities surely played a major role in their rise.

Their explosive diversification and dispersal in the late Cretaceous and early Tertiary nearly parallels an equally momentous rise and diversification, or speciation, among insects. An intimate interdependence between these two groups, therefore, must be considered a relatively recent geological phenomenon—one initiated about 70 million years ago. Most earlier relationships probably were essentially haphazard, and at first contributed little to the eventual physiognomy and requirements of either plant or insect.

Nothing in nature is constant but change. In a nutshell, organic evolution, expressed in terms of phylogenies, is "descent with modification" as a function of perpetual genetic variation, mutation, and natural selection. Those traits that do not contribute to species survival are gradually discarded. In rare cases, according to fossil evidence, species remain in a state of what might be called "evolutionary suspension," or *status quo*.

That is, they have changed relatively little either because their environment has not changed, or because they are adapted to a wide range of environments. Cases in point are such examples as the "living fossils" of the Dawn Redwood (*Metasequoia*), the Katsura tree (*Cercidiphyllum*), Ginkgo, and, to some extent, the roses. Among the insects, the dragonflies, roaches, and silverfish fit into the living fossil category.

Essentially, then, over a long period of time, each gradual and minute change in both flower and insect morphology was triggered by subtle genetic and environmental changes. In some cases, our interpretation of an early phylogeny is only as valid as the fossils on which it is based, and additional fossil material that comes to light frequently causes us to reconsider and modify previous conclusions. This procedure permits a flexibility to amend previously established plant—insect relationships according to new evidence.

Some of the early insects of the Carboniferous and Permian periods were doubtless carnivorous or omnivorous, while herbivorous forms were destructive to plant life. It has been suggested—as a hypothetical beginning to plant—insect interdependencies—that there gradually appeared cells that secreted nectar.

These surrounded the reproductive (flower-like) regions, and the feeding activities of insects began to be channeled toward such central lures. This, then, may have resulted in the preservation of vegetative structures and, concurrently, promoted the transfer of microspores and insured pollination. From such convergent developments arose a mutualism that was beneficial—in some cases a survival factor—for plants and insects alike.

It is sometimes assumed that flower types were determined by their exploiters, the pollinators. To the uninitiated, the examples are superficially plausible when we say that "bumblebee flowers," such as clover, arose in response to specific structures of the bumblebees; that "bird flowers," such as cardinal-flower or trumpet-vine, arose in response to parrots and hummingbirds; and that "bat flowers"—*Kigelia* (the tropical sausage tree) or certain cacti—developed in response to bats. Such reasoning assumes an independent "leading" development of the fauna with a resulting adaptation of flowers to their visitors. It can hardly be argued that the insects are the sole exploiters of flowers for food as long as the flowers equally "exploit" insects for pollination—an exchange that results in harmonious relationships to each other's advantage.

Many instances of insect behavior toward plants may be traced to a functional structure, color, or fragrance in plants, and one might be tempted to say that insects "behave" in a particular fashion because of these attributes. Whether the insects directed or caused the evolution of flowers to conform to their own entomological specializations is certainly not decipherable from the geological record.

But if they did so, it would follow that flowers must always have lagged behind insects in their optimum adaptation. It is more realistic to assume that reciprocal adaptation among primitive and advanced flowers and insects was achieved contemporaneously or in a limited, seesaw fashion.

Secretion of nectar, for instance, is physiologically linked with flower maturity, the opening of stamens, and the discharge of pollen. Nectar glands, or nectaries, of the earliest flower types attracted primitive arthropods of the class Insecta. Floral parts provide a region where sugars, starch, albumen, fats, and vitamins accumulate to be incorporated into developing fruits and seeds. Nectar and pollen thus furnished, then as now, a concentrated and palatable food for the arthropod visitors who "learned" to utilize these sources for their daily needs.

Today, insects are divided into two great subclasses, the wingless Apterygota and the winged Pterygota. Wingless insects entered the geological stage during early Devonian time and have remained relatively unchanged and primitive to this day. They contain the present orders of Thysanura, with bristle tails or caudal filaments and chewing mouthparts (silverfish), and the Collembola, minute—or even microscopic—vegetarian springtails. Only one in a thousand of all insect species is wingless today.

Pterygota appeared somewhat later, but the exact time of derivation from the wingless forms is problematical, and the morphological changes that resulted in wing development are still not clearly understood. The attainment of winged flight precedes that for reptiles and birds by nearly 100 million years. This mode of locomotion obviously unlocked new vistas and evolutionary potentialities for the whole group. Nearly 200 million years elapsed before the greatest number of insect species took to the air, which coincided with the early Tertiary ascendancy and diversification of the flowering plants.

Among the Pterygota, the Palaeoptera, or ancient-winged, had outstretched, non-flexible wings that could not be folded flat over their bodies. Dragonflies and mayflies are examples.

They were mainly predaceous, and evolved into the largest insects that ever lived. Fossil dragonflies with a wingspread of 30 inches are not uncommon. Toward the end of the Carboniferous the Palaeoptera gave rise to the new-winged insects, the Neoptera, which could fold their wings. Today, 90 per cent of all existing orders of Pterygota, comprising 97 per cent of all species, belong to this group.

For the first 100 million years following the origin of the winged insects, the air was shared with no other organism. They reigned supreme until the Jurassic and the advent of the birds, which soon turned the insects into a source of food. Neopterous insects of the Upper Carboniferous, related to our cockroaches and stoneflies, had an incomplete, or hemimetabolous, metamorphosis, in which each stage was larger than the previous until adult size was reached.

Insects with a complete, or holometabolous, metamorphosis—that is, those that go through grub and pupal stages—appeared in Lower Permian times and comprised about 5 per cent of all insects. Today, 88 per cent of all insects are in this group; their larvae and caterpillars differ from adults in structure and in food and ecological requirements.

One of the keys to the successful evolution of pollinating relationships is the faculty of insects to distinguish diagnostic forms of basic flower types that consist of a definite number of sepals and petals. Such structures are termed "numeral" patterns and were first recognized in "flowers," or fructifications, in the cycad-like plants of the early Mesozoic. Gradually, through various evolutionary lines, they attained their present angiospermous complexity of ovary-enclosed seeds.

Perhaps one evolutionary system of flower "type classes" may be equated with a complementary sensory evolution of pollinating insects. During the progress of natural selection, each previous sign of "recognition" on the part of the insects was augmented by a new, more adaptive, and more successful sign. Primitive, spirally arranged, beetle-pollinated magnolias of the Cretaceous, for example, gave rise to some modern species (*Magnolia denudata, M. coco*) with a radial symmetry and a more efficient pollination by Hymenoptera.The following botanical terms in their evolutionary sequence telescope these trends:

Amorphic, or "structureless," extinct flower types usually possessed dense whorls of leaves or bracts, often without definite form, and growing beneath the floral reproductive structures. Some of the Triassic ancestral cycads with amorphic "flowers" (*Wielandiella*) attracted certain primitive beetles and roaches, and possibly some wingless insects that learned to associate the pollen and secretory food supply with the density of whorls of bracts or basal leaves of the flowers.

Haplomorphic, or "simple," flowers, in their often "spiral" arrangements of variously colored petals and sepals (the perianth), trace their ancestry to the middle and late Mesozoic Era. Magnolias, the tulip tree, and waterliles belong in this group, which is considered to be one of the most primitive among the angiosperms.

Actinomorphic, or "regular," flowers have emerging "radial" and twodimensional symmetries. Roses and peonies, which have all floral parts and nectaries on one level, are examples. Here are associated composite, radial forms of from three- to six-perianth segments that probably are perceived by the insect as a unit. Together with color and fragrance, these flower types represent a food attractant for certain flies or bees and a few other well-adapted pollinators, including beetles, that need no specific ability to gather the abundant pollen offered. No flowers of the truly radially symmetrical types have been reported from the Cretaceous Period.

Pleomorphic flower types (the root word means "more than one"), with a four- or five-part recognition pattern, were found in Cretaceous beds, but they became abundant only in Tertiary floras. Some lily-like compressions of the Tertiary suggest bee pollination, as in modern, colorful monocotyledonous flowers with a three- or six-part perianth. Flowers of the mustard and dogwood families were first recognized in the Oligocene Ruby shale deposits of southwestern Montana, of some 35 million years ago. Bees (including bumblebees) and flies (including flower flies) were abundant by then, and may often have chosen specific flowers on which to feed.

Stereomorphic, or three-dimensional, often tubular, flowers with a symmetrical three-, five-, or six-parted perianth, are a late Tertiary derivative from pleomorphic ancestry. They include gentian, bellflowers, columbines, primroses, phlox, and daffodils.

Their corollas protrude considerably above the deep-set and often concealed nectaries. Butterflies and hawk moths, some of which have refined adaptations of mouthparts (a long, curled tongue, for instance), or certain bees (which, with long tongues or very small bodies, are able to force entry into the tubes) feed on and pollinate stereomorphic flowers.

Zygomorphic, or bilaterally symmetrical, floral development represents the highest and most intricate evolution of morphological structures in response to, or contemporaneous with, a very limited number of pollinators whose sensory organization is attuned to a highly specialized perianth. These perianths offer complicated recognition patterns together with nectar and fragrance that must be grasped in their entirety, as was experimentally shown, no matter how ornamented their symmetry may be.

Zygomorphism crosses the boundaries of many families (orchids, milkweeds, legumes) and demonstrates what may be the ultimate in floral specialization. Orchids and their pollinators pose the greatest morphogenic and taxonomic problems to an investigator. They do not always conform to botanical criteria that are considered valid for other families. For instance, they possess unique pollen structures, the pollinia, which adhere as units to the visiting insect (often flies and moths), and are thus assured of delivery to a receptive stigma of the same species. This is made easier because the highly adapted pollinators regularly visit the same kind of flower.

Throughout evolutionary time each flower type, from the amorphic to the zygomorphic, has corresponded to a concurrent level of sensory adjustment in those pollinating insects that had the greatest reaction to a particular plant. Such adjustments continue to evolve, if imperceptibly, without the sacrifice of previously existing, advantageous flower—insect relationships.

The story of the present adaptation of insects to their flower hosts, and the intricate endowment of flowers to lure, attract, trap, deceive, and feed pollinators to their own ends is a source of wonder. Even birds and bats entered the competition when specialized structures and functions, serving the pollination of the host, arose with equally efficient features for food requirements of their vertebrate visitors. This resulted in an advantageous, mutual dependence.

After these more or less theoretical considerations, let us now look at some specific aspects based on relatively recent discoveries of plant and insect compressions in our own western backyard. During the summer of 1947, I found the first Tertiary plant compressions in the upper Ruby River basin of southwestern Montana. I could not foresee, however, the significance of that find until I showed a shoe box full of fossil leaves and assorted insects to Dr. C. A. Arnold of the University of Michigan. He immediately recognized the scientific potential of the flora and encouraged me to expand the collection. That suggestion resulted in seven additional field trips during the next fifteen years, and an accumulation of about 12,000 specimens, including some 200 species of plants and a great number of insect remains.

Among paleoentomological regions, the Oligocene Florissant sites of central Colorado had been the most lucrative and best-known in the United States, and this fauna was exhaustively described seventy-five years ago by S. H. Scudder. The Ruby plant sites rank second only to those of Florissant in richness of insect species and may surpass them in detail of preservation.

The Ruby River valley is 70 miles from Yellowstone National Park in a basin of the Rocky Mountains formed by geological faulting. It is flanked by the Ruby and Gravelly ranges. During most of Tertiary time, the entire valley was occupied by a lake in which a sequence of sediments was deposited. A vegetation much denser and more varied than that of today supplied leaves and other plant debris that were carried into the lake by numerous streams and buried in the mud. A high mineral content of the water and a possible absence of decay organisms provided favorable conditions for the preservation of organic remains, including insects that had been washed or blown into the lake.

Fossil sites exposed along the walls of dry washes consist of laminated, light-colored shale that separates along some of its bedding planes. Where such rock is exposed to annual sub-zero temperatures and to baking during the summer, it splits into progressively thinner slabs. These are the "books" of shale with their "pages," or layers, so thin that the slightest breeze will waft them into the air. It is this "paper shale" that contains the fossil compressions of most plants and insects at the Ruby sites.

We rightfully assume that the large, varied Ruby River basin insect fauna was attuned to the equally varied angiospermous flora of trees, shrubs, and herbaceous plants, and that a dependent relationship existed between these two groups. This image is recreated from fossils nearly fifty times as old as mankind itself—from plants and insects that left nothing but films of carbon, which are sharp, clear replicas of their former selves.

The Ruby fossil plant assemblage provides an eloquent basis for local divisions of vertical life or climatic zones, each with its adapted entomological fauna. In most cases, diagnostic criteria for fossil insect identification are sufficient to place them in their proper orders, families, subfamilies, tribes, and even genera. Their taxonomic status is therefore accurate except on the species level. The entomological sample of the Ruby shales probably represents but a small fraction of the genera and species that actually existed. Recognizable fossil insects thus indicate a reliable, if not a specific, correspondence with then-existing plant relationships and with plant—insect interdependence.

In light of applied evolutionary principles, the botanical and entomological participants of the Ruby flora give us a glimpse into antiquity that allows us a sharp focus on present phenomena. Our limited picture can only dimly suggest the meanderings of the evolutionary path, the immensity of geologic time involved, and the astronomical numbers of species and individuals that fell by the wayside as links between past and present.

## CROP PLANT EVOLUTION

In considering crop plant evolution it is appropriate to begin with the time scale. In the Old World the beginnings of agriculture, which include the beginnings of the culture of the major Old World cereals, the wheats and barleys, can now be identified with some confidence in northern Iraq and neighbouring regions in Iran, Turkey and Jordan. They have been dated at between 7000 and 6000 B.C. Godwin has assembled the evidence for the belief that the practice of agriculture reached north-western Europe about 3000 B.C.

Thus the period during which the evolution of our oldest crop plants has gone on, is about 9000 years. In the early part of the period various plants went through the early stage of domestication and were then abandoned—*Chenopodium, Polygonum, Camelina*—and some late comers were adopted and successfully established—rye and oats.

These, having been more recently domesticated, have reached their present state in a shorter time. For the New World the most recent evidence indicates that the beginnings of agriculture probably date from about 5000 to 3000 B.C., giving for the New World a maximum evolutionary span of about 7000 years.

Of the crops discussed in these lectures, wheat and barley date from the beginnings of agriculture, and oats and rye from a later period. Doggett has suggested that the domestication of *Sorghum* may have been undertaken by people who brought primitive wheats with them when they migrated into regions where wild *Sorghums* occur, and *Sorghum* must therefore be grouped with the secondary crops.

Maize and potatoes are New World crops, and considerable evidence is now available on the beginnings of maize cultivation, giving the earliest date as about 5000 B.C.

in Central America. There is no evidence as yet on the antiquity of potato cultivation, but from what is known of the date of the beginnings of agriculture in South America, it seems likely that potatoes have been cropped since 3000 B.C.

It appears, therefore, that the whole vast complex of cultivated plants has evolved in about 9000 years, and some of our advanced and successful crop plants have been developed from their wild ancestors in half that time. The rate of change involved must be greater than in any other group of organisms, and it is not too much to say that for the student of evolution, the crop plants provide the most rewarding material to be found in all biology.

There are not only the ancient crops with 9000 years of history, but also the recently adopted crop plants, such as the forage plants where the wild progenitors and the cultivars are still to be found side by side, and even such an evolutionary phenomenon as Hevea rubber, of which some individuals of the original domestication are said to be still alive.

Godwin has commented on the wide range of plants that contributed to the diet of early agricultural man, and it is interesting to speculate on the factors that governed the choice of plants for domestication, and determined their success in more advanced agricultural systems. The food needs of mankind may be classified under three heads, carbohydrates, oils and fats, and proteins.

For his carbohydrate supplies the wide range of plants with which he has experimented has been narrowed down to two main groups, with only a very few plants that do not fall within them. The most important group comprises the grain-producing species of the Gramineae. Second in importance to these is the very diverse group of dicotyledonous roots and tubers.

Among the grains, such dicotyledons as *Chenopodium* in the Old World and *Amaranthus* and *Chenopodium* (Quinoa) in the New have given place to the cereals. And such exceptional carbohydrate suppliers as the banana are giving place to roots and tubers, such as yams and sweet potatoes. One might hazard a guess that among grain producers, the highest ratio of grain to vegetative matter is achieved in the monocotyledonous Gramineae and that an equally favourable ratio of edible to inedible material is only achieved among the dicotyledons among those that produce roots and tubers.

Vegetable oil production comes from monocotyledons and dicotyledons alike, from coconut and oil palm on the one hand, and groundnut, sunflower and the *Brassicas* on the other. In the provision of the other great human need, protein, the Leguminosae occupy the position held by the Gramineae in the supply of energy.

They have a subordinate position in agriculture, partly because the cereals also provide substantial amounts of protein, and partly because they are not as productive as the cereals. Here again, as seed bearers the dicotyledons are not as productive as the monocotyledons.

The choice of wild plants for domestication depended upon their immediate attractiveness. We can see the process going on in the collection and study of grasses and legumes for pasture improvement in Australia, in many developing tropical countries, and even in western Europe and North America. Of those initially selected, only a few are successful and spread, and these only persist so long as they maintain their place in competition with other crop plants.

Thus even so long established a forage crop as Sainfoin has almost disappeared, since it did not respond as well as alternative fodder crops to the changing circumstances of British agriculture. A striking feature of the accounts of crop plant evolution here recorded is the great importance of the continuing genetic contact between the crop plant and its wild relatives. Dodds has described the association between cultivated and wild forms of the potato in South America.

Mangelsdorf has drawn attention to the importance of teosinte in the evolution of modern maize. Bell has reported on the persistence of wild relatives of wheat as weeds in wheat fields, and on the opportunities for hybridization and polyploidy that have followed. Doggett has described the situation in the *Sorghum* crop, where weed forms persist in the crop throughout its African range.

In *Sorghum* in particular, the existence and significance of hybridization between the crop plant and its wild relative is apparent to the careful observer, and the effectiveness of gene exchange is evident from the similarity between weed and cultivar throughout the African range of the crop. The maintenance of the distinction between them is an excellent illustration of the strength and effectiveness of 'disruptive' selective forces in a field crop.

*Sorghum* appears to have reached India uncontaminated with its weedy relative, and it would be an interesting comparative study to enquire whether the rate of change and the extent of adjustment to agricultural circumstances have been any greater in India without the weed than in Africa in its presence. That change and differentiation are not dependent upon the differentiated gene pool established under disruptive selection is shown by the cottons, in which the crop has evolved in isolation from its wild relatives, and disruptive selection situations are very rare.

One of the major consequences of domestication is an enormous increase in the area of distribution of the species. This is most striking in the spread of crop plants formerly confined to one hemisphere. The spread of wheat and barley to the New World, to South Africa and the high altitude tropical latitudes, and to Australia, and of maize throughout the tropics, subtropics and warm temperate regions of the Old World has all taken place in the last four hundred years.

Similar vast extensions in distribution have taken place with most of the important crop plants of the world, and indeed in many cases the chief areas of production are beyond what was the range of the species until quite recent times. Cacao production is concentrated in West Africa, rubber in South-east Asia, bananas in the New World, and cotton in regions of North America well beyond the limits of the species at the time of Columbus.

These vast increases in range of the major crop plants could only have been possible in very plastic organisms. Limits can be discerned beyond which a crop plant cannot be established. Well known examples of such limits to distribution are those to which continental tropical or sub-tropical crops such as maize and soy beans are subject. All attempts to establish them in the agricultural system of the mild, damp, cloudy British Isles have failed, whereas in the drier and hotter summers of continental Europe the selection of well adapted strains has not been difficult.

Though limits can be recognized, it is not the limits but the range and variety of the climates in which adaptation is successful that is remarkable. Probably the most remarkable of all is the example of the annual cottons.

Six hundred years ago all cottons were perennial shrubs, and since they are all frost susceptible, the crop was entirely confined to frost-free tropical countries. In each of the four cultivated species, however, the range of variability in morphology was such as to permit of the selection of forms that fruited early enough to give a worth-while crop in the first season. These early fruiting cottons were grown in countries with hot summers but cold winters, and the annual habit was imposed on them by frost in the winter.

Selection for high productivity completed the process, and now obligate annuals make up almost the whole of the world's crop, and are grown not only in areas with cold winters but also in countries with hot dry seasons which likewise limit the success of the more primitive perennials. Thus in no more than 600 generations, the whole habit of the plant has been changed in response to the limitations of a climate to which the early cultivated forms were quite unsuited.

A similar, though less extreme, change in habit has been established in a much shorter period following the introduction of *Sorghum* into the agriculture of the United States. The common African forms, 8 ft. to 12 ft. tall, were quite unsuited to mechanized agriculture, and since the produce is a coarse grain used for stock feed, low cost was essential for successful production. Short forms were known, but these also were too tall for convenient mechanical harvesting. In quite a short period, however, *Sorghums* were bred that were short enough for full mechanization, and on them the whole of the present American crop is based.

Their genetic structure is well understood, the commercial types carrying a small number of major dwarfing genes, together with a polygene complex that magnifies the major gene effects. It is instructive to note that whereas the major gene effects can be studied conveniently within the American *Sorghum* population, crosses between American dwarfs and Indian agricultural types underwent such complex segregation that the effects of the major genes were lost.

These enormous increases in the distribution of our crop plants have been followed by extremely rapid differentiation of geographical races. The nature and consequences of the differentiation have varied with the circumstances of the particular crops.

The nature of the sample of the population from which the original supply of seed was taken has almost always left its mark on the race developed in the new area. Ellis has recently shown that the maize of Nyasaland can be sorted into a group of races that matches fairly closely the races that have been described from north-eastern South America, and the evidence is consistent with the view that the crop is descended from Portuguese introductions from Brazil.

Among the cottons of American Upland descent established in the Old World, those of India were derived from the medium staple New Orleans types of the 1860's, and their modern derivatives are shorter in staple than current American types. Those of Africa, on the other hand, came from introductions of a small range of long staple American varieties of the first decade of the present century, and they are still longer in staple than the bulk of the American crop.

The sample that was originally introduced was the raw material of the new race. The race was then fashioned by the selective forces of the new environment, and these have had the most striking effects, in a surprisingly short time. The selection of hairy leaved types in American cottons both in India and in Africa has given rise to a highly distinct phenotype, resistant to the jassid pest that is ubiquitous in the Old World tropics. In cotton, these changes can be fairly accurately timed, and it appears that the development of jassid resistance by natural selection in the fields of the Indian cultivator took no more than about fifty generations.

Under selection by plant breeders in southern Africa, the establishment and spread of jassid resistant types was accomplished in a much shorter time. Similar racial differentiation is evident in the potatoes. The European and North American forms have become widely distinct from the types still to be found in the home of the potatoes in South America, even though the number of sexual generations involved must be quite small. Moreover, improvement in South American potatoes now depends heavily on the return to their original home of the modern advanced cultivars developed in the new areas of cultivation.

This complex pattern of spread and differentiation depends upon the existence of a pool of genetic diversity out of which new genotypes emerge under the selection of the diverse environments to which man subjects his crop. In considering the differentiation of the cottons one cannot fail to be impressed by the contrast between the genetic uniformity of the wild species and the enormous genetic diversity of the cultivated species.

It seemed necessary to postulate some process by which diversity could arise within a species in the course of domestication and development as a crop plant. Fisher's demonstration that an expanding population would become more variable through the survival of a larger proportion of the naturally occurring mutants provides a statistical basis for increasing variability, and there is good reason to suppose that in other crop plants as well as cotton, the gene pool has been augmented in this way.

Nevertheless, it appears that in other crop plants there is not the same disparity between the variability of wild and cultivated species as there is in cotton. Bell has remarked on the variability of the weedy relatives of wheat. Doggett's wild *Sorghums* match the cultivated forms in diversity, though some at least of this may be ascribed to gene exchange. Most important of all, Cooper's forage plants at the very beginning of domestication have within their current limits all the variability that one is accustomed to find in a wide ranging crop plant.

Moreover, recent work with *Drosophila* makes it clear that superficial uniformity may conceal an extensive gene pool in a wild species—if indeed *Drosophila* can be regarded as a wild species. Consideration of the time scale discussed above suggests that many stocks of *Drosophila* must have been bred in captivity for as many generations as some of our crop plants, and we ought perhaps to class *Drosophila* with the honey bee in a small category of domestic insects.

The structure of the crop population and the distribution of the variability within it, depends upon the breeding system and the impact of selective forces. Breeding systems in crop plants range from free gene exchange throughout large populations through wind pollination, as in maize, to virtually complete genetic isolation through sterility, with clonal propagation, as in bananas.

The wide range of gene exchange through wind pollination is associated with high individual heterozygosity, and the twin phenomena of inbreeding depression and hybrid vigour. A more limited degree of gene exchange occurs in crop plants that are partly selfpollinated and partly cross-pollinated often by insects.

Cotton falls in this category, and while the extent of heterozygosity as indicated by segregation on inbreeding, is less than in maize, gene exchange is adequate to ensure the occurrence of the recombination on which progress under selection depends. It serves as a warning against overconfidence in plant breeding techniques to note that the gene exchange from which the modern bacterial blight-resistant Upland cottons of West Africa arose, would not have taken place if the early agricultural botanists had taken the precautions against cross-contamination that we now regard as part of the plant breeder's routine.

Many crop plants, including wheat, barley and oats, are self-pollinated in those parts of their range where the greater part of their crops are produced. They have become virtually mixtures of pure lines, the genetic diversity being partitioned between individuals, and hardly assorted within the genotype of the individuals at all. In the most extreme case, where sexual reproduction has been replaced by clonal propagation, genetic change may be restricted to that which arises by mutation within the clones, and re-assortment of the gene material has then ceased altogether.

In all crop plants the flower structure is such that cross-pollination must have occurred in the past. The Old World cereals have flowers suited to wind pollination. The legumes have the flower pattern characteristic of insect-pollinated plants. Moreover, it is now apparent that at least some of the self-fertilizing crop plants are normally cross-pollinated in their original areas of distribution. The Howards showed that cross-pollination occurs in wheat in India. Rick reported cross-pollination in tomatoes in South America and demonstrated the relationship between spread beyond the range of the pollen vectors and the development of a flower type that facilitated self-pollination.

Moreover it now appears that the classical Vavilov type of distribution of variability in a crop plant, with a progressive decline in diversity from centre of origin to periphery, is associated with the development of self-pollination. Crops such as wheat and barley lose in variability with every move to a new area since any pure line not represented in the mother stock for the new crop is lost forever.

Crosspollinated plants such as maize and cotton do not exhibit the Vavilov effect, but on the contrary develop new centres of diversity wherever the crop is established. Indeed it may be that it was this versatility of a crosspollinating crop that led Anderson to read the Maize story backwards, from Asia to America.

The loss of sexual reproduction must bring evolution in a crop plant as in anything else, almost to a halt. Some increase in clonal diversity will come about by mutation, as is to be seen in the bananas. But with the end of segregation and recombination, there is an end also to the spectacular evolutionary progress characteristic of the crop plants. Only in the banana has the degeneration of the sexual cycle gone to the point where seed production has completely ceased and even in this crop, breeding progress is still possible. The banana breeders have developed an ingenious technique, first proposed by Dodds, whereby they breed superior genomes in seeded, inedible bananas, and then by hybridization add an improved genome as a complete unit to the genome of a sterile, parthenocarpic edible banana. A splendid example of doing good by stealth!

Root and tuber-producing crop plants do not necessarily become seed sterile, and though seed production may be reduced even to vanishing point in commercial crops, it can generally be stimulated sufficiently to make progress by plant-breeding methods possible. There is thus a range in the potential for adaptive response, depending on the breeding system.

Outbreeding crop plants may be expected to change in response to changing environments by natural processes alone. Inbreeders will be much less responsive, and adaptation and improvement is only likely through the deliberate intervention of the plant breeder. For those crop plants that are no longer propagated by seed, improvement depends upon the intervention of the plant breeder to re-establish the sexual cycle, though such plants differ from inbreeding seed producers in that the clonal material is generally highly heterozygous, and a seedling progeny consequently segregates widely.

We are well aware of the fact that long continued selection to meet the needs of man results in the loss of characters that are essential for survival in the wild. What is perhaps not generally recognized is the further consequence that the breeding system may be so altered—by the development of obligate self-fertilization or by the loss of sexual reproduction—that any further evolutionary changes may also be dependent on the active intervention of man.

The nature of the genetic changes that have gone on during the development of the advanced crop plants is of great importance for an assessment of future prospects. These can be classified under three heads, changes in ploidy, major gene changes, and changes in polygene complexes. Polyploidy is very common among crop plants, but it is evident that increases in chromosome number are by no means necessarily related to crop plant improvement. The most important species of cultivated wheat and oats are polyploid, and polyploidy is common among vegetatively propagated crop plants. Sugar cane, bananas and potatoes are outstanding examples.

But development in polyploid wheats has been no greater or more successful than in diploid barley, maize or rice. The tetraploid New World cottons are larger, more vigorous, more productive, and yield produce of higher quality than the diploid species of the Old World.

But among the cultivated species of *Phaseolus* the New World species excel those of the Old World in ways that are almost exactly parallel, and the species of both hemispheres are diploid.

It seems probable that the occurrence of polyploidy depends more on the breeding structure of the crop plant than on the selective forces of domestication. In genera of the Gramineae where crosses are possible over a wide range of related species, polyploidy is very common.

Moreover, it is common in wild species and in species such as forage grasses that are in the early stages of domestication, as well as in advanced crop plant species such as *Triticum* and *Avena*. In the Leguminosae, on the other hand, crossing between species is difficult, and often impossible, even between those that appear to be closely related, and polyploidy is rare.

The incidence of polyploidy depends not only on the chances of occurrence but also on the prospects of the polyploid after it has arisen. Riley's demonstration of the genetic nature of the stabilization of meiosis in polyploid wheats is of the greatest importance in elucidating the steps that must occur in the establishment of a successful polyploid. Kimber's analysis of the data on meiosis in polyploid cottons is sufficient indication that the system of genetic stabilization demonstrated in polyploid wheat may occur in other genera also.

Where vegetative propagation is possible, a polyploid may be successfully established, though its meiotic cycle may be unstable or even very irregular. Thus what has come to be known as 'nobilization' in sugar cane breeding involves exploiting the occurrence of unreduced gametes to increase chromosome number by the addition of a whole set through crossing. The resulting seedling is then propagated solely by cuttings.

In such material, progress through plant breeding depends on exploiting the genetic variability in the simpler, sexually normal members of the species or genus, and producing commercially valuable clones by crossing with their complex polyploid relatives. It is then important for the breeder to maintain a clear distinction between his breeding stocks and these commercially acceptable clonal products, the latter being 'dead end' stocks of no further use for breeding.

In considering the importance of major genes in crop plant evolution, it is necessary first to make it clear that all the evidence indicates that genes do not fall into two categories, large and small, but cover the whole range in magnitude of effect. A 'major' gene is no more than a gene having an effect that is large in comparison with the variation due to environmental causes, and hence one that can be identified and studied individually.

In this respect, Knight's analysis of the genetics of blackarm resistance in cotton is important, since in the and climate with irrigated agriculture in which he worked it was possible to identify a number of genes with effects of different but measurable magnitudes, and to demonstrate the existence of yet others that had too small an effect to be studied individually. Moreover, in the variable climates of the rain-fed regions of Africa, even Knight's major genes became 'minor' genes, in the sense that they could not be individually identified.

There is, nevertheless, a real sense in which changes in major genes have been an instrument of evolutionary change. It is difficult to conceive of the persistence of an unstable polyploid long enough for meiotic stability of the kind demonstrated by Riley to be built up by the selection of a constellation of genes of small individual effect. The major changes which made possible the development of modern maize involved two loci.

The origin of spinnable lint from the seed hairs of the wild relatives of the cottons can be accounted for by a change in a single gene. And the difference between brittle and tough rachis in cereals is in general simply inherited.

Turning to recent breeding improvement, the short stemmed *Sorghums* bred for combine harvesting in the United States depend basically on a few major dwarfing genes. It is easy to see that the occurrence of a large mutation would provide an effective stimulus for human selection, whereas the existence of a range of variation in the same character due to minor genes might not.

In this way, the establishment of disease resistance, awnlessness in cereals, some forms of pest resistance, besides such quality characters as grain colour and texture, has been brought about in large measure by selection of major genes. A major change in any important character, however, is rarely adequate in itself.

That the American dwarf *Sorghums* have acquired minor genes as well as major dwarfing genes is indicated by the complexity of segregation for height in crosses with tall Indian strains.

Major genes for disease resistance seem to be particularly liable to a major responsive change in the parasite, and plant breeders are turning increasingly to minor gene controlled 'field resistance' as more likely to give consistent resistance. In fact the greater part of the genotype of any organism, and of the genes controlling any character, consists of genes of small individual effect.

In crosses between related types there may be clear segregation of a single gene, but if the relationship is not close, or if for other reasons there is wide diversity in the segregating populations, the effect of the same major gene may be obscured by the segregation of many other genes having minor effects on the character.

To take an example from animal breeding, the transfer of the hornless character from a polled to a normally horned British cattle breed is a simple Mendelian exercise.

The establishment of the polled character in a heterogenous African race of cattle is a much more complex operation, involving minor genes for horn size, differences between the sexes in expression of the character, and changes in dominance relations, all indicative of the segregation of genes with widely different effects.

It follows that whatever major changes may arise from simple gene differences, the establishment of a new and superior stock is unlikely to occur unless the minor gene constitution of the material is also subjected to selection, but given the initial improvement conferred by the major gene, further progress of great significance may well follow from selection in the polygenic variation.

The efficient manipulation of polygenic variation is only now becoming possible, and in breeding programmes with all kinds of crops, statistical analyses of heritability are likely to lead to much greater exploitation of genes of small individual effect.

## Chapter 30

# Neurological Ideas in Psychology

In this chapter we are going to see the ideas about the nervous system have been a major influence on both psychological and philosophic thought, with distorting effects even for those who deny any interest in neurology. If an influence of this kind exists, it should be recognized, to keep it under control. Subjective idealism is the prime example, but we also look from this biological point of view at other classical formulations, and conclude by asking how ideas about the brain, and "reductionism," may be used with a clarifying instead of a restricting and distorting effect.

The pervasive influence of neurological ideas in psychology may mean an unavoidable contamination of philosophy. In the theory of knowledge the two disciplines come together, and Descartes' great contributions to philosophy were after all presented in the context of a theory of behavior-which in turn was a theory of how the nervous system works, so "contamination" began early. Descartes' was truly a theory of behavior since it gave an account of how the mind controls the body's movements and how it gets the sensory information needed to do the controlling.

Though the resulting picture of the nervous system as an input-output system only was grossly incomplete, this was a brilliant performance and so effective that the theory lasted almost into the present century, with changes in detail but little in principle.

Better ideas were obtained of how a nerve pathway transmits energy, for example, and better knowledge of the course taken by each pathway to and from the cortex, but there was still a very Cartesian concentration on sensory input to the cortex and an apparently independent motor output from it—independent, that is, in physiological terms. Linkage was a nonphysiological mental function. The first modification in principle was made by Edward Lee Thorndike in 1898, and by Ivan Petrovich Pavlov independently in 1902. This we will return to shortly.

The effects of the Cartesian scheme are still evident in psychology. In philosophy, major lines of thought appear to have been attempts to live with or to escape the implications without renouncing the scheme itself. Parallelism for example escapes the anomaly of an interaction between substances totally unrelated—mind and matter—by endowing the material brain with all the powers of mind: a ridiculous solution.

Philosophic idealism and the solipsism that Berkeley escaped by the skin of his teeth (in an appeal to divine intervention) are incompletely worked out Cartesian products; they can be shown to be either internally contradictory or, followed to the logical conclusion, wholly meaningless.

## NEUROLOGICAL IDEAS IN PSYCHOLOGY

Until 1930 or thereabouts psychology was frankly neurological in an innocent sort of way; William James for example talked about habits as pathways worn in the brain, Wilhelm Wundt called his a physiological psychology. With Thorndike and Pavlov, the latter phrase took on real meaning, and a class of learning theory developed that permitted attack on a much wider range of phenomena than had ever been dealt with before, and began to show the real dimensions of the psychological problem.

Unfortunately, these theories still had an essential lack, and Karl S. Lashley in a brilliant series of experiments between 1920 and 1930 showed that their neurological assumptions had little connection with the real world. Mostly I think because of Lashley's work, psychologists stopped casting their theoretical explanations in neurological terms but—the point I make here—this did not mean the end of the neurological influence. One's thinking may be constricted by ideas whose source one does not recognize and which one might reject if one did.

Figure is a diagram of what is implied by the Cartesian theory of behavior as it was held, say, in the early nineteenth century. Incoming information to the brain from the sense organs is delivered to the "sensorium," a sort of bulletin board to be read by the mind, which may then initiate response via the "motorium." In this scheme, an immaterial mind plays a necessary part in adjustment to the environment (apart from the automatic, noncognitive reflex responses, controlled at lower levels in the nervous system).

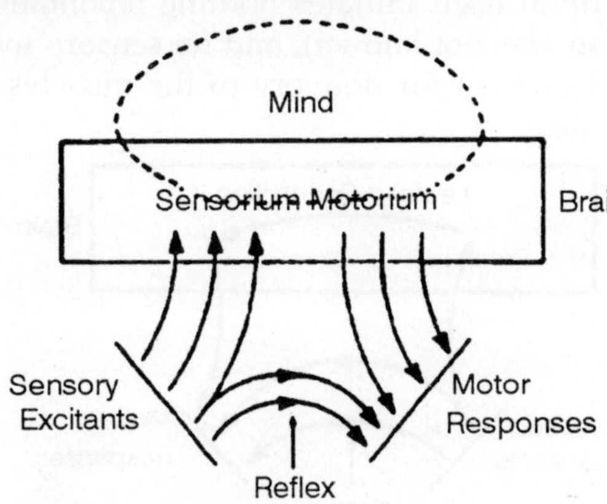


**Fig.** Diagramming the Relation of Mind to Brain in the Cartesian Nervous System after Bell's Separation of Sensorium and Motorium

An immaterial mind had a necessary role in any cognitive response, as the link between input and output. This did not apply to reflexes, which have their own built-in cross connections

Thorndike was radical; he said there was no need to invoke any mental process to explain a cat's learning, and proposed instead that a "satisfying state of affairs" following response might act directly on the cat's brain to strengthen (i.e., by synaptic change) whatever pathway was active in producing that response, after it had occurred by chance; conversely when response is followed by an unsatisfactory or painful state of affairs. This opened a new era in psychology and determined the main lines of theoretical debate in the next 50 years or so. By that time it was clear that although Thorndike was wrong in major respects—cats demonstrably have intelligence, and there is more to learning than his "law of effect"—he was asking the right questions.

Looking again at Figure, one may see that no provision was made in this scheme for any independent activity by the brain; it proposed that the special function of the cortex is to make new connections possible but that all connections are ones that go straight through. At any given stage in learning, what the brain does is completely controlled by the sensory input., The brain itself initiates nothing (spontaneous firing by the neuron was not known), and no sensory message can be reworked or held for delivery to the muscles at a more appropriate time.

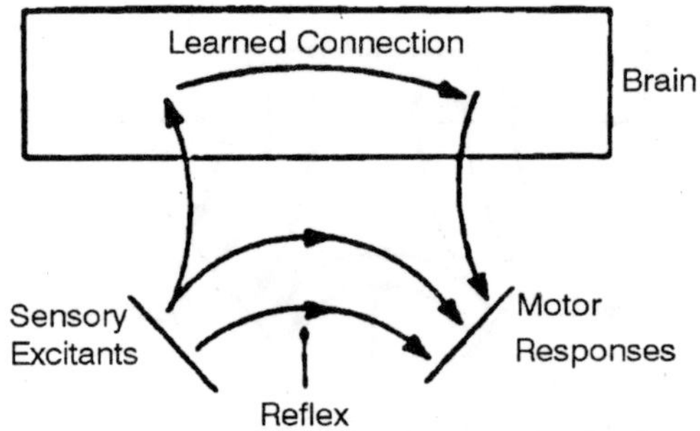


**Fig.** The Cartesian Nervous System, but with Mind Removed and Transcortical Connections Substituted therefor

Also, no connection between two incoming messages is possible: That is, all connections are S-R (stimulus-response), none S-S (stimulus-stimulus or sensory-sensory), which means that two sensations happening together cannot be associated; each may be associated with the same *response,* which is a different matter.

The denial of S-S associations nicely demonstrates the neurological influence. It was early discovered and emphasized in physiology that the neuron is polarized and conducts only in one direction. Hence two neurons parallel to one another at the same level in transmission cannot become connected—that is, cannot become able to excite one another—because the sending end of one cell (the axon) cannot excite the sending end of the second cell.

Thus if cell A in the figure is excited by one sensory event, B excited by another, simultaneous, event, the two cortical processes cannot become connected (though they may be conditioned to both excite the same motor response). Exciting one cannot excite the other—though behavioral evidence clearly shows that this is wrong, as we will see.

All this is still very Cartesian in denying that the nervous stystem by itself has any capacity for the essential creativity of thought. It was also consistent with—indeed, required by—neurophysiology of the early 20th century. In 1920 the cerebral cortex was solely a variable transmitter: in the common analogy, a mere switchboard.

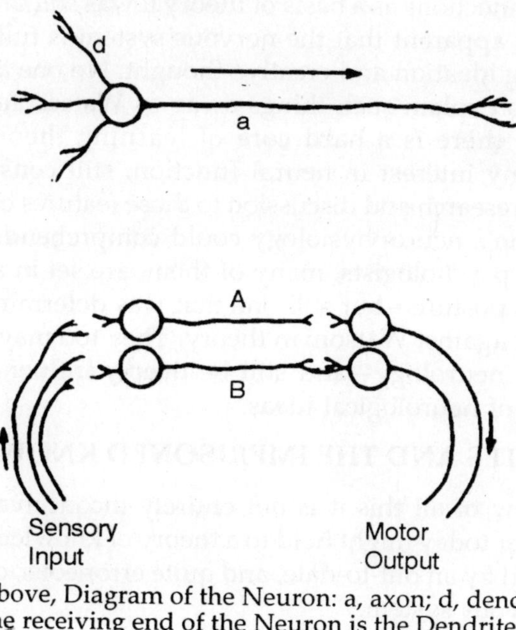


**Fig.** Above, Diagram of the Neuron: a, axon; d, dendrites. The receiving end of the Neuron is the Dendrites; the Axon sends but does not Receive

Below, two neurons, A and B, receive sensory stimulation and excite motor output; being at the same level in sensorimotor transmission, they cannot excite one another, lying axon to axon instead of axon to dendrite. Watson had full scientific warrant for denying any purely internal activity such as ideas or imagery, set or delayed response (in which excitation is held for short periods), or perception in the form of an elaboration of sensory input.

It is understandable that a psychologist might maintain such views, courageously, when well-established neurological conceptions leave him no alternative; but how can they be held after one has renounced all concern with the nervous system in one's theorizing? It seems however that Watson's ideas, fused with Pavlov's, took on a life of their own, idependent of their source.

In 1920, it was *unscientific* for the Behaviorist to talk about images or set or the association of ideas, and when in the 30's the "learning theorist" or nco neoBehaviorist had given up cortical connections as a basis of theory it was *still* unscientific. Today it is apparent that the nervous system is fully capable of handling ideation and creative thought. No one any longer attempts to explain such things away as Watson felt obliged to do, but there is a hard core of learning theorists who, denying any interest in neural function, still conspicuously limit their research and discussion to those features of behavior that Watson's neurophysiology could comprehend.

Other psychologists, many of them, are set in a different theoretical posture—but still one that was determined by the long battle against Watsonian theory. They too may deny any interest in neurology—and still be deeply influenced by an earlier set of neurological ideas.

## DESCARTES AND THE IMPRISONED KNOWER

In view of all this it is not entirely inconceivable that a philosopher today might hold to a theory of knowledge largely determined by an out-to-date, and quite erroneous, conception of the nervous system.

Descartes did not originate the neurologizing—the brain was already regarded as the "seat" or "organ" of mind, and Augustine for example had wondered that the soul should know so little of the brain, to which it was closest, compared to other things farther removed—but Descartes achieved a new sophistication with his theory of how the nerves worked. (He accounted even for the phantom limb, a remarkable insight). Their mechanical twitches and minute hydraulic pressures, as means of neural transmission, would reinforce existing ideas of the mind as essentially different from bodily functions, but the theory had a further and most important implication.

Between mind and external reality was now interposed a crude communications system that in effect shut the mind off from the external world, which could be known only at secondhand. The consequence was that theory from Locke to Titchener—200 years—was concerned almost entirely with the sensory rather than the motor problem and, more fundamentally, with the plight of a mind imprisoned in the skull and informed only by a low-grade signaling system.

A parallel in today's terms: a prisoner shut in a dungeon with five TV screens—the five senses— that picture a lively intricate world outside, but one that the prisoner has never been directly in contact with. How can he know whether the programs are live, or merely taped: how can he be sure there is any world such as that that appears on TV screens?

It was George Berkeley in 1710 who considered the predicament in which Descartes had left the soul and drew the conclusion that there is no reason to suppose that an outside world exists—at least, it is a matter of inference, theory rather than fact.

It was a dazzling intellectual performance by a 24year-old: dazzling to those who followed, so that they failed to see that the argument contained within it the seeds of its own destruction; and dazzling presumably to the author himself. It is understandable that a young man finding himself the possessor of an idea that appears to turn metaphysical philosophy inside out may not be impelled to search for fundamental flaws, especially if those whom his conclusions outrage are unable to refute them.

The argument opened the door to solipsism, the theory that only the thinker exists and the external world is a set of ideas within the thinker's own mind. In less extreme form, it appears in the question, whether there is any sound when a bell rings in the desert if there is no one there to hear it. It is the idea parodied by Lewis Carroll, when Tweedledum warns Alice that she was only something in the Red King's dream and if he awoke she would go out-bang!—just like a candle. Solipsism has been said at times to be logically inescapable, though obviously absurd.

It has thus been an embarrassment to philosophy. But it is not inescapable, and nothing could illustrate more forcibly the hidden presence of neurological assumptions underlying philosophic thought: For solipsism, and related forms of idealism, are products of a particular conception of the nervous system, as merely a transmitter of information inward and of motor instructions outward. It is this conception that makes *knowing* a relation of mind to brain instead of a relation, primarily, of the whole organism to its environment. Abandon that Cartesian-Berkeleyan idea, and any logical necessity for solipsism disappears.

However, much more interesting is what happens if we do accept Berkeley's argument but follow it to its conclusion. The result is that it ends up in contradiction or, when contradiction is avoided, becomes meaningless. For the moment then let us accept the conception of a mind that knows only its own sensory content, not any external object or event. We have seen what reason there is to doubt that there is such a state of affairs, but let us look at the consequences of supposing it to be so.

We then have a mind that knows itself and its content. No outside world is known to exist, though it may be guessed at, and we now consider the proposition that all apparent information from outside is illusory. The green of the grass, the hard chair on which I sit, the other persons in the room, all exist as ideas only.

But—ideas where? Not in a mind that is resident in this physical body that I call mine, for that body is itself only an idea. This is the point at which the argument leaves the rails in the implicit retention of the physical identity of the solpsistic thinker. The whole case depends on the idea that sensory information corresponds to no external existence, and the thinker's body must be illusion as much as any part of the universe depicted on the thinker's sensorium (on the TV screens of the imprisoned knower, in the analogy proposed above). Otherwise, contradiction.

That contradiction, dimly realized, is what makes for absurdity: for solipsism otherwise loses meaning instead of being absurd. Solipsism is selfcontradictory, and thus absurd, when it is taken to mean that this chair and room and other persons in the room exist as ideas within the head of this sensorily known observer, myself, for the observer as a physical person does not exist either.

Popper has said that the theory that the world "is just my dream" cannot be refuted though it is obviously false; but all he considers is a form of Samuel Johnson's refutation, which of course was no refutation at all (Sam kicked a stone, to show that it was really real). But the theory is readily refuted by pointing out to the "dreamer" that he is a puppet in the dream too, so the dream cannot be his. The world, and all physical existence, may indeed be a dream, but not the dream of any human thinker. Now let us try the argument without the contradiction.

The mind of the solipsist comprises all existence, and thus cannot itself be comprised by anything else. Also, it cannot be a function of or dependent on an individual human body, the solipsist's. It is instead coextensive with the physical universe, and in effect is the universe. Within that universe no distinction of mind from matter is possible—all is mental.

Yet that latter statement has little meaning, for in this context the significance of "mental" lies in a contrast with "physical" and that contrast is no longer possible. Any feeling that this conceptual product of the original argument is an *idealism* as distinct from materialism is merely a hangover, an emotional residual, a preference for one word rather than another as a result of a formerly held but—according to the theory—erroneous distinction. One can only say that the upshot of all this intricate argument is that one ends up in a monism instead of dualism; otherwise, no change.

For we are back where we started. The original problem has reappeared, in the question of how to deal theoretically with the thought and behavior of the (ideal) men and women in relation to their (ideal) environments, and even how to understand the behavior of the (ideal) physical self. In other words, the mind-body problem has reappeared, with the sole change that one may if one wishes consider both mind and body of one of these individuals to be items in an all-encompassing mind, thus constituting a monism—and clearly, opening the door to the identity theory of the relation of mind to body.

What difference then remains between such idealism and the equally extreme theory of materialism? In each case, denial of an opposed mode of existence robs the identifying adjective ("ideal" or "material") of its meaning in this context, and we are left with a monism. No qualification: monism is monism, and "materialism," for those who dislike the word, can mean that matter and mind both are ideas in the mind of God, if there is a God.

It is unlikely that anyone ever took solipsism seriously, except as an embarrassment. But it still has modern echoes that are encountered every now and then, the writer taking the Cartesian neurology for granted in the assumption that what one can know is one's sensations. This can lead to some surprising conclusions.

Two earlier examples will make the point. The first is the conclusion of Ernst Mach that psychology and physics have the same subject-matter. As an eminent physicist (his name preserved in "Mach-1," "Mach-2," etc., with respect to air speeds) he might be thought to know what he was talking about. How could he come to such a conclusion? But consider the thinker bottled up in his own field of consciousness, with the problem of decoding the signals presented on the sensorium. *Those signals* are what the thinker deals with; whether they originate with human beings or with the hardware of physics, the scientific material is the same. An absurd conclusion, but arising from the same source as solipsism.

The second example is Bertrand Russell, also eminent, in philosophy as well as in mathematics. In his book *Philosophy,* Russell affirms that Watson could not see one of his rats in the maze, nor a physiologist see a living brain: for direct observation of an external object is not possible, and what one sees is in one's own brain (the sensorium?). The nature of external events "can only be inferred by the help of physics." Russell freely states that any other conclusion would be preposterous, for "the event which constitutes seeing comes at the end of a series of events" that begins with the object that one seems to see.

This stands the truth on its head, for almost any other conclusion would be more sensible than Russell's. The argument is self-consuming; he takes for granted the physicist's results in order to cast doubt on what the physiologist sees, but if the latter can discover things in the outside world only with the aid of physics, where did physics get the aid necessary to make its own observations in the first place?

The conclusion is silly, hardly worth disputing except to show, once more, how distorting the effects of an underlying neurological assumption can be.

Any suggestion by an idealist that makes doubtful the direct knowledge of an external reality is nonsense, for it ultimately derives from taking for granted the physical existence of our sensory equipment, thought of as intervening between the knower and the all physical existence; which is a contradiction.

## PARALLELISM, EPIPHENOMENALISM, INTERACTIONISM

It thus appears that the argument of subjective idealism leads to a monism which differs from materialism only as a form of words. If one feels better by thinking of all existence as ideal, items in the mind of God or in some other allencompassing mind, well and good, but that view hardly permits any further belief that mind and body are not of the same kind of stuff. Quite the contrary. For the biological scientist, neither theory nor experiment is affected by adopting such harmless philosophic views. Much the same thing may be said of parallelism, epiphenomenalism, and one form at least of interactionism: the one that regards mental processes as lawful, so that mind and body, though unlike, may operate together as a deterministic system.

Parallelism is the theory that mental events and brain events run side by side, perfectly correlated but not causally related: in the old analogy, like two clocks that stay perfectly in step but not because either influences the other. It has been highly regarded as a way of avoiding commitment to an interaction of mind and body—or even worse, identifying them—while recognizing how closely they are related. But this is evasion rather than avoidance, for it is an empty solution.

It maintains the separation between mind and body only by allowing the brain to have all the powers of the mind. It makes Shakespeare's mind something entirely different from his body—but it seems to be overlooked that it was his body, his brain and peripheral nerves and hand, that wrote the plays. Parallelism says that the actors in the theatre, representing anger and fear and love in Shakespeare's plays, did so with no guidance from their conscious minds; whatever thought there may have been in those entirely separate minds, the *bodies* functioned on the stage as selfprogrammed robots.

To account for the writing of Newton's *Principia*, the movements of the hand and fingers that produced the manuscript without guidance from Newton's thought, one must suppose that his brain had equivalent skills.

Apart from the evident absurdity, a more general question is raised. What meaning, what real significance, is there in saying that brain and mind, as conceived of in this way, are separate? Similarly, epiphenomenalism also regards human awareness as entirely distinct from the brain, but only as a kind of by-product of its activity—a kibitzer that does not affect the play of the cards.

The brain must have all the powers of human thought without being itself conscious: Here again, what can it mean to think of such a mind as distinct from the brain's cognitive activities? A brain that functions in every respect like a mind is a mind; are there then two minds here? In the famous two-clock comparison with parallelism, the two clocks are separate entities by virtue of their separation in space; if in addition to being identical in function they also occupied the same space, as mental activity and brain activity appear to do (visual perception, auditory perception, verbal fluency, verbal comprehension, and so on each relating to particular parts of the brain), they would be one clock. To paraphrase C.S. Peirce's principle of pragmatism: Consider what practical effects such a distinction of mind from brain may have; then our conception of those effects is the whole conception of the distinction. The practical effects are null, the distinction is merely verbal.

A closely related argument applies to what I referred to earlier as a lawful interactionism. Here the mind acts on and is acted on by the brain according to rules that it is the business of psychological research to discover. If as we saw in earlier chapter the activity of mind is not known directly—if I know the minds of others by inference from their behavior, and know my own mind from judgements based on external phenomena—then the situation is this:

The objective evidence tells me—the dualist—that something complex goes on inside my head; part of this I know is the activity of sensory and motor nerves and the brain, but part of it—I believe—must have a delicacy and complexity that no nervous system is capable of; I conclude therefore that *something else* is active also.

It is perfectly evident that the conclusion is hasty, at a time when the limits of delicacy and complexity of brain function are not yet known, but the more general question again is this: Supposing it exists, what are the criteria by which the *something else* is classed as nonphysical?

It is known from its effects only, and even if its nature is mysterious it is no more mysterious than magnetism or gravity or other things that the physicist knows only by their effects. In any other field of thought, what has lawful physical effects is regarded as part of the physical universe; on what basis is another conclusion justified here—apart from unthinking preconception? This whole question has been re-opened by the "split-brain" patients of Bogen, Gazzaniga, and Sperry. In these patients the left and right hemispheres of the brain were disconnected surgically for the relief of otherwise intractable epilepsy. After operation the activity of the two hemispheres could be very different and a patient might seem to have two minds, a left-hand and a right-hand mind. Sensory and motor connections in the nervous system are crossed, so the right hemisphere gets its information about the outside world from the left side of the body (and the left half of the visual field), while the left gets its information from the right side.

In these patients, as in the majority of the population, the language centers were on the left side, so the left hemisphere could talk to the experimenters but the right could let them know what was going on, internally, only through its control of the left hand. (The right hemisphere does however have a low-level comprehension of speech and of printed language). In these circumstances one hemisphere can be given information the other lacks, by projecting it to one side while the patient fixates the center of the projection screen; or the two hemispheres can be given different instructions and carry out their different tasks at the same time—among a group of objects out of sight, for example, the left hand searches for and finds one object, the right hand a different object.

So—two minds in one head? It would be embarrassing for the dualist if he must conclude that an immaterïal mind can be cut in two by the surgeon's scalpel. Sperry has dealt with the situation by proposing that the mind is a continuously emergent product of the brain's activity, emphasizing both the separateness and the inseparability of "mental forces" and "physiological forces" as they interact. For example: The subjective phenomena of conscious experience are conceived to be direct emergent properties of brain excitation, inseparable from the brain process and its structural constraints, but nevertheless different from and more than the sum of the electrochemical and physiological activities.

This is not epiphenomenalism, though it may sound like it, for the mental processes have a physiological action ("consciousness does things"). The emergent idea is a slippery one, its meaning hard to pin down precisely, but here it allows Sperry to recognize the presence of two minds in the split-brain patient without having to suppose that the mind can be chopped up as a brain can be.

Sir John Eccles has taken a different tack, based on a more classical approach to dualism. Originally he minimized the conflict of interactionism with the law of conservation of energy by arguing that very little energy would be expended by the will and the conscious mind if they acted on the brain by influencing it at the level of subatomic particles and only within the range of the uncertainty principle. More recently he has adopted Sir Karl Popper's indeterminism and therefore can view small violations of the conservation laws with equanimity. He can then deal specifically with the split-brain problem simply by denying the existence of the second mind on the right side. In Popper and Eccles he concedes that the right hemisphere continues to carry out complex functions, but since it cannot talk it gives no evidence of consciousness and so does not have a mind.

But is that a test of consciousness? What are the signs of consciousness in the year-old baby, in the deaf-mute, or for that matter in the porpoise and the chimpanzee? The activity of the disconnected right hemisphere includes memory, perception, and purpose.

It lacks speech, but by way of compensation it has better pattern perception than the left hemisphere-better recognition of faces, for example. It is capable of feelings of propriety. A young woman patient was shown a picture of a nude, presented in the left visual field for perception by the right hemisphere. She blushed and showed embarrassment, but could not say why.

It seems obvious that the embarrassment was a right-hemisphere event, which adds to the evidence of consciousness on the right side of the brain. To avoid that conclusion, Eccles argues that the visual information must somehow have reached the left hemisphere, via uncut pathways at the base of the brain. It seems likely that this might happen where intensity is concerned, but all the clinical evidence is opposed to the idea that it could occur with visual pattern, and in any case it is difficult to accept the idea that the left hemisphere was embarrassed by the presence of an object it could not recognize.

All such difficulty is avoided by Sperry's emergentism, but here again we must ask: What meaning is there in saying that the emergent consciousness is not simply an aspect or property of the physical brain and its activity, as magnetism is a physical property of moving electrons? By what criteria is it distinguished from the more molar aspects of that activity? What specific properties has it that mean it is not part of the physical universe? Until that question is answered, with respect to Eccles 'formulation as well as Sperry's, the separateness of mind and brain is verbal only and can have no empirical significance.

## NEUROLOGIZING: USEFUL TOOL BUT HANDLE WITH CARE

Let us now consider the proper use of neurological ideas, and the meaning of "reductionism." This is a topic on which misunderstanding is easy, and it may be vain for me to hope that misunderstanding can be avoided here—but one must try.

Properly done, relating psychological things to neurological things—the mental to the anatomical and physiological—has value both as stimulant and as corrective. It has a bad name with philosophers and psychologists (but not all of them), due in the first place to having often been done badly (and dogmatically), and in the second place to being not understood when done right.

Properly done, reductionism does not substitute neurophysiology for psychology; when it assumes that pain or ecstasy consists of neural firing it recognizes the reality of pain and joy at the same time; it does not try to explain them away. It is the method of theoretical analysis followed resynthesis, whose validity depends strictly on whether the result accords with the psychological as well as the neurological evidence.

However, its value is that it may do this by showing the need of re-examining one or other of the two bodies of evidence and the conclusions drawn from it. Really, it can destroy nothing but error, by psychologist *or* neurophysiologist, and its positive effect may be to show an experimenter where to look for new evidence or a theorist how to look at the available evidence from another aspect.

A different failure of understanding is to suppose that identifying mental processes with physiological processes is mere translation, done only to sound scientific and impressive. A critic reviewing a book of mine said that "a connection between cell-assemblies" is no improvement on "an association of ideas," if ideas are cell-assemblies. But there are constraints on the functioning of cell-assemblies; if the identification is right, these constraints affect the theory of ideas. For example, it becomes necessary to conclude that what one usually thinks of as an idea must be a complex of simpler ideas (i.e., a constellation of cell-assemblies).

One can readily see in that case how the idea of the self, as complex, may consist in part of the same components as the idea of another. It also suggests a mechanism of Freudian symbolism, if the idea of a church steeple and the idea of an erect penis are complexes that share a common core (a sub-idea of verticality); it relates to James' notion that the same idea can never recur in thought—an overstatement apparently, but cell-assembly theory does suggest strongly that the "same idea" occurring in different contexts may have the same core but differ in fringe components; and so on. The identification of "idea" with "cell-assembly" or "complex of assemblies" is not vacuous but affects theory. Ideas have developed new properties.

The stimulus of reduction can be in either direction. Our concern here is its value for the development of psychological concepts, but I note in passing that there is no superiority of one source of evidence over the other. Psychological facts can show the need of revision of neurological theory. A clear example is Hunter's delayed-response experiments, which showed that there was error in the current conception of brain function which Watson accepted. What Hunter was saying, and Watson might have said, was that the neurophysiologists were wrong. Unfortunately—for them as well as for psychology—neurophysiologists of the day did not give much weight to the behavioral evidence, and the indicated revision of physiological theory had to wait.

What has given reductionism a bad name is the conclusion, after a theoretical analysis of a mental variable has been made and it is "reduced" to some pattern of neural activity, that the mental process in effect no longer exists. This is the nothing-but fallacy: Mental activity is a myth, what *really* exists is something in the brain. Such an attitude is absurd, but in certain circumstances it seems easy to accept. I once listened to a debate between a hard-boiled physiologist and a clinical psychologist who was evidently less hard-boiled, since he did not defend himself well.

The question concerned anxiety and how it should be treated. Said the physiologist, "Either you think anxiety is something in the soul, which is superstition, not science, or you must agree that something has gone wrong in the patient's brain—so it is the brain you have to treat, not the mind." What he meant was, stop the psychotherapy and find out what drug to give.

This is the nothing-but fallacy: If anxiety and thought and memory are identified with neural activities, it may be said that there is nothing there but those neural activities; anxiety and thought and memory have been replaced as impalpable and unreal, illusory.

But this *is* fallacy. It is comparable to concluding, in physical science, that sunshine is a myth, all that really exists is a bombardment of photons; or that the desk at which I sit is illusion, the reality being only a complex of molecules—or indeed, that the reader of this passage also is illusion and mirage.

Obviously this is all nonsense; when a complex is reduced, theoretically, to its component parts, the whole still exists. Anxiety must be a pattern of firing of neurons in the limbic system, but the pattern is as real as the individual neurons.

For all this there is a clarifying analogy that allows us to see that there *is* a limited sense in which the thing analyzed, the whole, does not exist—but only at another level of analysis. At its own level it is still there, as shown in the following.

An engineer designing a bridge must think at several levels of complexity. His conception of the bridge as a whole is very molar, in terms let us say of a center span, two side spans, two piers, and two abutments. Now when he turns to the design of the center span he begins to think in terms of lowerorder units such as steel beams, rivets or welding, and masses of reinforced concrete. However, these items are still very molar. Any engineer if asked would say that a steel I-beam is just a special arrangement of atoms or of electrons, neutrons, and so forth.

At this last level of analysis a bridge *is nothing but* a complex constellation of atoms, and at this level of analysis a steel I-beam is merely a convenient fiction, a crude concession to the limitations of human thought which would be incapable of planning in more microscopic terms. *At this level of analysis* there is indeed nothing but atoms or atomic particles.

But there are other levels of analysis; from the point of view of a practical man, all this stuff about atoms may be fine in theory (and may even be useful when one thinks about improving the quality of steel) but when it comes to bridge-building it is no more than theory. At this level of analysis the I-beam is an elementary unit, obviously real and no fiction. Reality now is steel and concrete.

So from one point of view, reality is the atom, the steel beam being a convenient way of dealing with large numbers of atoms in a particular pattern; while from another, the steel that is heavy and cold and resistant to distortion is reality, and atoms are theoretical items only. For different modes of thought, different realities: "reality" referring evidently to the mode of being that one takes for granted as the starting point of thought.

Now the behavioral problem. At a certain level of theoretical analysis there is no reality but the firing of single neurons: emotions, intentions, and consciousness are convenient fictions. But human thought does not restrict itself to any one level of analysis when seeking explanations, and those conceptions (of emotions, intentions and consciousness) are, like that of a steel I-beam, necessary as well as convenient for other levels of analysis. They are fiction only at more microscopic level; when one is dealing with the phenomena of everyday experience and of the clinic, fear and anger and grief are only too real.

For scientific purposes also these conceptions or others like them will always be necessary. Even if we could identify the part played by every one of the 9 or 10 or more billion neurons in the brain, the human mind of the scientist is obviously incapable of thinking of the whole activity in such terms.

It is not possible to follow the varying patterns of the firing of these cells as individual units. What one must identify is the larger patterns of the activity, and it is just such larger patterns that we know as anger or fear—or love, or mirth or pride. The wood is as real as the trees, a storm is as much a reality as the raindrops and gusts of wind of which it is composed.

Thus reductionism is not a means of abolishing psychological entities and processes but a way of learning more about them. In principle, the attempt to identify mental events with brain processes is an attempt to see them in a different context. Emotion as distinct from emotional behavior is a theoretical entity; when one looks for a physiological process to identify it with (again theoretically), one may find that one of its supposed properties cannot be included in the physiological translation—but that some other property, not thought of before, is suggested at once.

The question then is whether the modified idea of what emotion is or how it takes place makes behavioral sense, as one returns to the actual observation of human behavior. With luck, the theoretical conception has been sharpened, and the agreement of theory with fact improved.

That is a very general statement: Let me make it more specific with an example. In my attempt some 30 years ago to develop a general theory of behavior, including a theory of thought, it was evident that some of the mystery might disappear if thought could be regarded as a series of activities in small groups of neurons in the brain ("cell-assemblies"). For example, attention and set, which were wholly without explanation at the time, would at once become intelligible; psychologists need no longer avoid such conceptions on the ground that there was something wrong about them since they lacked rational explanation.

So far so good: clearly the theory being worked out would be useful—and although it was reductionistic, its immediate force was *not in the direction of getting rid of a mentalistic conception but just the opposite:* It restored one that had been lost because no one could see how it worked. A further effect however is of more interest here. The new theoretical conception was incompatible with widely accepted views of perception, including my own view.

It suggested a quite different basis of perceived identity from what I had thought was established fact; but when I now looked at the experimental evidence again, from the point of view of the neurological conceptions was trying out, it was plain that there were large gaps in the evidence.

The new idea of perception was distasteful, to me, but not contrary to existing fact, and because of the explanatory power of the neurological conceptions in other areas I went ahead and published. Then a series of experimental studies supported this general line of thought, and most explicitly confirmed the new and originally distasteful theory of perception.

It is not suggested that reductionism always has such useful effects; I have much more often been wrong than right, and here describe a case in which I was lucky and reductionism paid off. My point is that attempting to identify the neural bases of psychological variables is a way of probing and testing and—with luck—improving one's perceptions of reality at the level of observable behavior.

## Chapter 31

# Evolution of Mind and Behavior

The argument of this chapter is that the behavioral signs of mind and consciousness are evident only in the mammals, with the possible exception of some of the larger-brained birds; that relatively small-brained mammals like the rat or the hamster may have very small minds (like the penguins of Anatole France's *Penguin Island,* to whom the Lord gave souls but of a smaller size)—but still minds, whereas fish and reptiles, and most birds, seem to be reflexively programmed and give little evidence of that inner control to which the term mind refers.

The best evidence of continuity, in the development from lower to higher mammals, is to be found not in their intellectual attainments, their capacities for learning and solving problems, but in their motivations and emotions. Man is evidently the most intelligent animal but also, it seems, the most emotional.

## THE NEURAL BASIS OF DEVELOPMENT

Our knowledge of the way the nervous system developed is obviously theoretical. Not only is the fossil record incomplete: At best it can only tell us about the size and shape of the brain, or the brain and spinal cord, and it is a long step from the size of the brain to knowing how it functioned. A bigger brain does not necessarily mean a higher level of performance. In practice, we must also assume that we can tell what the course of evolution was from some animals now living, as far as learning and intelligence are concerned. We may for example assume that among mammals the following represent behavioral levels like those through which our own ancestors progressed: rat, cat, dog, monkey, ape, man.

This is only an assumption, but it is more reasonable than one might suspect, because of the fact that the brains of all living vertebrates are built on the same master plan, and that the brains of mammals in particular (including those listed above) are identical except in size and the relative sizes of the component parts; they are extremely complex, and years would be needed before a student could become able to identify the internal structures of, say, the monkey brain and master what is known about their interrelations—but having done so, a week or two might be enough to find the same structures in the brain of another mammal; and as for the interrelations, he would know them already.

Now it seems evident that the nervous system must have evolved in the first place as a communications network, a role it still plays. With the emergence of thought in the mammalian brain, it has taken on a further role, but the primitive reflexive connection of receptors (sensory cells) with effectors (muscle and gland) is as vital for the mammal as for any other animal.

It is no good for the eyes or nose of a hungry animal to be excited by food unless the word gets to the legs and the muscles of the jaws, so the food is reached and eaten; and every step the animal takes requires a coordination of muscles that is achieved by the nervous system. Breathing and blood supply are regulated by neural connections, and so with the processes of digestion. The nervous system of the higher animal is still a communications network, integrating the activities of the organism and harmonizing them with the normal environment by the transmission of sensory information.

G.H. Parker, from his studies of living animals, concluded that the first step toward a nervous system was a specialization of some cells for contraction-that is, as muscle cells—following a course represented in Figure. The development begins with cells in the outside wall. These contractile cells then migrate inward while others, still external, specialize in excitability and become elongated, to connect with the inner-lying muscle cells and excite them. A further stage is the appearance of interneurons entirely inside the organism that connect neuron with neuron or neuron with muscle; and finally, the massing of interneurons in the central nervous system, like the central switchboard of a telephone exchange. There is a long step from an early nerve-net organization of which *D*, might be a part, to the far more efficient operation of the central structure.

As far as I can discover, the course of that later development in unknown, but it is evident that the development took place twice, independently, in vertebrates and invertebrates, for the two follow quite different plans. In the vertebrate the central nervous system forms a single mass of brain and spinal cord dorsal to the alimentary canal—that is, next the back, in the spinal column—but in the invertebrate is ventral and consists of a series of separate masses (ganglia) connected by relatively thin neural strands.

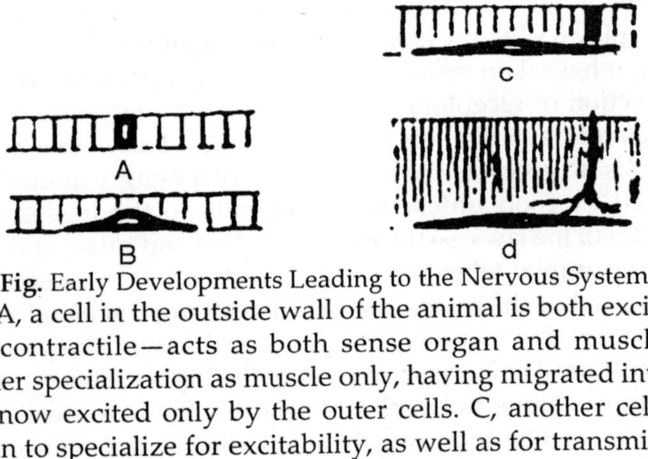


**Fig.** Early Developments Leading to the Nervous System

A, a cell in the outside wall of the animal is both excitable and contractile—acts as both sense organ and muscle. B, further specialization as muscle only, having migrated inward and now excited only by the outer cells. C, another cell has begun to specialize for excitability, as well as for transmission toward the motor cell. D, the primitive nerve-muscle combination. B represents the level of the sponge, D that of amore efficient coelenterate

In considering the evolution of the vertebrate brain the first obvious point that is discussed is its size in different species: ranging from a small fraction of a gram to about 7000 grams. It is immediately evident that size alone does not determine the brain's level of function: does not, that is, determine intelligence.

The elephant has a brain almost four times as large as man's (5000 as compared to 1300 grams), and if there is one thing certain in comparative psychology it is that the elephant is not four times as smart as a human being. Within the human species, again, brain weights of normally functioning individuals vary from about 1000 grams to about 2000, and the differences do not correlate with estimated level of intelligence; some persons with large brains have been very capable, like Samuel Johnson and Georges Cuvier (over 1800 gms), but others have been far less so, and some with small brains like Anatole France (about 1100 gms) have nonetheless had outstanding ability.

But size must still be important, and there have been persistent attempts to explain why the elephant, for example, or the bottle-nose porpoise, should have a larger brain than man. Such attempts assume that it takes a large brain to control a large body. The notion is apparently that only the excess of brain tissue, over and above what is necessary for that control, is available to provide a basis of intelligence.

Thus one may calculate the ratio of brain weight to body weight to put the elephant in his place: his brain is only 1/600th of the total weight of the animal, whereas man's is about 1/45th of his weight. So we conclude that man has more brain to spare for thought and intelligence. But the fact is that the elephant, despite that very low ratio of brain to body weight, is a very intelligent animal, so there is something wrong with the argument; and worse, the mouse has as good a ratio as man and the squirrel monkey a much better one (1/12).

The brain-weight: body-weight ratio has fascinated anthropologists and students of evolution, and various formulas have been devised to make it work better. We need not go into these complicated schemes because they provide no rational basis for an index of intelligence, which is what they are meant to do.

It may be that one of these can provide valid evidence that *Australopithecus* is a collateral ancestor of modern man, by showing that he was proportioned similarly, but it provides little evidence of the level of intelligence more than is provided by the relatively large brain—apart from the size of the body. It is true that a larger brain tends to function at a higher level, and it is also true that brain size and body size are correlated. This last fact however is not because it requires a big brain to manage a big body.

There are other reasons why the two should be related in size. A big body is needed to carry the skull that contains a big brain, and the female must be large enough to have a pelvis that permits delivery of the relatively largebrained infant at term. As for the neural control of a big body, heavily muscled, a thick spinal cord may be required and therefore a large motor cortex and cerebellum, though that is not a self-evident proposition.

Twice as many muscle cells do not need twice as many neurons to control them unless there is also some element of increased agility or intricacy of movement. In that case we would be dealing with a brain that is larger because it is functioning at a higher level, and not merely with one that is big because the body is big.

The conformation of the brain is a different matter, as is the relative size of its different parts. In the example preceding, of motor skill, the presence of well-developed motor structures in the brain might be interpreted as having nothing to do with intellectual capacity; but a moment's thought reminds one of the dexterity required for making and using tools and weapons, which by common agreement must have been a main avenue of man's intellectual advance.

One might then expect that a large motor cortex would be a requirement for reaching the level of *Homo sapiens*. The other main avenue, also agreed on, was language, and it is now clear that we must regard language as a highly developed motor skill, as well of course as a perceptual skill on the part of the listener.

On the perceptual side, the crucial question of intelligence and thought is the extent to which there is an autonomy of cerebral function: how far the brain is dominated by the present sensory input, how far it is capable of an internal activity that transcends the present situation. We will return to this point shortly; I raise it here to show the theoretical significance of the relation of the mass of association cortex to that of sensory cortex, which I have referred to as the A/S ratio.

If sensory cortex is extensive and association cortex small, there is little prospect of the occurrence of any activity in the brain other than that determined by the present sensory environment. *Sensory cortex* here means the total mass of the several sensory projection areas; *association cortex* is the old name for the regions that are neither sensory nor motor, on the reasonable assumption that the association of sensory events must take place here, though of course other functions must be based here too. In particular, this must be the key to the thought process.

The essential character of thought involves imagery and other, high-order, representative activity, corresponding not to the present situation but to others of the past or the possible future. This cannot take place in sensory cortex, which is of course under the full control of the present sensory environment.

Also, as above, if association cortex is small compared to sensory cortex it too will be dominated, and thus a small A/S ratio means a low level of thought and intelligence. A large A/S ratio does not guarantee intelligence, but it is a requirement (necessary but not sufficient).

This is not a very practical index, it is true. Even with existing mammals there are no quantitative values available for the extent of primary sensory cortex; with a good deal of work they could be obtained from serial sections of the brain, but there is no such possibility with a fossil brain. Nevertheless, it is interesting to find Holloway emphasizing the indications of a small visual area in the australopithecine brain compared to an enlarged parietal and temporal cortex that must have been mostly association cortex. From this and other indications Holloway concludes that the human brain was already distinctive, divergent from the main primate stem, as long as three million years ago.

Man's disproportionately large A/S ratio we may see as an essential factor in his intellectual development, but it can hardly be the whole story. We know also that something else occurred to make his brain distinctive, although it is not something that would be detectable in the fossil record: the lateralization of function, the specialization of left and right hemispheres for different roles, which is not known in any other species. Presumably this is an essential part of the whole development. But it seems clear that there is still something else, whose nature we do not know but which makes the human brain more efficient, gram for gram or kilogram for kilogram, than any other.

Efficiency here means the capacity to process information and record it, to abstract from it and to generalize, and to perform the creative syntheses that so far outdo the analogous syntheses of other animals: all this as shown in human behavior.

The size of the brain is evidently important, up to some threshold value perhaps of 1000 grams, after which some other variable is increasingly important. It is this variable that permits a relatively small brain like Anatole France's to make its owner an outstanding figure among his literary contemporaries, and lack of it that accounts for the fact that some much larger brains have been possessed by morons.

What can this other variable be? A first suggestion is likely to be that the human superiority is due to the possession of language, as a by-product of the lateralization of function in the human brain, but this does not account for the possession of language itself. As we will see later, language, though an important adjunct, cannot be the main mover in thought. A possibility is some metabolic development in the human nervous system that makes the individual neuron more efficient, perhaps because of a better blood supply. This seems unlikely.

A more likely theoretical possibility is found in the finer structure of the brain and the greater frequency of short-axon cells than in other species, on the assumption that a good proportion of these are inhibitory. In a large brain there must be a problem of "noise," the activity of neurons firing spontaneously, or continuing to fire with inappropriate timing.

The effect on thought is returned to later. Here I note that Rakic has drawn attention to the frequency of short-axon cells, or what the conference from which his report originated preferred to call local circuit neurons (LCNs), in the human brain. In posterior cortex these presumptively inhibibitory neurons increase with phylogenesis: visual cortex, rabbit, 31% LCNs; cat, 35%; rhesus monkey, 45%; and man, about 75%.

Rakic quotes Ramón y Cajal, that brilliant observer: "The functional superiority of the human brain is intimately linked up with the prodigious abundance of the so-called *neurons with short axons*" and "intellectual power does not depend on the size or number of cerebral neurons but on the richness of the connective processes" implying also that it does not depend on brain size except as this may make the richness possible.

As the mammalian forebrain increases in size there is also an increase of inhibitory LCNs, although presumably without a close correlation between the two variables; a large brain, with a relative lack of LCNs, may be outperformed by a smaller one with more LCNs. This may account not only for the human superiority to the elephant, but also for the low correlation between brain weight and intellectual performance within the human species.

## THE EVIDENCE OF EXISTING BEHAVIOR

Now let us consider evidence from existing animals, whose brains can be examined directly and whose behavior can actually be seen. The evidence is more direct but we must keep in mind that it is only theoretically relevant to the course of evolution, since it depends on the assumption that the levels of behavioral complexity that we know correspond to the stages through which evolution passed, at least roughly.

First the question of where mind and consciousness appeared. Evolution is a continuous process; if consciousness is an attribute of the active human brain, this might be thought to mean that consciousness is present in some degree wherever there is neural activity.

But this need not follow. I have already invoked the physicist's idea of a critical mass in arguing earlier that we need not attribute consciousness to computers. There are degrees and degrees of consciousness, but even the simplest or most primitive may have occurred in evolution only when a certain very large number of neurons made possible a certain pattern of complexity of neural firing.

Thus there may be so to speak a quantitative continuity—a steadily increasing number of neurons-resulting in qualitative discontinuity. The existence of consciousness in the higher animal does not mean necessarily that it exists in some degree in lower ones.

I have already proposed that consciousness is a product or accompaniment or form of thought, and I now point out that no one attributes thought or consciousness to the body of a chicken with its head cut off, despite its activity for some seconds after the head is cut off, nor to the lower part of the spinal cord of a human paraplegic though it continues to serve its reflexive functions.

Consciousness can hardly be identified from any simple motor index, for it is a variable state that can produce complete immobility, as in listening, or extreme activity as in running to catch a bus. It is not all-or-none, for it is present but impaired after taking in alcohol, or in one who can talk while recovering from a knockout but later remembers nothing of the conversation. It is qualitatively different in the vivid rememberable dream from what it is in crossword-puzzle solving, or in anger or in anxiety. A variable state, as thought patterns are variable.

But identification becomes possible if, as above, we assume that it is a function of the thought process. We lack evidence of thought, and thus of consciousness, in most birds as well as in fish, but this may well change with further work (it has not really been looked for, experimentally, except in mammals), and it may be that thought and consciousness are present but not to the extent characteristic of even the small-brained mammals with their prominent cortices (the cortex is rudimentary in birds).

Also, in attributing consciousness to the mouse or the hamster, we need not suppose that it is the fully developed consciousness of the large-brained mammals, including a conception of *self* and *other* that is implied by the occasional manifestations of altruism and empathy.

Thought in birds, if it exists, must be a sort of one-track process compared to thought in the mammal. The method of training a falcon is to capture an adult bird and then by close confinement in darkness, hand-feeding, and later feeding while the bird is still held by a sort of leash, convincing it that food is to be had only at the hand of the falconer, so that the falcon can then be turned loose to catch prey—other birds—and bring them to the falconer.

Earlier learning, from the period when the falcon fed itself on its own prey, appears to be completely over-ridden by the later experience in a way that is quite impossible with a mammal. Consistent with this, while not in itself decisive evidence, is the readiness with which even as clever a bird as the crow can be imprinted on a human being, taken early from the nest and looked after by man; the crow thereafter avoids other crows and consorts with its keeper. Something of the sort is possible with some mammals such as a deer, but the imprinting is not as complete and takes much longer to establish.

Birds are efficient learners, and one might think that this allows us to place them in the hierarchy of evolution. But learning is not a key that can be used for such a purpose. When one looks at the kind of learning to which each species is adapted one can find rapid, efficient learning at all levels of development. For example, A. P. Baerends has demonstrated complex onetrial learning in the solitary digger wasp. The wasp digs a hole in the ground, stings a caterpillar and drags it into the nest and lays an egg on it. The egg hatches and the larva eats the caterpillar.

The wasp inspects the nest each morning to see how much more caterpillar is needed. She may thus supervise two or even three nests, alı at different stages of development, and the provisioning is guided by learning for the wasp first visits all the nests, then goes hunting. An earthworm can learn a Y-maze (Y-shaped, with a single choice point) in 20 trials, about what the laboratory rat needs for the same task. On the other hand, the rat can learn a more complex maze about as fast as human subjects.

It should not really be surprising that higher and lower animals learn equally well when each is presented with a task to which it is suited. (The higher animal's superiority is in learning many more simple things, ones that require more complex responses, and ones depending on more complex relations). Learning ultimately reduces to synaptic modification, and there is no apparent reason why the modification should take place more slowly in a system of few synapses than in one with many.

In fact, the noise problem mentioned above suggests that it should be quicker in the simpler nervous system, and it also appears that the only significant difference in rate of simple learning is that it is slower in the higher animal during infancy: theoretically because experience must first act to organize the transmission paths of the infant brain.

If the rate of learning is not the key to evolutionary level, what about problem-solving? That after all is directly related to intelligence, and intelligence is what we are concerned with. But there are difficulties here too. The different species differ in sensory and motor equipment as well as in brain capacity. If a dog is better than a rat in a maze-like situation and solves the problem immediately, it may be because of his better vision.

The tests that Wolfgang Köhler used to demonstrate insight in the chimpanzee are not suitable for dogs or cats, because they require the primate hand. There is some reason to suspect that the big-brained porpoise, *Tursiops truncatus,* is more highly developed intellectually than the chimpanzee, but there is no possible way of making such a comparison with the two animals in captivity (and the porpoise at least is not likely to be systematically followed and observed in his native habitat, the sea).

Rough comparisons of course are possible, and I do not doubt that the dog is smarter than the rat, or the ape smarter than the dog. Köhler for example showed in *The Mentality of*

*Apes* that the chimpanzee readily sees the possibility of using a box or other object as a stepping-stone to reach a banana hung from the ceiling, but no such behavior has ever been reported for the dog. Though a dog lacks hands to pick up a chair he could push it with a shoulder or drag it with his teeth to where it would allow him to get food otherwise out of reach.

This might be awkward, but my point is that the dog does not even see a possibility that is obvious to the chimpanzee. It is a perceptual lack, not motor, and from this point of view it is relevant here to observe that there is a little evidence concerning the development of perceptual capacity in evolution. The laboratory rat, trained to seek food behind a triangle (and not behind a square), does not recognize and respond to the triangle when the figures are rotated from the training position, nor when the figures are black on white instead of white on black as in training.

The cat recognizes a rotated triangle, but not one with the color reversal. The chimpanzee recognizes the triangle despite either change—but not when it is made up of six small circles, though a two-year-old human child recognizes this one too. This and other evidence can be cited if it should be necessary to show that rat, cat, ape and man form an ascending series intellectually, but this is possible only because they are well spaced, separated by long steps.

We have no basis for comparing animals closer together in ability, such as cat and dog; such evidence as could be cited would involve learning, which as we have seen has serious limitations—in this case, for example, because dogs have been bred for thousands of years for their readiness to learn. Perhaps dogs are smarter than cats, but it has not been shown to be so. We have no test methods ready by which to appraise the comparative intelligence of the horse for example if that question should be raised, and no basis at all for comparing intelligence in ape and elephant.

Neither of these two is at his best in confinement: The Hayes's report of the behavior of their home-reared chimpanzee, Vicki, is enough to show that the cage-reared adults of the Yerkes colony (from which Vicki came) were grossly impaired by being so reared, and reports of the elephant's intelligence, for example in the handling of timber in the Burmese forest under the general direction of his mahout, describe a kind of behavior that could not be studied in a zoo or even in the circus.

The classical test of a capacity for thought in animals is the delayedresponse test used by Hunter; as we will see shortly, it does clearly show the presence of representative processes in some animals, but it too is unsatisfactory as a basis of phylogenetic ranking. The length of the delay period, which might have been an index of evolutionary level, does not correlate well with other evidence of ability, small variations of procedure make large differences in the results, and it is not clear how some of these differences are to be interpreted.

The delayed-response procedure however has a fundamental importance. Historically, it was a direct result of behaviorism and Watson's denial of the existence of ideas and imagery. Carr and Hunter appear to have asked themselves what these things mean in terms of behavior. The answer was, acting as if an absent object is still present. When one looks at an object something happens in the brain; if that *something* happens when one is not looking at the object, it is an idea.

Thus, Hunter devised a procedure in which the experimental subject was given a sign (a light) to show where food was; the sign was removed (the light was turned off) and after a longer or shorter delay the subject was allowed to go to the food. If he went to the right place the animal was acting as if the light were still shining. Thus he had an idea. Hunter later simplified and improved the procedure by just letting the animal see the food put in one of two containers, then lowering a screen for the delay period to prevent the animal from simply keeping his eyes fixed on the goal. This is the procedure that has been used since that time.

The best performances on Hunter's test, by an animal subject, have been obtained from monkeys and apes. In one experiment particularly the evidence of ideation—of the visualization of an object seen earlier but now removed from sight—was obtained in an experiment in Tolman's laboratory by Tinklepaugh. The food his monkeys worked for was sometimes lettuce, which they liked, and sometimes banana, which they liked better. When a monkey saw lettuce being put in one of the food cups and later, when he lifted the lid, found the lettuce, he took it and ate it.

But when he saw banana being put in the cup and then found lettuce—Tinklepaugh having made a slight change while the screen was lowered—the monkey's behavior was disturbed. He might leave the lettuce and search around and under the cup looking for the missing piece of banana (no other interpretation makes sense). On occasion he had a temper tantrum. This is unambiguous evidence of a representative process, the search and the tantrum resulting from a conflict between expectancy and perception.

There is other evidence of thought in the infrahuman animal. The classical studies of insight by Köhler during the first World War when he was interned on the island of Tenerife are well known and I will not describe them here. Some of his most interesting observations were incidental to his experiments, and the clearest evidence that I know of, of thought and planning by adult chimpanzees, was also provided informally by certain members of the Yerkes colony who liked to spit water on visitors. (The caretaking staff were immune, for water hoses were at hand with which to drench the offender in return).

Even though a visitor was warned to stay at a distance, the chimpanzees looked so innocent and friendly that he would drift insensibly closer as he went from cage to cage, and sooner or later would come within the six-foot range.Now if one was stationed where the chimpanzees could be watched when company was coming one would see a very interesting course of events.

An animal who detected the approach of visitors would slip over to the water supply at the back of the cage, fill his mouth, and return to the front of the cage and—when the visitors came near—press himself close to the cage-wire as if inviting contact.

There is no doubt of the purposiveness of this behavior,

nor or the chimpanzee's enjoyment at the upset behavior of the visitor on being soaked. In other situations the chimpanzee shows capacity for deceit, acting in a way that is normally an invitation to friendly contact until the caretaker or visitor is within reach, then attacking viciously.

This is two-stage planning: invitation followed by attack. The water-squirting has *three* stages, getting loaded, attracting the visitor, and soaking him. In the demonstrated abilty to think of doing one thing while actually doing another, or to simulate one attitude while actually having an opposite one, the chimpanzee gives his most convincing evidence of approach to man in mental level.

Let me observe that all this comparative evidence also illuminates the behavior of human beings, quite apart from trying to trace an evolutionary history. From the human data alone one could not guess what an intellectual feat it is to recognize a triangle as such, to see the possibility of using a box as a stepping stone, or still more to think of piling one box on top of another.

By human standards the chimpanzee's mechanical achievements are primitive indeed—in piling boxes for example the chimp has no idea of having to center the upper box over the lower, but places it anyhow and may actually climb a structure that he has to hold together as he climbs, his weight on top of the second box being enough to hold it in place—but in comparison with the dog, whose behavior really is at a high level in complexity of function, the chimpanzee takes on a new stature: and man a still higher one, by comparison with the chimpanzee. The comparative perspective gives one respect for the intellectual capacity of the most ordinary of human beings.

## EMOTIONS AND MOTIVATIONS

The intellectual gap between man and any other now-existing species is enormous, so much so that if it were not for the anatomical evidence, and the behavioral evidence we will now consider, one would have to be very skeptical indeed about the idea that man and ape or man and elephant share a common ancestor—no matter how far back. The evidence to be considered concerns emotion and motivation and shows a far closer relation between man and ape for example and appears to be so directly connected with an individual animal's intellectual level that we might almost use it as a measure of intelligence.

For this purpose we can again assume that rat, cat or dog, monkey, ape, and man form an ascending series in ability. First, consider anger. In the rat, there is nothing that justifies being classified in this way, as a special form of aggression. The rat will bite if attacked or frightened, and in the case of the female to protect her pups, but there is no sign of the transient attitude of aggression toward another member of the same group that, in higher species, may follow even a minor injury or no injury (think of the tantrum of Tinklepaugh's monkey on finding lettuce when he expected banana), and no sign of jealousy or the peculiar behavior called sulking (whose peculiarity consists in refusing to accept a desired object that had been refused).

With the pet dog, *anger* is at times a meaningful term; he is capable of snapping at (not necessarily biting) the child that teases too long, with whom at other times he is on a friendly footing, and he occasionally shows signs of jealousy and even of sulking. When we come to the ape however the term takes on its full meaning. The anger is almost indistinguishable from man's in the variety of causes, in the different ways in which it can be shown, and in its apparent intensity.

In addition to the characteristically sudden violent aggression even in response to apparently trivial provocation, the chimpanzee shows jealousy, sulking, and that most interesting phenomenon, the infant temper tantrum-instinctive if there is anything instinctive at this phylogenetic level, since it appears in characteristic form without prior practice and certainly without guidance.

In the Yerkes colony, Kambi always sought attention from any staff member, but would spit at him if he was unwise enough to pay attention to a second animal nearby. Fifi was refused milk (when offered it she had been in the habit of taking a first mouthful and, like a winetaster, rolling it over her tongue and spitting it out before drinking, and the feeder tried to cure her of the wasteful practice), had a tantrum, and then refused for three weeks to take any milk from the one who had refused her, though she would accept it from others after a day or two.

A pure fit of the sulks: even a human being could hardly improve on such a specimen. The infant temper tantrum may be recognized at once by any human mother, the baby chimp banging his head on the floor, attempting to pluck handfuls of hair from his body, and apparently choking to death—all the while keeping an eye on mother to see if this is having the desired effect.

In the adult animal, rage can be provoked by any of a long list of things that might cause anger in man: persistent noisiness, being startled or teased, importunate begging, a perceived attempt on the part of another to annoy (e. g., Bokar was enraged when Dick, from the next cage, spat water at him even when it did not hit him). Another male had a tantrum repeatedly when a sexually receptive female in a separate cage would not sit where he could gaze on her, the lovely creature; still another had a tantrum at seeing a member of staff take a vaginal smear from a female in heat, though in fact he had often seen the procedure, a routine in a continuing study of the chimpanzee's menstrual cycle.

Finally, Bimba liked human beings and was liked by them, but she was also short-tempered and had a way of suspecting that she was being teased if there was any delay in handing over the food at lunch-time and when the feeder's hand came within reach would attack instead of accepting the food. In these causes of anger and in the variety of manifestation there is no apparent evolutionary discontinuity between ape and man.

Nor is there any such discontinuity in the causes of fear, when allowance is made—as we must of course with anger also—for the ape's intellectual level and his lack of language; in man a potent precipitant of both fear and anger. Irrational fears and abhorrences are common in both man and chimpanzee, and remarkably similar in the two species. Man, especially "civilized" man, consistently arranges his affairs so as to limit exposure not only to the possibility of real injury but also to situations that encourage imaginative fears as well as ones that are horrifying or simply nasty. It is extraordinary how readily we forget that fact and how persistently it is left out of account in the discussion of human nature and human society.

The chimpanzee we regard as a wild animal, man as in a sense domesticated, tame, unemotional; but this is ethnocentric blindness. The chimpanzee on his own ground is not perceptually excited, as Jane Goodall's work shows; and his "wildness" in captivity amounts to objecting to physical restraint, as well as readiness to use force if opportunity offers when he wants his own way. None of this makes him different from man, and in certain respects we may— almost—use him as a mirror in which to see certain characteristics of our own species.

Adult chimpanzees fear snakes when they have had no injury from them, not even physical contact, and of course they have not been told that snakes are dangerous. The infant fears strangers (this "shyness" begins at about four months, about six months in the human baby) and if he is brought up in the nursery and is used to three or four persons only, the disturbance may be great.

The adult avoids small animals or, if escape is not possible, attacks viciously. An adult female was greatly disturbed at finding a worm in the biscuit she was eating. These various susceptibilities clearly show a kinship with man. Now we come to one in which the relation may not be evident at once—but only because we keep on forgetting the existence of certain human characteristics.

In an accidental observation I found that the adults of the Yerkes colony were panic-stricken at the sight of a model of a chimpanzee head. The model was in clay, about half life-size, made only to discover whether the chimpanzees would be more interested in a three-dimensional representation of a familar object than in a two-dimensional one. I was astonished at what I saw when this innocent object was presented to their view.

Some of them screamed, defecated, fled from their outer cages to the inner rooms where they were not within sight of the clay model; those that remained in sight stood at the back of the cage; their gaze fixed on the model held in my hand. None would approach.

Subsequent tests showed that this was a generalized response to any object recognizably representing a part of the chimpanzee or the human body, including an actual chimpanzee head that had been preserved in formalin; a model human head, very lifelike, from a display dummy and a human hand taken from the same source; and a plaster cast from a death-mask of another chimpanzee that died about this time.

With repeated testing the strength of the response decreased, but none of the animals ever approached and touched one of these objects, and a doll—a representation of a naked human baby—was put in the same cage with some of the adults and, despite the well-known tendency of the chimpanzee to investigate destructible objects by taking them to pieces, the doll remained untouched.

A further point, also of great interest. The emotionality is a function of age. One and two-year-old animals paid no attention to the model head. They simply begged to be picked up when a familiar attendant approached with it in his hand. Five and six-year-olds however were fascinated by it—but unlike their elders, wanted to get a close as possible and would undoubtedly have taken it to pieces if they had got a chance.

As I have said, the emotional response of the adult chimpanzee seems at first sight to be mysterious, unlike anything we have known about, but if one considers the intellectual level of the chimpanzee and still more what the exciting objects represented—namely, parts of the body—there is a parallel to be found. This is in the attitude of human beings brought up in a sheltered environment (as the chimpanzees in question also were) to the sight of surgical operations and dismembered bodies.

Watching even a color film of an open-heart operation causes fainting. The dead body has some of the same emotion-provoking properties; no one in this society is likely to admit being afraid of the dead, but actual fear is evident in the elaborate ritual that in many societies is or used to be necessary to propitiate the spirits of the dead, and if that fear does not exist in our own society we still take remarkable care to limit exposure to the dead body and to ritualize the procedures for dealing with it.

The parallel between chimpanzee and man is more marked when it is recalled also that the human young, like the young chimpanzees, tend not to be disturbed by the sight of flowing blood, talk about broken bodies in air crashes, and gruesome tales of witches who eat little children and wolves swallowing grandma.

As adults we read these "children's stories" to five-yearolds without a qualm, but, it seems, only because we ourselves became familiar with them at a less sensitive if not more bloodthirsty age. It also reminds us that the usual adult sensitivity does not appear when the child has grown up exposed to bloodshed, just as the child who plays with snakes and continues to do so during later childhood is likely to be immune to the otherwise universal horror of harmless as well as poisonous snakes.

The chimpanzee data clearly show that the fear need not be learned; on the contrary, it is *not* fearing snakes that must be learned. In all this, it is evident that there is a close relation between intellectual level and the emotional susceptibility that we have been considering.

It is the higher animal that is more disturbed and vulnerable to the wider range of provocations; and, in a given species, the older rather than the younger. This runs counter to common sense, which sees the human five-year-old as the excitable one, not his father or mother. But there is an answer to such objections.

The child's emotionality appears mostly when he is being made to conform to adult requirements: the adult's lack of emotionality is mostly dependent on living in an environment that is particularly suited to adult sensitivities, and the adult has also had long practice in not giving free rein to the emotions that do occur even in that sheltered environment. How is a relation between intellectual level and susceptibility to emotional disturbance to be understood?

Obviously a contributing factor is the greater ability of the higher animal compared to the lower, or of the adult compared to the child, to detect danger, to perceive another's malice or intent to obstruct, or to realize that one has made oneself ridiculous. A closely related factor is the greater capacity for imagining danger or malice or ridicule where it does not exist.

Thus shame is not something known to the two-year-old, for example, and fear of the dark tends not to appear till the age of three and is stronger at four and five when the child is more capable of imagining monsters in the bedroom shadows. But this is still not the whole story.

In a simple analogy, we might think of emotional disturbance as a momentary breakdown of the orderly normal functioning of the brain (often followed, it is true, by well-organized flight or attack, but momentarily disruptive nonetheless). The more complex a piece of equipment the more ways its smooth functioning can be impaired, and so with the more elaborate thought process of the adult higher animal. This is analogy only, but there is a fundamentally noncognitive, unreasoning element in some of these emotions. A good example is the powerful fear or horror aroused by snakes. An adult can be convinced that a nonpoisonous snake is harmless and still find it so abhorrent that he cannot bring himself to touch it, let alone pick it up.

There is no rational basis for that avoidance. In some way the limbless writhing, the lack of hair or feathers, and the mysterious movement over the ground without benefit of legs—between them, it seems that these set up an intrinsically *perceptual* conflict, or conflict of percept with preexistent concepts of animals and animal movement. Similarly, the emotional upset at the sight of a surgical operation has nothing cognitive as its cause.

One might argue that the equally strong upset at the sight of an autopsy or the dissection of a cadaver is the result of knowing that we all must die—that it is in other words a fear of dying, somewhat displaced. But this cannot account for the reaction to the surgical operation, whose function is to maintain life, and it certainly cannot account for fainting on seeing a badly lacerated hand being sewn up or for the frequently emotional reaction to the sight of a badly disfigured face—or even the stump of an amputated arm or leg.

A physiological clue to this surprising relation between intellect and emotionality comes from the discovery about 1950 of the role of the ascending reticular activating system (ARAS), or arousal system, low in the brain stem, in the maintenance of thought and consciousness. This also has the value of relating the negative or disruptive emotions that I have talked about so far to the positive and integrative emotions such as joy and love.

My emphasis on anger and fear, in the preceding, is not because they are more important but because they provide most of our evidence. They are easier to observe and record because they often appear as a sharp break in the ongoing flow of behavior. I have no doubt that friendship, regarded as a nonsexual, nonparental attachment to another member of the species, is a mark of the higher animal and of the older rather than the younger animal, but there is little evidence to cite in support, apart from the fact that it is prominent in human behavior and the fact that strongly established friendships have been observed in chimpanzees and porpoises.

The arousal system, deep in the brain stem, is relevant because as Lindsley has pointed out it is essentially involved in emotional behavior. There are cognitive elements in emotion also, and the limbic system is involved as well as the arousal system, but it is the arousal system that puts the energy into an emotional response. Its activity is necessary if the cortex is to function at all, and it is not unreasonable to suppose that its role increases in importance as the cortex becomes bigger and its operations become more complex. On this basis we can understand a close relation between intelligence and emotionality. A relation between positive and negative emotions also appears. It seems that there is an optimal level of arousal; below this level, the subject tends to act so as to raise it, above it he tends to act so as to reduce it. The positive emotions are found in the lower and middle parts of the range, the negative emotions in the upper part. The positive emotions are states in which the subject tries to maintain contact with the stimulating object or situation, to achieve closer contact, or to make repeated contact.

The stimulation, in short, is desirable. The negative emotions are reactions to undesirable stimulation. They are states in which the subject tries to change the stimulating conditions (by attack, in rage, or by avoidance), or ones like grief in which the subject has no control over the stimulating conditions (absence due to death, e.g)..

Oversimplifying somewhat, a low level of arousal is boredom and the subject seeks excitement; a moderate level is the norm in waking behavior and tends to maintain itself and the concurrent behavior; and a high level tends to take a form of rage or fear and interrupt (at least momentarily) the behavior leading up to it.

A qualification is that different kinds of behavior may tolerate, so to speak, different levels of arousal: a 100-yard dash, for example, is itself a form of violence, not requiring any subtlety of cortical function, and so arousal may be high without doing anything but support the behavior. It also seems clear that there are different forms of arousal to be taken account of. Another qualification is that very high levels of arousal may result in a prolonged impairment of thought (i.e., more than seconds in duration) and even the motor impairment of the so-called paralysis of terror.

How the arousal system has its psychological effects is still conjecture. We know that its activity is necessary for transmission of sensory messages through the cortex, and for the interaction of cortical systems. We may suppose then that in an unexciting situation any stimulation that raises the level of arousal will be responded to selectively, maintaining the arousal; also, the cortex can excite the arousal system, as well as vice versa, so an exciting thought can take control of behavior and result in a seeking out of excitation. However, when arousal is still higher, cortical transmission and excitability may be too much increased.

Cortical processes that interfere with or inhibit one another may occur simultaneously and disrupt the behavior that led up to that situation. As for "paralysis," this is something that is sometimes observed in persons who suddenly and unexpectedly find themselves in a dangerous situation calling for immediate action. It may take the form of an actual inability to move, or an equally fatal inability to think clearly or to decide on a course of action.

Different forms of arousal are seen, for example, in anger and in laughter, and it appears that these result from different patterns of firing in the arousal system. In fear and anger the sympathetic nervous system is excited, and the aftereffects of the arousal last for some time.

In laughter the momentary effect may be just as strong ("helpless with laughter") but there is little of the autonomic aftereffect. The arousal system consists of some 28 separate pairs of nuclei, which makes possible different patterns of activity, all with the same function of activating the cortex but with different properties otherwise.

## ALTRUISM

Finally, let us consider altruism as another mark of the higher animal and a product of evolution. It is a distinguishing mark of the human species. As we will see, it is not unique to man, but it is common in all human societies, ranging from the very frequent acts of small assistance to occasional risk and even loss of life. A widely held view is that human generosity is really selfishness (due to fear of social censure, or the reward of social approval), or possibly a closely related form of stupidity (having been punished as a child for greed, the adult does not fully realize that he is now free from such control).

Another basis for asserting that generosity or kindness is really selfish is to argue that the giver gets pleasure from giving, the rescuer from rescuing, and so on. This is nothing but a play on words. It leaves unaffected the real distinction between helping and not helping, between those who can get their kicks in this way and those who cannot. For our present purposes, it means only that we redefine altruism as a tendency to help others without *external* reward (and of course not because of fear of being punished if one does not help). It is most easily understood as a product of empathy, the partial identification with another that has already been discussed.

There is unambiguous evidence of a disinterested (but not *un*interested) concern for others in the porpoise and the chimpanzee, though the evidence is more complete for the chimpanzee reared in captivity since his history can be known from birth.

The story of the porpoise's contacts with man is fascinating and depressing: fascinating in itself and because what had been thought to be ancient myth, uncritically recorded by Aristotle, Pliny and others of the classical era, has suddenly been made factual; depressing because human promoters have found out how to cash in on the fascination by imprisoning this intelligent, friendly and sensitive animal in small pools all over the country, for human entertainment and promoter's profit.

The story is recorded by Kellogg, together with a report of the extraordinary ability of the porpoise, like that of the bat, to perceive and identify objects by hearing or "echolocation"; and by Alpers, who records in detail the story of Opo, the young female porpoise that came into the shallow waters of the beach at Opononi, New Zealand, in 1955, to make friends with a young lady named Jill Baker and to play with her and the other bathers: full confirmation of Pliny's tale of similar friendships in the Mediterranean in classical times.

The evidence of concern for others includes help given by porpoises to other porpoises, which is not disputed, but also help given to human swimmers in trouble, whose meaning at least *is* disputed.

First, the help to other porpoises: A female giving birth in the pool of Marine Studios in Florida was surrounded by other females in a clearly protective way. They warded off an aggressive male and a shark attracted by the blood in the water. When the newborn infant began a first slow gradual ascent to the surface to breathe, a second female accompanied the mother in swimming just below the infant in a position to give support if it had failed to reach the surface.

There is no doubt that the porpoise understands the need of other porpoises to reach the air (the animal's own rising to the surface is not merely reflexive and noncognitive).

If an infant is sick or does not swim, the mother will support it at the surface, continuing to do so for some time, and it is known that adults will give similar support to other adults.

Siebenaler and Caldwell have recorded two examples. In one a porpoise in a school was stunned by the explosion of a stick of dynamite nearby; in the other a captured porpoise was accidentally struck on the head as he was being lowered into a holding tank.

Both times two other animals came immediately to the stunned animal's assistance, one on each side under a pectoral fin to hold him up till he could swim again. It is of interest to note that such aid from a single animal would have been no good, since it would simply tilt the unconscious animal over so that the blow-hole would be submerged; the dual effort constitutes a higher level of intelligent cooperation than anything recorded for that animal genius, the chimpanzee.

This evidence of help spontaneously and promptly given to others in trouble, with its evidence of an understanding of the need of air by other porpoises at least, should be kept in mind when we consider the reports of aid given to human beings in trouble in the water.

The classical tales of rescue in the Mediterranean may have originated with real events (in view of the two modern incidents reported in the following) but they show signs of fanciful elaboration due to word-of-mouth transmission.

They have become folklore (unlike Pliny's circumstantial accounts of boy-porpoise friendship), but the fact that they exist at all and were widely believed by those who knew the animals may carry some weight.

However, Kellogg provides two modern instances of help by porpoises to human swimmers. A woman bather was caught in an undertow and was swept under water, could not regain control, and was losing consciousness when she was bumped violently and found herself in shallower water where she could stand up.

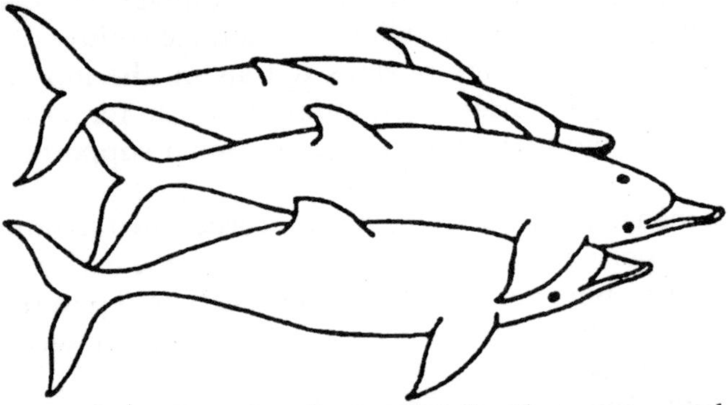

**Fig.** Sketch of two Porpoises Supporting Another, one on each Side

A man who had seen this told her that when he arrived she looked like a dead body in the water and that she was shoved ashore by a porpoise (which was still present, near the beach). Another woman fell overboard at night and was twice nudged away from adverse currents by a porpoise that stayed close till the swimmer could stand up "and then took off like a streak on down the channel."

An undue fear of anthropomorphizing—reading human attitudes and purposes into animal behavior without justification—can easily lead to the opposite error, of refusing to recognize the plain meaning of infrahuman behavior that happens to go beyond the observer's narrow view of what an animal is capable of. The difficulty, in the two cases above, may be either a low estimate of porpoise intelligence or a disbelief in the possibility of altruism toward members of another species, that is, human beings. In view of the whole record of porpoise behavior, neither is a good basis of skepticism.

For the chimpanzee, the evidence of altruistic behavior is clear. I do not of course suggest that altruism is necessarily more frequent than selfishness, whether in man, porpoise, or chimp, but argue only that it does occur. In the Yerkes colony it had been observed that an animal possessing food might give some of it to a hungry neighbor, and Nissen and Crawford put this tendency to experimental test. They found that when two chimpanzees were friends, the one with food would share freely; when they were not friends the "rich" animal might still find the other's begging irresistible, and so give less-but still give.

A more dramatic evidence of concern for another's welfare appeared in an experimental study of the chimpanzee's attitude toward human beings, and the effect of a visitor's boldness or timidity. The experimenter when disguised as the Bold Man, wearing heavy gloves and safely outside the cage wire, simply acted as one with no fear of chimpanzees. If the chimp was friendly, so was he, but he took no nonsense from others.

In one cage were two close friends, Lia and Mimi. Lia was wary of the Bold Man but Mimi attacked (i.e., tried to scratch or bite). The Bold Man replied in kind, catching one of Mimi's fingers through the wire mesh and bending it back slightly, enough to hurt. Mimi screamed with rage, beat on the cage wire, and tried again to catch a finger of the Bold Man.

Lia was also angry when Mimi was hurt and joined in the attack, but was frightened off by the apparent vigor of the Bold Man's responses. Having retreated to the back of the cage, evidently afraid, she nevertheless returned—not to re-enter the fight but to try to pull the reckless Mimi away and prevent her from attacking the Bold Man. When the test was repeated in later weeks Lia showed less emotional excitement but continued to work at getting Mimi out of trouble.

Observers thoroughly familiar with the animals of the colony were left with no doubt: Lia was afraid, but also feared for Mimi and was willing to run some risk to get her out of the danger zone.

It is extraordinary how strong the tradition is that selfishness is the only credible motivation—extraordinary, that is, in view of all the human experience to the contrary. The attitude long antedates Darwin, but it was reinforced when evolution came to be regarded as a process of "survival of the fittest" (and by implication, survival of the most selfish). Sharing food in times of famine, or trying to rescue a drowning swimmer, can only make one's own survival less likely. Such behavior occurs, however, and is not really the exception in such situations.

With the development of genetics in this century the problem becomes more specific. One may say that it is not the individual that is the significant variable in evolution, but the family or tribe or species, and the gene pool; a gene pool that throws up some proportion of individuals prone to help others may survive better than one that does not, which is obviously true where parental behavior is concerned. Without that instinctive pattern of behavior no bird or mammal would now exist.

The question is, however, what evolutionary mechanism can it be that produces such a situation in the first place? Apart from parental behavior, one might see how with learning cooperative tendencies might be maintained once established, but how would that state of affairs originate?

For part of the problem a relatively direct mechanism is available from genetic theory. This deals with parental behavior and supportive behavior between siblings. Relatives have genes in common, so the support given to one's kin, though it does not favor one's own survival does mean that those genes of one's own that produce such behavior will survive, transmitted by kin—so the behavior reappears in the next generation. But this does not apply to supporting behavior given to ones outside the family. Here there is another variable for the geneticist to take account of.

Wilson in *Sociobiology* as well as others have confused the issue by failing to consider the differences of behavioral control in the worker ant and in the higher mammals. The "altruism" of the ant is reflexive, noncognitive, genetically programmed given a normal environment. The altruism of chimpanzee and man is cognitive, conscious, and idea-controlled, and it is with respect to this behavior in the higher animal that the problem of true altruism arises. A directly genetic explanation of the behavior of the sterile ant's supportive behavior can be accepted (and also presumably for the strongly instinctive and more or less compulsive maternal behavior of bird and mammal) but we can look elsewhere for explanation of the evolution of true altruism. The idea in short is that this is a by-product, not something with direct evolutionary value but a consequence of something else that does have such value—namely, a high intelligence and the capacity for thought.

We have seen that there is strong behavioral evidence for the existence of empathy, the identification of self with other, in chimpanzee as well as in man. It was also seen that there is a theoretical rationale for such a state of affairs, to be found in the fact that the infant must first develop an idea of a person from the perception of others and develop an idea of self only on that basis.

Thus the idea of the self and the other have a common core. In favorable circumstances, then, learning that is based on self-interest may with suitable modification transfer to support for another.

This is of course not inevitable, just as altruism is not a universal feature of human behavior; the necessary "suitable modification," the fact that the actual response is not the same in helping another as in helping oneself, may account for the fact that the response is often not made. But the fact also is at the cognitive level that "sympathy," and the *idea* of help, is very common indeed.

That it is often compelling is evident in human society; and the same thing can be true of the chimpanzee, when an animal with food is subjected to continued begging by another and finally gives in—by throwing the food violently at the beggar, apparently unable to resist. One still encounters that assertion that true altruism in human beings is not possible, whatever may hold for the ant and the bee. It is implied that generosity is a kind of artificial addition made in childhood; and by the time maturity is reached the human being has done some complex learning, making it practically impossible to show that this is not really so.

When one is faced with the flat assertion that altruism is not possible however, the animal data are there to rebut the assertion. In the case of the chimpanzee at least the evidence is clear, that altruism can appear spontaneously, and we need no longer shut our eyes to the fact of kindness—sporadic kindness at least—as an outstanding characteristic of the human species. Meanness and cruelty are species characteristics also, it is true, but they are not the whole story.

## HEREDITY, THE IQ, AND ATTENDANT CONFUSIONS

In this modern and enlightened day a chapter on the heredity-environment question may look like flogging a dead horse. Unfortunately, the horse is not dead. I hope with luck to make this chapter short but, long or short, it is still needed.

There are vital social and educational as well as theoretical issues that are widely misunderstood, not because they are so complex in principle but because of the persistence of old habits of thought. An unfortunate result is that well-meaning people, concerned to support the deprived, white as well as black, sometimes take positions that tend to injure istead of support. In this biological-psychological problem area even the biological scientist may become confused in dealing with behavior, clear as he is about morphology. As for the social scientist, his simple preference seems to be to abrogate all the laws of genetics as they apply to differences between human beings.

William James once referred to a certain misconception about selfknowledge as "the psychologist's fallacy." The psychologist's fallacy today is to suppose that all behavioral problems are problems of learning, and to forget—or deny—the significance of heredity and constitution.

There are social scientists, including many psychologists, who believe and are prepared to state publicly that, barring birth injury, all members of the human race are born not merely equal but identical, apart from such trivia as muscular strength and agility, sex organs, skin color, and the like; the differences of ability and of propensity or motivation that are so marked at maturity are all products of experience during growth.

Even more, it has become fashionable to deplore research in this area, clearly with the problem of race relations in mind and the entirely false notion that the research has been injurious to the American black. As we will see, the truth is exactly opposite, certain current aberrations to the contrary notwithstanding.

The idea that all is learning and that human differences are all a question of an acquired personality—the motivations and self-image that originate in the experiences of childhood—that idea is to be found in the discussion of matters as widely separate as obesity and Women's Liberation. There is research that blames obesity on a love of food—that is, on greed—with no apparent recognition of a continuing need produced by the underlying physiological disorder; in general, being overweight is regarded as the result of eating too much and exercising too little—with no recognition of the genetic variable that makes it impossible for some of us to gain weight and makes it inevitable for others. Likewise, alcoholism is really regarded as self-indulgence, despite lip service to the idea that it is an illness.

There is a constitutional variable here too. In effect some people are no more capable of becoming alcoholics than a skinny woman is of getting a little fatter, much as she might like to; for others the addiction can be established only too readily.

The difference presumably appears in the immediacy and degree of pleasure that is derived from alcohol, and the beginner who finds great enjoyment in a drinking party might do well to regard that fact as a warning. He is not likely to do so however if he takes for granted that all men are born equal (constitutionally, with respect to alcohol) and, seeing others drink freely without becoming drunks, takes for granted that he can too.

From this same egalitarian point of view, *mental illness* is a misnomer, and there are psychologists, many of them, who are indignant at any suggestion that neurosis or psychosis is a medical problem, to be treated in the first place by one with medical training—that is, by a psychiatrist. The idea instead is that such disorders arise from experience and constitute only a problem of reeducation.

Now if there is one thing in this field that has been established in the past 30 years or so it is that there are genetic factors in the common psychoses, which must have a biochemical basis, and that both neurosis and psychosis are better treated with chemotherapy (tranquillizers, antidepressants) than without. But let us not go to extremes. That statement does not deny importance to psychotherapy and behavior-modification methods. Not all disorders have a special biochemical component, and in those that do psychotherapy may still have great value though by itself it does not cure.

The egalitarian tendency in feminism shows up, uncontrolled, in those writers who deny any inborn psychological differences between male and female. The differences between adult men and women are all products of social pressure. The tendency is also evident in those who recognize the existence of a genetic component but are concerned to keep it to a minimum and put the burden of proof on the hereditarian. No great weight is given to biological evidence from other species, even when it favors the female: leadership by the female contrasted with dominance among males, in deer, for example, or the apparently greater social intelligence and courage of the female chimpanzee.

## THE BIOLOGIST'S FALLACY

The biologist has a fallacy too, though I hasten to say that just as the psychologist's fallacy is shared with sociologists and other social scientists, so the biologist is far from being in sole possession of his. The biologist's fallacy is to ask whether a given character is inherited or acquired. In that form today it is rare except when dealing with behavior and even then is more likely to take the form of asking *how much* some aspect of behavior depends on heredity and how much on environment, or *how important* heredity or environment is for the behavior. But either question is bad; the only answer is that any behavior depends fully on both variables, that both are of hundredpercent importance. The relation between them is not additive, so that as one becomes more important the other becomes less; instead, one depends fully on the other.

In an analogy used elsewhere, heredity and environment are related to behavior as length and breadth of a field are related to its area. To say that human intelligence is 80% heredity, 20% environment—or that it depends more on heredity than on environment—is to make the biologist's fallacy. It is like saying that the area of the field depends more on its length than on its width. Heredity by itself, or the fertilized ovum at the moment of conception, can produce no intellligence whatsoever. Its potential can be realized only with the conjoined environmental action of the uterus until birth and a supporting postnatal environment for some time after that. And *vice versa*: The environment can do nothing without a fertilized ovum to work on.

There is also a tendency to forget how varied the different modes of environmental action are and even, when dealing with intelligence, to reduce the all effects of environment to learning—and at that, a kind of learning that is known only in the half-grown or adult organism. As we will see, the infant's first learning is of a different kind and needs to be taken account of separately.

The accompanying table is a means of keeping things clear. It is a mnemonic rather than a logically tight analysis (e.g., the genetic factor in behavior is the DNA carried by the ovum, not the ovum as a whole) but precise enough to serve its purpose. Factor I then is (roughly) heredity. Factors II and III are the normally supportive physical environment, distinguished because Factor II, prenatal, is sometimes forgotten but should not be: Malnutrition in the mother for example can have serious effects, and so can her use of drugs.

Factors IV and V are both experiential, but here again a distinction is important. Factor IV, exposure to the normal sensory environment of the young of the species, has very often been forgotten, and some of the learning it induces is different from, and prerequisite to, the learning that occurs later in development.

**Table: Classes of Factors in Behavioral Development**

| *No.* | *Class* | *Source, Mode of Action, etc.* |
|---|---|---|
| I | genetic | physiological properties of the fertilized ovum |
| II | chemical, prenatal | nutritive or toxic influence in the uterine environment |
| III | chemical, postnatal | nutritive or toxic influence: food, water, oxygen, drugs, etc. |
| IV | sensory, constant | pre- and postnatal experience normally inevitable for all members of the species |
| V | sensory, variable | experience that varies from one member of the species to another |
| VI | traumatic | physical events tending to destroy cells: an "abnormal" class of events to which an animal might conceivably never be exposed. unlike Factors I to V |

This is true both theoretically and empirically. *Theoretically,* the first learning is the slow development of cellassemblies which, once developed, make possible the faster—often onetrial—learning of maturity, as a changed relation between assemblies. *Empirically,* infant learning is slower and completely lacks insight; it must begin in very early infancy and cannot be replaced by later learning, at least not fully; without it, adult personality and intelligence are grossly defective (e.g., when dog or monkey is reared in isolation in the laboratory, or the human baby is reared in what amounts to isolation in certain orphanages).

The codification of these factors in development has the immediate value of making it impossible to think of behavior as determined by any one variable. It also clarifies terms. Thus, reflex behavior is a product of Factors I to III; instinctive behavior, especially in the higher animal, brings in Factor IV as well (instinctive behavior is no longer thought to be wholly unlearned but depends also on the learning that comes from simple exposure to the common environment). Intelligent behavior likewise is a product of Factors I to IV, but whereas instinctive behavior is evoked by features usual in the common environment, intelligent behavior is evoked by the unusual, and so involves the stimulation of Factor V.

Factor V is also the source of the learning that differs from one animal or person to another, but it must be emphasized that Factor V alone can do nothing: For all cognitive learning Factor IV is necessary, and Factors I to III are necessary for any learning whatsoever.

Here it may be seen also that behavior may be unlearned, in the usual sense of the word, and still be dependent on prior learning. An insightful or an instinctive action is unlearned since it does not result from practice or instruction or observation of another's performance, and yet it requires the Factor-IV learning normal to the species.

## THE IQ AND ITS VALUES

The in thing today in the discussion of intelligence is to talk about *the heritability of the IQ*, but not always with comprehension of the limits of the meaning of that statistical conception. Heritability is the proportion of the variance that results from differences of heredity. It is not easily determined for the IQ, so estimates differ. It has been common to arrive at a figure of.80 (80% of the variance), but on the other hand one egalitarian has tried to show that it is about zero—concluding that one's heredity has nothing to do with one's intelligence.

That extreme view, and at the other extreme, the idea that high heritability would mean that environmental modification can have little effect on the IQ—specifically, on the mean IQ of U.S. blacks—is adequately dealt with in the outstanding monograph by Loehlin, Lindzey, and Spuhler. The book is outstanding in its treatment of race and the uncertainties that enter into the interpretation of a figure for the heritability of the IQ, and what heritability means or does not mean in the comparison of black and white.

On one point I believe they should be clearer: in the reference of that term *intelligence,* which in the psychological literature sometimes means, approximately, the goodness of the infant brain (or at an even earlier stage, the quality of the genetic contribution thereto), and sometimes means the level of intellectual function that is achieved at maturity.

That is, the term sometimes refers to genotype, sometimes to phenotype. In the past I have distinguished between the two meanings as *intelligence A* and *intelligence B,* and I believe to avoid confusion the distinction is still necessary. Intelligence in sense *A* cannot be measured; it is an even more inferential entity than other psychological variables, which are also inferred from behavior.

Intelligence in sense *B* however is quite measurable, with relatively short steps of inference by tests which in effect sample a cultural product, to determine the extent to which the person being tested has mastered the ideas, modes of thought and of solving problems, ways of perceiving, and the store of information characteristic of the society for which the test is standardized. Thus the idea of a culture-free test is a contradiction, and in fact the test items that predict best are "Binet-type" items: ones so culture-loaded that they offer no difficulty for those who can succeed with them at all.

They are not problems because, in effect, they merely ask for solutions that the subject has already worked out, and what he has already learned in the ordinary course of experience. Thus the test does not pose a set of new problems to be solved here and now. Solving new problems might seem more logical for an intelligence test, but in fact it works poorly; what Binet really discovered was that the level or quality of mental function *in the past* is the best predictor for the future.

The so-called culture-free tests are neither culture-free nor predict well. The Raven Matrices test for example makes ingenious use of a certain kind of logical operation, nonverbal and independent of any specific past experience, but it is not the test by which to find out how well a child from the slums is prepared for school. On the other hand, the idea of a culture-

*fair* test is simply confused. In what respect can a Stanford-Binet or Wechsler test be unfair? Only as an index of intelligence *A* in cases where the environment has been deficient—that is, as an index of native ability. But no test can measure intelligence *A*. The question is simply, How well does the test predict? *The same conditions that impair the IQ also impair clerical and academic aptitude,* so a Wechsler rating would predict about as well for a Stone-Age Indian as for a New Yorker if one wanted to know how well either would do in a business office in the city or at Columbia University.

The IQ is in bad odor today. It is said to be racist and unjust and a means of further oppressing the oppressed, all of which is completely wrong. If the results are abused the abuse should be corrected, but the intelligence test is a tool of the greatest potential importance, for the disadvantaged student particularly. It should be recognized that the IQ is essentially an academic *diagnosis*, an estimate of the child's probable rate of learning the kind of thing that a school wants him to learn. The diagnosis has a known probable error, but the error is least where the diagnosis is most important—that is, with low IQs—and it helps in adjusting the course of study to the needs of the individual.

Too often the slower learner who is not obviously destined for a special school for the retarded is simply dropped into the main stream under pressure to keep up with the average, which he cannot do. He becomes frustrated and unhappy and learns to hate everything connected with school. Formerly, he failed examinations annually and repeated grades and quit school as soon as he could manage it; today, examinations (like the IQ) are considered discriminatory, and in the name of egalitarian principles the slower learner is likely to be promoted year after year without examination and so ends up in high school unable to read—when with some adjustment of the curriculum he might have got at least the beginning of an education.

Similarly, the bad name of research on intelligence is undeserved. Those liberal thinkers who would dissuade us from it clearly think it will turn out badly—that is, that it will support the idea that the black citizen is genetically inferior. But there is no reason to expect that result, and the whole force of research has been to the contrary.

Consider the U.S. Army data of 1918. There is little doubt that in the nineteenth century the black was regarded as intrinsically inferior; to be convinced one need only look at some of Abraham Lincoln's talk about how one might deal with the newly freed slaves when the Civil War would be over. The first big dent in that attitude was made by the report of Robert M. Yerkes and his colleagues on the result of testing American recruits in the first world war.

Instead of finding the black scores to be a separate distribution, massed below the white, they found a great overlap, a significant proportion of blacks making better scores than a significant portion of whites. A myth at that point was shown to be nothing but a myth—by psychological research.

The research then went on to make a further contribution by showing how dependent test performance is upon the subject's social background. This seems to have been generally regarded as a defect of the available tests, and it was at that stage that various attempts were made to develop a culture-free intelligence test—not realizing that the environmental effect might go deeper and affect intelligence itself (intelligence *B*, that is) as well as the IQ.

That further realization came with animal studies in which rearing conditions could be completely controlled, using as experimental subjects the laboratory rat and Scottish terriers. This, and perhaps also the provision of a theoretical rationale—the necessity of experience to develop cell-assemblies that are the basis of intelligence—threw a new light on human data that already existed.

Gordon and Asher had reported losses of 25 to 30 points of IQ after the age of about 5 years, the former in the canal-boat children of England, the latter in the Kentucky Mountain children: The data imply that the environmental stimulation was adequate for cognitive growth in the first years of life, inadequate thereafter.

William Goldfarb's unique data in particular were

decisive. He showed that three years in an orphanage, about the ages of one to three, resulted in a mean IQ of 72 in comparison with a mean of 95 for a matched group placed at the same ages in foster homes, the test being given some years later between the ages of 10 and 12. At this time Beth Wellman and her colleagues were trying to convince others that improving the environment could raise the IQ of deprived children, and met ridicule. The animal data changed that situation completely.

All this work should have made it clear that the IQ and intelligence *B* are both vulnerable to the effects of an impoverished environment, and where the blacks are concerned that their lower mean IQ—10 to 15 points lower—may be low for that reason only. But it should also be clear that intelligence *A* is removed by too many steps of inference from the observed IQ for us ever to be able to compare it in any two socially distinct groups in the country.

## HERITABILITY

It is evident that some psychologists (and it seems some biologists) still think in an unacceptable older way about heritability; if heritability is high, little modification of the phenotype will be made by manipulation of the environment, and the phenotype mirrors the genotype.

The falsity of such ideas is emphasized by Loehlin, Lindzey, and Spuhler, and even more vigorously by Feldman and Lewontin. But the inertia of that older way of thinking about heredity and environment is great, and the argument may become more effective if it is translated into concrete terms. Hence the two following thought-experiments.

T-experiment A: Let us suppose that some miraculous new development of genetic engineering makes it possible to provide all infants of the next generation with a first-class, high-quality set of the genes that contribute to the development of intelligent behavior.

When this is done, all having the same genetic constitution in that respect, everyone's intellectual endowment will be the same but, since the children must be brought up by their present parents and in very different environments, their behavior will vary considerably.

What about the heritability of the IQ once these children have grown up and society consists of them and their children? Heritability is the proportion of the variance that is determined by variations of heredity, but there is no variation of heredity, so the result of a profoundly important hereditary modification is to reduce heritability to zero—but this obviously does not mean that the importance of heredity is reduced.

T-experiment B: Mark Twain once proposed that all boys should be brought up in barrels and fed through the bung-hole till the age of 12. Let us imagine that his advice was taken and has been followed since then. The result today is that there is a big gap between the intelligence of men and women; and the women, who now staff the universities and run the country, ask themselves scientifically what to do about the abysmally low male IQ.

In the best academic tradition they determine its heritability first. This turns out to be very high indeed, since the environment during growth is practically identical for all males and cannot contribute to the variance of the male IQ. If the female scientists are among those for whom high heritability means that the environment is not an important cause, they will see no reason not to go on rearing boys in barrels.

Heritability is a descriptive statistic that has limited value as an analytical tool with which to investigate the complex relation of heredity to environment in the mechanisms of behavior. High heritability of the IQ does not lessen the importance of the environment as a determinant of behavior, nor show that the IQ in a segment of the population has not been depressed by adverse conditions during infancy.

On the other hand, it does show that the genetic variable is also a source of variability of the IQ. Put in the crudest terms, it shows that some children inherit better brains than other children, and it is really this crude proposition that is offensive to the egalitarian attitude to which I referred in the early part of this chapter. In the attempt to refute it, some extraordinary illogic has been indulged in by otherwise intelligent people. One pair of critics for example has argued as follows: all the children in a certain clinic come from the same ethnic group (Italian, let us say), but the ones from middle-class homes are brighter than those from working-class homes. Since they all come from the same gene pool, the difference between them must be of environmental origin, due to the kind of homes they are reared in.

This is absurd. It begs the question by assuming what the authors set out to prove, namely that all members of an ethnic group have the same genetic endowment, and does not even consider the logical alternative that middle-class homes are middle-class because the parents have superior genetic endowments that they also pass on to the children. The argument has no significance at all.

Another critic denies that the IQ measures something that determines how well a child will do in school. When social level is held constant the IQ has little predictive value, so social level is the true determinant of success in school. This argument too has a large hole in it. It overlooks the possibility that the IQ, determining the levels of school success in preceding generations, is in effect a main determinant of social level. The argument could as well have gone like this: Holding IQ constant, social level is a poor predictor of school success, so it is the IQ that is the real determinant, not social level.

Or try this analogy: I have a theory that age is a determinant of weight in the growing child. I determine the correlation of age and weight between the ages of 2 and 12, and it is high, which supports my theory. You however object, saying that height is an important variable, so you recalculate the correlation holding height constant. You find that age has little effect in children of the same height—so age is not a factor in a child's weight, it is all due to his height. It is evidently still difficult to think logically in the heredity-environment business.

## Chapter 32

# Physiological Theory

The theory of cell-assemblies is a theory of thought and a general theory of behavior, but it began at a more modest level. My problem was to understand some of the effects of brain operation, and in particular how it was that a patient might have a large chunk of brain tissue removed without much effect on his IQ or his intelligence as it seemed to his family. How could there be no loss of intelligence after surgical removal of both prefrontal lobes (about a sixth of the mass of the cerebrum) or removal of the entire right half of the cortex?

The only answer seemed to be that intelligence, as measured, somehow concerned the kind of concepts the patient had developed while growing up and the way he had learned to think about the common situations of everyday life, and that these ways of thinking, once established, were not easily changed. But such an answer only moved the problem over into another area. What is a concept? How is a way of thinking established, and why should it last?

There are complications here that still must be reckoned with by anyone who would really explain human behavior. In particular, there is a crucial difficulty in reconciling the theory of learning with the theory of perception, and it seems to be avoided both by the learning theorist and the perceptionist. ("Learning theorists," self-named, have in common a narrow view of learning that derives ultimately from Watson's stimulus-response proposals and his denial of mental processes; and I coin the term *perceptionist* to refer similarly to one with a narrow view of perception, without regard to the problems of learning).

In both cases a black-box approach is taken, which permits one to avoid the constraints originating in the structure of the brain. Much can be done, and is done, with such nonphysiological theory, but the theory must ultimately be translatable into or concordant with the language of the anatomist and the physiologist. Physiology is not the whole answer by any means, but "physiologizing" may at times reveal difficulty that would otherwise remain unsuspected.

Showing how the theory of cell-assemblies was developed is one way to show what the difficulties are, or some of them, and perhaps is the clearest way to explicate the theory itself. This also has some interest as an example of theorizing in the real world, with its doubts and difficulties and false starts, unlike the idealized and simplified cases to be found in the textbooks of logic.

Logicians prefer to discuss cases of outright success (good theory) or outright failure (bad theory). It is surely of some interest to see how an author could get himself into the position of having to publish a theory of perception that he actively disbelieved, but published nevertheless because it was a necessary part of a larger theoretical scheme that did have value (and eventually got independent support). But the experiments that showed that the theory was right—more or less—also showed that it was wrong: a situation that will be clear later in this chapter. It appears that the theory was, and is, on the right track or aimed in the right general direction, but still has a long way to go.

The theory is physiological in the sense only that is uses the available physiological evidence and attempts to say in physiologically intelligible terms what kind of activity goes on in the brain. But it is still theory, and it is important to make clear that the problem must be dealt with theoretically; and that this is a *psychological* problem. There is no possibility whatever of solving the problem of the over-all patterns of activity in the brain by physiological methods. There is no possibility for example of putting in enough electrodes to record the patterns of neural firing in thought. Ten thousand electrodes could not do the trick.

Thought must be known as theoretically as the chemist knows the atom. Physiological methods can deal with part systems: for example, with input from retina to striate and peristriate cortex or output from Betz cells to spinal cord and thence to muscle; with the effects of damage to the red nucleus or the behavioral effects of removing the amygdala; and so on. All such information is fundamental, but a further fundamental feature is missing, namely, how these part systems are coordinated in the ordinary behavior of the intact unanesthetized animal.

We know much of how visual excitations reach the cortex, of how the motor cortex interacts with cerebellum and the nuclei of the midbrain to control spinal cord activity and thereby the activity of the muscles, but there is no physiological account available of how Hubel and Wiesel's visual edge detectors determine motor output when a cat's eye is caught by a mouse running across the floor.

There is no prospect of getting such an account, or of learning by physiological means what goes on in the cortex when human eyes are engaged in reading a novel. If present theoretical answers to such questions are poor, it is nevertheless only by means of developing better theory that we can hope for better answers.

Understanding the integrative action of the whole brain is a psychoneurological or a neuropsychological problem, a highly theoretical enterprise that depends on bringing together the results of anatomical, physiological and behavioral analysis. That is what the theory of cellassemblies tries to do.

## THE PROBLEM OF PERCEPTION

The facts of perception represent the first great hurdle for theory, now as in 1945 (when cell-assemblies were first thought of, especially to deal with this problem). A conflict with the theory of learning is due to the apparent fact that the perception does not, or need not, depend on the excitation of any particular input lines, and hence the apparent necessity of concluding that a perceptual process is not localized—may not have to be an activity of any particular cells in the brain.

But learning must depend on changes in particular cells; and any response to perception must be in particular muscles and so must be determined by particular motor pathways from brain to cord to muscles. If the locus of a perceptual activity is variable, if it is now an activity of one set of neurons, now of another, how can a perception be affected by learning, and how can it excite the proper output path and not a lot of other paths?

Learning is a changed pattern of conduction in the brain that results from experience and makes a change of response potential. In the simplest case (the formation of an S-R connection) excitation from the stimulus reaches a muscle or gland that it did not reach before, making a new response. In the more complex case of cognitive learning there is often no response at the time the learning occurs; a new ideational system is organized or a connection is made between two such systems, for example, and it is only later when, in the right circumstances, some change of response shows that the learning did occur.

If this changed pattern of conduction persists it is memory; and the persistence, it seems clear, means that the learning must be a change related to particular pathways, or particular neurons. In other words it must be a localized effect.

But this seems not to be true of perception, which—appparently—may be excited by different sets of input lines every time it occurs and so cannot be localized. So it seems, at least with familiar objects.

A book or a cow or a car is recognized not only in different parts of the visual field—you don't have to look right at one of them to perceive it—but also when it is seen from different angles and at different distances. Lashley and Köhler both believed in fact that what matters in perception is not what sensory receptors are stimulated but only the *Gestalt*, the form or pattern of stimulation.

However, no one was able to show how learning could take place with repeated trials, if the input never followed the same input lines and the same central pathways. The effect of a second trial cannot sum with the effect of the first if they are in different places in the brain. And this Gestalt-Lashleyan theory never succeeded in showing how a perception can excite a specific response. When a learned response is made to a perception, at *some* point the excitation from the stimulating event must get on a specific set of motor pathways to produce that response.

Köhler never faced the problem (at least not in his published work). Lashley recognized it and tried to deal with it by his theory of interference patterns, but concluded that the theory was unsuccessful (he was given to saying that he had destroyed all theories of behavior, including his own).

A related modern idea is the hologram, which it seems likely Lashley could not have approved any more than his own interference patterns.

By analogy with the optical hologram of the physical scientist, the brain hologram is supposed to be widely dispersed but to contain all the information in each of its parts; thus it does not matter where a perceived object is in the visual field, or by which fibers in the optic nerve the information is transmitted.

But essential difficulties remain. First, different perceptions result from the same sensory input, though the hologram must be the same. Figure is the classic example, from Edgar Rubin, the Danish psychologist. Without eye movement, so that the sensory input remains the same, one sees at one moment a vase or goblet, at another moment two faces in profile staring at one another.

This variability of perception is not limited to such ambiguous figures—they serve only to provide clearcut examples—but is universal. Depending on what is going on in your thought at the time, when you look at a chair what you may perceive is its size, its color, a distinctive feature of its shape, its distance from the table—or all these and other features one after another in close succession.

The difficulty for hologram theory is to account for this influence of one's attitude (essentially, one's preexistent mental activity) upon what is perceived, and the resulting Variability of perception. Another difficulty is the constancy of the visual world, and the fact that the other end of the room does not look different as your eyes move, or when you move from one point to another, although the brain hologram is modified with each such movement.

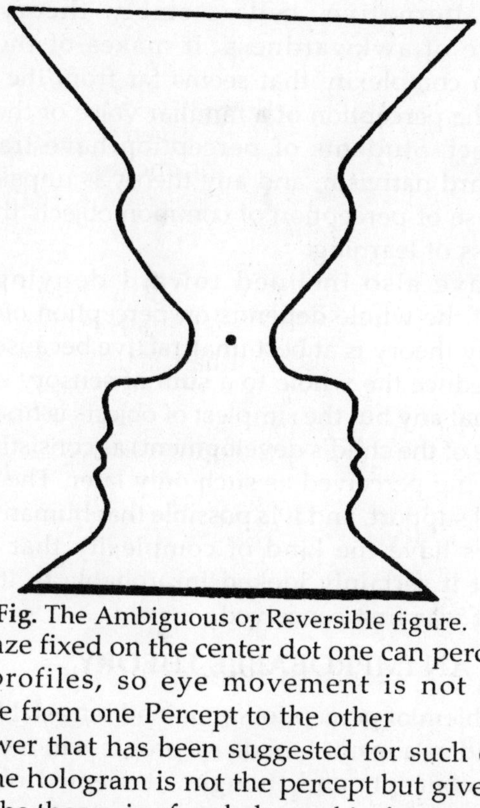

**Fig.** The Ambiguous or Reversible figure.

With Gaze fixed on the center dot one can perceive either goblet or profiles, so eye movement is not necessary to the change from one Percept to the other

An answer that has been suggested for such criticism is to say that the hologram is not the percept but gives rise to it. In that case the theory is of no help, and in fact is not a theory of perception at all, until we know what the activity is that it gives rise to and how it does so.

The original idea was ingenious and attractive, but the attraction is superficial until we see how to deal with the ambiguous figure and the variability of the percept despite an unchanging sensory input; with the selectivity of attention and the influence of the immediately preceding cerebral activity; and with the fact that the process of perception is characteristically sequential and not like a single photographic recording, except possibly with highly familiar objects. There are other difficulties—visual completions, subjective contours, perceptual learning—and until some at least of these are dealt with the hologram remains at the level of analogy.

As an alternative, cell-assembly theory has the disadvantage of awkwardness. It makes of the simplest perceptions a complexity that seems far from the directness and ease of the perception of a familiar voice or the sight of a familiar object. Students of perception have traditionally inclined toward nativism, and any theory is unpalatable that makes the ease of perception of common objects the result of a long process of learning.

They have also inclined toward denying that the perception of the whole depends on perception of parts, and thus assembly theory is at best unattractive because, although it does not reduce the whole to a sum of sensory elements, it does argue that any but the simplest of objects is first perceived (in the course of the child's development) as consisting of parts, the whole being perceived as such only later. The theory has experimental support, and it is possible that human perception actually does have the kind of complexity that the theory requires, but it certainly looked improbable to its author—me—when it was first conceived.

## DEVISING AN IMPROBABLE THEORY

The problem of perception remained intractable for about five years and as a result I made no progress in my attempt to understand concepts and thought. It seemed obvious that concepts, like images, must derive from perception, and I could think of no mechanism of perception that corresponded to my ("perceptionist") preconceptions.

In fact, by 1944 I had given up trying to solve the problem. What happened then was that I became aware of some recent work of Lorente de Nó in conjunction with some observations of Hilgard and Marquis, which led me to think of a possible solution for a different problem: the problem of set and attention. And this brought me straight back to concepts and the thought process, but now from a different point of view.

In 1940, set, in various manifestations and known by various names such as attention, intention, attitude, expectancy, and so forth, constituted an entirely unsolved problem.

As Gibson showed, the effect, whatever produced it, was ubiquitous in psychological investigation, freely recognized by the cognitivist but also by some hard-line experimentalists who would rather not; on the other hand, not even a suggested mechanism was available by way of explanation. Attention and set were a mystery, until Hilgard and Marquis's references to an activity of the brain that is relatively free from sensory control showed where the trouble must lie: In that throughtransmission nervous system represented in Figure, with a brain that can do nothing but act as errand boy in the delivery of sensory messages to the muscles.

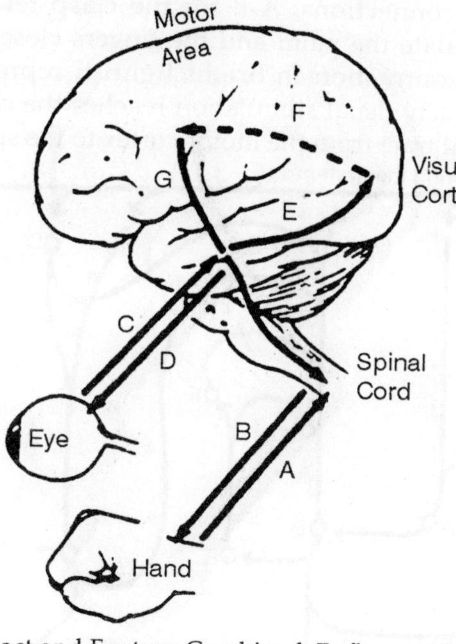


**Fig.** Fact and Fantasy Combined: Reflex and learning

Set was a mystery because it meant that the brain was doing something that—it seemed—the brain could not do; but if on the contrary the brain can act independently of the sense organs, then set and attention and attitude could be understood. Figure diagrams two kinds of transmission in the nervous system. One is built in, one acquired. The first is the basis of the unconditioned reflex, a function of the spinal cord or brain stem (the stem is the extension of the cord into the skull), but not of the cortex.

The representation is schematic, but sound enough in principle. The second is a longer route, from sense organ to cortex, to another point in the cortex, and thence to the spinal cord and out to the musculature. Like the first pathway, it is a straight-through sensorimotor connection; unlike the first, it has no basis in reality, as far as the crucial link *F* is concerned. This was supposed to be formed in the course of learning, and Lashley showed that such direct connections do not exist.

The solid lines A, B, C, D, E and G schematically represent built-in reflex connections, A-B for the clasp reflex of the newborn (stimulate the palm and the fingers close), C-D for the pupillary contraction in bright light. E represents the pathway by which visual stimulation reaches the cortex, and G the pyramidal tract from the motor cortex to the spinal cord.

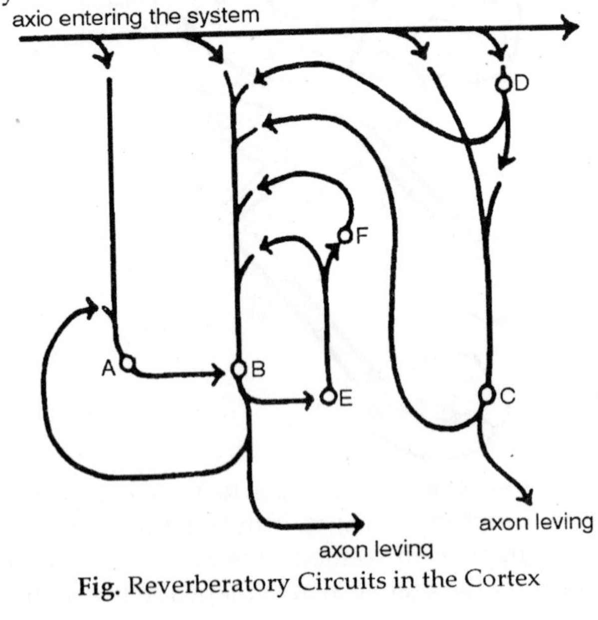


**Fig.** Reverberatory Circuits in the Cortex

Surgical stimulation of the latter path will make a patient's hand move whether he wants to or not. All these effects are automatic.

The broken line F is a theoretical idea of an earlier day, concerning learning: the idea that visual cortex gets a new connection direct to motor cortex whenever a visual stimulus is conditioned to arouse a motor response (as diagramed, a movement of the hand). But such a connection is now known to be fantasy only.

The essential basis of an alternative view was provided by Lorente de Nó, who showed that the cortex is throughout its extent largely composed of enormously complex closed or re-entrant paths, rather than linear connections only between more distant points. A cortex so organized offers a very different theoretical situation.

A single fiber from outside the system excites four neurons, A, B, C, D. A excites B, which re-excites A as well as exciting E. E re-excites B directly and also indirectly via E1, and so on. Two of the four neurons send their axons out of the system, so that the internal activity will have effects elsewhere. Based on a drawing by Lorente de Nó. From Hebb. By permission.

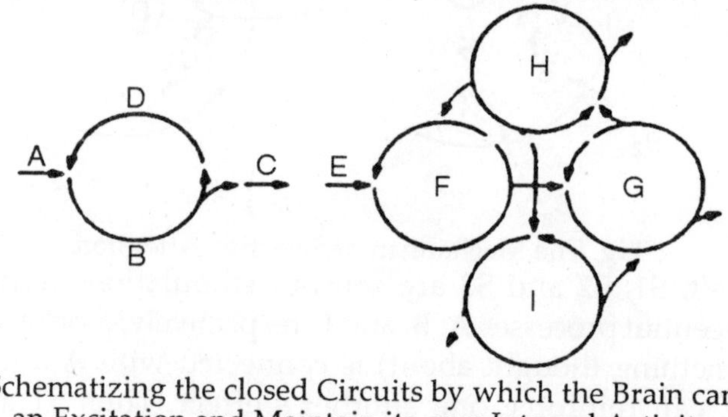


**Fig.** Schematizing the closed Circuits by which the Brain can hold an Excitation and Maintain its own Internal Activity.

Left, the simplest case: a pathway A excites reverberation B-D-B-D-..., which permits a single volley from A to result in prolonged firing by C. Right, interaction of self-re-exciting systems ("cell-assemblies"), with reverberation of reverberations.

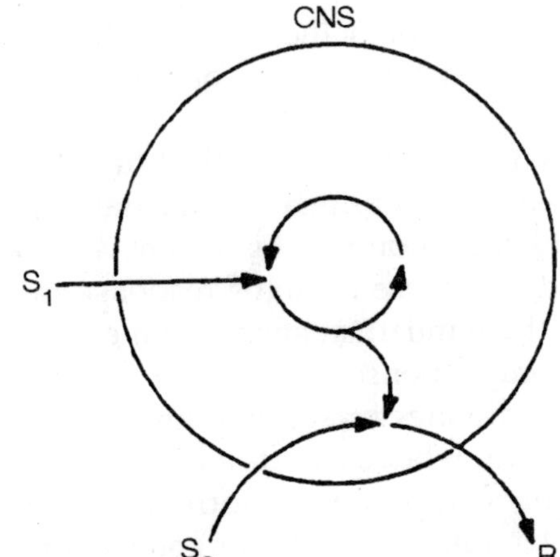


**Fig.** Schematizing the Mechanism of Set

The first of two stimuli, S1, excites a central activity, which by itself is not sufficient to determine the response but maintains its activity until S2 is presented, exciting an immediate response R.

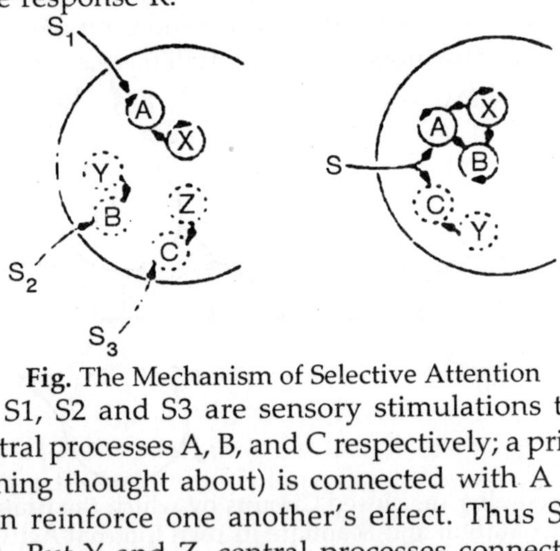


**Fig.** The Mechanism of Selective Attention

Left, S1, S2 and S3 are sensory stimulations tending to excite central processes A, B, and C respectively; a prior activity X (something thought about) is connected with A and so S1 and X can reinforce one another's effect. Thus S1 arouses percept A. But Y and Z, central processes connected with B and C, are inactive, provide no support, so S2 and S3 fail to have any effect (do not enter consciousness). Right, the same sensory event can arouse different perceptions (e.g., at sunrise, one may perceive the sun moving upwards or the earth rolling downwards). In the diagram, stimulation coincides with a central activity X; with that support, S excites the percept or thought process A-B. If Y had been active instead of X, S would have excited C, a different percept.

When an excitation reaches the cortex, instead of having to be transmitted at once to a motor path, or else die out, it may travel round and round in these closed paths and may continue to do so after the original sensory stimulation has ceased. A delayed response to stimulation is now possible and, fundamentally, a mechanism of thought now becomes intelligible, as a complex activity entirely within the brain in which one closed system excites another, that one exciting a third or re-exciting the first, and so on. These central processes of course continue to receive sensory inputs also; most of the time thought processes are in touch with the sensory environment and are influenced by it.

Set and attention then become intelligible also. In the clearest case, set occurs when two stimuli are presented one after the other and the response to the second is controlled or modified by the first. The first "sets" the subject to respond to the second. In the hundred yards' dash, the *ready* signal prepares the runner for the *go* and produces a faster start. The instruction to the subject to add (or subtract) is the first stimulus, the presentation of digits such as 6 and 2 is the second; the response "eight" (or with the other set, "four") is obviously determined by both stimuli, not by the second alone. The effect of the first must be held till the second is presented. The nervous system had no way of holding, and so set was a mystery.

Another mystery was banished at the same time. In 1939, Brogden showed that dogs were capable of associating two sensory events. This too was impossible for the nervous system which is designed to permit sensorimotor associations, but not sensory-sensory ones. Brogden exposed the dog to a light and a sound simultaneously; he then conditioned the light to a response, and having done so tested the dog with the sound—and obtained the same response, though the sound and the response had not been paired.

I have already pointed out, in an earlier discussion, that although the learning theorists had forsworn any interest in the nervous system, they continued in effect to deny the existence of any psychological phenomenon that was inconsistent with the nervous system as it was known to Watson in 1920.

Brogden's results could not be denied, but for twenty years or more there were persistent attempts at explaining them away—although by now it had become easy to see how the central representative processes could become associated, if the two stimuli excited activity in reverberatory closed pathways and not mere straight-through transmissions.

I need hardly say that such ideas were very interesting, with their suggestion of solutions for long-standing problems of behavior. But serious difficulties showed up at once, including my old difficulty with perception, and I did not feel that I had made a "discovery," or that the problems were really solved.

However, the ideas *were* attractive, and I worked with them as a sort of game, to see how far the difficulties might be reduced. The first point was that one of these self-re-exciting systems could not consist of one circuit of two or three neurons, but must have a number of circuits, both because of "subnormality" or fatigue (when one circuit fatigues temporarily another can continue) and to be able to deliver a sufficiently massive excitation to another such system to fire it. I could assume that when a number of neurons in the cortex are excited by a given sensory input they tend to become interconnected, some of them at least forming a multicircuit closed system.

Eventually such a system was called a cell- assembly, so that term can be used here. The idea then was that a *percept* consists of assemblies excited sensorily, a *concept* of assemblies excited centrally, by other assemblies. And this brought me back to the problem of the localization of a percept. The fundamental idea of the theory as it was later published was that repeated exposure to a given sensory stimulation will organize an assembly, and the difficulty I have already referred to now became inescapable. A familiar object is recognizable at different distances, at different angles, and in different parts of the visual field—left, right, above or below the direction of vision.

In each case, the retinal stimulation is different and so a different cortical excitation results. In each case, it seemed, a separate assembly would be required, amounting to a large number for each recognizable object and meaning that visual recognition would have to be separately learned for each direction, distance, and aspect of the object as seen. An improbable conclusion, to say the least. Regretfully, I saw that the idea would not work. It was too interesting to give up entirely, but I could not take it seriously.

The game then was to see whether the general idea could be made less improbable. For one thing, suppose that it is not the whole object that one must learn to see separately in its different aspects but the component parts. The fundamental work on perception in animals had been done with simple figures such as square and triangle, so I asked what the effect would be of supposing that what one learns to perceive is not a triangle as a whole but the three angles or the three sides. At once there was a simplification of the problem.

A large triangle and a small triangle of similar shape have identical parts as far as foveal (central) vision is concerned. The same thing applies when the same triangle is seen at different distances, so it would not be necessary in the first place to have to learn separately to recognize it for each size and each distance.

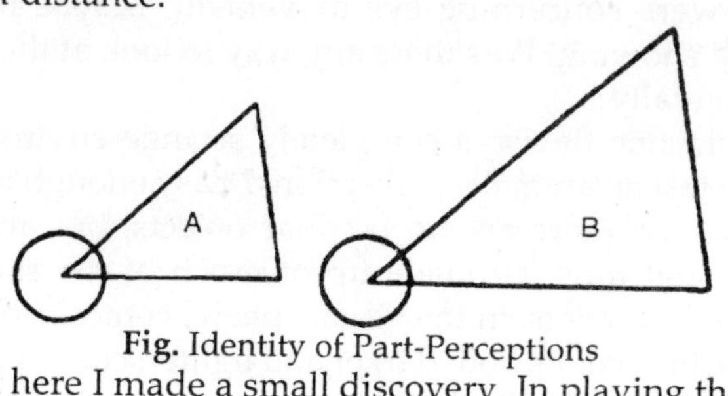


**Fig.** Identity of Part-Perceptions

And here I made a small discovery. In playing the game I was looking at the experimental data from a new point of view and I suddenly realized that the whole literature was biased by the tacit assumption that eye movement is unimportant in visual perception (except, of course, when fine detail must be examined closely).

The small and large triangles, as wholes, cannot excite the same retinal cells at the same time but with separate fixations can have a series of identical effects in central vision (represented by the circles).

Now it is true that the tachistoscope, which presents the visual object for too short a time to permit eye movement, shows that perception of a familiar object is possible without looking at its parts one after another, but this is not true of strange objects. It cannot be true when visual perception is first developing; for the newborn, all visual objects are strange. Even with things as apparently simple as a triangle the infant may have to fixate the three angles repeatedly, one after another, before ever managing to perceive the triangle as a whole.

It was a surprise to see how much the problem was changed by taking account of eye movement and a variable fixation. The cell-assembly notion began to look less absurd. But this was still a game. It still implied that a baby must slowly learn even to see the world, quite apart from learning what things mean or how they behave: I thought, a most unlikely idea. It implied too that an adult would not perceive strange objects when not even their parts were familiar: still more unlikely. However, I had just discovered how biased my notions were concerning eye movement. Maybe my biases were still showing. Was there any way to look at the question experimentally?

Could one devise a completely strange environment in which to test an animal's perceptions? Easy enough to prepare one in which there are no familiar objects, but any livable environment must be made up of much of the same set of perceptual elements: in the visual sphere, contours of variable slope, various colors both darker and lighter, curved and plane surfaces and their intersections; noises of various pitch, complexity, intensity and temporal pattern; roughness, sharpness or roundedness of edges and points, vibrations, and so on: These are components of which any environment is made up and must be familiar, strange as the combinations in which they occur may happen to be.

As I thought about this aspect of the question, the special case of vision came to mind. It is possible to rear an animal in darkness and at maturity expose him to an environment that in one respect at least is totally strange. Nine years earlier in fact I had reared rats in darkenss myself and tested their vision, as the basis of a Ph.D. thesis. When I recalled that work I recalled too a feeling that there was something peculiar, something remarkable, about a book I had read at the time. The book was von Senden's valuable compilation of all the published case reports of people who were born blind, with cataract, and who were later made able to see by removing the cataract. I looked at von Senden again and was astonished at what I found.

His subjects were in effect blind at first, as the game I was playing suggested. They could distinguish and respond to colors but had practically no perception of shape or pattern. (All that was "peculiar" about these reports, earlier, was that they flatly contradicted my beliefs). A prolonged learning process was needed after operation before the patient began to see the world in the way that a normal person does. The learning was so arduous that most of the patients eventually gave up the attempt and returned to living as blind persons.

*Now* my attitude had to change. What I had thought too unlikely to be considered looked as though it was true, and the game had to become a serious investigation. The change was reinforced when Austin Riesen, entirely by coincidence, chose that time to bring out into the light two young chimpanzees who had been in darkness from birth.

We were both working in the Yerkes Laboratories of Primate Biology and the baby chimpanzees had been reared in darkness with the general expectation that perception would be unimpaired by the lack of visual stimulation. They showed exactly the opposite, and to a degree that was astonishing to us all-me least perhaps, because I had been rereading von

Senden, but I was astonished too. For the two animals showed no sign of awareness of the new stimulation to which they were exposed in the normally lighted laboratory. Von Senden's human patients were verbally prepared for what was to happen.

They were immediately responsive when the bandages were removed from their eyes, could perceive that something was there when an object was held in front of them, and could even tell sometimes that a circle and a square were in some way different. There was no sign of this in the chimpanzees. Visual reflexes were strongly marked, but there was not the slightest suggestion of any other response. There was no effect of the visual stimulation on the behavior of the whole animal. There was no attempt to explore visual objects tactually, and no evidence of any emotional reaction to this completely strange experience.

The two young animals, I thought, had formed no visual cell-assemblies whose activity would be perception and which would have permitted thought about the new phenomena and an investigation of them. My scheme had suggested that the genuinely unfamiliar could not be perceived or thought about and now it seemed that I had confirmation, full and running over.

It is entirely fitting, in this story of stumbling discovery, to report that the evidence that converted me to belief, without which I might well have dropped the whole thing, was later shown to be bad. It was certainly not the crucial evidence that it seemed.

The two young chimpanzees were not merely facing a strange environment, they were facing it with defective equipment. Chow, Riesen and Newell eventually found that rearing in darkness causes the death of some of the cells in the chimpanzee's retina (although not in the rat's: K.S. Lashley, interested in the structural basis of rat vision, examined the eyes of the rats I had reared in darkness and found no degeneration).

Again, the eye of a kitten covered with translucent plastic, and thus deprived of patterned stimulation during growth, loses some of its central connections. Either or both of these defects may have been present in von Senden's patients. At the time however I had no suspicion that this was so: There were those who thought than an unstimulated cell might eventually die, in what was called "trans-synaptic degeneration," but their views at the time were not widely accepted in physiology. Thinking that the case for the theory of cell-assemblies was now established, I set out in earnest to see what its other implications were.

There were of course still difficulties. The next one unexpectedly widened the enquiry to take in the topic of emotion. Here I was still concerned with the problem of strangeness. At the earlier stage I might have rejected the ideas with which I was working because they meant that a truly strange object could not be perceived—and then Riesen's evidence and von Senden's showed me that this might really be true, peculiar as it seemed. Now the question concerned the effect of strangeness on the stream of thought.

If thought is a series of cell-assembly activities, these must ordinarily be excited both sensorily and centrally.

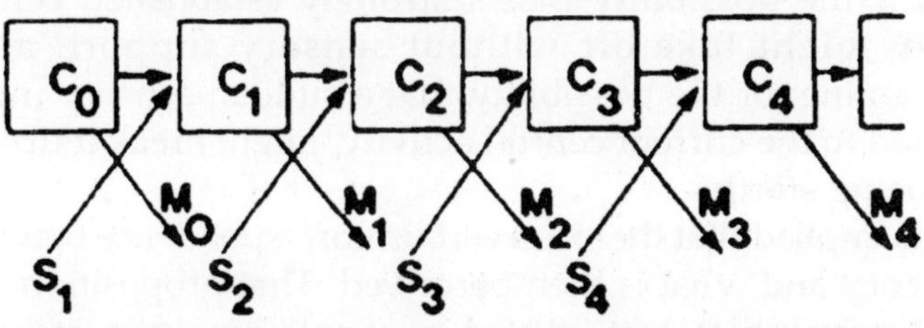


**Fig.** Relating Sensory input (S) to Central Processes (C) and Motor outflow (M). C1 to C4 Represent the whole Activity of the Brain at four Successive Moments in Time.

At any moment that central activity is determined both by the immediately preceding activity and by sensory input; motor response follows. Obviously the response is not determined by sensation alone

When event *A* in the familiar environment has always been followed by event *B*, as perceived, the corresponding cell-assembly a will have been regularly followed by assembly *b*, and therefore will have connections with it and tend to excite it; or if *A* instead has been followed at various times by one or other of a set of events *C, D,* or *E, a,* instead of exciting b strongly, will excite *c, d,* and *e* somewhat less strongly. In this scheme, the facilitation from *a* is the basis of expectancy.

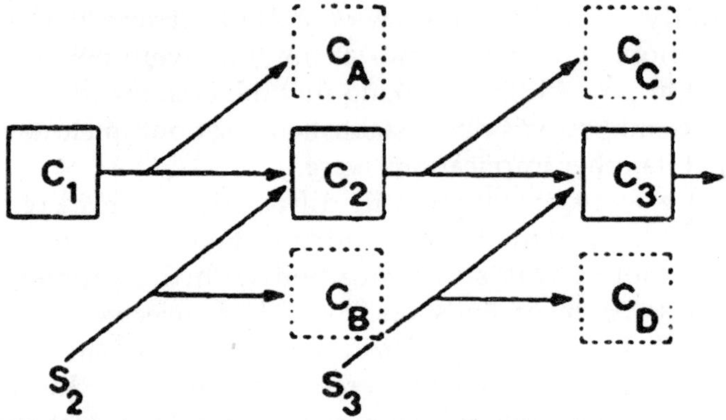


**Fig.** The Relation between Sensation and thought. C, central Process; S, Sensory input. C1 might give rise to Various Sequels; these are Represented by CA and C2. S2 also tends to Excite Different Central Activities; these are Represented by C2 and CB.

Both excite C2, which therefore is the upshot. What is not shown is the possibility that a strongly established central activity might take off without sensory support, as in daydreaming; or the possibility that a sudden sensory input, unrelated to the current central activity, might break it up, the effect being startle

It is implied that there is a certain correspondence between expectancy and what is then perceived. That proposition was quite acceptable, but it had a corollary. In a strange environment, thought ought to be disrupted, or might be disrupted, by the lack of correspondence between central facilitation and sensory input. In short, thought might be disorganized in a really strange environment.

Here, "strange environment" does not merely mean a place where one has never been before, but one in which the ordinary concordances are absent. It must not be too strange, for it was implied that the problem would arise when things were perceived—but perceived in the wrong relation to each other. Again it seemed that the theory was pointing in the wrong direction. One might have trouble adjusting to such an environment but surely one would still be able to think.

Once more, the chimpanzee to the rescue. In the preceding months my experimental work (as distinct from playing theoretical games) had been a study of personality and emotion in the animals of the Yerkes colony. I had stumbled on the fear aroused by a model of a chimpanzee head and related objects, and was completely puzzled by it. The phenomena were obviously significant and must be reported, if only as a reminder of the human reaction to sight of a major operation or to contact with a dead body: a reaction that is persistently avoided or forgotten in the psychological textbooks. I prepared a report of the facts with no attempt at explanation.

The paper was ready for dispatch to the Psychological Review when suddenly I realized what it was describing: a series of situations in which strangeness, a departure from day-to-day experience, was disruptive of behavior (and presumably of thought). The strange situations—a head lacking a body, a model that looked like an animal but obviously was not, the inert body of a chimpanzee being freely handled by the human staff—were not quite what I had been thinking of, and the disruption of behavior was not what I had thought of either: but disruption there was, and plain to be seen.

Now I had a theoretical explanation: Emotion is a disruption of cellassemblies. The explanation had gaps and minor inconsistencies, but it needed only a slight extension to offer, in general terms, a rational and comprehensive account of fear and its near relatives anger, shyness, and depression. Nothing of the kind existed in the literature. Another bonus was that it drew attention to the extent of the phenomena of emotional disturbance, of which the literature at the time gave no hint.

The theoretical account turned out to be grossly incomplete, since I knew nothing at the time of the arousal system nor of the limbic system, both of which are intimately concerned in emotion. It may be that the disruption of cell-assembly activity is only a precipitant of emotional disturbance instead of being itself the emotion, but this theory still offers the only relatively complete account of the conditions giving rise to fear, anger, and so on, and the only suggestion about what they have in common. In addition to the conflict between expectancy (including desire) and what one perceives, the simple absence of accustomed sensory input might disrupt cell-assemblies that are excited centrally but are dependent also on sensory excitation for stable functioning.

This would account for a number of "deficit reactions," such as the three-year-old's fear of the dark and the depression of loneliness or homesickness in older subjects. Finally, malnutrition and endocrine change could disrupt, a changed blood chemistry disturbing time relations in the assembly. Here would be comprised the neurotic reactions of U.S. servicemen in the second World War who did not eat enough salt to compensate for heavy sweating in the tropics, the psychotic fears described in cases of pellagra (lack of B-complex), and the depression of the menstrual cycle. None of these conclusions could be rigorously deduced from the theory, but they were possible inferences, and they introduced order into a topic where no order existed previously. At this point the theory seemed to be ready for launching.

## EXPERIMENTAL TESTS: THE EARLY ENVIRONMENT

The main validation of the theory was found in three sets of experiments. It is fine to have found a rationale for set and attention and to discover how the varied phenomena of emotional disturbance might be fitted into an organized theoretical scheme, but the case for the actual existence of cell-assemblies became much more firm with the support of new experimental data.

The first of the three investigations concerned the effects of the early environment on intelligence and thought, the second, its effects at maturity, and these two are discussed here. The third, which concerned perception, will be left till the following chapter.

The theory proposed that thought processes (and therefore

the operations of intelligence) consist of an activity of cell-assemblies, and that these assemblies are organized in the first place by sensory stimulation from the young animal's normal environment. The first experiment therefore was a test to see if animals that had had a good environment during growth would be better problem-solvers at maturity than ones from a poor environment.

A preliminary test was made with two litters of pet rats, compared to cagereared littermates. From the point of view of the theory, rearing in the small laboratory cages must fall short of an adequate perceptual stimulation; the pet rats on the other hand, taken home to be reared by my two young daughters and having the run of the house much of the time, had a much wider exposure and should be more intelligent.

At maturity both the laboratory group and the pets were tested with a rat-intelligence test that had been developed by Kenneth Williams and later improved by Rabinovitch and Rosvold. The results of this preliminary work looked good and a more formal investigation was set up by Bernard Hymovitch. He devised for some rats a radically restricted environment (but with all the comforts of home, apart from boredom) and an enriched or "free" environment for others, in a large wooden cage containing a number of objects to make what he called a Coney Island for rats. The principal results agreed with the theory: The rats reared in restriction were greatly inferior to the free-environment rats. But two further results were of almost equal interest.

Hymovitch also showed that the environmental effect occurs only during the period of growth. One group of rats, reared in restriction, were put for an equal period of time into the free environment, as adults; another, reared in the free environment, were restricted as adults.

Both were then tested and the second group were much better than the first. In other words, the effect of the early environment on intelligence was not reversible. This closely parallels the human data. The environmental influence on the IQ is limited essentially to the period of growth, ending somewhere between the ages of 12 and 18. A favorable environment thereafter cannot reverse the damage that has been produced by an unfavorable environment earlier.

The other extra in Hymovitch's experiments was a demonstration of the extent of visual-perceptual learning that the rat is capable of. One group of animals was reared in small wire-mesh cages that let their occupants see the outside world but permitted no locomotion. The rat could turn around and stand up or lie down, no more. The cages were placed in daily rotation in a series of places in the experimental room and in the free-environment cage so as to give the animals an extensive visual experience of the environment but no experience whatever of exploring it.

The wholly surprising result was that these animals were as good in solving maze problems as the free-environment group (which could explore but did not have as wide a visual experience) and of course much better than the restricted. Skeptical of this result, Forgays and Forgays repeated the experiment and showed that the mesh-cage group were not quite as good as the free-environment group, but still confirmed Hymovitch's results of extensive benefit from a purely visual experience.

The dog is nearer man in psychological complexity. If the results of the rat experiments were to be relevant for man it was of value to see if the same results would be obtained with the higher animal. Thompson and Heron and Melzack and Scott reared Scottish terriers in two ways: in restriction, and with the degree of freedom that a pet dog usually has. The restricted dogs during the whole period of growth were in small single cages where they could hear and smell other dogs and the human occupants of the lab, but could not see or touch or be touched by them.

Whereas an adult dog would be miserable in such circumstances, being used to more freedom, these dogs knew nothing else and in the words of an experienced dog handler were happy as larks. Their littermates were given as pets to members of the university community and their friends, with the understanding that they would be recalled on occasion for testing in the laboratory. The effects of restriction were more extreme than in the rat experiment. Thompson and Heron showed that the restricted dogs were inferior to the pet group in the solution of maze problems and in a number of insight tests devised especially for the investigation. They were untrainable, abnormal in personality and social behavior in ways that were hard to describe, and could not be bred, though physically they were splendid animals.

Melzack and Scott's research dealt with the most striking peculiarity of all. The dogs reared in restriction were unmindful of pain. The reflex response was normal but the emotional or motivational component was missing. It may be recalled that after lobotomy, for intractable pain, the human patient may say that the pain is the same—but doesn't bother him any longer. Thus pain must have two components, a recognizable sensory quality and, for normally reared dogs and normal people, an accompanying strong emotional reaction.

The restricted dogs acted as if lobotomized. One of them saw a lighted cigar fall on the floor and thrust his nose into the glowing coal to investigate it. He pulled back reflexively, but touched the coal with his nose twice more. In experimental tests, electric shock elicited the same reflexive withdrawal, but produced no subsequent avoidance of the electrically charged object. Clearly, the motivational and emotional qualities of pain stimuli are a function of a normal infant experience.

## THE ADULT ENVIRONMENT

The animal experiments showed in brief that the normal growth of the mind requires the sensory stimulation of the normal complex environment. The next study, using human subjects, showed that the integrity of the mind at maturity continues to depend on that stimulation.

This was the so-called sensory-deprivation experiment—which, since it did not deprive the subject of sensory stimulation but did prevent perception, might better be thought of as a perceptual-deprivation or perceptual-isolation experiment.

The experiment was done by Woodburn Heron and

collaborators, W.R. Bexton, B.K. Doane, and T.H. Scott. They hired college students to do nothing—or very little, such as answering a few questions at testing time—for 24 hours a day. The subjects were well paid to lie quietly on a comfotable bed; wore translucent but not transparent goggles; heard a constant slight buzzing sound from a small speaker in the foam-rubber pillow except when the experimenter was on the line with a test question; and wore long cardboard cuffs from the middle of the forearm to beyond the fingers.

Thus they received much sensory stimulation, but all of it unpatterned except in the tests or occasional requests for information. Excepted also are the breaks in which the subject ate or went to the toilet, although he wore the goggles over his eyes throughout. It was estimated that on the average 22 hours a day were spent in full restriction.

The main results: The subject's ability to solve problems "in his head"-mental arithmetic problems, anagrams and the like that could be attempted while in restriction—declined from about the end of the first day onward, and coherent thought became harder to maintain. Some of the subjects entered the experiment with the intention of planning future work or reviewing mentally for examinations. The reports were consistent that this soon became impossible, and there were even complaints that it was hard or impossible to daydream. Beginning about the third day there were increasingly complex visual hallucinations as well as a few somesthetic ones.

For some hours after emerging from the experiment, visual perception was also greatly disturbed. The experimental subjects were more vulnerable to propaganda than the control subjects who were hired to listen to the same stuff, advocating belief in ghosts and poltergeists as well as clairvoyance and telepathy. (The message was that scientists who don't believe in ghosts or ESP are biased and refuse to consider the evidence; both groups were affected by it, but the experimental subjects much more). And finally, the experimental subjects showed a 10% slowing of the EEG, a most surprising result, demonstrating a physiological change of brain function as a result of a mere lack of pattern in the input from the sense organs.

As far as theoretical predictions are concerned, the prediction that intelligence would be impaired by restriction in infancy was reasonably specific and was confirmed.

The personality change in dogs was not predicted, nor the relative insensitivity to pain, but these results strengthen the more general proposition, quite consisent with the theory, that the early environment has a vital organizing function in mental activity at maturity. Similarly, the specific results of the perceptual-isolation experiment were not predictable, but their general tenor supported the prediction that was made, that such conditions would have a disorganizing effect on thought.

But it is important to say also that in some respects the theory was contradicted by the results. The students who entered the isolation experiment showed no impairment of problem-solving ability for some 24 hours, and other abnormalities appeared even later. The theory has no way yet of accounting for that delay. Similarly, the subjects showed a good deal of unanimity in reporting after isolation that it took 24 to 36 hours to recover motivation and to resume normal work patterns.

Together with the results of the stabilized-image work reported in the following chapter, these experiments show that the theory is of the right kind but has far to go. The experiments with stabilized images reported in the next chapter show, I think conclusively, that cell-assemblies exist, but the isolation experiment described here shows that there is much that we do not understand about how they function.

# Chapter 33

# Perception and Cognitive Development

The classical conflict between ideas about perception and ideas about learning. It seemed that a perceptual process could have no specific locus in the brain; but perception is a source of learning and establishes memories, and for this reason, it seemed, *must* be localized. Also, perception gives rise to specific responses and so must transmit excitations to specific efferent pathways.

It seemed therefore that some way had to be found to localize the perceptual process, theoretically, and this necessity was a main determinant of the proposed characteristics of a cell-assembly.

I have also said that the result was distasteful and violated my pre-existing ideas about perception, although I could find no conclusive evidence to oppose the new theory or support the pre-existing ideas.

Now I come to evidence that shows that cell-assemblies do exist and function in perception as the theory proposed. Defects in the theory also show up, but suggest new possibilities for cognitive learning.

The result is to relate learning and perception more intimately. To simplify, perception in the early stages is a consequence of a primitive learning process, but thereafter learning becomes a function of perception—the learning, that is, that is most characteristic of the adult human being—and a function of cognitive structures that originate in perception.

## THE STABILIZED-IMAGE EXPERIMENTS

My doubts about cell-assembly theory were abolished by a brilliant set of experiments by Roy Pritchard and Woodburn Heron that opened a new page in the study of perception. They put a decisive limitation on certain holistic ideas from Gestalt psychology but also gave new and stronger support for other Gestalt propositions.

The background of the research is as follows. There is always some tremor of the muscles that control the movement of the eyeball in its socket, and consequently some slight movement of the image, on the retina, of what is looked at, even though one tries to keep one's gaze fixed. R. W. Ditchburn and L. A. Riggs independently discovered about 1952 that this slight movement of the image on the retina is necessary to normal vision, for when they managed to stabilize the image of a line crossing the visual field the line promptly disappeared, reappearing intermittently.

When I heard about this, I saw that it was an obviously important discovery concerning visual sensation, but it was not related, apparently, to my ideas about cell-assemblies. But then I heard several years later of an experiment done by Pritchard in Ditchburn's laboratory in England. A somewhat more complex object was looked at instead of a single line, and it did not disappear all at once; instead, different parts disappeared and reappeared, and what part disappeared was affected by what the observer attended to.

*This* was a different matter, with a direct bearing on the theory. The theory said that attention is a selective support for some sensory inputs and not others, thus determining what is perceived at any particular time. Disappearance of part of the object must mean that the assembly concerned, or group of assemblies, is fatigued or inhibited and has stopped firing. *When* this happens must be affected by the supporting activity of other assemblies.

Theoretically, the support might prolong activity, or by increasing the activity within the assembly might hasten the onset of fatigue (my informant had reported that the part of the object attended to disappeared sooner; in our own subsequent research this effect was sometimes obtained, but more frequently attention had the effect of keeping the part-object visible longer). Here was a way of getting direction information about assembly activity and the interaction of assemblies.

When Pritchard joined our research group at McGill he and Heron made a further development of a technique that Pritchard had already used in England. The target—the object to be observed—was attached by a short stalk directly to a contact lens on the observer's eye with a tiny collimating lens to let the target be seen as if at a comfortable distance. The targets were mostly line diagrams; a drawing was photographed and the transparency was trimmed to a size that would fit into the optical system on the contact lens.

The main problem of adaption for the subject was to learn not to be surprised when an object or part of an object in clear view suddenly disappeared or when something equally suddenly reappeared all at once—not gradually, but as a whole—since the result of surprise was a jerky movement of the eyes that disturbed the position of the contact lens and so destabilized the retinal image. I will not try to report all the dramatic results that Pritchard and Heron obtained before I was invited by them to have a contact lens fitted so I could observe them myself.

**Fig.** Sketch of Contact lens and Optical Apparatus Mounted on the Eyeball of the Reclining Observer

The wire taped to the forehand of the subject connects with a Tiny lamp to illuminate the Target A sample of those that gave direct evidence of the activity of cell-assemblies is found in Figure. With outline drawings, straight lines without intersection acted, in perception, as units, disappearing or reappearing all at once.

This is in full agreement with the proposition that the perception of a single line is the activity of a cell-assembly, as a system that is active as a whole or else inactive. This essentially is the evidence that demonstrates the existence of cell-assemblies.

In Figure on the other hand appear results showing that the theory of how assemblies work is not satisfactory as it has been presented. For example, the completion in perception of an incomplete triangle is stronger support for *Gestalt* ideas of this kind than any previously available, but the fact that the completion is thinner, less vivid, than the rest of the line as perceived, conflicts with the idea that the perception of a line is the activity of a cell-assembly, without qualification.

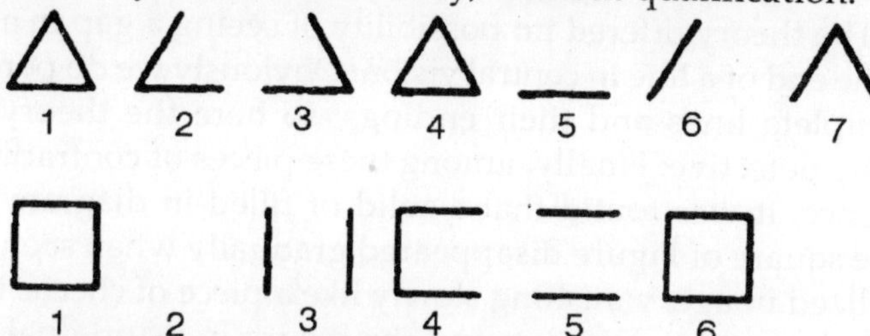


**Fig.** Two sample Sequences of Patterns Perceived with Stabilized Images; above, Triangle as target; below, Square.

The sequence of course is variable from one time to the next. The blank space in position 2, below, signifies complete disappearance of the square, followed by partial regeneration

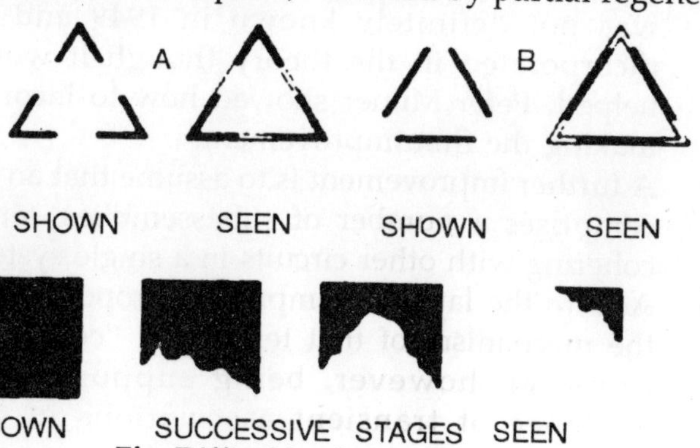


**Fig.** Difficulties for the Theory

Above, A, the target triangle with incomplete sides (left) is perceived complete but the completions are fainter than the rest of the triangle; B, similarly when it is the apices of the triangle that are missing. The completions agree with theory, but the faintness does not. Below, the square disappears gradually; this too is inconsistent with the theory. The postulate of subassemblies may remove the difficulties.

If the activity is all-or-none the line as perceived cannot vary in faintness or vividness. The fact of completion supports the cellassembly idea, since the idea was that the assembly activity meant perceiving a line of indefinite length running continuously across the field, but the loss of vividness does not, and now another difficulty appeared: How can an *incomplete* line be seen?

The theory offered no possibility of seeing a gap in a line, nor the end of a line in central vision. Obviously we do perceive incomplete lines and their endings, so here the theory was grossly defective. Finally, among these pieces of contradictory evidence, it was found that a solid or filled-in diagram such as the square of Figure disappeared gradually when seen with stabilized image, vanishing slowly like a piece of cheese being nibbled away by a mouse, and the theory in its original form had no way of dealing with that effect.

However, the theory is improvable. The defects are not fatal.

- In the first place, the existence of a neural inhibition was not definitely known in 1949 and was not incorporated in the theory though it would have helped. Peter Milner showed how to incorporate it, making the first improvement.
- A further improvement is to assume that an assembly comprises a number of subassemblies, temporarily cohering with other circuits in a single system;
- Add to the latter assumption a proposal concerning the mechanism of that temporary "coherence"—the proposal, however, being supported by other evidence of transient associations in cognitive learning.

In this respect, the need of improving the theory with respect to perception leads to a way of dealing with a long-standing but disregarded problem of the learning and memory that result from momentary stimulations. We will return later to the first point concerning inhibition, which may be as essential to intellectual activity as Sir Charles Sherrington showed it to be for reflex activity. It seems that the difficulties just discussed can be dealt with by assuming that the perception of a line in the visual field depends on the excitation, by the line, of an elongated narrow set of subassemblies in the striate cortex (the primary visual area), as well as excitation of peristriate circuits activated by that particular combination of subassemblies and helping in return to maintain their activity.

It must be assumed, if this is to account for the facts, that the subassembly has a very small extent in terms of visual angle, bordering on the microscopic in relation to the extent of the visual field. Variations in vividness or intensity of the line as perceived result from differences in the proportion of subassemblies that are active within the cortical region that the line excites.

The perception of finer or coarser lines becomes intelligible in the same way. And so with the perception of an imcomplete line, with associated implicit motor activation of the eye muscles or actual eye movement. And, last of the difficulties listed above, the gradual fading of solid figures such as the square of Figure becomes intelligible, as the density of excitation—the proportion of subassemblies that are active in the stimulated area—decreases with inhibition or fatigue.

The third point is the mechanism by which subassemblies and the peristriate circuits cohere to form a single system: transient, but momentarily functional. The proposal in brief is that any discrete cortical or corticosubcortical organized activities excited together—assemblies or subassemblies —tend to be reciprocally exciting, forming one larger pattern of activity because transmission of excitation at the synapses connecting them tends to keep the synapses "open" and active for some short period of time. On this basis, subassemblies that have often been active together, and have excited peristriate circuits connecting them, may constitute a unified cell-assembly whose activity is ended only by inhibition.

## COGNITIVE ASSOCIATION AND INHIBITION

Now to consider that third point—the immediate but temporary cohering of subassemblies—and some related phenomena of cognitive association. The commonest form of human learning is the immediate one-trial association of one perceptual event with another or with the ongoing cognitive background activity such as an awareness of where one is. The associations that psychology has emphasized so much—and to which, as a basis of higher behavior, such strong objection has been made—are ones that depend on repeated exposure, as in conditioning.

The one-trial associations we are concerned with here are quite different in that respect and must depend on a different mechanism. A glimpse of a clouded sky or a word or two from a radio weather report is enough to let one know that there is no hope of sunbathing, and one does not have to rehearse a verbal response ("it's cloudy, it's cloudy, it's cloudy,...") to establish the learning.

In the news is a reference to a person or event of no interest at the time, but a few minutes later someone tells you something that makes it of interest and the reference is recalled. Learning took place. Or you see a car and later can report its color or the direction it was being driven. The information mostly is not retained long, but perception is constantly providing you with *knowledge,* transient or lasting.

Obviously, it is not setting up specific response tendencies—you are not learning then to *do* something—but makes a modification of cortical function that means, in turn, modification of the way you will respond in any of an indefinite number of possible situations in the future. As I have said, this is the commonest form of adult human learning. It occurs constantly during waking hours, and it has been totally disregarded by learning theory.

Laboratory-style control of the phenomenon is possible with the repetition of digits, which is not as well understood as the lack of theoretical interest in it might let one think. Ask an experimental subject to repeat a set of digits such as 1-2-3-6-4-5 and he does so with ease; but how does he do it, in view of all his past practice with 1-2-3-4-5-6? That counting sequence must be highly overlearned, the synaptic modifications on which it depends must be strongly established, and yet the new sequence is organized instantaneously, though usually not permanently, of course. Apparently in the same way, you see a pencil lying on the shelf and associate the two perceptions, of pencil and of shelf, so that ten minutes later when you hear someone ask "Where's the pencil?", you are able to answer.

The repetition of digits in a unfamiliar order may be possible only because the cell-assemblies concerned became temporarily connected when they are excited in close temporal contiguity, and remembering the whereabouts of the pencil similarly because of a connection between assemblies constituting the percept of the pencil and those whose activity is the awareness of its locus, solely because they are simultaneously active. What explanation is there?

In 1949 it was considered that a neuron fired once would be harder to fire a second time, and consequently, that transmission at the synapse would increase in difficulty with repetition. The use of microelectrodes, directly recording activity of individual neurons in the central nervous system (a great body of research that began in the early 1950s), showed instead that a neuron fired once is more likely to fire again, and synaptic transmission temporarily facilitates further transmissions.

This makes intelligible, physiologically, an immediate temporary coalescence of activity in two or more cell-assemblies or assembly groups. It is implied that there must be multiple synaptic connections available, ready to be activated, but this must be the result of a complex early experience. It is consistent with the fact that the transient associations are possible only with familiar materials (it is easy to repeat a few words or the names of digits in your own language, but not when foreign words are used).

This makes a fundamental change in the concept of the cell-assembly itself. There is a new problem to deal with if the assembly, once activated, tends to go on firing. Here inhibition enters the picture, with a contribution to make to the level of cognitive functioning.

The same microelectrode methods that showed continued firing by the neuron also showed the existence of inhibition as a direct suppression of activity in one neuron by the activity of another. At least some of the closed loops of the cortex must have the function of suppressing activity in the system, not maintaining it. I have already referred to the possibility that the high efficiency of the human brain, in comparison with its weight, may be the result of the great frequency of Ramón y Cajal's short-axon (or local-circuit) neurons that appear to be inhibitory.

They may have the function of promptly shutting off a cell-assembly once it has acted to fire the next assembly of an internal sequence or, once it has made its contribution to a motor act, controlling the duration of firing in the separate sets of efferent fibers leading to motor centers. A maintained psychological state or process (an image or a concept) is more likely to be a function of a closed cycle or reverberation among a set of assemblies than of reverberation within the assembly itself.

Thought is a sequence of activities. When one of the events in the sequence has done its job of exciting the next it should cease promptly; otherwise it must be disruptive, like an actor who keeps on repeating his speech after it has provided the cue for the next actor's speech. If *A* is to fire *B* and *B* fire *C*, and *A* keeps on firing *B* after the time for *B*'s function has passed, it must prevent a smooth flow in the sequence of events.

At the least, it must make thought woolly and inefficient but at the worst—if this happens widely throughout the cerebrum—must grossly disorganize the thought process. A built-in inhibition to shut a cell-assembly off as soon as it has fired will prevent that possibility, and it becomes intelligible that the remarkable power of the human brain is accounted for by the"prodigious abundance" of short-axon neurons that so impressed Ramón y Cajal, if these neurons are inhibitory or if many of them are. It is also possible, incidentally, that these small unmyelinated neurons are more vulnerable than others to anoxia, especially at birth, and that the syndrome of "minimal brain damage" or "dysfunction" (where behavioral disorder in the growing child makes birth injury seem likely but no specific loss can be demonstrated) is one result. It is suspected that eidetic imagery is a product of a slight degree of brain damage, and eidetic imagery is the vivid reinstatement of a percept, briefly, after the object perceived has been removed from sight.

I also find personally that with advancing age I am bothered more by the phenomenon of words "running through my head," in the familiar phrase, or the persistent return of my thought to a topic I would as soon forget till I can do something about it. This is a common phenomenon but it has become worse with age, and that may be the result of a selective loss of inhibitory neurons: possible evidence that they are more fragile.

## COGNITIVE DEVELOPMENT

The newborn infant starts out with a well-developed reflex capability, which may give a false impression of conscious awareness. Fantz showed for example that when the baby's eyes are exposed to a diagram in an otherwise blank background, he will focus on the diagram rather than the blank area and with different diagrams will focus oftener and longer on one than on the other.

This looks as though the baby is perceiving, and prefers some diagrams to others. But that does not follow. The eyes move reflexively in response to retinal stimulation and the behavior does not necessarily show anything except that a diagram is a more effective reflex excitant than a plain surface, and that some patterns are more exciting than others.

For some psychologists the emphasis on learning in cell-assembly theory is unacceptably empiricist, with too much talk of learning and too little recognition of the innate. But that emphasis is a strategy, not the expression of a conviction. With reasonable certainty, one can say that eventually we will know as much about the operation of genetic factors in the determination of behavior as we do about the operation of experience, but that day has not yet come. An essential source of information about the innate factor is finding out within what constraints learning operates, what is learned easily as well as what needs no learning, and what can be learned only with difficulty or not at all.

Cell-assembly theory supposes that cognitive development begins at zero, at birth or conceivably somewhat before birth though to no great extent. But it does not give experience a blank slate to write on (or to develop the metaphor, if blank, the slate has grooves and indentations that make the appearance of some letters inevitable, other letters impossible).

In the case of visual perception, what assemblies develop and become the basis of perception is fully dependent on an innate property of the organism, the reflex responsiveness of the eye muscles, as well as on the innately determined structure of the striate and peristriate cortex which—according to the theory—is what makes the formation of cell-assemblies inevitable, given exposure to a normal visual environment. To take this further, learning is involved in the functional relation between parallel lines that makes it possible for larger and smaller similar triangles, or similar triangles in different parts of the visual field, to be perceived as similar—and to evoke the same naming response, for example—since the triangles are composed of three sets of parallel lines.

But this is not empiricism any more than nativism. The learning is inevitable in view of the responsiveness of the eye to peripheral stimulation, and the fact that the resulting eye sweep causes parallel rows of excitation across the retina wherever there are points of light in the visual field. According to the theory, it is this relation between the organism and its environment that produces a functional identity of horizontal lines anywhere in the field, or an identity of vertical lines.

Certain kinds of visual learning are determined in advance, by the way the organism is constructed, and recognizing this puts no more emphasis on experience than on innate properties. Another example of great interest has to do with the development of language. Condon and Sander have shown that the newborn baby is selectively responsive to the rhythms of human speech, innately prepared for and impelled toward a form of learning that, we know, is inevitable in normal conditions: that is, learning to talk. Exposed to the sound of speech, there is a correspondence in "microanalysis" (taking behavioral segments of about 1/10 sec duration) of fine movements by the baby with spoken language, whether English or Chinese. No such correspondence was found with the sound of tapping or isolated vowel sounds. The baby is built to react to verbal stimulation, *predisposed* for learning language.

Such predispositions are bound to be hard to demonstrate directly. Holloway's paleological evidence suggests an early divergence of the evolutionary path of brain structure toward the distinctive human pattern, and the known facts of that pattern—lateralization and localization of verbal function, usually in the left hemisphere—clearly mean the existence of an extensive innate factor in language, but it would certainly not be possible to show this experimentally and there is little likelihood of being able to observe it naturalistically, by direct observation.

It is hard to see how remarkable things are that we all share. There is nothing extraordinary about language-any baby can learn it, where's the problem?—and it took a Chomsky to make us realize how great in fact the problem is. It might have been seen at once by a visitor from outer space, with a different cognitive structure; or analysis of the language-learning of human infants by human adults might have been made easier if the analyst had some babies from outer space for comparison, with aptitudes as complex but not the same as those that make language-learning easy for the human baby. If babies are all so to speak equipped with slates that have special grooves and slots for auditory experience—innate modes of receiving certain classes of sounds and fixing and relating them to other experience—then anything that experience writes on those slates will have certain features in common. But the babies' elders started out with the same slots and they must be hard to discover, for the learning they control is so "natural" that its peculiarities will be evident only by the light of theory, slowly developing.

Let us return to the cell-assembly. The same innate cortical structure and function that permits postulation of the formation, by a learning process, of the elementary assemblies that have been discussed so far make it possible also to postulate the existence of higher-order assemblies whose activity is the basis of abstraction and generalization in thought.

What I have talked about so far is the first-order assembly whose activity is the awareness of such perceptual elements as a contour in the visual field, a pressure on the skin at some point, a simple taste, a noise of particular pitch, and so on. When one of these events happens repeatedly, those neurons that are often excited by it will organize to form an assembly. (In vision, there is reason, as we have seen, to think that the assembly comprises subassemblies, and the same thing may occur elsewhere. For example, it could account for differences of intensity in the perception of a note of given pitch or of pressure at a particular point on the skin).

When the theory was in the course of construction, about 1945, I realized that the facts on which the cell-assembly was based also implied the possibility of what I then called a superordinate assembly. When a group of assemblies are repeatedly activated, simultaneously or in sequence, those cortical cells that are regularly active following that primary activity may themselves become organized as a superordinate (second-order) assembly. The activity of this assembly then is representative of that combination of events that individually organized the first-order assemblies.

After a baby has developed assembly activities for lines of different slope in the visual field and then is repeatedly exposed to some triangular object—to take the simplest possible example—the combinations of activity of the three assemblies may lead to the formation of a higher-order assembly whose activity is the perception of the triangular object rather than its three sides as such. This might be a second-order assembly if the object is seen always in the same orientation, or a third-order assembly formed in the same way by the repeated excitation of second-order assemblies, fired by seeing the object in different orientations. This elaboration of the theoretical scheme seems highly speculative, but it has physiological support. There is no point in trying to elaborate too much detail, however, because the possibilities are too varied.

For example, it seems that superordinate perceptions are likely to develop from the stimulations of such a familiar object as a toy animal or mother's face or hand before any such artificial idea as that of a triangular shape develops, although for purposes of exposition it was possible to be clearer with that example. My concern here is to show at least in principle the possibility that higher-order assemblies are the basis of abstract ideas. And in that attempt a most effective support comes from the work of Hubel and Wiesel, who have demonstrated physiologically a hierarchy of neurons in the visual system.

Simple cells are ones that are excited directly by a line or contour of a specific slope in a specific part of the visual field; complex cells are excited by simple cells; and hypercomplex cells are excited by complex cells. A complex cell is fired by any of a number of simple cells that vary somewhat in locus.

Thus its activity is generalization on a small scale, since it makes the same response to different sensory stimuli; and it is also abstraction, since there is transmission of only one kind of information about the sensory stimulation (i.e., slope of line is included, locus is omitted). The parallel of simple, complex, and hypercomplex (and at least one of still higher order) with first, second, third, and probably fourth-order cell-assemblies is obvious. Hubel and Wiesel's research makes the existence of higher-order cell -assemblies a considerably less speculative idea.

From all we know, this is the kind of complexity that is demanded of the brain of the growing child. Scott regards as conservative an estimate that one billion ($10^9$) cell-assemblies are formed in a normal human brain in the course of ordinary experience. That may be a conservative estimate, but it represents the formation of assemblies at the rate of one per second for 30 years, sleeping time included, or for 45 years counting only 16 hours per day. This it appears is what makes possible the normal development of intelligence and the human learning capacity, including the capacity for abstracting and generalizing, and being able to see the wood instead of the trees—seeing the larger pattern as well as the detail—that is fundamental to human thought.

## PERCEPTION IN THE DEVELOPMENT OF LANGUAGE

This is not the place, and I am not the author, to attempt any systematic treatment of language. Nevertheless, the theory of cell-assemblies has clear implications concerning language-learning and I must point to some of these and, if I can, go far enough into the topic to interest the psycholinguist in its possibilities.

Hebb, Lambert, and Tucker dealt with the question with the aim of avoiding the extremes of nativism on the one hand and empiricism on the other. What follows is based on that discussion, which built largely on perception and the memory image that results from perception.

Three main points: Elaborate perceptual learning must precede both comprehension and use of words; a general concept of a name (nounness) is intelligible on the basis of a relation between heard names and seen objects, and similarly with action words (verbness); and with imagery a basis of active-passive transformations becomes understandable.

*First,* in learning to name an object, the theory does not allow us even to consider the beginning of language as a direct association between the sight of an object and the motor response.

The child must first learn separately to *perceive* both word and object, with the development of auditory and visual cell-assemblies of lower and higher order. Now a connection becomes possible between the auditory perceptual complex excited when the mother says "doggie," for example, and the complex excited by the sight of a dog. This is not S-R but S-S learning—not stimulus-response but a sensorysensory or more accurately a perceptual-perceptual association.

But we must take account of the response capacity also. To provide for the child's ability to say "doggie" himself we must assume that the basis for motor learning is being laid at the same time.

As the child hears his own vocalizations, more or less random at first, connections are built up between the resultant auditory stimulations and the concurrent vocalizations: a threadbare idea but still possible. As the speech of adults becomes familiar and the child begins to perceive the constituent phonemes and phoneme combinations, that auditory stimulation must have a shaping effect on the child's own vocalizations and make them approximate closer and closer to the adult patterns.

The result is a tendency to repeat a spoken word and, because of the S-S association referred to above, to name a seen object—to the extent that object and name are already familiar. That circular-reflex type of explanation looks like the revival of an idea long dead and buried. All that saves it from banality is the accompanying notion that the process is effective only in combination with the proposed extensive perceptual learning.

This is the problem of imitation; instead of making imitation the key to language-learning, the proposal shows that imitation itself must be accounted for first, a product of a prolonged period of mostly latent learning whose first effect is comprehension, well before the appearance of talking.

*Second,* consider the concept of a name and what Slobin has called "nounness". Fundamental to this is the generalized idea of a thing, an object, a space-occupying and sense-stimulating *something,* as the activity of a higher-order cell-assembly made up of neurons that are usually or always active in the relatively small number of different situations of infancy when a visible, tangible object attracts attention.

Those neurons must be a small proportion of the total number excited on any one of such occasions but may still be a large number in absolute terms. The theoretical possibility of such an assembly is clear and the psychological support for its existence is also clear.

The highly abstract idea of something, a thing or event with little specification other than its existence and perhaps its function or effect, is a thoroughly familiar one and an important component of thought in many situations. It appears early ("What are those *things*, Mommy?") and is common at maturity; consider such terms as thingummybob, *je ne sais quoi*, and gadget or gismo, this last definable as an ingenious something with which to do something or get some effect.

Parallel with that idea of a thing is the idea of a name. Many neurons are excited every time the mother's vocalization coincides with the appearance of an attention-getting object or with some action by the mother that draws attention to a particular object. The small proportion that are common to these excitations, organized in an assembly and concurrent with the assembly complex that is the idea of a thing, form the idea of a name.

Now the child perceives the mother's vocalization both as a particular word related to a particular object, and as a name. In the same way the general idea of something happening—organized activity of neurons common to perceptions of things moving or changing—is paralleled by generalization from hearing the accompanying words, such as *run, cry, fall, drink*.

The child develops meaning for the individual words but also develops the higher-order concept of a reference to (observable) activity. This is the first stage of the development of nounness and verbness.

In the second stage there is a further generalization. Familiar with "Bobby is crying," "Bobby is running," and so on, the child hears a new word not associated with observable action but in a familiar verbal setting: "Bobby is sleeping," and *sleep* is heard as an action word even though no action is visible. Besides exciting the idea or image of the immobile Bobby, as seen, it excites the higher-order cell-assembly that has already been built up by *run, cry* and so on—the abstract idea of verbness, now generalized. Note however: The theory says that what assembly group is excited by a particular sensory stimulation is a function also of the context. In another verbal context ("Sleep is good for Bobby," for example) the same word will acquire nounness, and context will continue to be decisive in determining which property the word sleep has in future situations—which higher-order assembly it will excite.

Nounness and verbness at first sound like esoteric and fanciful ideas, remote from the world of science and the study of behavior, but in fact a physiological meaning can be assigned to them—a meaning that is at least intelligible and one that is possibly true.

*And third,* of these illustrative possibilities, is a theoretical basis for the relation of active to passive voice, which has been a problem. This is found in the fluidity of perception and memory image. As one surveys a scene of activity the focus of one's attention varies from moment to moment, and verbal report may vary accordingly. When agent *A* impinges on target *T* at a moment when attention is on *T* the perceptual sequence is *percept of T-impact from A.*

Verbal report may then tend to take the same pattern, "*T* was hit by *A*", but if instead attention had been on A, with perceptual sequence *percept of A—movement by A—impact on T*, report might tend instead to be "A hit T." This is not a rigid theoretical inference—other factors may operate—but it does show how the different speech forms, active and passive, are determined by different perceptions—of the same event. Similarly with the memory image:

When Johnny is called on to report playground violence, he may recall the scene beginning with the aggressor so the report takes the form "Billy hit Kathy," but if Kathy is the center of interest for him at the time of report, it would take the form "Kathy was hit by Billy."

These proposals, which attempt to show the relevance for psycholinguistics of cell-assembly theory with its emphasis on perception and representative process (image and idea, or assemblies of lower and higher order), must not be taken to mean simplification of the problems.

If the theory makes the idea of nounness intelligible physiologically or shows how another approach may be made to the difficult problem of active-passive transformations, what it does is substitute one complexity for another, and the new one is manifestly no less complex than the old.

The general conception relates the controlling processes of verbal and nonverbal behavior more intimately, and denies that psycholinguistics can pursue its problems successfully without regard for the theory of perception and of thought generally, but that complicates rather than simplifies the problem.

## Chapter 34

# Biological Inequality of Men

A first glance should scan the way in which the species, the race and the individual actually share in a vast common store of some ten to forty thousand *genes*—those agents that are present in every cell and which so largely determine the type of person that will issue from the egg. It is plain enough that the entire human species as it exists today contains a greater variety of genes, and of genes combined in special ways, than does any single one of its races. It is equally clear that each race contains more genes and special combinations of genes than does any individual of that race.

From this point of view one can see that an individual—a person—must develop from something progressively less than the total variety and arrangement of genes possessed by his two parents, by his clan, by his race, and by his species. An individual thus becomes a neat and shrunken pattern that almost everywhere incompletely fits the unfurled fabric that is his species; and the fit to his race is merely a little better.

Moreover, an individual must develop from only one of the many possible arrangements of a reduced and limited number of genes carried by two parents. Everyone starts and ends his career with such a strait jacket of genes. Though this jacket usually permits considerable variety of outcome, it is always worn and is always persuasive.

A slight correction or explanation belongs here. Genes usually or often exist within the race in several slightly modified (mutated) forms called alleles, and where the word "gene" is herein used it would usually be more accurate to say allele. Each individual, however, will be provided with only one of the several forms (alleles) of the gene by each of his parents.

He will thus be equipped with two alleles, and whether these two will be unlike or identical depends upon whether his father and mother gave him unlike or identical alleles of the same gene. When two unlike alleles are thus brought together one of them is commonly able to overcome, or cover up, the usual influence of the other. And one gene (allele) is sometimes, though apparently not always, able to affect more than one of the traits that later blossom in the man or woman.

Some outstanding facts relating to human individuality and biological inequality thus lie in the field now being previewed. First, the individual just discussed is provided with a special and unique assortment of genes that exists in no other person living or dead—except in cases of identical twins. Second, his physical and mental capacities, and a host of his defects, deformities and susceptibilities to disease, are either limited or definitely determined by the particular packet of genes from which he started his career as a fertilized egg.

Third, the child's father and mother each carried some or many genes that could not be put into each sperm or each egg. These genes went instead into different sperms and different eggs, though only one sperm and one egg may be used to start any one person. Therefore, since this distribution of at least many genes to each egg and sperm is based on chance, and since unions of sperm and egg likewise occur on an unpredictable or chance basis, the personal characteristics, limitations and destiny of the individual arising from this union are based broadly and essentially on chance.

Just here, at the introduction of the word "chance," both the reader and the writer face a special obligation. The reader should now note well the point at which chance enters into this account. It enters through the *way* unlike eggs are produced by the mother, unlike sperm by the father; and also through the further and unrelated fact that the union of a particular sperm with a particular egg is an unpredictable or chance affair. However, any human egg uniting successfully with any human sperm will surely produce a human being, not an individual of another species; and if egg and sperm are of the same race the individual produced by the union will surely be of that race.

It is *law* that governs these more basic matters; no unpredictability or chance relates to these points. But the genes for the personal or distinctive characteristics and endowments listed above are the precise items that whirl in the mills of chance. The writer's obligation is to make this point wholly clear, and also to note that the word "chance" requires a fuller definition, which can be given better later. Finally—and of first importance—since our talents, defects and personality traits rest upon chance, they do *not* rest on Providence—unless dice-throwing methods are used by that agency.

This preliminary view of the area included in this chapter sketches a most penetrating biological story, with one meaningful intellectual implication attached to it. This implication-the replacement of Providence by *chance* as the source of our very personal endowments and defects—has an educational value so pre-eminent that one can only marvel at the small margin by which it escapes complete elimination in the total education of everybody.

The writer's experience leads him to estimate that this particular implication is being pointed out perhaps to about one or two per cent of college *graduates* in the United States. Because of religion-controlled public opinion, it is usually unmentionable at all lower levels of public and private education.

## INDIVIDUAL AND RACE

Everyone has a valid, even it quite incomplete, understanding of what an individual is. But biologists and anthropologists need help, along with some of the liberty of arbitrary statement, when they attempt to define a "race" of mankind. Equal difficulties do not arise over whether mankind of the year 1950 represents more than one species; in other words, whether the few main types of surviving men should be called species rather than races. Practically all of the most competent opinion rejects the idea that these types have the rank of species. However, our look at the human family needs only to extend to the present records of fossil men to find some three to six true species of man.

While the matter of species of mankind is momentarily before us, it is useful to consider the notable variation or diversity that is present within each of the now known species of man. Modern man, our own species, is by far the most variable wild species of animal with which we are acquainted. Associated with this fact, and partly in explanation of it, is the circumstance that man has a wider geographical range than any other animal species, with the possible exception of his parasites.

Again, man's method of evolution, though leading to perhaps less extreme divergences than are found in some other species, has involved more intertwining or mixing of genes from all the extreme types that exist than is found in most other species. This, too, has resulted from the migratory nature of man. Over long periods of time the races of modern man have met and, in no small degree, have mixed. This has afforded unusual opportunities for highly diverse individuals to arise from the combination of genes carried in the most extreme types and races. It should be understood that the few primary races of modern man originated in a still earlier stage of prehistory, when isolation, not migration and mixing, was the rule.

When this question of the diversity of man is examined in some two to five fossil species of the human family, it is evident, first of all, that the whole of that unlikeness has not yet been found and studied. It is nevertheless clear that these fossil men also were a diversified lot, probably much split into races or types. Indeed, some trained students of these ancient men also assign some of these species as wholes to a genus other than Homo, or modern man. In any case, the species of fossil men were probably rich in quite dissimilar individuals and races.

Returning now to the question of races within our own species, it is notable that the word "race" has been variously used and abused. Historically, this word was first used some two hundred years ago to designate the few and chief biological types of existing men. It would seem that confusion is the principal result of using the word "race" to mean something really different from that. To be sure, the Swedish naturalist Linnaeus used the word "varieties" instead of races for his four main types of man. And he tried also to use mental, or temperamental, rather than physical traits as a basis for his classification. A bit later Buffon—and also Blumenbach, the founder of anthropology—made skin color the basis of his classification, and so arrived at five races—Caucasian, Mongolian, Ethiopian, American (red), and Malayan.

In the light of present knowledge, skin color alone is too slight a matter to constitute a principal or an important biological type of man. Dark or black skin is widespread among the Hindus, who are otherwise characteristically Caucasian; and "White Indians"—with white skins but non-Caucasian—live in Panama. But associated questions arise: Are other and more significant evolutionary differences nevertheless tied to, or primitively associated with, skin color in some of these five color types?

This seems probable. Are such significant evolutionary differences tied to something other than skin color? This is perhaps less probable, though the genes responsible for the so-called blood groups tell much concerning the lineage and relationships of the main types of mankind. When these and related questions are firmly answered, it will be possible to make a very short and more satisfactory list of races of modern man.

One need not doubt that human races really exist. The need is for further knowledge that will tell us which peoples best represent those races, and which peoples have resulted from the mixing of two or more true races or subraces. The human family, probably to its advantage, is already in no small degree a bulging realm of hybrids. And man now finds few questions more intricate than that concerned with his own not-too-recent ancestry.

Most anthropologists, utilizing color and several structural characteristics, now point to three primary races—

White or Caucasoid, Negroid, and Mongoloid. They are also measuring and defining a much longer list of primary subraces—including Alpine, Mediterranean, Nordic, Negrito and Eskimoid—along with some blended hybrid groups called composite races and subraces. Knowledge and use of the "blood groups" has not led to a radically different classification of men.

One authority ( Weiner) arrives at the three above-named races, or at five; another authority ( Boyd) finds indications of six primary races. Races are populations, originally isolated, that differ significantly in the frequencies of one or more of the genes they possess. Species differ from races in that the frequencies of a greater number of genes are usually involved; and also in the "reproductive isolation" of the population—for even when brought in contact two species rarely choose to interbreed. The number of pairs of genes that characteristically separate the Negro and White races has been estimated at from six to a few times that number. Man probably has ten to forty thousand genes.

Does race, as listed and defined above, have significance in relation to personality? Is race significant in relation to such physical items as strength, speed or endurance? Is it significant in the talents and skills of human handicrafts? Is it significant in the sharpness of one or another of the special senses? Is race of significance to things mental? Is it related to defects of development that arise along the path from egg to adult? Is race related to disease or to susceptibility to disease?

This list of questions may be left temporarily without answer. First of all, there is an obligation to point out that a difference-whether expressed in a race or in an individual—does not at once indicate whether this difference, or trait, is the result of genic difference (heredity, nature) or of some particular environment (nurture). Endowment with unlike genes may and frequently does lead to the development of biologically *unequal* men.

Dissimilar physical environments (nurture), and also sociological elements of culture, may give rise to highly important *differences* in men. Races or individuals, however, may accumulate and monopolize a gene or a group of genes in a sense that neither of them can monopolize any ordinarily encountered element of environment or of culture. It is genetic inequality only that is of primary concern in this chapter.

The large extent to which a trait may be and usually is modified by a different or special environment is a large subject not discussed here.

Nevertheless, present knowledge of the interrelations of nature and nurture, and especially our present ability to find or to trail a gene, frequently enables us to spot a difference that is hereditary.

This task in humans, one must admit, is sometimes frustrated by limitations on the biological tests that may be performed on human beings. Fortunately, the more thoroughgoing tests already performed on our animal relatives often supply the essentials of this missing information.

On the other hand, no species is comparable with man in the extent to which he has been routinely examined, treated and measured—medically and otherwise. Thus despite the relatively undeveloped state of human genetics, we have more than the beginnings of information on genetic inequality in racial and individual man.

With these considerations in mind, some answers to the above list of deferred questions can now be attempted; at least they may be approached. Does race have significance in personality? Two special and hitherto unmentioned items are involved in this question, and these far from negligible items must next be acknowledged.

First, personality is surely not a single trait but a combination of very many traits—essentially it includes the entire organism. Some of these traits may therefore have an hereditary basis that is perhaps practically the same in all human beings.

Second, the extent to which personality, as this is estimated by an onlooker, is a subjective rather than an objective matter is both confusing and important. To how great an extent does the "personality" of John Smith depend upon whether he is caught up by the eyes and ears of Mary Jones or by the coolie Chang Tan Wan? One may at least note that, if John is well bathed, Chang can smell him and Mary cannot.

To whatever extent personality is related to such items as stature, body build, skull and face form and proportions, skin color, hair form and distribution, and eye color, there is little question of its foundation upon genes whose frequencies vary from race to race. Indeed, these are among the principal items now used by anthropologists in the classification of races. With these remarks the writer may partly evade and longer delay an answer to a too complicated question. It is with the widely scattered components or fragments of personality that science is best prepared to deal, and items relating to these fragments will appear throughout this account.

The relation of race to such things as physical strength, speed and endurance is somewhat less complicated, although most of the data and measurements required to establish actual differences are not available. Physical strength is certainly much affected by stature, while stature in turn is largely controlled by genes. And measurements are scarcely needed to establish the fact that short and small subraces like the Negrito and Bushman have less physical strength than have much taller groups like the Nilotic Negro, the Nordic and the Apache.

Racing speed is likewise affected by stature and bodily conformation, which are known to vary genetically in tribe and race. Endurance, however, is doubtless built of several or many traits. It is not clearly associated with either stature or with genes now adequately identified. This, however, is rather far from suggesting that endurance rests upon genes whose frequencies do not vary from race to race.

The extent to which race is significant in the talents and skills required in human handicrafts must be left largely unanswered. Here again not one but many separate traits are involved. For these several traits the relative roles of nature and nurture—the latter always includes the potency of practice, training and education—is unmeasured and unknown. It would be strange and unexpected, however, if the known and considerable structural and other differences of primary races failed to support some of these talents and to hinder others. And it is certain that several of these differences—structural and other—are based upon the special or characteristic share of each race in the total store of human genes.

Discussion of the extent to which the special senses differ in the different races will be limited to a single case. Several years ago it was found that a laboratory-made substance known as phenylthiocarbamide (PTC) is distinctly bitter to most people but quite tasteless to others. Family studies showed that the ability to taste this material is hereditary and due to a single gene. In America, about one person in four finds it tasteless, while three in four declare it is bitter. It was later found that chimpanzees can likewise be divided into tasters and nontasters of PTC.

This substance is not found in nature, but, as is suggested in the preceding sentence, the gene that permits some men to detect it may be much older than man. Another substance was recently found that does occur widely in nature, particularly in cabbage and turnip, and it too is bitter to most people and tasteless to others.

When several people were tested both with this substance and with PTC, it was found that every person reacted to the two substances in the same way—found them both either bitter or tasteless. The natural substance derived from cabbage acts as an antithyroid drug. In nature, therefore, the gene that gives ability to taste PTC is perhaps significant only as it directs the avoidance or the consumption of this other antithyroid drug. In the present account, PTC and the antithyroid substance serve as a quite simple instance of biological inequality in men. The presence of a single gene, obtained from either parent, determines that the child will be a taster.

It is further found that this gene is unequally distributed among various races of man. Whites show fewest tasters: Arabs, 63 per cent, and North Americans, 70 per cent; pure Nilotic Negroes, 96 per cent, and American Negroes, 91 per cent; pure American Indians, 94 per cent; Chinese, 94 per cent. It thus appears that both races and individuals are unequally provided with some of the sole means by which a human being may make contact with the external world and sense its properties. And this inequality of ability to learn clearly rests on chance, not on Providence.

It is easy to ask: Is race of significance to intelligence and mental qualities? But it is impossible to give an answer that is both informative and brief. In fact, what is known on and around this matter provides little more than the outline of an answer to this loose and composite question. A good grasp of that outline, however, would do more than provide a partial answer to this particular question.

It would provide a sharper view of the extent to which both nature and nurture contribute to the bundle of things we call personality; it is that bundle—not really divisible into mind and body—that we are everywhere and always actually compelled to meet and study. Whether one wishes to examine the foundations of tallness or of ability to learn, those foundations will be found interlocked with a tangle of genes and with countless items of unrepeatable individual history—growths, impressions, nutritions, movements, accidents, responses, and urges.

From a unique store of genes, an unrepeatable history, and the never-ending contacts and impressions of life, each personality is built. To identify or partially to isolate the components of personality is no slight task. If some genes did not assert themselves clearly and forcefully the task would be nearly hopeless.

Much effort has been directed to the measurement of the "intelligence" of individuals, races, and various social and ethnic groups. As between individuals of the same family, community, school, or social group, it is clear enough that the tests used in such measurements have shown that some individuals are superior to others in some or several of the items that enter into intelligence. When the tests have been applied to various races and ethnic groups, the differences found were usually small.

And almost always there has been a possibility or probability that the differences obtained could be assigned to environment or training, or to defects in the tests used in the measurement, and not to racial inequalities in genes. It seems that much improved tests, each directed toward one only of the several primary components of intelligence, according to Thurstone, are already being developed. Of the tests hitherto made, it is perhaps safe to say that actual racial differences averaging only five or ten per cent would not have been satisfactorily established by those tests.

Despite the difficulties and limitations that attend all such measurements, it seems clear that the *range* of many or all mental capacities is much the same in all of the three primary races and in nearly all of the subraces thus far studied. This is a very penetrating fact. It means that any and all of these ethnic groups (a very few peculiar subraces excepted) have a certain percentage of individuals—perhaps varying from a few per cent to fifty or sixty per cent—with a *higher* mental rating than has the lower fifty per cent of *every other* ethnic group. This, in turn, leads to the conclusion that the true home of mental inequality is the *individual*, not the race.

There is evidence for the existence of a gene that is essential to having or not having so important a thing as an instinct; and some evidence also that this gene is being shuffled about in our own mills of chance. The parental or maternal drive, a probable progenitor of love and altruism, is the instinct here in mind. Whether this gene is distributed equally or unequally among the several races and the smaller ethnic groups, however, is quite uncertain; and even for the two sexes this point is far from clear. It is nevertheless probable that this gene is missing in some women and present in an occasional man.

Some of the above conclusions regarding mental qualities tend to agree with the firmly expressed opinion of Confucius: "Men's natures are alike; it is their habits that carry them far apart." The role assignable to habits—and to the somewhat broader and related thing known as culture—is an imposing one. Culture is of unusual significance in what we call mentality. And again —however loose and temporary this association may be—culture, habits and manners are items usually associated with ethnic group, nationality and race. Cultural differences and cultural achievements—such as attainment of a low or high state of civilization—are not based mainly on genetic differences. The explanation of such differences lies mainly in the history of the cultural experience that each group has undergone.

Culture is at once the child, and often the long-term subruler or dictator, of a people, nation, ethnic group or race. So considerable is the weight of culture that it, and not the genes, probably often predominates and tips the balance for such composite and inclusive things as temperament, personality and character. If racial prejudices must persist, they may rest, with trace of justice, mainly on nothing more substantial than the alterable items of culture and manners. Yet the racial prejudice fostered chiefly in the Western world is now a threat to the security and welfare of all Western peoples.

The wholly surprising power of culture, as described above, is fully supported by the following statement by philosopher Edman:

It is hard to believe, but as certain as it is incredible, that the modern professional and business man, moving freely amid the diverse contacts and complexities pictured in any casual newspaper, in a world of factories and parliaments and aeroplanes, is by nature no different from the superstitious savage hunting precarious food, living in caves, and finding every stranger an enemy.

The last two of our list of questions are concerned with the possible relation of race to defective development and to susceptibility to disease. These two questions will be treated as one, and only a few of the known instances of this association will be cited.

Unlike other races, some Negroes carry a gene that, when obtained from one parent only, causes their red blood cells to assume irregular (sickle, holly-leaf) shapes when subjected to certain treatments. This condition is known as sicklemia and it occurs in about eight per cent of American Negroes. Such sickle cells have a normal span of life and apparently they do not inconvenience the individual. It seems, however, that when an individual obtains this gene from both parents, and thus has a double dose of it, these red cells have a very short span of life and they may assume the sickle or related shapes while normally circulating in the blood.

This condition is known as sickle-cell anemia, and individuals showing it either die young or live only briefly beyond maturity. Nearly two (1.8) per one thousand American Negroes are born with this chronic and eventually fatal defect, and this proportion should be higher in the various African tribes, in some of whom this gene is known to be more than twice as frequent as in American Negroes.

A related though different hereditary defect of the red blood cells is found in the peoples who now or earlier lived around the Mediterranean Sea. In this disease the red cells are reduced in size and otherwise differ from normal cells; more especially they provide the blood with somewhat less than the normal amount of its oxygen-carrying coloring matter, hemoglobin. The disease is called Cooley's anemia, or better, since it exists in two quite unequal states of severity, Thalassemia minor and major. In southern Italian or Sicilian stock now living in New York, the minor type was found to occur about once in 25 births; the major type occurs about once in 2,400 births.

Another gene or group of genes affecting the quality of the blood, and which is sometimes responsible for the death of the unborn child, is associated with the Rh factor. This factor, an antigen, was first discovered in rhesus monkeys and has since been found in all races of men. When the father contributes the appropriate gene the developing embryo will produce the Rh antigen. When the mother lacks this antigen she may develop the corresponding antibody, which, in turn, may diffuse through the membranes into the blood of the embryo—sometimes with fatal results to the embryo.

A first child of parents of this type is much more likely to survive than is a child from any later pregnancy. Only in very recent years has it been learned that, throughout the ages, unborn humans in notable numbers have been and now are being killed by this chance combination of genes—a chance combination regarding which the responsible parents could have no possible knowledge. About 85 per cent of North American Whites carry the gene for the Rh antigen. Considerably higher percentages are found among Indians, Chinese, and American Negroes.

Further brief reference will be made to one other series, but to one series only, of the "blood groups." These blood groups bring no defect or disadvantage to the individual, though they do become important to survival when transfusions of blood are made from one individual to another. The presence or absence of two antigens, called A and B, in your blood stream determines the blood group or type to which you belong—O, A, B or AB. People of Group O have neither of the two antigens, those of A and B have the corresponding antigens, while those of Group AB have both antigens. Three different genes provide the basis for the four blood groups. Your own blood group is thus genetically determined; it becomes established months before birth, and no known environmental circumstance—including age, disease or climate—can change it.

The higher apes have four blood groups, which correspond to those in man. Human races vary greatly in the extent to which they share the four blood groups. According to Weiner, the Indians of Peru are 100 per cent O. Chinese are 30 per cent O, 25 per cent A, 35 per cent B, and 10 per cent AB. Negroes of the United States have corresponding percentages of 47, 28, 20 and 5. For North American Whites these values are 45, 41, 10 and 4. The significance of the blood groups in an interpretation of racial distinction and in human migrations has been earlier noted.

This account will be concluded with a list of some commonly known diseases that owe their presence to a gene or to genes. In most of these cases it is not known whether or to what extent the genes involved are shared unequally by the various races. Two such diseases, which relate to the use or disposition of food products, are sugar diabetes and gout. The nervous system thus suffers from a form of idiocy, from epilepsy, from ataxia, and from Huntington's chorea. The blood is thus subject to bleeder's disease, whose victims—because of the constant threat of the scratch that may prove fatal—can have neither an occupational nor an emotional life that is normal.

The skin is exposed to hereditary urges resulting in rubber skin, in scaly skin, in purple birthmarks, and in the absence of sweat glands. One by one, the muscles of some boys at ages three to six may begin to fail and weaken, and in this response to a gene from the mother their life ends before their twentieth year. The skeleton may be predisposed to fragile bones, to legs and arms of disproportionate length, and to shortened fingers and toes. Much of deaf-mutism is hereditary. The eyes are thus subject to a three-way attack—by degeneration of the choroid and retina, by deformation of the lens, and by color blindness. And the body generally is genepropelled to albinism, to dwarfism, and to allergy.

Races certainly share some diseases unequally. Negro children are relatively immune to scarlet fever; Whites are highly susceptible to yellow fever. The foregoing excursion has scanned parts of the evidence for the definite fact that the total store of human genes is shared unequally by the various races and subraces of men. Also that individuals within each race or subrace share very unequally and often most tragically in that common store of genes. Again, both races and individuals share not only the desirable genes but also those that lead—effectively and practically—to crippling defect, disease and premature death. And distribution of these fateful genes that point to triumph or to tragedy is based on chance. Every human being is a monument to the sway and tides of chance.

## CONSTITUTION AND PERSONALITY

There are two approaches to the problem of personality and the individual. The approach through genetics has been followed in the preceding pages. It remains to examine the other approach, which is through some segments of anatomy, psychology and physiology. It must not be inferred that these latter sciences are excluded from the preceding account, nor that genetics is excluded from the point of view that will next be presented. The facts are otherwise.

## ANATOMIC AND PSYCHOLOGIC ASPECTS

It is easy to understand what we are doing when we collect information on the "constitution" of each of a group of living men. To be sure, these men should wear no clothes while their physical traits are being observed and classified; and it is best to observe them with care and at some length—at work, relaxed, and at play—in order to assess the items of their temperament. But a child of ten years would have little difficulty in making a very crude assessment of constitution in each one of a small group of men.

This child would label one as short, fat—also jovial and lazy; another as thin, big-fisted—also obstinate and courageous. Trained specialists can now set down many items for the structure and the temperament, and even for the peculiarities of the internal chemistry, of an individual. To study the pattern of these several items in an individual is to study constitution.

The thing called personality is an actually achieved pattern. It is ordinarily understood and approximately described in terms of items of difference from other personalities. Reasons for constitutional studies have varied among such aims as gaining a better knowledge of man as an animal species, the selection of the right man for the right job, and insight into the relation of constitution to disease.

This latter relation was already partly recognized in the earliest medicine of India, Persia and Greece. And medicine—psychiatry in particular, at this moment—has provided much of the incentive to more and better investigation of constitutional factors. Finally, throughout the ages those disciplines which our universities call the humanities have recorded human inclinations, frailties, achievements and traits, thus mirroring the sort of creature that man was or has now become. History, literature and art all display the fascinating variety of the special qualities and disposition of particular persons and peoples.

There is now no generally accepted way of putting together, of classifying, or of describing the physical and temperamental items of the well-measured individual. Some of the classifications now in use relate especially to psychiatry or medicine rather than to human biology and psychology. The present brief consideration of this central topic will be based on the method of classification introduced (and still being developed) by Sheldon and his associates. An outline of that method is as follows:

*Group 1:* A relative predominance of structure associated with digestion and assimilation—conserve energy and easily put on fat (called *endomorphy*). The temperamental or behavioral traits associated here are relaxation, conviviality, and gluttony for food, for company, and for affection or social support (called *visceratonia*).

*Group 2:* A relative predominance of bone, muscle, and connective tissues—tend toward massive strength and muscular development (called *mesomorphy*). The associated behavioral traits are bodily assertiveness and desire for muscular activity—dominating, love of action and power (called *somatotonia*).

*Group 3:* A relative predominance of the skin and its appendages, which include the nervous system—small mass in relation to surface, height large in relation to weight (called *ectomorphy*). The associated temperamental traits are inhibition of the traits under Groups 1 and 2; also overconsciousness, love of privacy, and avoidance of overstimulation. (called *cerebrotonia*).

Of course, the descriptions provided above are sketchy and incomplete. They are meant to indicate with reasonable clearness the three major components of constitution within each of which several special structural and behavioral items naturally belong. It is not to be supposed that one has much chance of ever seeing a person built *wholly* of components of, for example, Group 1.

Heredity is involved here; and in compounding an individual the versatile artilleries of chance will rarely or never use only shells of a single type.

Larger or smaller amounts of the elements from all of the three groups will be found in every individual—often even within the head, torso or legs of an individual—and the extent to which this is true can be measured. This measurement is done by using for each group a scale, represented by the numbers 1 to 7.

With only three numbers set in the following order, 5-5-1, the constitution of an individual can be rather fully described; and thereafter it is easy to find that many people of various races throughout the world fit this same description. In this grouping of numbers, the first figure means that the qualities of Group 1 are present in slightly more than average amount, since 4 is the average on a scale of 7.

The second figure has a similar meaning in regard to the qualities of Group 2. The final figure indicates the lowest recordable amount of the qualities of Group 3. "These are squat, heavy organisms, built close to the ground. They don't put their necks out far."

Objective measurements can be made, of course, on any and every type of man. When these measurements, on both structure and temperament, have been classified and described according to this method, they bring into clear view many of the most remarkable facts now known about ourselves—about our constitution and personalities. Within each group it is found that structure and behavior are highly correlated. In this way the unity of the organism—the inseparability of body and mind-which elsewhere has been frequently observed, is again observed in man. On the other hand, components of personality that are seriously dissimilar and incongruous—as well as the more usual congruous components—are sometimes or often forced to integrate themselves into personalities.

So marvelous are the powers of purely biological integration that both grit and grain are used to make this "bread of life"—at least, to make a two-legged type that can eat and walk and feel. But some of these products prove to be impossibilities at the social level; they swarm inside and outside our charitable and correctional institutions and our hospitals. Finally, these penetrating studies in constitutional psychology, though made with aims partly similar to those of Freud, serve to question or to replace methods used by Freud. Freud's method seeks the keys to mental disorder by trying to get a look at the subconscious self. For these same keys the Sheldon school looks at the entire organism, and directly to measurable items of body structure and temperament.

Quite other studies have shown that at chemical and physiological levels of cellular activity numerous gene-controlled differences and inequalities exist. These may take such forms as addiction to drugs; absence of an enzyme or hormone; increased or decreased vitamin requirement of the cells; and perhaps a changed sensitivity of some tissues of the body to one or another of its own hormones.

In some strains of rats and mice it is now known that an appetite for alcohol is hereditary, and again that this appetite can be abolished by some unusual supplements to their diet. Here susceptible strains are ones in which parts of the cellular chemistry are so deranged as to require in their food extraordinary amounts of one or another vitamin, mineral or amino acid. It was said above that some personalities that are passably welded together biologically nevertheless fail at the social level.

Notable members of this or of a related group are those who lead an approximately normal social life for several years and thereafter develop a "split personality," or schizophrenia. Here the years of near-normality are those long years that humans spend in attaining maturity. And those same years manage to register much experience in forms commonly spoken of as conscious and subconscious. The person who becomes schizophrenic, however, is one who—among other things—does not readily or adequately accomplish maturity, and who ultimately also fails to deal suitably with his registered experience.

His world too much resembles the dream world. It is clear that an integration, a suitable merging of parts, has failed at one or another level in the development of these individuals. The role of heredity in this failure has been very hard to establish; possibly that has now been done ( Kallman). It is also probable that the functions of the suprarenal glands are deranged in this remarkable disorder. The psychiatrist's art can do much toward restoring normal personality to some but not to all of these victims.

Considered as a disease, schizophrenia provides one fourth of all hospitalized patients in the United States, and of course in no country do all such cases enter a hospital. After prolonged study of this elusive constitutional disorder, in which one or another item in the welding process fails, physiologist Hoskins has said: "It is doubtful that any mishap of existence causes in total so much human distress or entails so much unproductive expense."

Constitution and personality cannot be well understood without rather more than a passing glance at the system of involuntary (autonomic) nerves that presides, in belly and breast, over the main business of living and of procreation. All that occurs there, except for an occasional command of will to change the respiratory movements, is under the speedy and precise control of these nerves. Many of the keys to personality and to human inequalities are hidden in our deep internals—that very central part of our bodies that we may never have opportunity to see.

The ever-responsive glands of the body are touched and fingered by these autonomic nerves, as are also the deep-lying muscles within the body cavity and within the walls of the blood vessels everywhere. Even the muscle cells ruled by these nerves are wholly unlike other muscle cells; they are smooth or unstriped, while those large cells which are subject to the will are abundantly striped.

These nerves communicate with a very special part of the brain—an old and primitive part bearing the names hypothalamus and thalamus. To and from this old brain flow the nerve impulses that regulate the basic and busy traffic of life; and usually we are quite unconscious of the tangles and the regulation of that endless traffic.

The common and total effect of all these activities and precise adjustments is that we feel a state of well-being or that all is not well. Sometimes, however, the messages reaching the old brain are sufficiently intense to overflow and be carried somewhat indirectly by other nerve fibers to the newer brain, the cerebrum. There, in the cerebrum, such an overflow permits consciousness to intrude itself into the traffic problems, where it affects them and is affected by them. The importance of such overflows, with the resulting involvement of consciousness, lies in the fact that all this has an effect on personality and may even lead to self-imposed disease.

This item requires a word of further comment. Each individual has his own personal endowment of autonomic nerves; and this must have a genetic basis while exhibiting its many grades of inequality. At one extreme there is stability, the ordinary body functions proceeding calmly and automatically even when all sorts of exciting things are happening in the intellectual, emotional and physical spheres of the person's life.

At the other extreme is found such a lability, or fluid reactibility, that autonomic functions are chaotically disturbed by every external pressure to which the individual is exposed. In the words of Draper, "The functional quality of the autonomic nervous system, the speed of its reaction to stimulation, and its stamina in the face of long-continued strain are probably the basic factors in determining the physiological character of the individual." When an individual is described as having "guts," a really good autonomic nervous system is implied.

Within this area of nerve and brain two other items are worthy of notice. First, a word in regard to a home for the emotions in the old or primitive brain; and thereafter a note on the degree of fusion of older and newer parts of the brain. It seems to be fairly established ( Cannon) that the emotions and instincts are rooted in the old or primitive brain—the hypothalamus and thalamus. And it has been noted above that these elements of the brain are closely tied to the autonomic nervous system.

Evidently, therefore, the emotions are originated or modified by the impulses arriving from the deep-lying (visceral) organs, and those organs may in turn be affected directly by outgoing impulses, which the emotions themselves produce.

It is on this basis that we may understand the fact that worry may lead to ulcer of the stomach, and that fear or rage will release adrenalin from suprarenal glands by way of preparing the person for fight or flight. And likewise we may understand that things like alcoholism and peptic ulcer are sometimes exorcised or cured by an unexpected emotional experience, such as "getting religion" or inheriting a fortune. The main point of present interest, however, centers in the meaningful fact that, in man, reason and emotion tend to dwell in separate segments of the brain. In what ways, or to what extent, are these two segments united? What are the consequences to men of an imperfect union of these two parts?

Full answers to these questions would relate to limitations on human performance, to the illumination of human behavior in all its diversity, and to many items of human inequality. Only partial answers and a further citation of some related facts belong here. Hitherto unnoted is the certainty that the newer brain, the cerebrum, is the seat of reason and the area of the more acute consciousness.

This newer part has been added to the several parts of the earliest reptilian brain; and though a passable integration of new with all old parts of the brain has been accomplished—albeit by roundabout rather than by direct nerve-fiber connections in the case of the thalamus—human behavior can be and sometimes is dictated by the emotions, not by reason. In this fact lie enduring dangers to the race. "The voice of the intellect is soft," say Freud and others.

Nevertheless, it has recently been shown that some otherwise unassociated nerve fibers do pass directly from the thalamus to the cerebral cortex. And the discoverer ( Walker) of these fibers interprets them as quite recent arrivals, and as perhaps indicative of nature's continuing evolutionary attempt to achieve in man a better association (integration) of emotion and reason.

If this interpretation is correct, it provides an alluring—even if painfully distant—promise to the race. Moreover, some implications of this interpretation are intriguing. A recent origin, and a continuing increment, of these slender threads that expand the rule of reason in existing man provide a probability that our own species has thus advanced genetically over all·earlier species of man.

And they likewise suggest that the very unequal packets of genes carried by men now alive may include well-concealed inequalities of this hidden wave that slowly drifts toward reason.

## HORMONAL AND SEXUAL ASPECTS

The hormones have much to contribute to a complete story of constitution and personality. No complete story, however, is being told here. The genetic inequalities of man—our central subject—are observed most readily either at their starting point, the genes, or near their terminal point, the fairly established traits of the going organism. To a large extent, the hormones play their parts at points between. The potent and wandering molecules of hormone that are released from certain special cells are mates and cousins to other molecules released in all cells under genic action—all involved in the orderly building and maintenance of the organism.

However, in addition to or as a very special part of this widely shared task, these wandering molecules uniquely fraternize with the nervous system in making the many parts of the body into one, in the integration of the organism. It is perhaps largely because of the way our knowledge has developed that a few constitutional factors can now be best approached in terms of these intermediate agents, the hormones. The genes involved are in a dimmer background; the hormones earlier came within our view.

There is no uncertainty concerning the large extent to which most aspects of constitution, personality and behavior are affected by absence or by excess of certain hormones. Even the circumstance that a male fetus must develop within a female body, and is thus subjected temporarily to the female sex hormone of its mother's blood, sometimes produces adverse effects in that male child.

The female embryo runs no similar risk. The environment in general, as well as the genes, partly determines the amount of hormone that is actually produced. Absence of the thyroid hormone, thyroxin, from the developing embryo and child means that it will at best become a dwarf with an undeveloped mind (cretin).

Another type of dwarf, with normal body proportions and mentality (midget), results from a deficiency of hormone produced by the pituitary gland. And an overproduction of hormone by this same gland in more mature years may lead to overgrowth and even to a deforming giantism. Many of these cases relate to abnormality, and their bearing on studies of normal human constitution is not too clear at the present moment.

On the other hand, it is clear enough that in the case of a boy the general growth of the body is promoted by one or more growth hormones of the pituitary, by thyroid hormone, and by male sex hormone (androgen). In a girl these hormones act similarly except that the androgen is largely replaced by female sex hormone (estrogen). It is possible that the special ability of estrogen to stop growth at the ends of long bones is one of the reasons for a shorter stature in sisters than in their brothers; perhaps another reason is the smaller amounts of androgen produced by girls.

It is likewise clear that in boys the androgens assist the usual development of male genital organs, beard and change of voice; while in girls the estrogens similarly direct development of female genitals, uterus and breasts. There is fair evidence from humans for a gene that sometimes leads to incomplete sexuality (pseudohermaphroditism); for another that wholly prevents the development of an ovary in females; and for still another gene or genes controlling the time of onset of puberty.

The incomplete evidence for the three conditions last cited is greatly strengthened—and, in addition, proofs are provided for the existence in animals of a genetic basis for a wide variety of components of constitution—by tests that may be very briefly sketched. In a study conducted during a period of twenty-four years by the writer, a strain or race of pigeons was permanently established in which both true and pseudo hermaphroditism occur in every generation among the birds that should become males; some of these individuals produce both ova and sperm. Likewise, through selection, a race of doves was formed in which the females become sexually mature—lay their first eggs-at six months, and another race that matures only at eight or nine months.

The constitutional types thus established also included races with unusually large, others with unusually small, testes; races with large, others with small, thyroids; races with long, others with short, intestines; races whose tissues showed either a low or a high responsiveness to the hormone prolactin; and races with low, others with high, rates of heat production. This study was started with a careful selection of thirty-seven pairs of birds, for each of which the pedigree was known and, in some cases, certain "constitutional traits" were either known or suspected, and selection was continued during three generations.

These birds were chosen from a large bird colony known to be mongrelized comparably with the human population of a large American city. It is entirely probable that exactly similar tests made on humans—granting the centuries and the many impractical items of the test—would lead to similar results. Certainly these briefly described tests on animals prove that in them genes really exist for all of the above-cited elements of constitutional inequality.

A concluding word on hormones should note that in extreme cases these agents may express themselves frankly in at least three large areas of constitutional study—the structural, the psychological, and the chemical. We meet, or perhaps there comes into a clinic, a woman whose arm span is greater than her height.

That fact alone, as she enters the door, may be enough to tell the knowing student or the physician that here is a woman —and a body type called hypo-ovarian—in whom the ovary has for long been underdoing its work; in whom the gag reflex is diminished and the tendon reflexes are increased; and whose glands are now forming, and whose urine is now excreting, the sex-hormone products in reduced amounts.

Sex is an item of constitution so familiar that it would seem to need mention but not description. Nevertheless, it should be set in the same frame with other elements of constitution, and a further word should note its occasional incomplete development. Sex, too, is based on genes, but in an exceptional way. A certain group or block of genes occur singly in human males though they are paired in females. This results from the circumstance that two kinds of sperms—one with and another without this block of genes—are formed in equal numbers.

When a sperm deprived of this group of genes unites with any egg, a male child results; the other type of sperm adds this block of genes to that present in every egg, and thus yields pairs of all genes-and a female child. It is wholly a matter of chance which type of sperm will unite with the rare egg that may be fertilized, just as it is chance that decides for all other genes which of four grandparents will be represented in a particular egg.

Finally, sex, rather more than most elements of constitution, tends to get itself expressed very widely in the organism. Sex is a nearly wholesale endowment. There is little wonder that much of the subhuman world readily recognizes it.

Just as diversity of races is associated with some diversity of diseases, so the genes that decided your sex also apportioned you a larger or smaller share in certain diseases. Gout and the duodenal type of peptic ulcer belong mostly to males; while gallstones, osteomalacia and hyperthyroidism predominate in females. Even females of various types share unequally in some diseases. Dreaded pernicious anemia is frequent in florid, flabby blondes with rather large faces and blue eyes, but is rarely found in brunettes.

A unique thing about the abstract quality called sex is that it is observed and known only in terms of differences between two unlike types. In an all-male or an all-female world, our concept of sex could never arise. But once the idea and fact of sex difference was recognized by man, it became a dividing line between the two chief groups into which all humans are classified by themselves and by others. And here, indeed, there is no doubt of human biological inequality.

To this contrast of bodily form and function, however, nature herself seems content to assign mainly unlike and unequal shares in reproduction and in a few diseases; meanwhile, and in addition, humans themselves have assigned to sex almost everything on earth and in heaven. In the numerous cultures of our own day—barbaric and civilized-there is scarcely a category of performance that is not apportioned on a basis of sex. To round out this absurdity, the apportionments of one culture contradict those of another.

Just as the idea of, and the wish for, progress is either unknown or quite new among men, a declared and purposeful search for sound and unexplored ways of developing and utilizing the special qualities of each of the two sexes—and especially those of the female sex—is yet to be attempted in any culture anywhere. It is true, of course, that bars and taboos are being lifted in many countries; but tradition, masculine vanity, law, custom and religion all compete for rule within this exceedingly broad area. The divisive aspect of sex usually still has right of way over any concerted effort to exploit the rich possibilities of what can be made of two rather similar genetic endowments that are so flavored as to yield two (often) superb and complementary personalities.

Let it be understood that the individual's huge endowment of genes, apart from the minute fragment that tips the scale for sexuality, does not under all climes and conditions rest on clear favor to either sex—ever-present chance is taking good care of that. And the traits that may develop from those genic impulses usually share nature's urge to full maturity and existence in both sexes. Some differences in such things as interests, physical strength, and averages for skills and reaction time—along with temperamental quirks in females that to males may look like those of an alien species—do indeed arise during early or later life.

These differences are probably closely related to the dissimilar production of hormones in the two sexes. The bravery of a girl, however, is truly human bravery, whatever the tribe may call it. For the two sexes the overlap of traits, except the traits involved in sex itself, is almost complete and perfect; while for two distinct human races this overlap of traits, as noted above, is less complete, though even here the observed lack of fit seems to be of no great biological significance.

To those who have studied sexuality, that endowment has taken on hitherto unknown qualities during the past forty years. For much of a lifetime the writer himself has shared the privilege of joining his efforts to those of many others in a study of this intensely biological characteristic. It is now clear that, at or near its basis, sex is a rather fluid and a definitely reversible thing. But this innate reversibility of sex does not mean that it is everywhere practicable to change or reverse the sex—at any rate, not yet. And the group of mammals, including man, is precisely an area where this reversibility is not practical.

The more intimate, intracellular and functional nature of sexuality—through or by which the sex genes do their work and on which the fluidity and reversibility of sex is probably based—is a large and incompletely explored subject which lies beyond the range of this account.

But this brief reference to fluid and reversible aspects of sexuality, together with the enlightening fact that a basis for both male and female structures is laid down in all early human embryos, provides a background against which one should next scan a group of personality and sex problems of mankind. In all human societies there are some boys who do not develop into "normal" males, and perhaps as many other boys who fear that they do not or may not measure up to normal maleness.

Likewise among girls there are those who do not develop into "normal" females, and another number who have an unwarranted concern for their femininity. Among adults, both men and women, there are some who well know they have one or another sex abnormality or inadequacy; and there are some committed to homosexual practices or to countersex components of temperament, desire, voice or behavior that do much to color and complicate the whole business of living. Parts of this composite picture relate to biological inequality; parts of it relate to constitution; all of it relates to personality.

A first remark about this area might well note that both medicine and education can now be of service in several of these cases. It is next notable that in any society the current social ideal of male and female usually has a great, and often a quite unjustifiable, influence on an individual's thinking about his or her own sexual deficiencies.

This whimsical social ideal may be at variance with biological fact, and an awareness of this fiction may be all that is needed to throw off a fear or an illusion of inferiority. Among whiskerless tribes a whisker is not an accepted part of maleness; and perhaps the maleness of the fat man is better viewed in contrast to the femaleness of a fat woman than in comparison with the maleness of an athlete.

A further word on the possible relations of constitution and sexuality may be helpful in presenting parts of a biological view of both sexual normality and abnormality. If the bases of sexuality—viewed solely as an endowment of genes—can be fitted to male and female bodies as different as those of monkey, orang and chimp, these bases should also rather easily fit into the various human races and the several constitutional types—endomorph, mesomorph, ectomorph—and the many blends of these.

To the truly genetic bases of maleness and femaleness there may have been no single addition or loss between early monkey and modern man. It is probable that occasionally a gene is displaced from the small group of genes that normally tips the balance for sex, and that this displacement provides the basis for some very rare cases of genetically based sex abnormality. In vastly greater numbers of cases, however, other sources probably account for both an unequal expression of a type of sexuality in different body regions and for sex deviations in the temperament of an individual.

For example, the genetic trend to maleness in the skin and breast regions might not usually unfold itself as favorably in an endomorphic-viscerotonic as in a mesomorphic-somatotonic type of man. In other words, the prevailing local endowment or total constitution in a particular body region may make a difference in the degree to which a type of sexuality is expressed in that region. Sex is always a part only of the total constitution in any body region, and sex expression is a highly modifiable trait. Masculinity, in both sexes, is increased by hormones formed in adrenal tumors.

Again, a lush variety of activities and introspections—fetal conditions, parental errors, early and protracted tribal comment, sex ideals, customs, taboos, nutrition and adventitious physical contacts-may adequately explain the great majority of the temperamental sex inversions and failures that actually occur. As already noted, environment is so potent in the mental and temperamental area that the usual role of genes in it is especially difficult to measure and declare. Mentionable, too, is the generally overlooked fact that at the end of the last century psychologists Krafft-Ebing and Freud studied and described cases of sex reversal of temperament long before biology was prepared to admit sex reversal in the bodily structures of vertebrate animals.

Of these temperamental sex inversions of mankind one further notes that certain groups within any current crop of victims may hopefully appeal to expensive psychiatry. Always, however, the fuller solution lies in prevention. Thus this huge and enduring task lies largely in the lap of education. The appropriate education can become fully effective, however, only when it is already a part of the training of parents and most other adults.

And if this formidable educative task can be even started without teachers trained in, and permitted to teach, the best of evolutionary biology and some fragments of psychology, the race possesses some marvelous endowment that is entirely overlooked in this chapter and quite unknown to this writer.

## THE IMPLICATIONS OF BIOLOGICAL INEQUALITY

The preceding account records several of the now known facts of human inequality. Facts do not easily lend themselves to dissent and disagreement. In contrast, however, any remarks on the implications of these facts may easily meet with resistance. Certainly a human factor, a thing that is not a part of the biological facts presented above, enters into any discussion of the implications of these facts. The present writer's outlook—his conditioning, his inhibitions, his prejudice, his frame of reference—may be wrong, timid or warped. In that case, what does the reader find to be implicit in the foregoing facts? Perhaps what the reader himself has found in them, and already accepted as usable in thought and practice, is all that matters.

But the reader is not in position to set down his thought here, nor even to indicate whether his thought on these matters is satisfactory to him. To fail to attempt here some guidance to that casual reader who has patiently read this account, but is nevertheless uncertain of its several implications, seems a crime hardly less than treason to a partner in this adventure. Certainly, such a failure would represent an unpardonable reluctance to discuss, activate and disseminate a science—leaving it floating and free of useful application among men. The author therefore—hoping that his guards are up—readily takes this further look into and around the facts of human inequality.

One conclusion that follows an examination of the way genes are distributed to every individual has been indicated near the beginning of this chapter. The genes that so largely predispose or determine our small constellations of talents, defects, diseases, terms of life and personality traits are distributed at random-in such a way that it evidently is not Providence but chance that is involved in their distribution. The word "chance," if undefined, may not best convey the intended meaning to many persons. "Random" and "unpredictable" are words that help to give that meaning. Certainly Priestley's redefinition of chance as "the observation of events arising from unknown causes" may be clear enough to an electron-conscious few.

But chance thus defined will not carry the intended meaning to most readers. To them the "how" of gene distribution—or, rather, of the distribution of certain particular genes—is far better pictured by the full, honest and thought-inviting statement that the events with which we are here concerned are unpredictable and random, and that the human mind finds it difficult or impossible to imagine a way of shuffling genes that would better exclude Providence or God from a share in it.

Though familiar human inequalities thus rest on chance, an incautious reader may not project this statement into a declaration that the human species likewise and equally rests on chance. The agents of species formation, of evolution, were elsewhere recorded as mutation, recombination, hybridization, genetic drift and natural selection—the last named and mutation being the key items. Whatever the part played by chance within a particular agent—and largely involved is chance plus selection—these agents, laws or principles direct the formation of species.

The true diversity of humankind, we now know, includes the diversity expressed in living men plus that much more imposing diversity that results from the inclusion of those several earlier and now extinct species (or genera?) of men. Within this total of human diversity there is, unquestionably, unfamiliar and staggering human inequality; this must spread from highest man to another species, which is or was more like ape than like earliest man. From this perspective one may wonder whether our account of the differences between existing races of men, as this was presented early in this chapter, does not somehow minimize those differences.

On this point, and at this moment, the several now existing races should congratulate themselves on having presented so much evidence that their inequalities, though real, are neither shameful nor relatively very significant. For, in any case, the area of tragic inequality is the population within each race. The many households of any one race, through their sons and daughters, roll up the really impressive score of human diversity—inequality of interest, capacity and defect. And here, too, belong the more extreme inequalities associated with sex.

To the various peoples of this planet the many consequences of this human variation are, in some respects, similar. And the fact of human inequalities is, of course, partly or dimly recognized by all normal people. But it is extremely important that the reader now become fully aware of the very narrow and rather fruitless limits of that recognition.

This general public, which speaks lightly of all inequality as "difference," is unaware of its sources and extent and innocent of most of its far-reaching implications. All this is truly reminiscent of the pre-Darwin day. Then, too, men looked at apes and dimly recognized many resemblances to man; but wholly and fatally missing from their understanding were the implications of those similarities—the firm link of ancestry, kinship and descent, which, once fully acknowledged, transformed the world of thought.

From the whole field of biological science, only the principle of organic evolution has greater social significance than the principle of the biological inequality of men. No principle whatever has equal personal significance. This principle is basic in sociology, essential in psychology and ethics, and abrasive and subtractive in religion. In all these areas our momentary social and political progress depends partly upon widespread and full understanding and usage of this principle.

Yet everywhere this supremely meaningful biological principle is almost avoided, and is skirted or touched with one finger in the halls of education. Is this restriction at all associated with the bearings of this and related biological fact on religion? And the implications of this affirmative answer doubtless merit whatever action that answer may arouse in any reader seriously concerned with the welfare and future of the race.

While still fully conscious of the extreme range of human diversity, necessarily including a long history, one may digress for a moment—as one must in this book of dual purpose—to an item near absurdity.

The so-called revelations of the several religions claim a high and soul-saving importance, and their proponents usually speak and think of them as given to *all* men. The perspective supplied by these chapters clearly outlines the fact that all of the now extant revelations were entirely withheld from all but *one* species of man! And from that one remaining species they were withheld during much more than 90 per cent of its history to date. Words that came from above fell chiefly upon the dust of men who, ages ago, had made the long and unaided fight. The few who thus late were shown a road to heaven had already built a wheel that rolled quite well on earth.

Once it becomes clear that genes bring into human society large amounts of mental disorder, physical abnormality, inefficiency and disease, many extremely important questions arise. Do the carriers of these genes reproduce themselves to a usual or to an unusual extent? How does substitution of civilization for natural selection affect the ratio of sound to unsound genes in the race's store of genes? Within the United States there is a recognizable barrier, even though it be called a slight one, to getting the general public to give serious thought to human inequality.

Certainly an appreciable amount of resistance to this concept is fostered by schoolbook familiarity with words of our Declaration of Independence —"that all men are created equal"—but by unfamiliarity with the intended sense and usage of these words.

If an acceptance of reality provides any advantage to social survival of a people, it would appear that education may need to place unusual stress on this principle of biological inequality within the borders of the United States. This involves no reversal of early American aims and ideals.

Few human laws or institutions can be expected to persist successfully and indefinitely against biological reality. And men of good will wish that democracy may not be merely obtained but that it may also endure. The essence of democracy would seem to lie in continuous and effective efforts to increase the opportunities and capacities of all men everywhere.

Those human capacities can be neither known nor fully nurtured apart from a recognition of the principle of human inequality; and sustained efforts directed to that end can be best supported by the truth. It is the Soviet Union that now denies and rejects outright the principle of genetic inequality of men.

Democracies, if sufficiently educated and fortunate, may take truth as they find it; apparently they alone may hope to build their futures around it.

It is diversely unequal men and groups of men—as partly described in the preceding pages—that biology delivers to the attention, care and direction of sociology, economics, education, law and government. Within these latter areas—though not in biology—there is firsthand acquaintance with what men become as members of society and with the rules by which a human society passably maintains itself.

If a biologist crosses these thresholds his first purpose should be to insist that biological man does not there become the forgotten man. In the often regrettable forms already pointed out, unequal men are in all of the woodpiles. As noted immediately above, one great government—perhaps one almost equally well called a religion—specifically denies the only type of man or race that biology knows. A doctrine of complete equality of all races or individuals, wherever it appears, is dogma rather than science—a yoke rather than a crutch.

Childhood conditioning and the chance-born pool of genes propel a lifetime to strengths and failures. Philosopher George Santayana was born and passed his early years inside a thickwalled city in religion-haunted Spain. His special kit of genes could build in him the keenest grasp of thought and the finest art of expression.

But the total of those pressuring beginnings of youth and personality failed to form a man who could either like the modern world or find sympathy for the fate of contemporary men.

The many inequalities of men clearly mean—as this was pointed out by Davenport when genetic science was only ten years old—that there can be no *impersonal* science or art of medicine, hygiene or education, nor of any similar agency that deals with human beings. Only the genic jacket carried by the patient, plus the mental hazards he has encountered, will sometimes clarify the nature and the treatment of his disease. However we approach or are approached, personality clings to us as does our shadow. There are always questions of the practicability of obtaining this highly personal treatment in medicine and in education, but there is never a doubt about the need.

In the handling of the criminal and the insane, civilized peoples already partially recognize this advance in our enlightenment. With respect to the extreme criminal, however, there is something further to remark. The worst of human beings may or may not be vastly worse than jackal or wolf.

Their access to human companionship, however, certainly renders these human derelicts infinitely more harmful and dangerous to man—they can and do maliciously kill, torture and plunder.

When society's disposition of these extreme derelicts is dictated by one or another religious sediment or sentiment, rather than by values born to clean biological fact and social health, the race is left with another wound to lick.

Without a further and specific statement about the things that may now or later be done to overcome the frequently unwelcome pressures of genes, the preceding picture is darker than the reality it should faithfully represent. Most, though not all, human traits are known to be somewhat plastic; some are extremely so.

And the arsenals of organized knowledge have begun to bulge with instruments for counterpressures that transform disease to health; sometimes even abnormality to normality. The gene for diabetes is already partly countered with insulin; deforming mongolism of the embryo may be normalized by feeding the mother with vitamin A; pernicious anemia is successfully treated with vitamin B12; and the gene-slit harelip is glossed by the surgeon's skill.

Again, as earlier noted, the exhibition of many capacities hinges merely on circumstance; thus culture fully bridges the gap between a savage and the professional man who happens to be your neighbor.

To how many things will all these long arms of help extend in a later day? Surely the hand and brain of man are shaping tools that man can trust. And here as elsewhere it is not Providence but man himself that acknowledges nature and softens its many blows.

That person who now only lamely trusts expanding knowledge and unaided human performance has indeed been robbed.

There remains for further remark only the forming, growing, inconstant thing called individuality or personality—a thread that runs through all of the preceding account. For this further look at personality it is more convenient to assign a name and dates to a fairly typical and entirely possible individual history.

Thus we may see John Smith on a June day in 1950. He then seems tangible and definite enough as a person and personality. But the actually blurred and protean nature of the total personality of John Smith—in terms of 1870 or of 1900—will be grasped only when the simplest facts about him at these different dates are brought to mind.

Early in 1870, he passed from egg cell to fishlike embryo, and in that autumn was a newborn infant—certainly with numerous elements of body and personality as yet unformed, and just as certainly he was then capable of assuming a variety of alternative physical and mental traits. In 1900, many of those traits had blossomed; an outstanding person was actively meeting the world.

Here was a sensitive student easily, eagerly and rapidly adding to our knowledge of a difficult subject. Seeing him in 1950, we can find only a man who has already spent some years, in rebellion and violence, in a mental hospital.

It seems clear enough that nothing of human inequality hitherto discussed in this chapter at all approaches the personal and private inequality that—from egg to end of life—is progressively unfolded by each of us, while acquiring and pursuing a human existence.

Personality is unrivaled in the extent to which it is a new and forming thing that is nevertheless incapable of remaining the same thing. In one of its ranges, only a step or an instant separates the comic from the tragic. Its beginnings—its foundations-rest on elements that may not be reckoned extremely simple, but they are certainly extremely unlike a personality.

These beginnings, we have seen, trace to *two* quite small and quite independent bits of living matter, each bit—egg and sperm—enclosing several thousands of special kinds of molecules (genes) that in their entirety are either compatible with or complementary to the molecules (genes) enclosed by the other.

And each of these bits of living matter, like everything else that is alive, somehow carries *time* as a fourth dimension. Time it is, too, that permits the merging of egg and sperm to become complete; it is time that permits further slow growth, along with the actual formation of new parts with new reactions.

It is time that favors the origin of mechanisms for registering all, and for expressing much, that has occurred; it is time that permits also no end, during life, to these processes of registration and expression—that is, no end to changes in personality, no end to the molding and tempering of the individual.

At every point in all this sea of change the mind and body exist as a single unity, not as something separable. The two are a unified and common growth; "there are no mental states, only mental acts" ( Herrick). Yet as late as 1900 some textbooks of physiology used in reputable American colleges spoke of mind and *soul* as interchangeable items; and, for purposes of certain creeds at scattered points on earth, it seems that some such texts are still in use.

The early theology that gave man a "soul" also took from man his immense fourth dimension—time. While promising man a future, it filched from him a past that alone makes a present possible.

Surely those who today speak of a soul have need to provide it a home. If soul does not in some way represent personality, just what does it represent?

If it *is*—or if it relates to—personality or individuality, to which end or cross section of that flowing panorama does the detachable soul attach? Can this be at the very beginning of the individual—the *two* things, the unmerged sperm and egg? If it enters at the instant of merging—according to recent Catholic dogma—who can possibly identify such an instant in an unbroken chain of life?

To ask these questions is to indicate how thoroughly the concept of soul is now rejected both by fact and by reason. This entire survey of living matter and of personality has neither met that agency nor found a time or place for its entry or abode.

# Index

## C



## D

## Q



## R



## S

## T



## U



## V



## W



## Y